Total-Reflection X-Ray Fluorescence Analysis and Related Methods

CHEMICAL ANALYSIS

A SERIES OF MONOGRAPHS ON ANALYTICAL CHEMISTRY
AND ITS APPLICATIONS

Series Editor
MARK F. VITHA

Volume 181

A complete list of the titles in this series appears at the end of this volume.

Total-Reflection X-Ray Fluorescence Analysis and Related Methods

Second Edition

Reinhold Klockenkämper

Alex von Bohlen

Leibniz-Institut für Analytische Wissenschaften – ISAS – e.V.
Dortmund and Berlin, Germany

Published by John Wiley & Sons, Inc., Hoboken, New Jersey
Published simultaneously in Canada

Library of Congress Cataloging-in-Publication Data:

Klockenkämper, Reinhold, 1937- author.
Total-reflection X-ray fluorescence analysis and related methods.—Second edition / Reinhold Klockenkämper, Alex von Bohlen, Leibniz-Institut für Analytische Wissenschaften-ISAS-e.V., Dortmund und Berlin, Germany.
pages cm
Includes bibliographical references and index.
ISBN 978-1-118-46027-6 (hardback)
1. X-ray spectroscopy. 2. Fluorescence spectroscopy. I. Bohlen, Alex von, 1954- author. II. Title.
QD96.X2K58 2014
543'.62–dc23

2014022279

Printed in the United States of America

10 9 8 7 6 5 4 3 2 1

CONTENTS

FOREWORD

This second edition of the first and only monograph on total reflection X-ray fluorescence (TXRF) is thoroughly revised and updated with important developments of the last 15 years. TXRF is a universal and economic multielement method suitable for extreme micro- and trace analyses. Its unique and inherent features are elaborated in detail in this excellent monograph. TXRF represents an individual method with its own history and special peculiarities in comparison to other XRF techniques, and is well established within the community of elemental spectroscopy. In particular, TXRF has been realized and understood as a complementary rather than competitive instrument within the orchestra of ultramicro and ultratrace analytical instrumentation. In different round-robin tests, TXRF demonstrated its performance quite well in comparison with methods such as ET-AAS, ICP-OES, ICP-MS, RBS, and INAA.

Total reflection XRF is widely used in the analysis of flat sample surfaces and near-surface layers. Here, it may be applied as a nondestructive method especially suitable for the quality control of wafers in the semiconductor industry. It can be used for the determination of impurities at the ultratrace level and for mapping of the element distribution on flat surfaces. In addition to the composition, the nanometer-thickness of thin layers can be determined by tilting the sample at grazing incidence. Direct density measurements are a special and unique feature of TXRF after sputter-etching.

The authors have built a successful and well established team in the field of TXRF for about 25 years. In the first edition of this book, R. Klockenkämper described the principles and fundamentals of TXRF, the performance of analyses, and its applications. After his retirement, he cooperated with A. von Bohlen in order to examine the latest developments and to place TXRF in a leading position of analytical atomic spectrometry.

Several new sections of this second edition demonstrate the essential progress of TXRF. The new generation of silicon drift detectors, which are cooled thermo-electrically, is highlighted. About 80 synchrotron facilities around the whole world are listed—with work places that are dedicated solely to TXRF offering an extremely brilliant and tunable radiation. The previous fields of applications are enumerated and diversified, contamination control of wafers is shown to be standardized, and many new fields are represented especially in the life sciences. Combinations of different methods of spectrometry, such as NEXAFS and XANES, with excitation under total reflection build a trend and have been presented as future prospects. The worldwide distribution of TXRF's instrumentation and its different fields of applications are evaluated statistically.

This articulate monograph on TXRF with several color pictures provides fundamental and valuable help for present and future users in the analytical community. Many disciplines, such as geo-, bio-, material-, and environmental sciences, medicine, toxicology, forensics, and archaeometry can profit from the method in general and from this outstanding monograph in particular.

Geesthacht, May 2014

PROF. DR. ANDREAS PRANGE

Helmholtz-Zentrum Geesthacht
Institute for Coastal Research
Head of the Department for
Marine Bioanalytical Chemistry

ACKNOWLEDGMENTS

The authors are grateful to all the colleagues of our TXRF community for their laborious and important investigations and for manifold publications that build the basis of this monograph. Special thanks go to the attendees of the last conference on TXRF, who took part in the survey described in Chapter 6.

We also wish to thank Mrs. Maria Becker for carefully adapting the first edition in a readable word document, and for the diligent compilation of all references and all the data of synchrotron beamlines. Furthermore, we thank our former colleague Prof. Dr. Joachim Buddrus for proofreading chemical terms and formulas. Scientific and technical assistance of the Leibniz-Institut für Analytische Wissenschaften – ISAS – e.V., represented by members of the Executive Board, Prof. Dr. Albert Sickmann and Jürgen Bethke, is gratefully acknowledged. ISAS in Dortmund is supported by the Bundesministerium für Bildung und Forschung (BMBF) of Germany, by the Ministerium für Innovation, Wissenschaft und Forschung of North Rhine-Westphalia, and by the Senatsverwaltung für Wirtschaft, Technologie und Forschung, Berlin.

It is a pleasure for the authors to thank our friend Prof. Dr. Andreas Prange for providing a felicitous and penetrative foreword. The authors are also obliged to the publishers John Wiley and particularly to Bob Esposito and Michael Leventhal for their reliable assistance, and to Dr. Mark Vitha for his great care in editing the manuscript. We also pay tribute to the printers for the excellence of their printing, especially to our project manager, Ms. Shikha Pahuja, for the diligent organization.

LIST OF ACRONYMS

AC	Alternating current
ADC	Analog-to-digital converter
AFM	Atomic force microscopy
AITR	Attenuated internal total reflection
ALS	Amyotrophic Lateral Sclerosis
AMC	Adiabatic microcalorimeter
ANNA	Activity of Excellence and Networking for Nano- and Microelectronics Analysis
APS	Advanced photon source *or* American Physical Society
ASTM	American society for testing and materials
ATI	Atom institute
AXIL	Analytical X-ray analysis by iterative least squares
BB	Black body
BCR	Breakpoint cluster region (protein or gene) *or* British Chemical Standard - Certified reference material
BESSY	Berliner Elektronen Speicherring Gesellschaft für Synchrotronstrahlung
BRM	Blank reference material
CAS	Chemical Abstracts Services
CCD	Charge-coupled device
CHA	Concentric hemispherical analyzer
CHESS	Cornell high-energy synchrotron source
CMA	Cylindrical mirror analyzer
CMOS	Complementary metal oxides
CMOS	Complementary metal oxides semiconductor
CRM	Certified reference material
CVD	Chemical vapor deposition
CXRO	Center for X-ray Optics and Advanced Light Source
DC	Direct current

DCM	Double-crystal monochromator
DESY	Deutsches Elektronen Synchrotron
DIN	Deutsches Institut für Normung
DMM	Double multilayer monochromator
DORIS	Doppel Ring Speicher
EDS	Energy-dispersive spectrometry or spectrometer
EDTA	Ethylene-diaminetetraceticacid
EPMA	Electron probe microanalysis
ESCA	Electron spectroscopy for chemical analysis
ET-AAS	Electrothermal atomic absorption spectrometry
EXAFS	Extended X-ray absorption fine structure
FAAS	Flame atomic absorption spectrometry
FCM	Four-crystal monochromator
FEL	Free-electron laser
FET	Field effect transistor
FPS	Flat panel sensor
FT-IR	Fourier transform-infra red
FWHM	Full width at half maximum
GC-MS	Gas chromatography-mass spectrometry
GeLi	Ge(Li) detector; Germanium drifted with Lithium ions
GE-XRF	Grazing exit X-ray fluorescence
GF-AAS	Graphite furnace-atomic absorption spectrometry
GI-XRD	Grazing incidence X-ray diffractometry
GI-XRF	Grazing incidence X-ray fluorescence
GIE-XRF	Grazing incidence/exit X-ray fluorescence
GLP	Good laboratory practice
HASYLAB	Hamburger Synchrotron Strahlungslabor
HOPG	Highly ordered (oriented) pyrolytic graphite
HPGe	HPGe detector; high-purity Germanium
HPLC	High-performance liquid chromatography
HS	Humic substances
IAEA	International Atomic Energy Agency
IC	Integrated circuit
ICDD	International Centre for Diffraction Data
ICP	Inductively coupled plasma
ICP-MS	Inductively coupled plasma-mass spectrometry
ICP-OES	Inductively coupled plasma-optical emission spectrometry
IDMS	Isotope dilution-mass spectrometry

IEEE	Institute of Electrical and Electronics Engineers
IFG	Institut für Geräteentwicklung
IMEC	Interuniversity Microelectronics Center
INAA	Instrumental neutron activation analysis
IR	Infrared
IRMM	Institute of Reference Materials and Measurements
ISO	International Standard Organization
ITRS	International Technology Roadmap for Semiconductors
IUPAC	International Union for Applied Chemistry
JCPDS	Joint Committee on Powder Diffraction Standards
JFET	Junction Gate FET
KFA	Kernforschungsanlage
LED	Light emitting diode
LINAC	Linear accelerator
MBI	Max-Born Institut
MCA	Multichannel analyzer
MRI	Magnetic resonance imaging
MRT	Magnetic resonance tomography
NEXAFS	Near extended X-ray absorption fine structure
NIES	National Institute for Environmental Studies
NIST	National Institute of Standards and Technology
NSF	Nephrogenic Systemic Fibrosis
NSLS	National Synchrotron Light Source
PES	Photoelectron spectrometry
PGM	Plane grating monochromator
PIN	Positive-intrinsic-negative
PIXE	Proton or particle induced X-ray emission
PMM	Primary methods of measurement
PTB	Physikalisch-Technische Bundesanstalt
PVD	Physical vapor deposition
QM	Quality management
QXAS	Quantitative X-ray analysis system
RBS	Rutherford backscattering spectrometry
RMS	Root mean square (of the mean squared deviations)
ROI	Region of interest
RSD	Relative standard deviation
SAXS	Small angle X-ray scattering
SD	Standard deviation, absolute value

SDD	Silicon drift detector
SDi	Strategic Directions International
SEM	Scanning electron microscopy
SGM	Spherical grating monochromator
SiLi	Si(Li) detector; Silicium drifted with Lithium ions
SIMS	Secondary ion mass spectrometry
SOP	Standard operating procedure
SPM	Suspended particulate matter
SQUID	Superconducting quantum interference device
SR	Synchrotron radiation
SRM	Standard reference material
SSD	Solid-state detector
SSRL	Stanford Synchrotron Radiation Laboratory
STJ	Superconducting tunnel junction
STM	Scanning tunneling microscope or microscopy
SW	Standing wave
TDS	Total dissolved solids
TES	Transition edge sensor
TR	Total reflection
TRIM	Transport and range of ions in matter
TR-XPS	Total reflection XPS
TR-XRD	Total reflection XRD
TR-XRR	Total reflection XRR
TXRF	Total reflection X-ray Fluorescence
UCS	Ultra-Clean Society
UHV	Ultra-high-vacuum
ULSI	Ultra-large-scale integration
UPS	Ultraviolet photoelectron spectrometry
USB	Universal serial bus
UV	Ultraviolet
VAMAS	Versailles Project on Advanced Materials and Standards
VLSI	Very-large-scale integration
VPD	Vapor-phase decomposition
WDS	Wavelength-dispersive spectrometry or spectrometer
XAFS	X-ray absorption fine structure
XANES	X-ray absorption near-edge structure
XPS	X-ray photoelectron spectrometry
XRD	X-ray diffractometry

XRF	X-Ray fluorescence
XRR	X-ray reflectometry
XSW	X-ray standing waves

Chemical Compounds

APDC	Ammonium pyrrolidine dithiocarbamate
DNA	Deoxyribonucleic acid
h-BN	hexagonal form of boron-nitride
HMDTC	Hexamethylene-dithiocarbamate
mQC	murine Glutaminyl cyclase
MIBK	Methyl isobutyl ketone
NaDBDTC	Sodium dibutyldithiocarbamate
PEDOT:PSS	Polyethylenedioxythiophene: Polystyrene sulfonate
PEG	Polyethylene glycol
PFA	Polyfluoroalkoxy (polymers)
PEI	Polyethylenimine
PEO	Polyethylene oxide
PP	Polypropylenes
PTFE	Polytetrafluoro-ethylenes
PMMA	Polymethyl methacrylate
RNA	Ribonucleic acid
ROS	Reactive oxygen species
TEAB	Triethylamine borane
TMAB	Trimethylamine borane
TMB	Trimethylborazine

LIST OF PHYSICAL UNITS AND SUBUNITS

A	ampere
a	year (annum)
°C	°Celsius *or* centigrade
C	coulomb
cm	centimeter
d	day
eV	electronvolt
F	farad
ft	foot
GHz	gigahertz
GeV	giga-electronvolt
g	gram
h	hour *or* hecto
hPa	hectopascal
Hz	hertz
in	inch
J	joule
K	kelvin
keV	kilo-electronvolt
kg	kilogram
km	kilometer
kPa	kilopascal
kV	kilovolt
kW	kilowatt
l	liter
m	meter *or* milli
mA	milliampere
MeV	mega-electronvolt
min	minute
ml	milliliter
mm	millimeter
mol	mole
mrad	milliradian
N	newton
nl	nanoliter
nm	nanometer
Pa	pascal
pl	picoliter
rad	radian
rpm	revolutions per minute
s	second
sr	steradian (squared radian)
T	tesla
V	volt
W	watt
kΩ	kiloohm
μl	microliter
μrad	microradian
Ω	ohm
%	per cent (10^{-2})
‰	per mill (10^{-3})
ppm	parts per million (10^{-6})
ppb	parts per billion (10^{-9})
ppt	parts per trillion (10^{-12})

LIST OF SYMBOLS

Symbols for Physical Quantities (in general they are unambiguous; in exceptional cases their meaning becomes clear by their individual context; for a detailed definition and distinction they can have indices)

α	Glancing angle of incident primary beam
α_{crit}	Critical angle of total reflection
α_{d}	glancing angle determined by the detector's field of vision
α_{f}	Sommerfeld's fine structure constant *or* glancing angle determined by the footprint
α_{k}	glancing angles of Kiessig maxima
β	Imaginary component of refractive index *or* ratio of electron velocity and light velocity *or* take-off angle of the fluorescence radiation
γ	Lorentz factor
δ	Decrement *or* real component of refractive index (*or* sometimes difference)
Δ	Path difference *or* interval
ε	Efficiency of a detector
ζ	Vertical coherence length
η	Efficiency
θ	Polar angle of an electron's position (in plane of the orbit)
Θ	Tilt angle around horizontal x-axis (corresponds to α)
λ	Wavelength
λ_{C}	Compton wavelength
λ_{cut}	Longest wavelength of radiation refracted at a given angle
μ/ρ	Total mass-absorption coefficient
ν	Frequency *or* index
ξ	Horizontal coherence length
ρ	Density of an element *or* material
ρ_m	Radius of curvature of the circular electron orbit

σ	Shielding constant *or* roughness
σ/ρ	Mass-scatter coefficient *or* cross-section of X-ray scattering
τ	Dead time
τ_{dead}	Dead-time *or* shaping time
τ/ρ	Photoelectric mass-absorption coefficient
υ	Phase velocity *or* velocity of light in a medium
φ	Phase difference
Φ	Angle of rotation around vertical z-axis *or* work function of a spectrometer
χ	Tilt correction around horizontal y-axis
ψ	Azimuthal angle of an electron's position (vertical to the orbit)
Ψ	Angle of deflection
ω	Angular frequency (Larmor frequency) *or* fluorescence yield
Ω	Solid angle
a	Distance *or* period *or* axis *or* acceleration of a particle *or* lattice constant
A	Atomic mass of an element *or* absorption *or* ordinate offset *or* detector area
b	Axis *or* constant of Wien's displacement law *or* lattice constant
B	Slope of calibration straight line *or* absolute sensitivity *or* magnetic field strength
c	Concentration *or* molar ratio of an element in a sample *or* lattice constant
c_{A}	Area related mass of an element (area density)
c_{v}	Volume concentration of an element
c_0	Light velocity *in vacuo*
C	Particular constant
C_{m}	Material constant determining α_{crit}
d	Thickness of a sample or a particular layer *or* interplanar spacing of a Bragg crystal
D	Dead-time loss *or* thickness of a stack of layers
e	Elementary charge of a single electron *or* energy necessary for a special atomic reaction
E	Energy of photons *or* amplitude of the electric field strength *or* energy of radiation
E_{binding}	Binding energy of an electron within an atom
E_{crit}	Characteristic (central) photon energy of synchrotron radiation
E_{cut}	Cut-off energy of refraction
E_{el}	Kinetic energy of an electron (beam energy)

E_{kin}	Kinetic energy of a particle
E_{min}	Minimum photon *or* electron energy required *or* critical excitation energy
E_{max}	Maximum photon *or* electron energy accepted *or* photon energy for maximum brilliance
f	Absorption jump factor *or* frequency *or* length of the footprint *or* parameter of fading
F	Fano factor *or* Lorentz force *or* formfactor (fading coherence)
g	Relative emission rate
h	Planck's constant *or* height
$\hbar$	Planck's constant over 2π
h, k, l	Miller indices
i	Current *or* index
I	Intensity *or* current
j	index
k	Particular constant *or* Boltzmann's constant *or* order of Kiessig maxima
K	Calibration constant *or* Bessel function *or* undulator parameter
l	Length
L	Distance of two points
m	Matrix element *or* mass *or* order of Bragg's reflection
M	Matrix, two-dimensional
M	molar mass of ions *or* atoms
n	Count rate *or* refractive index *or* number density
N	Number of photons *or* layers *or* oscillations *or* net intensity
N_A	Avogadro's constant
P	Level of significance *or* probability *or* electrical power
q	Charge of a particle
Q	Auxiliary quantity of mass absorption
r	Radius *or* distance from the origin *or* absorption jump ratio
r_{el}	Classical electron radius
R	Reflectivity
R_a	average roughness
R_∞	Rydberg's constant
S	Relative spectral sensitivity *or* Poynting vector
t	Time *or* live time *or* thickness of a layer *or* student factor
T	Acquire time *or* transmissivity *or* tilt center *or* temperature
U	Voltage

υ	Small volume
V	Volume
υ/V	Dilution factor
w	Width *or* spiked volume
w_{beam}	beam width
W	Radiant energy *or* window distance
x	Lateral movement *or* axis
X	Addenda of trinomial expression of fluorescence intensity
y	Lateral movement *or* axis
z	Depth in a sample normal to its surface *or* vertical shift
z_{n}	Penetration depth of radiation in a sample normal to its surface
Z	Atomic number of a chemical element
z_{fade}	damping constant of fading

CHAPTER

1

FUNDAMENTALS OF X-RAY FLUORESCENCE

X-ray fluorescence (XRF) is based on the irradiation of a sample by a primary X-ray beam. The individual atoms hereby excited emit secondary X-rays that can be detected and recorded in a spectrum. The spectral lines or peaks of such a spectrum are similar to a bar-code and are characteristic of the individual atoms, that is, of the respective elements in the sample. By reading a spectrum, the elemental composition of the sample becomes obvious.

Such an XRF analysis reaches near-surface layers of only about 100 μm thickness but generally is performed without any consumption of the sample. The method is fast and can be applied universally to a great variety of samples. Solids can be analyzed directly with no or only little sample preparation. Apart from the light elements, all elements with atomic numbers greater than 11 (possibly greater than 5) can be detected. The method is sensitive down to the microgram-per-gram level, and the results are precise and also accurate if matrix-effects can be corrected.

Total-Reflection X-ray Fluorescence Analysis and Related Methods, Second Edition.
Reinhold Klockenkämper and Alex von Bohlen.

For these merits, XRF has become a well-known method of spectrochemical analysis. It plays an important role in the industrial production of materials, in prospecting mineral resources, and also in environmental monitoring. The number of spectrometers in use is estimated to be about 15 000 worldwide. Of these, 80% are working in the wavelength-dispersive mode with analyzing crystals; only 20% operate in the energy-dispersive mode, mainly with Si(Li) detectors, and recently with Si-drift detectors. At present, however, energy-dispersive spectrometers are four times more frequently built than wavelength-dispersive instruments due to the advantage the former provides in fast registration of the total spectrum.

A spectrum originally means a band of colors formed by a beam of light as seen in a rainbow. The Latin word "spectrum" means "image" or "apparition." The term was defined scientifically as a record of intensity dependent on the wavelength of any type of electromagnetic radiation. The "intensity" is to be interpreted as a number of photons with particular photon energy. Today, a spectrum can also be a record of a number of ions according to their atomic mass or it can demonstrate the number of electrons in dependence of their electron energy. The visual or photographic observation of such a spectrum is called *spectroscopy*. The term is deduced from the Greek verb "σκοπειν," which means "to observe" or "to look at." On the other hand, "μετρω" in Greek means "to measure" so that *spectrometry* is a quantitative photoelectric examination of a spectrum.

1.1 A SHORT HISTORY OF XRF

The foundations of spectrochemical analysis were laid by R.W. Bunsen, a chemist, and G.R. Kirchhoff, a physicist. In 1859, they vaporized a salt in a flame and determined some alkaline and alkaline-earth metals by means of an optical spectroscope. Today, optical atomic spectroscopy has developed a variety of new analytical techniques with high efficiency, such as atomic absorption spectroscopy (AAS) with flames (FAAS) or electrothermal furnaces (ET-AAS), and the inductively coupled plasma technique (ICP) combined with atomic emission or mass spectrometry (ICP-AES and ICP-MS). These techniques do entail some consumption of the sample, but they are highly suitable for ultratrace analyses of solutions.

Nearly 40 years after the discovery by Bunsen and Kirchhoff, in 1895, Wilhelm Conrad Röntgen (Figure 1.1) discovered a remarkable, invisible, and still unknown radiation, which he called X-rays. This name has been adopted in the English-speaking areas; only in German-speaking parts is the radiation called "Röntgenstrahlen" in his honor [1]. In 1901, Röntgen was awarded the first Nobel Prize in Physics. The great potential of X-rays for diagnostic purposes in medicine and dentistry was immediately recognized worldwide. Furthermore, different researchers clarified the fundamentals of X-ray spectroscopy and developed the methods of XRF (X-ray fluorescence) and XRD

Figure 1.1. Wilhelm Conrad Röntgen in 1895 (reproduced with permission of the "Deutsches Röntgenmuseum" in Lennep, Germany).

(X-ray diffraction) applicable to material analysis. Table 1.1 enumerates well-known and renowned scientists. Most of them came from Great Britain and Germany and almost all of them won the Nobel Prize in physics.

Hendrik Lorentz found the dispersion of X-rays and studied the influence of magnetic fields on rapidly moving charged particles by the "Lorentz force," which 50 years later has built the basis for beamlines at synchrotron facilities. Lord Rayleigh detected the coherent scattering of X-rays, and Philipp Lenard investigated cathode rays while Sir J.J. Thomson verified them as negatively charged electrons. Lord Ernest Rutherford created his well-known model of atoms containing a positive nucleus and several negative electrons. Max von Laue, Friedrich, and Knipping showed the diffraction of X-rays by the lattice of crystalline copper sulfate [2] and hereby proved both the wave nature of X-rays and simultaneously the atomic structure of crystals.

In 1913, Sir William Henry and William Lawrence Bragg—father and son—built the first X-ray spectroscope as demonstrated in Figure 1.2 [3,4]. It consisted of a cathode-ray tube with a Mo anode, a goniometer with a revolving rock-salt crystal in the center, and a photographic film on the inside wall of a

TABLE 1.1. Important Scientists, Mostly Nobel Laureates who Established the Fundamentals of XRF and TXRF

Scientist	Country	Life	Research or Discovery	Nobel Prize
Wilhelm-Conrad **Röntgen**	Germany	1845–1923	Detection of Novel Invisible Rays	1901
Hendrik Antoon **Lorentz**	The Netherlands	1853–1928	Lorentz Force, Time Dilation; Dispersion	1902
Lord John William Strutt **Rayleigh**	United Kingdom	1842–1919	Coherent Scattering of X-rays; Gas Densities	1904
Philipp **Lenard**	Germany (Austria/ Hungary)	1862–1947	Discharges in Cathode Ray Tubes	1905
Sir Joseph John **Thomson**	United Kingdom	1856–1940	Electrical Conductivity of Gases; the Electron	1906
Lord Ernest **Rutherford**	New Zealand, UK	1871–1937	Radioactivity; Rutherford's Modell of Atoms	1908
Max **von Laue**	Germany	1879–1960	Diffraction of X-rays by Crystals	1914
Sir William Henry **Bragg**, sen	United Kingdom	1862–1942	Determination of Crystal Structures by X-Rays	1915
William Lawrence **Bragg**, jun.	United Kingdom	1890–1971	Bragg's Law: $m\lambda = 2d \cdot \sin\theta$	1915
Henry **Moseley**	United Kingdom	1887–1915	Moseley's Law: $E = k \cdot (Z - \sigma)^2$	–
Charles Glover **Barkla**	United Kingdom	1877–1944	Characteristic X-rays of Elements; Polarization	1917
Max **Planck**	Germany	1858–1947	Energy Quanta or Photons	1918
Albert **Einstein**	Germany	1879–1955	Photoelectric Effect; Theory of Relativity	1921
Niels **Bohr**	Denmark	1885–1962	Atomic Structure examined by Radiation	1922
Karl Manne Georg **Siegbahn**	Sweden	1886–1978	X-ray spectroscopy; M series	1924
Arthur Holly **Compton**	United States of America	1892–1962	Incoherent Scattering; Total-Reflection of X-Rays	1927
Peter **Debye**	The Netherlands	1884–1966	Powder Diffractometry by X-Rays	1936
Kai Manne Börje **Siegbahn**	Sweden	1918–2007	X-ray Photo-electron Spectroscopy	1981

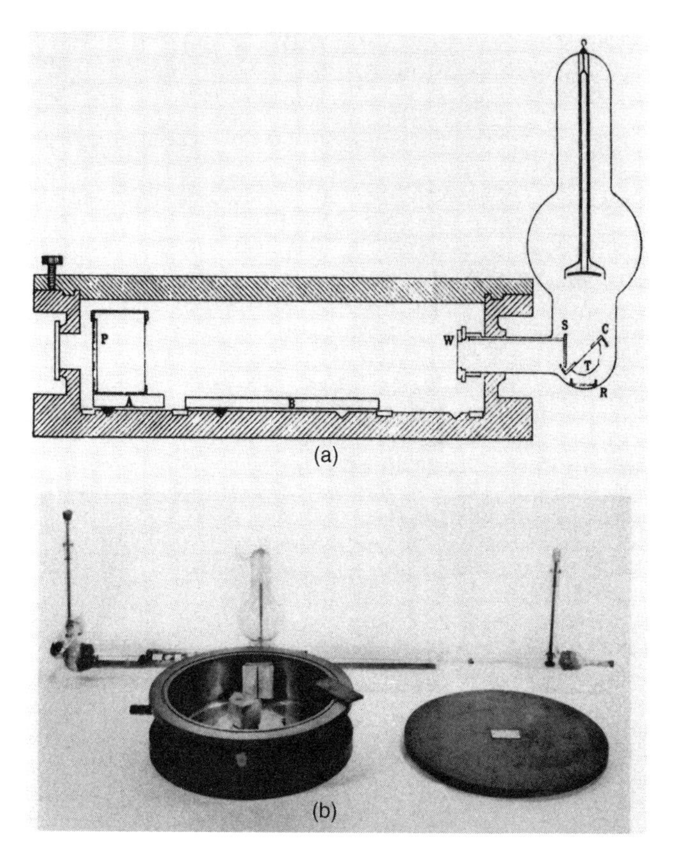

Figure 1.2. First X-ray spectroscope used by Moseley in 1913. (a) X-ray tube with T = metal target that can be exchanged; S = slit; W = window; goniometer with B = base for the crystal; P = photographic film. (b) A metal cylinder in front of an X-ray tube. The cylinder with slit and rotating crystal in its center can be evacuated. Figure from Ref. [3], reproduced with permission from Taylor & Francis.

metallic cylinder. The Braggs explained the diffraction of X-rays at the three-dimensional crystal as their reflection at parallel planes of the crystal lattice and determined the wavelength of the X-radiation according to the law later called Bragg's law. Furthermore, the interplanar distance of different other crystals had been determined. Then, in 1913, Moseley established the basis of X-ray fluorescence analysis by replacing the Mo anode by several other metal plates. He found his well-known law [3], which relates the reciprocal wavelength $1/\lambda$ of the "characteristic" X-rays to the atomic number Z of the elements causing this radiation. Moseley probably missed a Nobel Prize because he was killed during World War I at the Dardanelles near Gallipoli when he was just 28 years old (Figure 1.3b).

In 1904, Barkla had already discovered the polarization of X-rays, which is a hint to their wavelike nature [5]. Ten years later, he bombarded metals with

Figure 1.3. (a) Arthur Holly Compton in 1927 deriving his famous formula. Photo is from the public domain, © is expired. (b) Henry Moseley with an X-ray tube in 1913. Photo is from the public domain, © is expired.

electrons, which led to the emission of X-rays as "primary" radiation. Barkla excited the materials by this primary X-rays and together with Sadler he found their characteristic X-rays as "secondary" radiation [6]. He showed that the elemental composition of a sample could be examined by X-radiation and was awarded the Nobel Prize in 1917.

In contrast to the wavelike nature, Max Planck recognized the corpuscular nature of X-rays appearing as photons and Albert Einstein explained the photoelectric effect by means of such photons. Niels Bohr depicted the model of atoms consisting of a heavy nucleus with several protons and with an outer shell containing the same number of electrons. These electrons were assumed to revolve around the nucleus on several distinct orbits. The periodic system of the elements was discovered by Dimitri Mendelejew and Lothar Meyer in 1869. The naturally existing elements ordered with increasing atomic mass had got the place numbers $Z=1$ for hydrogen (the lightest element) up to $Z=92$ for uranium (the heaviest element). It could be explained now that Z is not an arbitrary number but the number of protons in the nucleus and the number of electrons in the outer shells of an atom. And the three anomalies of potassium, nickel, and iodine could be cleared up by the different atomic mass of their isotopes. Furthermore, six new elements could be predicted and had indeed been discovered in the next 20 years: the rare elements technetium, hafnium, rhenium, astatine, francium, and promethium.

Manne Siegbahn got the Nobel Prize for his discoveries of X-ray spectra. He determined the wavelength of characteristic X-rays with high accuracy by their diffraction at mechanically carved gratings under grazing incidence [1]. Arthur Holly Compton detected the incoherent scattering of X-rays. In 1923, he also discovered the phenomenon of external total reflection for X-rays [7]. He found that the reflectivity of a flat target strongly increased below a critical angle of only about 0.1°. In 1927, Compton was awarded the Nobel Prize in Physics (Figure 1.3a). Ten years later, Debye won the Prize in chemistry for his investigation of X-ray powder diffractometry. And finally, Kay Siegbahn, son of Manne Siegbahn, received the Noble Prize for the discovery of X-ray photoelectron spectroscopy in 1981.

The years of fundamental discoveries were gone now and the time of industrial applications began. Already in 1924, Siemens & Halske (Germany) had built the first commercially available X-ray spectrometer with an open X-ray tube, revolving crystal, and photographic plate. Coolidge developed a vacuum-sealed cathode-ray tube as shown in Figure 1.4. Samples could easily be excited now by X-rays instead of electrons. Soller built a collimator consisting of several parallel metal sheets just right for the collimation of a broad X-ray beam. In the 1930s, Geiger and Müller developed a gas-filled photoelectric detector, which allowed for direct pulse-counting instead of a complicated development of the photographic plate. This detector was replaced by a gas-filled proportional detector and by a scintillation counter in the 1940s. Simultaneously, different analyzer crystals were produced with various spacings and high reflectivity, for example, lithium fluoride and pentaerythritol.

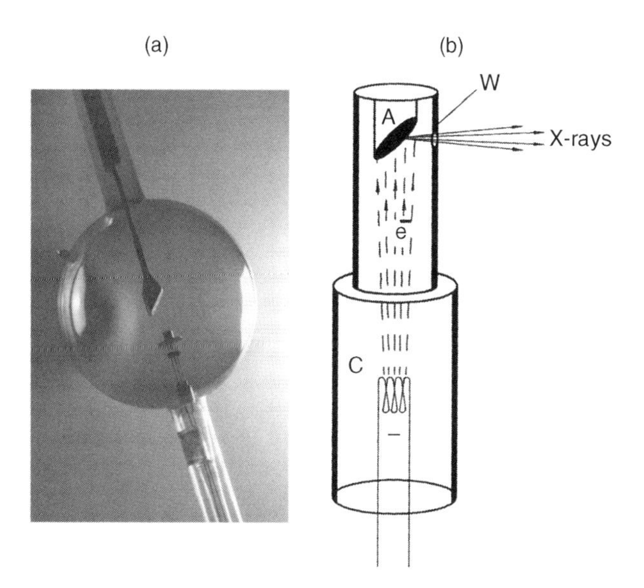

Figure 1.4. X-ray tube of the Coolidge type used as an X-ray photon source. (a) The vacuum-sealed glass bulb is an engineering marvel of glass blowing workshops from 1905. Photo of the authors, reproduced with permission from "Deutsches Röntgenmuseum," Lennep, Germany. (b) Sketch of today's X-ray tubes consisting of a metal–glass cylinder. C = tungsten-filament used as the cathode; A = metal block with a slant plane used as the anode; W = thin exit window. Figure from Ref. [8], reproduced with permission. Copyright © 1996, John Wiley and Sons.

After World War II, the first complete X-ray spectrometers became available, developed for example, by Philips, The Netherlands, by Siemens, Germany, and by ARL, Switzerland. In the 1960s, the spectrometers were equipped with hardwired controllers, servo transmitters, switching circuits, and electronic registration [4]. In the 1970s, X-ray spectrometers became computer-controlled and automated for a high throughput of samples. They were used for production and quality control in several branches of the metallurgical industry. Furthermore, X-ray spectrometers were applied in the exploitation of mineral resources and also in environmental protection. At this time XRF-spectrometers filled a whole lab, but in the 1980s the lateral dimensions decreased. In the decades since, XRF has developed into a powerful method of spectrochemical analysis of materials. However, classical XRF is not suitable for ultratrace analyses and it is notorious for producing matrix effects that may lead to systematic errors. Extensive efforts have been made to overcome these drawbacks, for example by matrix separation, thin-film formation, and mathematical corrections. Nevertheless, the new techniques of optical atomic spectrometry have surpassed conventional XRF in many respects.

From the start in 1895, X-rays were immediately applied to medical and dental diagnosis and later on for security checks at airports, for material analysis, ore mining, and pollution control. Furthermore, X-rays in astronomy have enlarged our view of the universe. In 1932, the "German Röntgen Museum" was founded at Röntgen's birthplace in Lennep, 50 km away from Dortmund, Germany. Today it is a global center of the life, research, and impact of Wilhelm Conrad Röntgen and presents numerous valuable original objects of the discovery, development, and application of X-rays [9].

1.2 THE NEW VARIANT TXRF

Simultaneously with the invention of semiconductor devices in the "silicon valley" after 1970, a new kind of an X-ray detector was developed. It could not only count the individual X-ray photons but could also determine their energy. Such a Si(Li) detector was called "energy-dispersive" instead of the "wavelength-dispersive" spectrometers used so far. The novel detectors were small and compact, did not need a goniometer with an analyzing crystal, and could collect the whole spectrum simultaneously in a very short time.

1.2.1 Retrospect on its Development

Additional important progress in XRF was made 50 years after the discovery of total reflection of X-rays by Compton. In 1971, Yoneda and Horiuchi [10] evolved an ingenious idea of using total reflection for the excitation of X-ray fluorescence. They proposed the analysis of a small amount of material applied on a flat, even, and totally reflecting support. An energy-dispersive Si(Li) detector, developed shortly before, was placed directly above the support for sample analysis. First, they determined uranium in sea water, iron in blood, and rare earth elements in hot-spring water. The theoretical basis and the experimental conditions were subsequently investigated. In Vienna, Austria, Wobrauschek wrote a PhD thesis on the subject [11], and together with Aiginger, they reported detection limits of nanograms [12,13]. In Geesthacht near Hamburg, Germany, Knoth and Schwenke found element traces on the ppb-level [14,15].

In the decade after 1980, a great variety of applications promoted a growing interest, and different instruments became commercially available (the "Wobi" module of the IAEA in Vienna, Austria; EXTRA II of Seifert in Ahrensburg, Germany; Model 3726 of Rigaku, Japan; TREX 600 of Technos, Japan; and TXRF 8010 of Atomika, Munich, Germany). The number of instruments in use increased to about 200 worldwide and the new variant of XRF turned out to have considerable advantages for spectrochemical analysis of different materials. At a first "workshop" in Geesthacht in 1986, the participants decided to call the new method "total reflection X-ray fluorescence analysis" and introduced the acronym "TXRF." A series of biannual international meetings followed. Table 1.2 lists the years, locations, and chairpersons. The papers presented were subsequently published as proceedings in special issues of scientific journals, mostly of Spectrochimica Acta [16–27]. The next conference will be held in 2015 as a satellite meeting of the Denver conference in Denver, Colorado.

In 1983, an angular dependence of the fluorescence intensities in the range below the critical angle of total reflection was first observed by Becker *et al.* [28]. It led to the nondestructive investigation of surface contamination and thin near-surface layers. This variant was also called "grazing-incidence" XRF. In 1986, the X-radiation of a synchrotron was first used for excitation by Iida *et al.* [29]. The high intensity, linear polarization, and natural collimation of this X-ray source were shown to be very useful and favorable in comparison to conventional X-ray sources.

In 1991, Wobrauschek, Aiginger, Schwenke, and Knoth (Figure 1.5) won the distinguished Bunsen–Kirchhoff Prize of the DASp (Deutscher Arbeitskreis für Angewandte Spektroskopie) for the development of TXRF. In the years after, first reviews and book contributions were published on the subject of TXRF (e.g. [30,31]). They enclose short surveys with some 10 to 50 pages. In 1997, this monograph at hand was published in a first edition, exclusively dedicated to TXRF. It was very well received on the market and within one year after publication, 450 copies of the book were sold. Today, it is still the only comprehensive monograph on the field of TXRF. Nearly 800 copies of the first edition have been distributed and nearly 350 different scientific articles have used the book as a reference, so it is the most cited item in this field of research. In 2002, the English edition was translated into Chinese and offered as a low-price book.

The development of TXRF and related methods can be read from the rate of peer-reviewed papers. Figure 1.6 demonstrates the publication rate of XRF

TABLE 1.2. Fifteen TXRF-Meetings Between 1986 and 2013

No	Date	Type	Location	Country	Chairperson	Proceedings
1	May, 1986	Workshop	Geesthacht	Germany	Michaelis	GKSS Report **86**/E/61 (1986)
2	May, 1988	Workshop	Dortmund	Germany	Klockenkämper	Spectrochim. Acta **44B** (1989)
3	May, 1990	Workshop	Vienna	Austria	Aiginger, Wobrauschek	Spectrochim. Acta **46B** (1991)
4	May, 1992	Workshop	Geesthacht	Germany	Andreas Prange	Spectrochim. Acta **48B** (1993)
5	Sept, 1994	Workshop	Tsukuba/Tokyo	Japan	Yohichi Gohshi	Adv. X-ray Chem.Anal. Jpn. **26s** (1995)
6	June, 1996	Conference	Eindhoven Dortmund	Netherlands Germany	de Boer Klockenkämper	Spectrochim. Acta **52B** (1997)
7	Sept, 1998	Conference	Austin/Texas	USA	Mary Ann Zaitz	Spectrochim. Acta **54B** (1999)
8	Sept, 2000	Conference	Vienna	Austria	Wobrauschek, Streli	Spectrochim. Acta **56B** (2001)
9	Sept, 2002	Symposium	Funchal/Madeira	Portugal	Maria Luisa de Carvalho	Spectrochim. Acta **58B** (2003)
10	Sept, 2003	Conference	Awaji-Island	Japan	Yohichi Gohshi	Spectrochim. Acta **59B** (2004)
11	Sept, 2005	Conference	Budapest	Hungary	Gyula Zaray	Spectrochim. Acta **61B** (2006)
12	June, 2007	Conference	Trento	Italy	Giancarlo Pepponi	Spectrochim. Acta **63B** (2008)
13	June, 2009	Conference	Gothenburg	Sweden	Johan Boman	Spectrochim. Acta **65B** (2010)
14	June, 2011	Conference	Dortmund	Germany	Alex von Bohlen	
23	Sept, 2013	Conference	Osaka	Japan	Kouichi Tsuji	Virtual issue Spectrochim. Acta (2014)

Figure 1.5. Four pioneers of TXRF analysis, from left to right: Peter Wobrauschek, Hannes Aiginger, Heinrich Schwenke, and Joachim Knoth were awarded the "Bunsen–Kirchhoff Prize" in 1991. Photo by R. Klockenkämper, private property.

(Figure 1.6a) and TXRF (Figure 1.6b) within the last 40 years. The number of all XRF papers started in 1970 at a level of about 100 papers per year, remained constant for 20 years, and exponentially increased after 1990 to a rate of 2500 papers per year. Between 1970 and 1985, TXRF papers appeared only sporadically. But in the years after 1986, their number grew explosively from some 3 to about 125 papers per year with large fluctuations. The impact of the special issues after every single TXRF conference can be recognized as special peaks, repeating every 2 years after 1989. Altogether, 1250 articles have been published in the field of TXRF. It is interesting to mention that only eight authors are connected with 30% of all published papers in this field.

The method of TXRF has been developed significantly and has become a high-performance variant of classical X-ray fluorescence. For a lot of elements, the detection limits are on the pg-level and even below. In general, all elements except for the light elements can be detected. TXRF analysis can be compared with ET-AAS, which is the high-power specialty of FAAS, and with ICP-MS, which even tops ICP-OES. TXRF ranks high among these competitive methods of element spectral analysis.

In the last 15 years after the first edition of this monograph, different review articles on TXRF have been published summarizing new developments and results [32–34]. Book contributions furthermore describe the subject with different aspects, for example, wafer analysis [35–37]. Specific articles deal with further developments, such as excitation with synchrotron radiation [38,39], with standing waves by grazing incidence [40,41], with biological applications [42], with sample preparation [43], and with portable instruments [44]. Today, TXRF is successfully applied all over the world

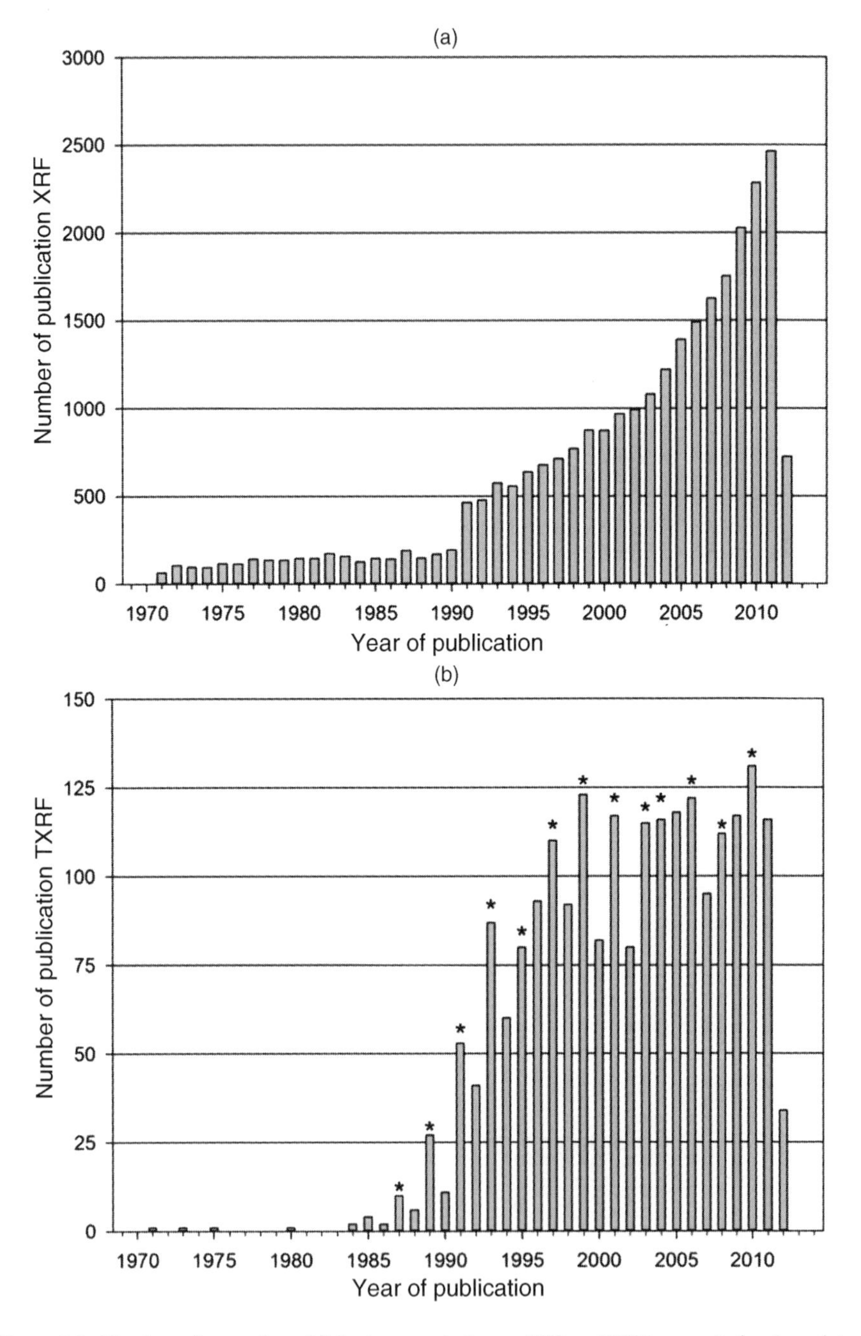

Figure 1.6. Number of annually published papers between 1970 and 2012 presented as bar plots. (a) For XRF in total. (b) Solely for TXRF. The data came from ISI Web of Knowledge, January 2012; http://thomsonreuters.com.

(see Section 6.3.3), and suitable equipment are installed and operated at several institutes and laboratories in a lot of countries: Argentina, Australia, Austria, Belgium, Brazil, Chile, China, Cuba, France, Germany, Great Britain, Hungary, India, Italy, Japan, Poland, Portugal, Russia, Spain, Sri Lanka, Sweden, Switzerland, Taiwan, The Netherlands, different states of the USA (CA, TX, IL, NM, ID, NY, MA, NJ, MD), Venezuela, and Vietnam. The users come from university institutes of chemistry and physics, from synchrotron beamlines at synchrotron facilities, and from chemical laboratories in industry, especially in the semiconductor industry—with particular interest in wafer production and control.

1.2.2 Relationship of XRF and TXRF

As is illustrated in Figure 1.7, TXRF is a variation of energy-dispersive XRF with one significant difference. In contrast to XRF, where the primary beam strikes the sample at an angle of about 40°, TXRF uses a glancing angle of less than 0.1°. Owing to this grazing incidence, the primary beam shaped like a strip of paper is totally reflected at the sample support.

Today, TXRF is primarily used for chemical *micro-* and *trace* analyses. For this purpose, small quantities, mostly of solutions or suspensions, are placed on optical flats (e.g., quartz glass) serving as sample support. After evaporation, the residue is excited to fluorescence under the fixed small glancing angle and the characteristic radiation is recorded by a Si(Li), or recently by a Si-drift detector, as an energy-dispersive spectrum. It is the *high reflectivity* of the sample support that nearly eliminates the spectral background of the support and lowers the detection limits from 10^{-7} to 10^{-12} g. Although this mode of operation does not permit the entirely nondestructive investigation of bulk material, it offers new challenging possibilities in ultramicro- and ultratrace analyses. Besides its high detection power, simplified quantification is made possible by internal standardization. This is because matrix effects cannot build up within the minute residues or thin layers of a sample.

A new field of application has been opened in the 1980s by *surface* and *near-surface layer* analyses. In 1983, the angular dependence of X-ray fluorescence at grazing incidence was investigated as already mentioned earlier [28]. This effect was used in the following years to investigate surface impurities, thin near-surface layers, and even molecules adsorbed on flat surfaces. Such examinations are especially applicable for cleaned and/or layered wafers representing the basic material for the semiconductor industry. The flat samples are examined either with respect to contamination of the surface or with respect to the setup of near-surface layers. However, this mode of analysis needs fluorescence intensities to be recorded not only at one fixed angle but at various angles around the critical angle of total reflection. From these angle-dependent intensity profiles, the composition, thickness, and even density of top layers can be ascertained. It is the *low penetration depth* of the primary beam at total reflection that enables this in-depth examination of ultrathin

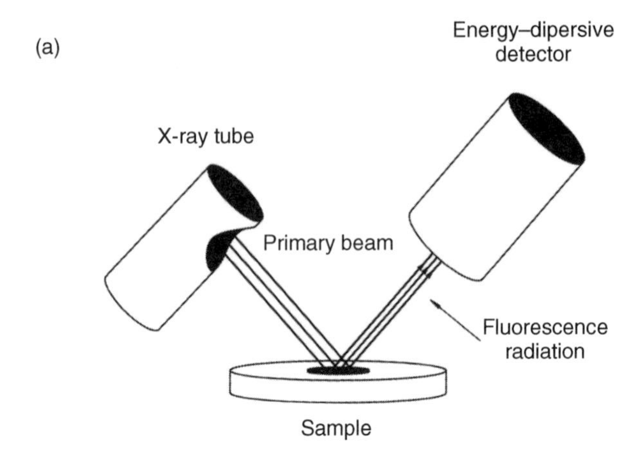

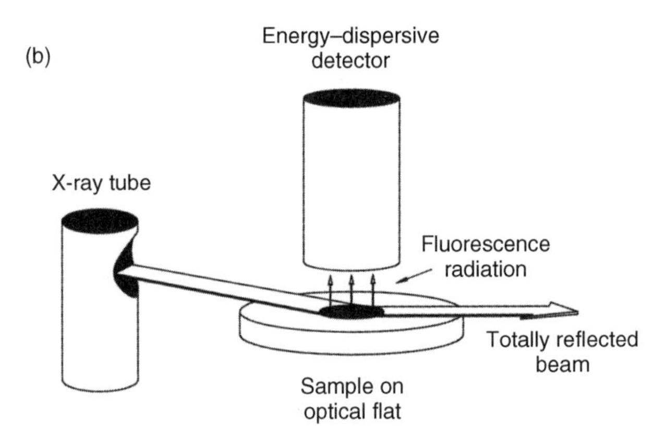

Figure 1.7. Instrumental arrangement for (a) conventional XRF and (b) TXRF. Comparison shows a difference in the geometric grouping of excitation and detection units. Figure from Ref. [8], reproduced with permission. Copyright © 1996, John Wiley and Sons.

layers in the range of 1–500 nm. The method is nondestructive and needs no vacuum—at least no ultrahigh vacuum (UHV).

In spite of the similarities in instrumentation, such as the X-ray source, the energy-dispersive detector, and pulse-processing electronics, the use of TXRF differs fundamentally from classical XRF. With respect to sample preparation and performance of analysis, it has a lot in common with AAS or ICP for trace element analysis and it is similar to X-ray photoelectron spectroscopy (XPS), Rutherford backscattering spectroscopy (RBS), and secondary ion mass spectrometry (SIMS) for surface and near-surface layer analysis. In fact, TXRF is able to compete, often favorably, with these well-established methods.

The main reason for this progress is the special geometric arrangement leading to total reflection of the primary beam. Accordingly, the totally reflected beam interferes with the incident primary beam and leads to standing waves above surfaces and also within near-surface layers. The unique role of TXRF is based on the formation of such standing waves and particular details can only be understood with regard to these standing waves.

The arrangement of grazing incidence is not restricted to XRF measurements. It can also be exploited for X-ray reflection (XRR) and X-ray diffraction (XRD). As early as 1931, Kiessig investigated the reflection of thin layers deposited on a thick substrate [45], and in 1940, Du Mond and Youtz observed Bragg-reflection of periodic multilayers [46]. It was not until the late 1970s that XRD at grazing incidence was developed. This monograph mainly deals with the technique of TXRF and excludes that of XRD. However, reports of XRR experiments are included when needed for a better understanding or even for complementary results. The usual TXRF instrumentation can simply be extended for such experiments.

1.3 NATURE AND PRODUCTION OF X-RAYS

Already in the seventeenth century, Isaac Newton described visible light as a beam of small corpuscles while Christian Huygens developed a picture of a beam with waves. In the corpuscle picture, all corpuscles travel at the velocity of light c. They follow straight lines that can be regarded as beams. In the wave picture, the light propagates as a wave showing crests and troughs. They follow each other with a frequency ν and at a distance λ called the wavelength and are always orthogonal to the direction of the respective beam. The speed of light in vacuum was shown to be nearly 3×10^8 m/s. Phenomena of reflection, refraction, diffraction, and polarization could be explained in the one or in the other picture, or even in both pictures. The wavelength of visible light was determined to lie between 0.4 and 0.8 μm.

In 1865, James Clerk Maxwell described light as an *electromagnetic* wave with electric and magnetic field strength. The photoelectric effect as a reaction of radiation with matter was explained by Einstein in 1905. Together with Planck he identified a light beam as an array of energy quanta called photons. A photon was defined as a corpuscle that carries an elementary energy unit E but has no rest mass. *In vacuo*, all photons travel at the velocity of visible light on straight lines. However, the dualism of the corpuscular and the wave picture was not dissolvable; neither corpuscles nor waves could have been seen directly, only the different phenomena have been observed.

1.3.1 The Nature of X-Rays

Shortly after their discovery by Röntgen in 1895, X-rays were assumed to be part of the electromagnetic radiation. Von Laue and the Braggs explained the diffraction of X-rays in the wave picture and measured wavelengths of about 0.1 nm. Such values are comparable with the spacing of crystal lattice planes. This value *d* was previously determined for simple crystals from Avogadro's number, the molecular mass, and the density, for example, for a rock salt crystal [1].

In order to describe the incoherent scattering of X-rays by electrons, Arthur H. Compton used the corpuscle picture. On the other side, in 1923 he detected the external total reflection of X-rays that again supported the hypothesis of the wave nature of X-rays. This dualism of a corpuscular and a wave picture was interpreted as a complementary nature of the electromagnetic radiation. It was overcome in 1927 by Niels Bohr in Copenhagen. Because of Heisenberg's uncertainty principle, the location of corpuscles cannot be determined with absolute certainty. The locus can be estimated by quantum mechanics only as a statement of probability expressed by a wave function.

X-ray photons have energies in the kilo-electronvolt range (0.01–100 keV) and wavelengths in the nanometer range (100–0.01 nm). Figure 1.8 demonstrates X-rays as a part of the wide electromagnetic spectrum including synchrotron radiation. Photon energy and wavelength are inversely proportional, according to

$$E = h\nu = \frac{hc}{\lambda} \tag{1.1}$$

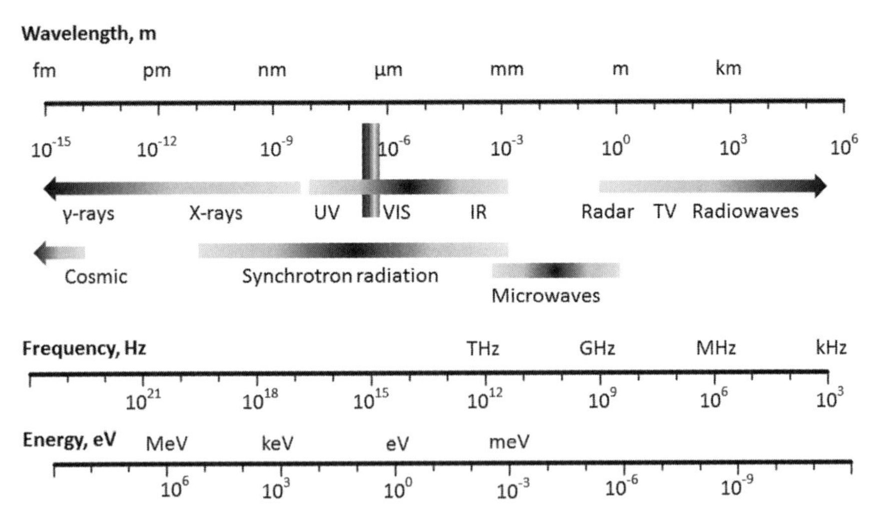

Figure 1.8. Spectrum of the electromagnetic radiation with wavelengths between 1 fm and 1000 km covering more than 21 orders of magnitude. The visible light appears in the very small region between 390 and 770 nm (gray ribbon VIS). X-rays span about four orders of magnitude far below the visible and the ultraviolet light (UV). The radiation of synchrotrons has a width of eight orders of magnitude.

where h is the Planck's constant and c is the speed of light ($h \approx 4.136 \times 10^{-15}$ eV·s; $c \approx 2.998 \times 10^8$ m/s).

The conversion of energy and wavelength of Equation 1.1 can be made by the simple relationship

$$E[\text{keV}] \approx \frac{1.23984}{\lambda[\text{nm}]} \tag{1.2}$$

Frequently used physical constants are listed in Table 1.3 with the latest numerical values from NIST (National Institute of Standards and Technology, Gaithersburg, MD). They are given in SI units, mostly with 9 to 11 digits and with a relative uncertainty of some 10^{-8} [47]. The values with SI units can be transformed into atomic units by the relationship $1\,\text{J} = 6.241\,509\,34 \times 10^{18}\,\text{eV}$. In the text, physical constants will be given with only 3 to 5 digits and atomic units.

1.3.2 X-Ray Tubes as X-Ray Sources

X-rays are originally produced by the bombardment of matter with accelerated electrons. Usually, such a *primary* radiation is produced by an X-ray tube of the Coolidge-type as mentioned earlier and shown in Figure 1.4. It consists of a vacuum-sealed tube with a metal–glass cylinder. A tungsten filament serves as hot cathode, and a pure-metal target, such as chromium, copper, molybdenum, or tungsten, serves as the anode. Electrons are emitted from the heated filament and accelerated by an applied high voltage in the direction of the anode. The high-energy bombardment of the target produces heat above all while the electrons are absorbed, retarded, or scattered. Finally, X-rays and Auger electrons can be produced. The heat is dissipated by water-cooling of the anode while the X-rays emerge from a thin exit window as an intense X-ray beam. Mostly, a 0.2–1 mm thick beryllium window is used. Reflected electrons, including Auger electrons, cannot escape from this window.

The X-ray tube is supplied by a stabilized high-voltage generator. High voltage and current applied to the tube determine the intensity of the X-ray beam. The voltage can usually be chosen between 10 and 60 or even 100 kV, the current between 10 and 50 mA, so that an electric power of several kilowatts can be supplied. However, only about 0.1% of the electric input power is converted into radiation and most of it is dissipated as heat. For that reason, such X-ray tubes have to be cooled intensively by water. A flow rate of 3 to 5 l/min is commonly needed.

The primary X-ray beam is normally used to irradiate a sample for analysis. By this *primary* irradiation, the atoms in the sample are generally excited to produce *secondary* X-rays by themselves. This effect is called X-ray fluorescence. The secondary radiation can be used as a color pattern of the sample as its *chromatic* composition changes with the *element* composition. The spectral pattern can be recorded like a barcode by means of an X-ray detector and constitutes the basis of XRF analysis.

TABLE 1.3. Some Frequently Used Physical Constants from NIST Reference Values [47]

Term of the Constant	Character	Numerical Value	Uncertainty	SI Unit	Relative Unc.
Avogadro's constant	N_A	$6.022\,141\,29 \times 10^{23}$	$0.000\,000\,27 \times 10^{23}$	mol^{-1}	4.1×10^{-8}
Boltzmann's constant	k	$1.380\,648\,8 \times 10^{-23}$	$0.000\,001\,3 \times 10^{-23}$	J/K	9.1×10^{-7}
Compton wavelength	λ_C	$2.426\,310\,238\,9 \times 10^{-12}$	$0.000\,000\,001\,6 \times 10^{-12}$	m	6.5×10^{-10}
Electric constant	ε_0	$8.854\,187\,817 \times 10^{-12}$	exact	As/Vm	
Electron radius, classical	r_e	$2.817\,940\,326\,7 \times 10^{-15}$	$0.000\,000\,002\,7 \times 10^{-15}$	m	9.7×10^{-10}
Electron rest energy	E_0	$8.187\,105\,06 \times 10^{-14}$	$0.000\,000\,36 \times 10^{-14}$	eV	4.4×10^{-8}
Electron rest mass	m_0	$9.109\,382\,91 \times 10^{-31}$	$0.000\,000\,40 \times 10^{-31}$	kg	4.4×10^{-8}
Elementary charge	e	$1.602\,176\,565 \times 10^{-19}$	$0.000\,000\,035 \times 10^{-19}$	C = As	2.2×10^{-8}
Fine structure constant	α_f	$7.297\,352\,569\,8 \times 10^{-3}$	$0.000\,000\,002\,4 \times 10^{-3}$	dimensionless	3.2×10^{-10}
Inverse fine structure constant	$1/\alpha_f$	137.035 999 074	0.000 000 044	dimensionless	3.2×10^{-10}
Magnetic constant	μ_0	$12.566\,370\,614 \times 10^{-7}$	exact $= 4\pi \times 10^{-7}$	N/A^2	
Planck's constant	h	$6.626\,069\,57 \times 10^{-34}$	$0.000\,000\,29 \times 10^{-34}$	Js	4.4×10^{-8}
Planck's constant over 2π	$\hbar$	$1.054\,571\,726 \times 10^{-34}$	$0.000\,000\,047 \times 10^{-34}$	Js	4.4×10^{-8}
Proton rest mass	m_p	$1.672\,621\,777 \times 10^{-27}$	$0.000\,000\,074 \times 10^{-27}$	kg	4.4×10^{-8}
Proton-electron mass ratio	m_p/m_0	1836.152 672 45	0.000 000 75	dimensionless	4.1×10^{-10}
Rydberg's constant	R_∞	$10.973\,731\,568\,539 \times 10^{6}$	0.000 055	m^{-1}	5.0×10^{-12}
Speed of light *in vacuo*	c_0	$2.997\,924\,58 \times 10^{8}$	exact $= 1/\sqrt{\varepsilon_0 \mu_0}$	m/s	

Contribution of the National Institute of Standards and Technology.

X-ray spectra generally show the intensity of radiation or rather the number of its photons in relation to the wavelength of radiation or the energy of these photons. Normally, X-ray spectra consist of two different parts, the line spectrum and the continuous spectrum.

1.3.2.1 The Line Spectrum

A line spectrum will be produced if a target or sample is irradiated with X-ray photons, as just mentioned, or is bombarded with electrons (or ions). In both cases a sufficient energy of photons or electrons is needed. The energy must exceed the binding energy of a bound inner electron of the target atoms, which therefore is called the *critical excitation energy*. The ensuing effects can be described best by Niels Bohr's atomic model, which supposes Z electrons revolving around a nucleus in different orbitals or shells and subshells, where Z is the atomic number of the respective element.

Owing to the high-energy impact, an inner electron can be ejected from the atom so that a vacancy is created within the respective inner electron shell. The atom with the vacancy is in an instable state of higher energy and tries to regain its stable ground state by two different processes. In both processes an outer bound electron fills the vacancy and the atom instantly emits either an X-ray photon, which is the basic process of XRF, or what is called an Auger electron. The energy of the X-ray photon must be equal to the difference of the previous and the subsequent energy state of the atom:

$$E_{\text{photon}} = E_{\text{previous}} - E_{\text{subsequent}} \tag{1.3}$$

The newly created vacancy in the outer shell can be filled in turn by an electron still farther out, and another X-ray photon can be emitted. These processes will follow each other successively and a series of photons will be emitted until a free electron ultimately replaces an outermost valance electron so that the atom has finally returned to the ground state.

Since the energy states of atomic electrons are quantized and characteristic of all atoms of an element, the X-ray photons emitted in this way have individual energies that are equal for all atoms of the same element but different for atoms of different elements. Consequently, these photons cause discrete sharp lines or peaks as intensity maxima in an X-ray spectrum that are characteristic for any single element of the sample target. Conversely, any element of the sample can be identified by its characteristic lines or peaks, comparable to a fingerprint or barcode. For this reason, the line spectra are also called *characteristic* spectra. Of course, line spectra of the same sample either produced by X-ray photons or by electrons are similar.

Although not every outer electron is permitted to fill an inner vacancy, there are a lot of allowed transitions according to the selection rules of quantum theory. The most important transitions are indicated in Figure 1.9

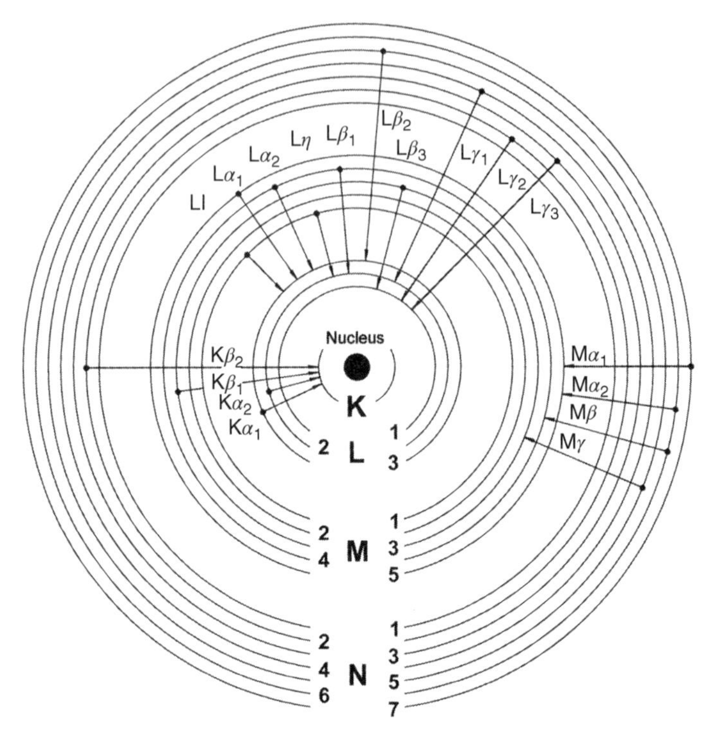

Figure 1.9. Electron transitions that are possible in a heavy atom and that produce the principal lines or peaks within an X-ray spectrum. Figure from Ref. [8], reproduced with permission. Copyright © 1996, John Wiley and Sons.

(see, e.g., Ref. [48]). They lead to the principal lines or peaks named here in the classical notation. There are three principal series, the K, L, or M series, which arise when the inner vacancy being filled is in the K, L, or M shell. A series contains several peaks named K, L, or M peaks, which mainly differ according to the origin of the outer electron. The most intense peak is called α, the next less intense peaks in descending order are called β, γ, η, and l. A further differentiation is made by an Arabic numeral added as an index, such as α_1 and α_2 for the α doublet. This classical notation proposed by Siegbahn after 1920 is not entirely systematic and indeed somewhat confusing. Meanwhile, the International Union of Pure and Applied Chemistry (IUPAC) has suggested a nomenclature that is solely based on the shell and subshell designation. Table 1.4 compares the K and L peaks in both notations [49].

From the preceding discussion it can be understood that the two lightest elements hydrogen and helium have no X-ray peaks at all because of their lack of inner electrons. However, all other elements have characteristic X-ray peaks. They appear in a spectrum with an intensity that depends on the energy of the primary X-rays or electrons, on the composition of the sample target, and on the efficiency of the detector. In the range up to 40 keV, normally each

TABLE 1.4. K and L X-ray Lines or Peaks in Siegbahn and IUPAC Notation

Siegbahn	IUPAC	Siegbahn	IUPAC	Siegbahn	IUPAC	Siegbahn	IUPAC
$K\alpha_1$	K–L_3	$L\alpha_1$	L_3–M_5	$L\gamma_1$	L_2–N_4	$M\alpha_1$	M_5–N_7
$K\alpha_2$	K–L_2	$L\alpha_2$	L_3–M_4	$L\gamma_2$	L_1–N_2	$M\alpha_2$	M_5–N_6
$K\beta_1$	K–M_3	$L\beta_1$	L_2–M_4	$L\gamma_3$	L_1–N_3	$M\beta$	M_4–N_6
$K\beta_2^I$	K–N_3	$L\beta_2$	L_3–N_5	$L\gamma_4$	L_1–O_3	$M\gamma$	M_3–N_5
$K\beta_2^{II}$	K–N_2	$L\beta_3$	L_1–M_3	$L\gamma_4'$	L_1–O_2	$M\zeta$	$M_{4,5}$–$N_{2,3}$
$K\beta_3$	K–M_2	$L\beta_4$	L_1–M_2	$L\gamma_5$	L_2–N_1		
$K\beta_4^I$	K–N_5	$L\beta_5$	L_3–$O_{4,5}$	$L\gamma_6$	L_2–O_4		
$K\beta_4^{II}$	K–N_4	$L\beta_6$	L_3–N_1	$L\gamma_8$	L_2–O_1		
$K\beta_{4x}$	K–N_4	$L\beta_7$	L_3–O_1	$L\gamma_8'$	L_2–$N_{6,7}$		
$K\beta_5^I$	K–M_5	$L\beta_8$	L_3–$N_{6,7}$	$L\eta$	L_2–M_1		
$K\beta_5^{II}$	K–M_4	$L\beta_9$	L_1–M_5	Ll	L_3–M_1		
		$L\beta_{10}$	L_1–M_4	Ls	L_3–M_3		
		$L\beta_{15}$	L_3–N_4	Lt	L_3–M_2		
		$L\beta_{17}$	L_2–M_3	Lu	L_3–$N_{6,7}$		
				Lv	L_2–$N_{6,7}$		

Source: From Ref. [49], reproduced with permission. Copyright © 1996, John Wiley and Sons.

element apart from H and He shows between 2 and about 10 intensive peaks, so that X-ray spectra in contrast to ultraviolet (UV) spectra can be regarded as fortunately poor in peak number. Figure 1.10 shows some examples for different pure elements excited at 40 keV. The lighter elements up to $Z = 25$ mostly show a Kα doublet, which is not resolved here, and a Kβ peak at higher energy. The heavier elements with $25 < Z < 57$ additionally have several L peaks mostly with an α doublet followed by a more energetic β and γ group. Heavy elements with $Z > 57$ are lacking in K peaks (their exciting potential is >40 keV) but show some M peaks in addition to the L peaks. In general, the most intensive K or L peaks are used for X-ray spectral analysis.

As mentioned earlier, the relationship of peak or photon energy and element was discovered by Moseley in 1913. He found that the reciprocal wavelength and consequently the photon energy are dependent on the atomic number Z of the elements. His well-known law can be described by

$$E_{ij} = k_{ij} \cdot (Z - \sigma_i)^2 \tag{1.4}$$

with certain constant values k_{ij} and σ_i for particular peaks or lines, j, of a series, i. This square law is demonstrated in Figure 1.11 by different parabolas, each representing a particular peak (see, e.g., Bertin [48]). As a consequence of Moseley's law, the atomic number of the elements could be ascertained beyond any doubt.

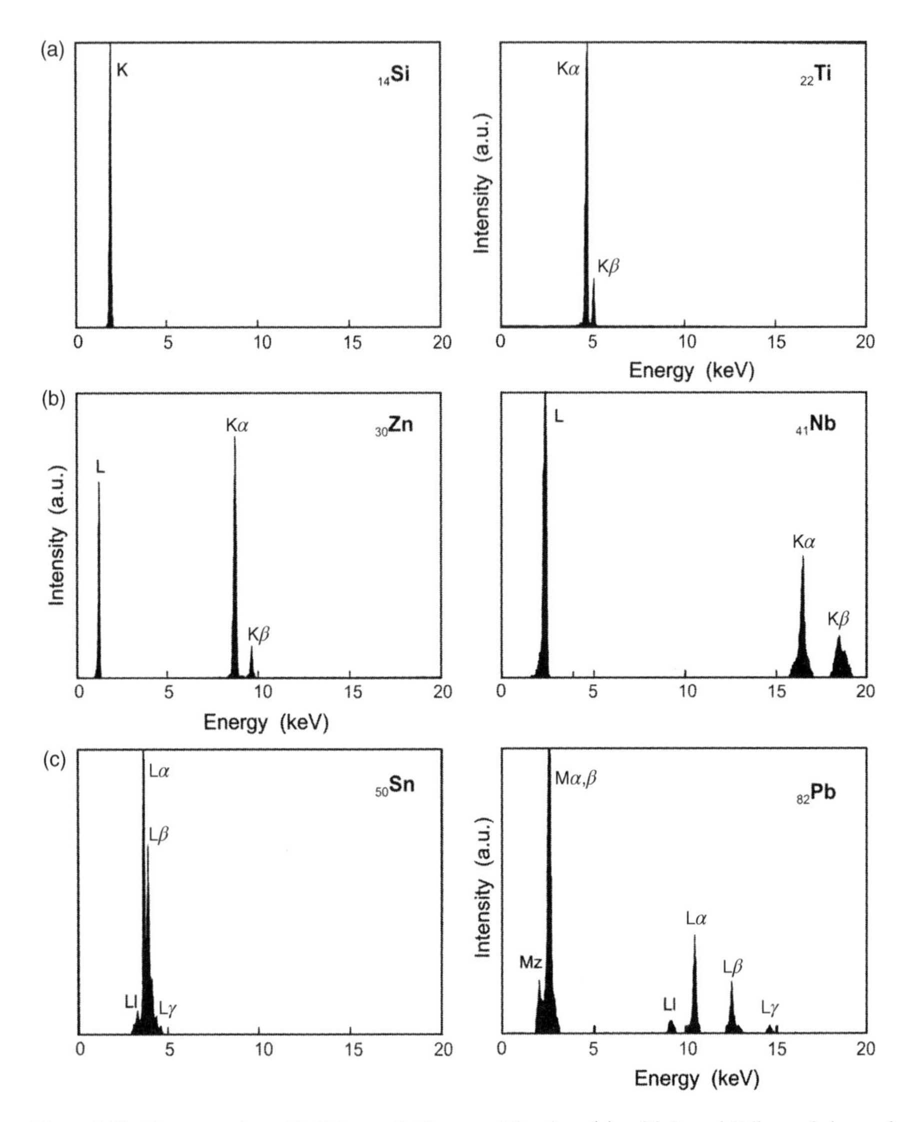

Figure 1.10. X-ray spectra with K lines of silicon and titanium (a), with L and K lines of zinc and niobium (b), and with L and M lines of tin and lead (c). The fluorescence intensity in arbitrary units is plotted against the photon energy in keV. Figure from Ref. [8], reproduced with permission. Copyright © 1996, John Wiley and Sons.

Quantum mechanics of atoms that are missing one single electron in an inner shell can explain Moseley's law. The values k_{ij} can be derived from

$$k_{ij} = R_{\mathrm{E}} \cdot \left(\frac{1}{n_i^2} - \frac{1}{n_j^2} \right) \tag{1.5}$$

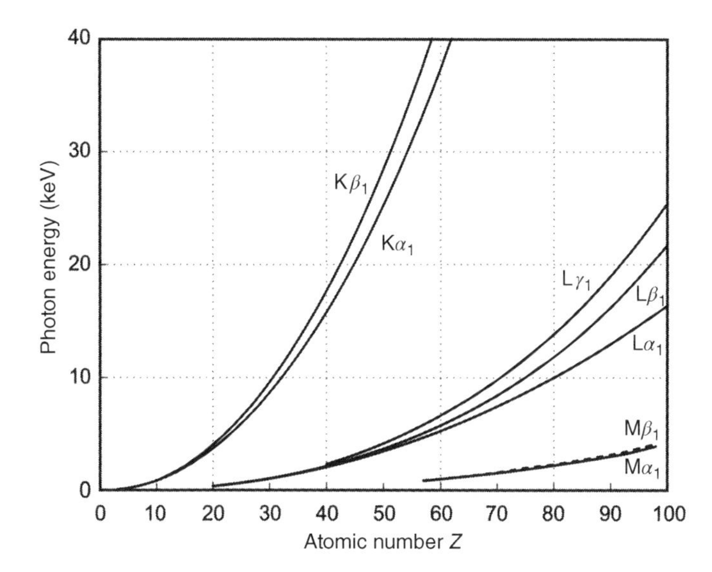

Figure 1.11. Moseley diagram of photon energies of the principal X-ray emission peaks dependent on the atomic number Z of the elements. Figure from Ref. [8], reproduced with permission. Copyright © 1996, John Wiley and Sons.

where R_E is bound up with the Rydberg energy, $R_\infty = R_E/hc$, which amounts to 13.606 eV (the binding energy of the 1s electron of hydrogen), and n_i and n_j are the principal quantum numbers of the involved inner and outer electrons, respectively, of the atom emitting the X-ray photon.

The quantities k_{ij} and σ_i characterize the different peaks. The factor k_{ij} for Kα_1 peaks is 10.20 eV, for the Lα_1 peaks is 1.89 eV, and for the Mα_1 peaks is 0.66 eV. The quantity σ_i can be interpreted as a shielding constant. From the point of view of an outer electron, it may be defined as that number of negative electrons by which the number of positive charges of the nucleus is reduced. $(Z - \sigma_i)$ may be defined as the effective nuclear charge. The σ_i values are not exactly equal for all elements but somewhat dependent on their atomic number Z. As listed in Table 1.5, experimental values for Kα_1 peaks are below 1.0; for Kβ_1 peaks, below 1.9; for Lα_1 peaks, below 7.4; and for Mα_1 peaks, at about 22.

The minimum excitation potential for the appropriate spectral series i can also be determined after Moseley's law. An inner electron of a target atom can only be expelled from its shell by an accelerated electron in the X-ray tube if its energy is above a minimum energy, E_{min}, also called critical excitation energy. During this process, the quantum number n_i of the inner electron changes to $n_j = \infty$, leading to

$$E_{min} = R_E (Z - \sigma_i)^2 \left(\frac{1}{n_i}\right)^2 \tag{1.6}$$

TABLE 1.5. Factors k_{ij} and Shielding Constants σ_i of Characteristic X-Ray Lines and Absorption Edges Experimentally Determined for Several Elements According to Moseley's Law

Quantum Number	$K\alpha 1$	$K\beta 1$	K Edge	$L\alpha 1$	L_3 Edge	$M\alpha 1$	M_4 Edge
n_j, initial	2	3	1	3	2	4	3
n_i, final	1	1	∞	2	∞	3	∞
k_{ij} [eV]	10.204	12.094	13.606	1.890	3.401	0.661	1.512

Atomic Number Z	σ_i for $K\alpha_1$	σ_i for $K\beta_1$	σ_i for K Edge	σ_i for $L\alpha_1$	σ_i for L_3 Edge	σ_i for $M\alpha_1$	σ_i for M_4 Edge
6–10	0.83						
11–15	0.93	1.64	2.11				
16–20	0.97	1.73	2.52				
21–25	0.97	1.80	2.85				
26–30	0.93	1.87	3.14	6.73	12.14		
31–35	0.86	1.87	3.38	6.95	13.26		
36–40	0.74	1.84	3.56	7.08	14.17		
41–45	0.57	1.76	3.66	7.19	14.94		
46–50	0.34	1.61	3.69	7.28	15.67		
51–55		1.48	3.67	7.34	16.37	21.35	32.93
56–60				7.39	17.08	21.46	33.92
61–65				7.40	17.76	21.65	35.28
66–70				7.39	18.43	21.85	36.87
71–75				7.35	19.07	22.08	38.31
76–80				7.31	19.67	22.29	39.67
81–85				7.25	20.20	22.47	40.86
86–90				7.20	20.64	22.61	41.85
91–95						22.68	42.39

For the K series with $n_i = 1$, σ_i takes experimental values of about 3; for the L series with $n_i = 2$, σ_i values lie around 17; and for the M series with $n_i = 3$, these values are around 39. E_{min} is also the minimum energy of photons, needed for the excitation in X-ray fluorescence when a sample is irradiated by an X-ray tube. And it gives the position of the absorption edges of the different K-, L-, or M-shells (see Section 1.4.1). The respective characteristic peaks of the elements always lie just below these edges.

Moseley's law is not very stringent since relative deviations of 0.1–0.2% for the lines and of 0.5–2% for the edges occur. Consequently, the exact positions of characteristic X-ray lines and respective edges are not calculated in practice by using this law but instead are obtained from tables or computer memories with measured values. They normally give the energies and wavelengths of the peaks and additionally their relative intensities within the defined K, L, or M series. The relative intensity of a certain peak in its series is determined by the probability of the electron transition causing this particular peak. The respective quantity is called emission rate g_{ij} and can be calculated from quantum mechanics. In general, the relative intensities are rather similar for most elements. For the K peaks, Kα:Kβ is about 100:15; for the L peaks, Ll:Lα:Lη:Lβ:Lγ_1:Lγ_3 is round 3:100:1:70:10:3; and for the M peaks, Mα:Mβ:Mγ is about 100:50:4.

The intensity of the total K, L, and M series is a function of the fluorescence yield ω_i. It gives the relative frequency according to which an X-ray photon and not an Auger electron is emitted after excitation of an atom. The relationship can be described approximately by

$$\omega_i = \frac{Z^4}{A + Z^4} \tag{1.7}$$

The constant A is about 9×10^5 for the K series; it is about 7×10^7 for the L series; and 1×10^9 for the M series.

The fluorescence yield for the K, L, and M series (see, e.g., Bertin [48]) correlates with the atomic number Z as shown in Figure 1.12. As demonstrated there, the X-ray photon and Auger electron emission are two competing effects, the frequencies of which sum up to 100%. The Auger process predominates for lighter elements, so that X-ray spectral analysis is not very effective for those elements with atomic numbers $Z < 20$ and especially for $Z < 10$. But for these elements, Auger spectroscopy is highly effective.

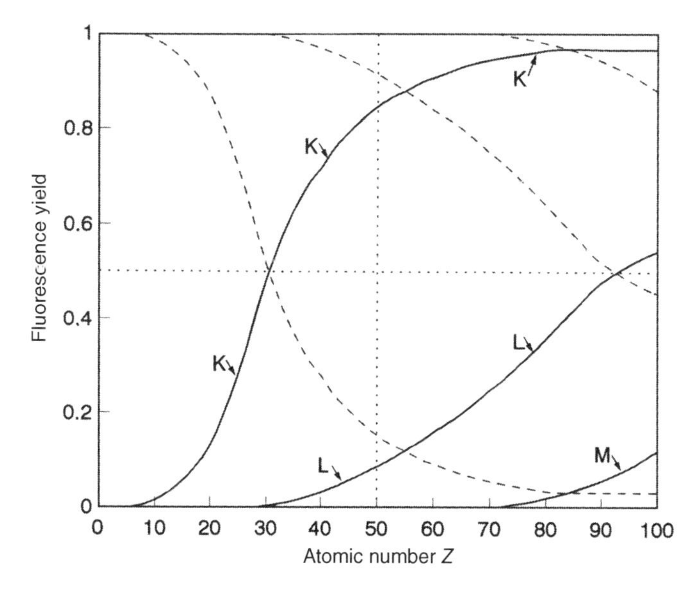

Figure 1.12. Fluorescence yield (———) and Auger electron yield (- - - -) as a function of the atomic number Z of the emitter. Figure from Ref. [8], reproduced with permission. Copyright © 1996, John Wiley and Sons.

The intensity of any characteristic line of a target element can be expressed in number of photons emitted per unit time or second. For lines in the primary X-ray spectrum, for example, produced by electrons of an X-ray tube, this intensity can be given by an empirical formula:

$$N_{\text{line}}\left(E_{ij}\right) = kIg_{ij}\omega_i \cdot E^m_{\text{min}}(E_0/E_{\text{min}} - 1)^m \tag{1.8}$$

where $N_{\text{line}}(E_{ij})$ is the number of photons with E_{ij}, k is a constant, I is the electron current, g_{ij} is the emission rate of the respective line, i, in its series, j, ω_i is the fluorescence yield of the target material, E_0 is the accelerating potential of the X-ray tube, and E_{min} is the critical excitation energy for the appropriate spectral series. The exponent, m, usually has a value below 2. It may be 5/3 for $E_{\text{min}} < E_0 < 2E_{\text{min}}$, and slowly decreases to 1.0 for E_0 values $>5E_{\text{min}}$. This equation for target lines is rather simple compared to the formula for a secondary X-ray spectrum produced by X-ray fluorescence of a sample.

There are some exceptions to the aforementioned selection rules. First, there are emission peaks that do not correspond to permitted transitions and therefore are called "forbidden" peaks. Second, there are additional peaks that arise from a double ionization by a simultaneous impact of photons or electrons on two inner electrons of the atom. As the energy levels of the doubly ionized atom slightly differ from those of the singly ionized atom, somewhat different peaks occur, which are called "satellite" peaks. Forbidden and satellite peaks are always weak, and satellites mainly appear in the K-spectra of lighter elements. Nevertheless, they must not be ignored in trace analysis if they are generated by a major component at energies close to small peaks of trace elements. Consequently, both forbidden and satellite peaks are also included in tables or stored in computers.

According to their energy position, the characteristic X-ray peaks are independent of the chemical bonding or state of the atoms. This advantage exists as long as only electrons from inner shells are involved in the X-ray emission process and as long as these electrons are not affected by the chemical vicinity of the atoms. In practice, this is the normal situation for the detection of higher photon energies and heavier elements. However, exceptions can appear for lower energies and lighter elements. If an electron from a valence or a near valence band is involved in the emission process, the respective energy level of the atom and the energy transition will be affected by the chemical state. Consequently, the characteristic peaks may be shifted for elements in different compounds. As the effect is in the range of a few electronvolts, it can be measured and used to get information on chemical bonding. However, other spectroscopic techniques are more efficient in this respect. For the usual X-ray spectrometrical practice, peak shifts are an exception but may be taken into account to avoid systematic errors.

1.3.2.2 The Continuous Spectrum

This kind of spectrum is defined by an intensity distributed continuously over a broad range of energy or wavelength covering about three orders of magnitude. For this reason, it is called "continuous" or "white spectrum." It is originally produced by energetic electrons or ions bombarding a target but actually not by X-ray photons themselves. However, if X-rays of a continuous spectrum are used to excite a sample, they will partly be scattered by this target and the original primary spectrum will be transformed into a somewhat modified spectrum that is likewise continuous but much smaller. Consequently, a continuous spectrum is present in any case, representing an inconvenient "background" that has to be eliminated from the analytical point of view. An example of a continuous spectrum produced by an X-ray tube is given in Figure 1.13. The characteristic L lines of the tube target are shown in addition.

A spectral continuum is produced by the fact that electrons penetrating into a target material are decelerated or retarded by impacts with the atomic nuclei of the target. The primary electrons lose their energy in these inelastic collisions, and this energy can be emitted as X-ray photons. A single electron can lose its energy completely in a single collision or stepwise in several consecutive collisions. Consequently, one single photon can be produced with the total electron energy or several photons with smaller parts of this energy. This is described in the corpuscular picture of X-ray photons with different energies.

In an X-ray tube operated at a voltage U_0 the electrons get the final energy E_0 according to

$$E_0 = e \cdot U_0 \tag{1.9}$$

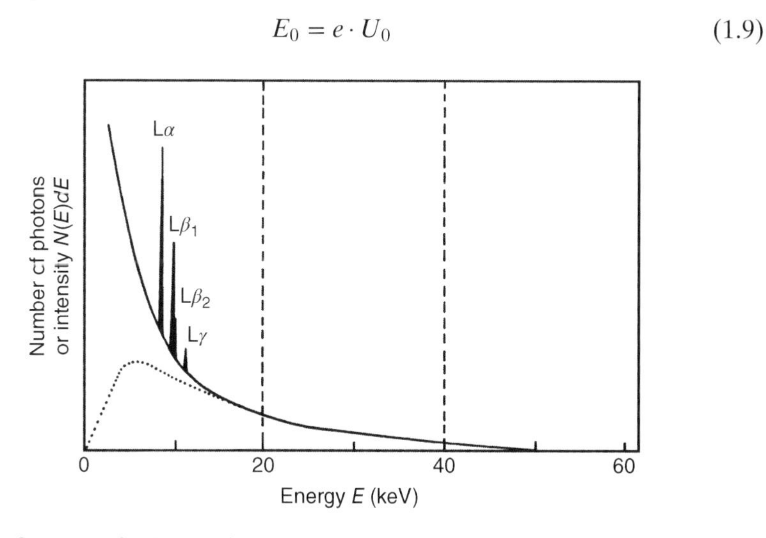

Figure 1.13. Spectrum of an X-ray tube operated at 50 kV and equipped with a thin target layer of tungsten as the anode. The spectrum is represented in the energy-dispersive mode. The K lines, with an excitation energy of 69.5 keV, cannot arise; but all L lines with a required minimum of 12.1 keV appear. The continuum of a thick solid target is illustrated by the dotted line. Figure from Ref. [8], reproduced with permission. Copyright © 1996, John Wiley and Sons.

where e is the charge of a single electron, called elementary charge ($e \approx 1.602 \times 10^{-19}$ Coulomb). Consequently, photons can carry away this maximum energy E_0 or lower energies down to zero. The spectrum covers the energy range between zero and the upper limit E_0 as shown in Figure 1.13. Since the retardation of electrons causes the continuous or white spectrum, this radiation is called *bremsstrahlung* in German (*bremsen* = to *brake and Strahlung* = *radiation*). The intensity distribution of the *brems*-continuum can approximately be described by

$$N(E)\Delta E \approx kIZ\left(\frac{E_0}{E} - 1\right)\Delta E \tag{1.10}$$

where $N(E)\Delta E$ is the number of photons emitted with energies between E and $E + \Delta E$, k is a constant, I is the tube current, and Z is the (mean) atomic number of the target. The formula shows the intensity or number of photons per second; it is inversely related to the energy of these photons, decreasing to zero when E approximates E_0. The graph of Figure 1.13 does not touch the energy-axis but it cuts this axis under the angle arctan ($-kIZE_0$). Furthermore, Equation 1.10 indicates that the intensity can be increased linearly by the tube current, the applied voltage (by virtue of E_0), and by the atomic number of the target material. Simultaneously, the line intensity is increased and the detection limits are lowered (see Section 6.1.2). For that reason, high-powered X-ray tubes equipped with a heavy-metal anode are usually applied in X-ray fluorescence analysis.

Equation 1.10 can be transformed into a wavelength-dependent equation known as Kramers' formula:

$$N(\lambda)\Delta\lambda \approx kIZ\left(\frac{\lambda}{\lambda_0} - 1\right)\frac{1}{\lambda^2}\Delta\lambda \tag{1.11}$$

where λ_0 corresponds to E_0 according to the Duane–Hunt law, $\lambda_0 = h{\cdot}c/E_0$. The relationship 1.11 is represented in Figure 1.14, which shows a sharp short-wavelength limit at λ_0, a broad hump with a maximum at

$$\lambda_{max} = 2\,\lambda_0 \tag{1.12}$$

and an extended long-wavelength tail [50].

The effects of target irradiation by electrons are highly complex and not completely understood. Consequently, the two spectral distributions of the continuum given by Equations 1.10 and 1.11 are only valid for thin target layers where electron backscattering and other effects can be ignored. For thick solid targets, they are approximations that are reasonably well in the high-energy or short-wavelength region (hard X-rays). For low energies or long wavelengths (soft X-rays), however, they are substantially modified by the self-absorption of

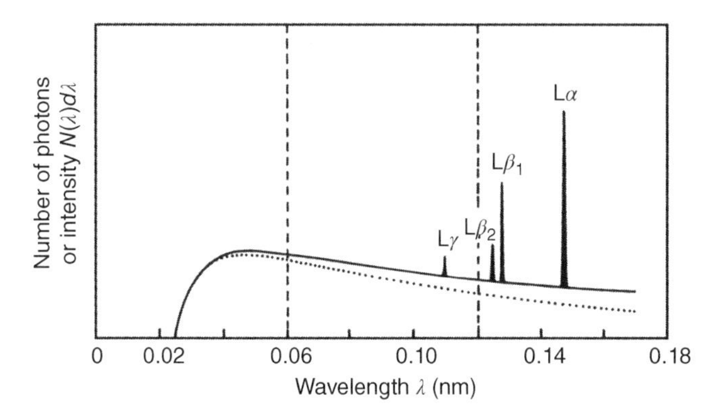

Figure 1.14. Spectrum of Figure 1.13 represented in the wavelength-dispersive mode. The X-ray tube operated at 50 kV may be equipped with a thin target layer (———) or a thick solid target (········) (see also Ref. [50]). Figure from Ref. [8], reproduced with permission. Copyright © 1996, John Wiley and Sons.

X-rays within the target itself. These absorption losses reduce the continuous spectrum, and this effect is further increased by the absorption of X-rays in the exit window of the X-ray tube. In the end, a quite different dotted curve results for the soft X-ray region. Figure 1.13 shows the corresponding continuum of a thick target with a maximum at E_{max} significantly below $0.5{\cdot}E_0$ whereas the respective distribution for a thin layer has no relative maximum at all. Figure 1.14 represents such a dotted curve for a thick target with a maximum λ_{max} above $2\lambda_0$.

The efficiency of X-ray production can be approximated for the continuum by integrating Kramers' formula (1.11). In relation to the input power of the X-ray tube P, the continuous radiation carries only a very small part, which is given by the empirical relation

$$\eta_{cont} \approx 1 \times 10^{-9}\, ZU_0 \tag{1.13}$$

In practice, η values between 0.001 and 0.7% can be reached. Most of the input power is converted into heat and dissipated by water-cooling.

1.3.3 Polarization of X-Rays

A wave is called "linearly polarized" if it oscillates only in one direction perpendicular to the direction of its propagation. That means that only transversal waves can be polarized. Usually, an original beam of visible light and also of X-rays is not polarized because all the different light or X-ray photons irregularly oscillate in different directions. However, shortly after the detection of X-rays, it became known that X-rays can be polarized like visible light; that means the electric and magnetic field vector of special X-ray beams oscillate only in a certain direction perpendicular to the beam direction.

In 1904, Barkla discovered the polarization of X-rays, which were emitted from an X-ray tube in a particular direction. He showed that X-rays are linearly polarized if the X-ray beam is perpendicular to the electron beam of the X-ray tube [5]. In this case, the electric field vector only oscillates in the plane spanned by both beams. Furthermore, if X-rays are scattered from a paraffin block they are linearly polarized since it can be shown that a second block can scatter these X-rays only in a particular direction [1].

The polarization has given evidence for the wavelike character of X-rays and moreover for the transversal kind of X-ray waves. Much later, synchrotron radiation was shown to be linearly polarized in the plane of the storage ring. The polarization of X-rays can be used to reduce the spectral background and thereby to improve the detection limits in X-ray fluorescence analysis.

1.3.4 Synchrotron Radiation as X-Ray Source

The first synchrotrons were constructed by Edwin Mattison McMillan in the United States and by Vladimir Iosifovich in the former Soviet Union in 1945, Synchrotron radiation (SR) was discovered in 1947 by General Electric in New York [51] when a bright arc of visible light was observed for the first time at an electron accelerator. Its closed electron tube was partly covered by a transparent instead of an opaque coating so that radiation became visible. Today, SR is obtained from a storage ring in which charged particles like electrons are stored in several bunches and maintained at high constant velocity or kinetic energy. The particles come from the actual accelerator ring called booster where they are accelerated by electric fields to almost light velocity. The relativistic particles are forced into a fixed circular orbit by several strong magnets. They are accelerated radially by the magnetic fields and hereby produce a brilliant radiation. Extended descriptions can be found in the literature [52–57] and online in the Internet [58–61].

Originally, the first users of synchrotrons constructed such large accelerators for particle physics. These machines called colliders were applied to high-energy collisions, for example, of electrons and positrons, as a new branch of science. SR was regarded as an undesired loss of energy that had to be compensated. Only several years later, scientists used the highly brilliant radiation emerging from those machines and recognized its incomparable potential for research offered to physicists, chemists, geologists, physicians, biologists, engineers, and art historians. The benefits of SR in all disciplines of application are unequalled.

A synchrotron facility usually consists of an electron or positron source, a first linear accelerator, a second circular accelerator, called a booster-synchrotron, and a storage ring that consists of a metallic tube with circular and straight sections with a total length from some meters up to several kilometers (Figure 1.15). In the booster-synchrotron, the electrons are accelerated by high-frequency (HF) amplifiers or clystrons to nearly light velocity. Around the storage ring, several dipole magnets, so-called bending magnets (BMs), are

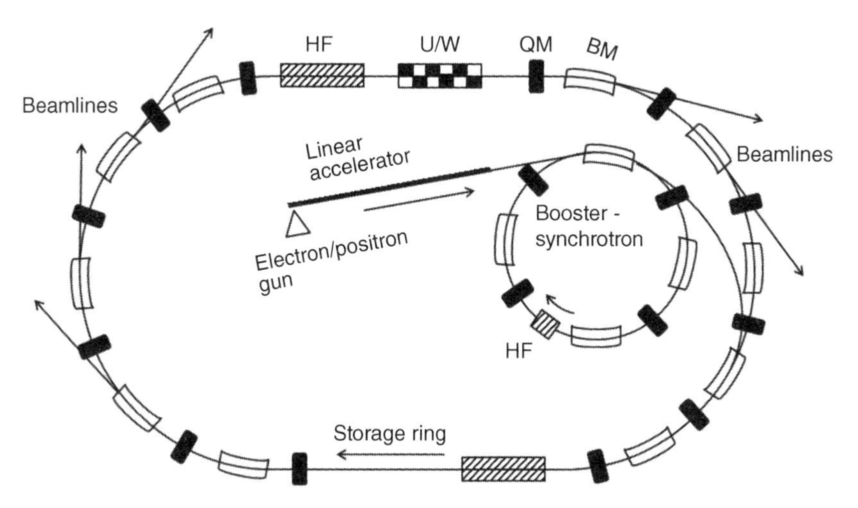

Figure 1.15. Schematic representation of a dedicated synchrotron facility. Essential parts are the electron or positron gun, the first linear accelerator, the second circular accelerator, called synchrotron or booster, the storage ring with a length of some 10 m up to several km. Different bending magnets (BM) and undulators or wigglers (U/W) are inserted for the production of the radiation. Quadrupole magnets (QM) are used for focusing the electrons. Several beamlines are arranged around the storage ring, always tangentially to the ring. A few electric HF fields provide energy in order to compensate for the radiation losses.

aligned, which force the electron beam into a fixed circular orbit. Furthermore, several quadrupole magnets (QMs) are used for focusing the electron beam again and again. The booster and the closed storage-ring are evacuated to ultra-high vacuum. Tangentially to the storage ring, several light pipes are attached at extra holes or switches and at the end of these beam lines different experiments are installed.

Electrons or positrons in the booster synchrotron come up to a relativistic velocity of more than 0.99999*c*. Furthermore, they get a high mass connected with a high kinetic energy of more than 100 MeV. In the storage ring, the fast electrons or positrons do not get any additional kinetic energy. Within the curved parts of the ring, the electrons are deflected radially by magnetic fields on account of the Lorentz force and lose kinetic energy in the form of SR (see Section 3.3.1). In the linear or straight parts of the ring, undulators and wigglers (U/W) are inserted, which also deflect the charged relativistic particles and produce SR (see Section 3.3.2). This radiation emerges tangentially to the orbit of the relativistic electron or positron bunches within the beamlines. The energy loss caused by the radiation of the particles is compensated by HF amplifiers in some straight sections of the ring.

Synchrotron radiation is also emitted in the first accelerator or booster but it is only used for diagnostics. Single electron bunches from the booster are filled into the storage ring one after the other when they have reached a certain relativistic velocity or kinetic energy. In the ring, they build an electron beam with bunches and gaps. This "train" uniformly rotates with almost light velocity in the ring. The temporary energy loss of the electrons according to the emitted SR is compensated again and again so that the original high velocity and kinetic energy of the electrons is kept constant here.

Synchrotron radiation covers a wide range of the electromagnetic spectrum from the infrared to hard X-rays with about eight orders of magnitude (see Figure 1.8). According to the electron or positron bunches, the radiation is pulsed. Furthermore, it is strongly collimated in a narrow cone in the forward pointing direction (fan with a vertical divergence of 0.1 mrad or 0.006° and a horizontal divergence of 5 mrad or 0.3°) and it is highly polarized. Linear polarization is ascertained in the orbital plane of the particles and elliptical polarization is observed at a small angle to that plane. Usually, a small bandwidth of the white radiation is selected to perform experiments with quasi-monochromatic radiation.

The scientific community using synchrotron light for spectrometry and diffractometry has been continuously growing since the 1970s. The high brilliance of SR sources exceeds that of conventional X-ray tubes by more than five orders of magnitude (see Section 3.3.3) and allows many experiments at the micro- and nanoscale with high lateral resolution and/or temporal resolution not known beforehand. Nowadays, synchrotrons are especially designed and constructed to produce this radiation. Facilities of the fourth generation will supply intense SR for a variety of experiments in a broad field of applications. The data (e.g., size and performance) of more than 60 machines spread over the whole world can be accessed online [60].

Synchrotron radiation is not only generated artificially by special electron accelerators but also appears naturally in astronomic objects. The radio emission of several galactic and extragalactic sources is caused by relativistic electron clouds trapped in strong magnetic fields. They can be found in jets of black holes, and in the nebulae of pulsars and quasars and can emit strong SR even in the X-ray region of the electromagnetic spectrum.

1.3.4.1 Electrons in Fields of Bending Magnets

Charged moving particles can be deflected in the magnetic field of a bending electromagnet. A particle with an electric charge q uniformly moving with a velocity $\boldsymbol{v}$ in a homogeneous magnetic field $\boldsymbol{B}$ is forced on a curved trajectory, for example, a circle, a spiral or a sinusoidal curve, by the so-called Lorentz force:

$$F = q[\boldsymbol{v} \times \boldsymbol{B}] \tag{1.14}$$

All vectors are bold-faced and $[\boldsymbol{v} \times \boldsymbol{B}]$ is the cross-product of $\boldsymbol{v}$ and $\boldsymbol{B}$.

For the synchrotron orbit spherical coordinates are chosen: r is the distance of a point from the center or origin (usually radius ρ_m), θ is the polar angle (between 0° and 2π or 360°), and ψ is the azimuthal angle of a position (0° within

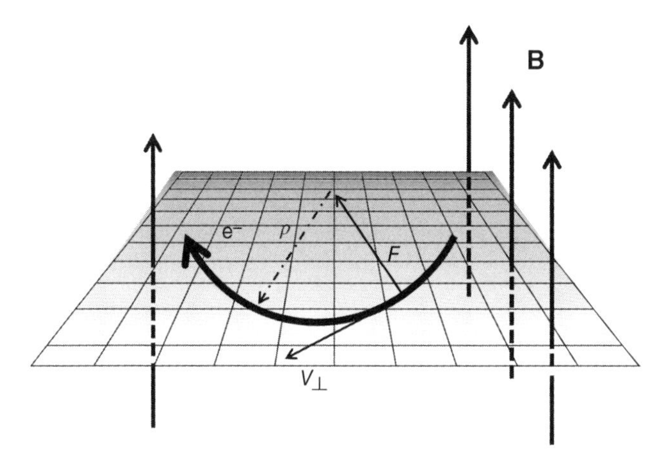

Figure 1.16. Trajectory of an electron e^- in a homogeneous magnetic field $\boldsymbol{B}$ perpendicular to its velocity $\boldsymbol{v}$. The electron is deflected by the Lorentz force $\boldsymbol{F}$ and moves uniformly on a circular orbit with radius ρ_m in the plane perpendicular to $\boldsymbol{B}$. The acceleration $\boldsymbol{a}$ is always directed to the center of the circle and is vertical to $\boldsymbol{B}$. Consequently, $\boldsymbol{a}_{||}$ is zero and $\boldsymbol{a}_{\perp}$ is the decisive quantity.

the orbit). The vectors $\boldsymbol{v}$ and $\boldsymbol{B}$ include the so-called pitch angle θ_p. If the magnetic field strength $\boldsymbol{B}$ is perpendicular to the velocity $\boldsymbol{v}$, the pitch angle is $\pi/2$ or 90°, and the charged particle will move on a circular curve within a plane perpendicular to $\boldsymbol{B}$ (Figure 1.16). The centripetal force for the particle and the Lorentz force can be set equal:

$$m_{\text{part}} \cdot \frac{v^2}{|\rho_m|} = q \cdot |\boldsymbol{v}| \cdot |\boldsymbol{B}| \tag{1.15}$$

where m_{part} is the mass of the particle and ρ_m is the radius of curvature of the circular orbit. Consequently, the radius can be calculated by

$$|\rho_m| = \frac{m_{\text{part}} \ |\boldsymbol{v}|}{q \cdot |\boldsymbol{B}|} \tag{1.16}$$

For relativistic atomic particles, Equation 1.15 has to be modified:

$$\frac{\mathrm{d}}{\mathrm{d}t}(m_{\text{part}} v) = q \cdot |\boldsymbol{v}| \cdot |\boldsymbol{B}| \tag{1.17}$$

The kinetic energy E_{part} and the particle mass are dependent on its relativistic velocity according to the Lorentz factor γ:

$$\gamma = \frac{m_{\text{part}}}{m_0} = \frac{E_{\text{part}}}{E_0} \tag{1.18}$$

where m_0 is the rest-mass of the particle and $E_0 = m_0\, c_0^2$ is its rest-energy. With $\beta = \upsilon / c_0$ we get

$$\gamma = \frac{1}{\sqrt{1-\beta^2}} \quad \text{and relatedly} \quad \beta = \sqrt{1 - \frac{1}{\gamma^2}} \tag{1.19}$$

Equation 1.17 can be rewritten as

$$\frac{\mathrm{d}}{\mathrm{d}t}(\gamma\, m_0 \upsilon) = q \cdot |\boldsymbol{\upsilon}| \cdot |\boldsymbol{B}| \tag{1.20}$$

Differentiation of the left-hand side leads to

$$m_0 \frac{\mathrm{d}}{\mathrm{d}t}(\gamma\, \boldsymbol{\upsilon}) = m_0 \gamma \boldsymbol{a} + m_0 \gamma^3 \boldsymbol{\upsilon} \frac{(\boldsymbol{\upsilon} \cdot \boldsymbol{a})}{c^2} \tag{1.21}$$

where $\boldsymbol{a}$ is the acceleration of the particle defined by $\mathrm{d}\boldsymbol{\upsilon}/\mathrm{d}t$. If $\boldsymbol{\upsilon}$ is perpendicular to $\boldsymbol{B}$, $\boldsymbol{a}$ is perpendicular to $\boldsymbol{\upsilon}$. Consequently, the scalar product $\boldsymbol{\upsilon}\, \boldsymbol{a}$ vanishes and we get

$$m_0 \cdot \gamma |\boldsymbol{a}| = q \cdot |\boldsymbol{\upsilon}| \cdot |\boldsymbol{B}| \tag{1.22}$$

For electrons or positrons, q is the elementary charge of about 1.602×10^{-19} C, m_0 is about 9.109×10^{-31} kg, and E_0 is nearly 511 keV. Table 1.6 gives typical β values for relativistic electrons, the corresponding

TABLE 1.6. Increasing β Values of Accelerated Electrons, their Lorentz Factor γ, their Energy and Mass, the Difference of their Velocity and the Light Velocity, and the Critical Energy of Emitted Photons

β	γ	E_{el} [MeV]	m_{el} [10^{-31} kg]	$(\upsilon - c_0)$[m/s]	E_{crit} [eV]
0.1	1.005	0.514	9.2	-2.7×10^7	9×10^{-8}
0.9	2.294	1.172	21	-3.0×10^6	1×10^{-6}
0.95	3.203	1.637	29	-1.5×10^6	3×10^{-6}
0.99	7.089	3.622	65	-3.0×10^5	3×10^{-5}
0.999	22.37	11.43	204	−29979	0.001
0.9999	70.71	36.13	644	−2998	0.03
0.99999	223.6	114.3	2037	−300	0.99
0.999999	707.1	361.3	6441	−30	31
0.9999995	1000	511.0	9109	−15	89
0.9999999	2236	1143	20369	−3.0	991
0.999999942[a]	2936	1500	26746	−1.7	2243
0.99999999	7071	3613	64413	−0.3	31333

[a] This value is realized at the synchrotron facility DELTA of the TU Dortmund.

Lorentz factor, the energy and the mass of the electrons, and the difference between υ and c. It can be recognized that γ, E_{el}, and m_{el} strongly increase with β while $\boldsymbol{\upsilon} - c_0$ strongly decreases, that is, $\boldsymbol{\upsilon}$ approaches c_0. The conditions for relativistic positrons are the same.

With help of Equation 1.18, with $q = e$, $\boldsymbol{\upsilon} = \beta\, c_0$, and $E_0 = m_0\, c_0^2$, the radius of curvature $|\rho_m|$ determined by Equation 1.16 can be converted into

$$|\rho_m| = \frac{\beta}{e c_0} \cdot \frac{E_{el}}{|\boldsymbol{B}|} \tag{1.23}$$

Since β is nearly constant in the booster and in the storage ring, this equation makes clear that the ratio of the kinetic energy and the strength of the magnetic field $E_{el}/|\boldsymbol{B}|$ must be constant in order to keep the radius of the trajectory constant and to fix the electron beam in a closed orbit. In the booster synchrotron, the field-strength has to be increased synchronously with the kinetic energy of the electrons. The name synchrotron is due to this condition. In the storage ring, $|\boldsymbol{B}|$ has to be kept constant because E_{el} should be constant.

1.3.4.2 Radiation Power of a Single Electron

The particles running in accelerators and storage rings undergo energy losses, which can be described combining the expression for the radiation rate of a relativistic particle with the expression for the acceleration of the particle in its orbit [52]. The acceleration, $\boldsymbol{a}$, is always perpendicular to $\boldsymbol{\upsilon}$ and to $\boldsymbol{B}$. Consequently, its component $\boldsymbol{a}_{||}$ (parallel to $\boldsymbol{B}$) is zero and the particle does not gain kinetic energy. However, the component $\boldsymbol{a}_{\perp}$ (perpendicular to $\boldsymbol{B}$) is not zero and determines the bending radius ρ_m of the electron trajectory, according to [58,62]

$$|\boldsymbol{a}_{\perp}| = \frac{q}{m} \cdot |\boldsymbol{B}| \cdot |\boldsymbol{\upsilon}| = \frac{\upsilon^2}{\rho_m} \approx \frac{c_0^2}{\rho_m} \tag{1.24}$$

Charged particles are deflected in the magnetic field according to $\boldsymbol{a}_{\perp}$ in the direction of the guiding center of a circle with ρ_m, so they are retarded radially inward. Because of this fact they lose energy as a kind of relativistic "bremsstrahlung."

The basic quantity for the radiation emerging from highly accelerated particles in perpendicular magnetic fields is the total radiation loss ($\Delta E/\Delta t$). It is a power quantity measured as energy per unit time emitted as photons with undefined energies. For a single relativistic particle the total radiation loss is given by [58,62–64]

$$\left(\frac{\Delta E}{\Delta t}\right) = \frac{2}{3} \cdot \alpha_f \cdot \hbar \cdot \frac{\gamma^4}{c_0^2} \cdot |\boldsymbol{a}_{\perp}|^2 \tag{1.25}$$

For a singly charged particle, the quantity α_f is Sommerfeld's fine-structure constant, defined by

$$\alpha_f = \frac{1}{4\pi\epsilon_0} \cdot \frac{e^2}{\hbar c_0} \tag{1.26}$$

where ε_0 is the electric field constant, and $\hbar = h/2\pi$ is the reduced Planck constant. α_f is dimensionless and amounts to 1/137. It determines the power of the electromagnetic interaction and is a measure of the probability for a photon to be coupled to an electron or positron. Consequently, it is decisive for the emission rate that equals the radiation loss $(\Delta E/\Delta t)$.

For relativistic particles with $\boldsymbol{v} \approx c_0$ and $\beta \approx 1$, we find from Equation 1.25 by use of Equations 1.23 and 1.24

$$\left(\frac{\Delta E}{\Delta t}\right) = \frac{2}{3} \cdot \alpha_f \cdot \hbar \cdot \left(\frac{c_0}{\rho_m}\right)^2 \gamma^4 = \frac{2}{3} \cdot \alpha_f \cdot \hbar \cdot \frac{q^2 E_{el}^2 |\boldsymbol{B}|^2}{m_0^4} \tag{1.27}$$

These equations show that the radiation loss induced by a single electron is dependent on $1/|\rho_m|^2$ and on γ^4, E_{el}^2, $|\boldsymbol{B}|^2$, and $1/m_0^4$. Large values of γ, E_{el}, and $|\boldsymbol{B}|$ lead to the high photon fluxes of synchrotron radiation while small values of $|\rho_m|$ and of m_0 also effect a strong radiation. Because of their low rest-mass, the radiation power for electrons is 10^{13} times stronger compared with that for protons. That is the reason why electrons or positrons are used as sources for synchrotron radiation in preference to protons or α particles. A single electron or positron causes power values of $>10^{-7}$ W. All electrons together in a beam of some 100 mA yield a radiation energy of some 30 kW [63].

1.3.4.3 Angular and Spectral Distribution of SR

The trajectory of charged particles crossing a homogeneous field is an arc of a circle as already demonstrated in Figure 1.16. The geometry of the radiation emitted by relativistic electrons observed in the rest frame of the relativistic electron, S', is that of a dipole. The angular distribution of the radiation with respect to the velocity vector in S' is

$$I_v \propto \sin^2\theta' = \cos^2\Delta\theta' \tag{1.28}$$

where θ' is the horizontal emission angle and $\Delta\theta'$ is its compliment to $\pi/2$ or 90°. I_v gives the probability distribution of photons that are emitted from a dipole leading to the isotropic dipole pattern of this figure with a width of π or 180°.

The aberration between this rest frame S' and the reference or laboratory frame S can be described by [58,62]

$$\sin\Delta\theta = \frac{1}{\gamma} \frac{\sin\Delta\theta'}{1+\beta\cos\Delta\theta'} \quad \text{and} \quad \cos\Delta\theta = \frac{\cos\Delta\theta' + \beta}{1+\beta\cos\Delta\theta'} \tag{1.29}$$

At the angles $\theta' = \pm\pi/2$, the intensity of the emitted radiation is zero in S'. According to Equation 1.29, the corresponding angles in S meet the condition of $\sin \Delta\theta = \pm 1/\gamma$ and $\cos \Delta\theta = \beta$. Consequently, $\Delta\theta$ is about $\pm 1/\gamma$. The radiation is beamed in the direction of the particles motion within $-1/\gamma < \Delta\theta < +1/\gamma$. It leads to the forward-pointing narrow cone of Figure 1.17 where the observation line and the velocity vector of the accelerated particles coincide within a horizontal emission angle of about $2/\gamma$. This is also valid for the vertical emission angle. The higher the speed of the electrons, the narrower the momentary emission cone of the photons become. Within a certain time interval the cone describes a fan that may be 10 times wider.

The emitted radiation of a single relativistic electron is a flux with a large number of photons. They all do not have the same energy but carry very

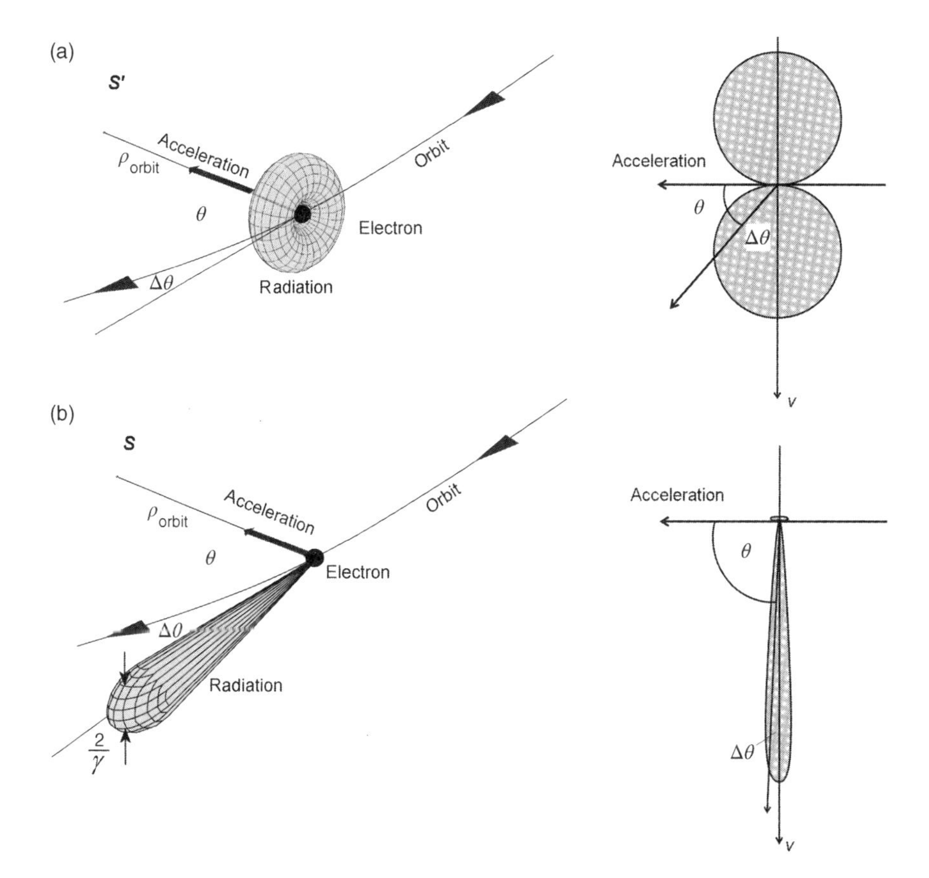

Figure 1.17. The emitted radiation of relativistic electrons deflected by a homogeneous magnetic field $\boldsymbol{B}$ vertical to the orbit plane. The rest frame S' of the electron (a) shows an isotropic dipole pattern while the relativistic frame S of the observer (b) gives a narrow cone in forward direction. Three-dimensional view (left). Cross-section of the orbit plane (right). θ and $\Delta\theta$ are emission angles between velocity $\boldsymbol{v}$ and Lorentz force $\boldsymbol{F}$ or acceleration $\boldsymbol{a}$. Its sum is 90° or $\pi/2$. Figure from Ref. [53], reproduced with permission from K. Wille.

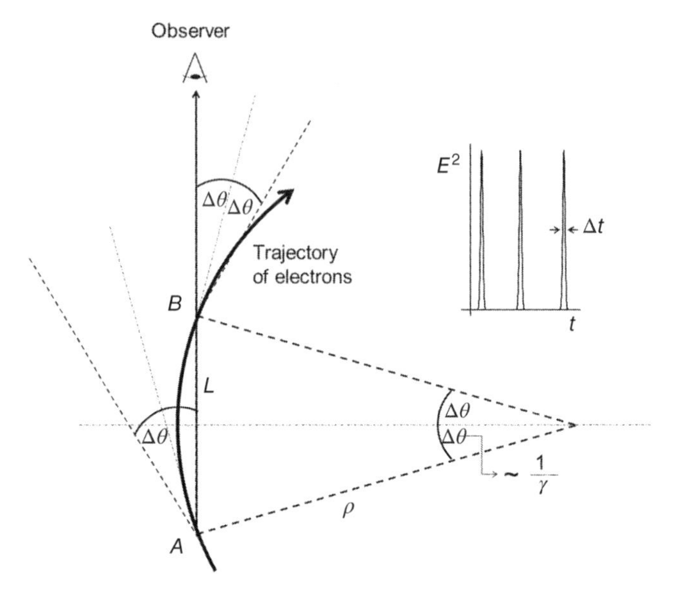

Figure 1.18 Observation of photons emitted from a relativistic photon by its deflection in a perpendicular magnetic field. The first photon emitted from point A moves to point B on the secant with length L at light velocity. The last photon is emitted from point B after the electron has covered the arc AB with the velocity $\boldsymbol{v}$. The observer detects an ultrashort radiation pulse with length Δt, which is equal to the difference between the time of flight of the electron and the photon. L is about 2 mm, $\Delta\theta$ is about 0.02°, and the bunch length is some centimeters. Figure from Ref. [53], reproduced with permission from K. Wille.

different photon energies. This spectral nature of the relativistic radiation emitted at a bending electromagnet in a storage ring can be illustrated in a simple way [52,53,63]. As demonstrated in Figure 1.18, a single electron moves on an arc of a circle with the velocity $\boldsymbol{v} = \beta\ c_0$ within a horizontal emission angle $2\Delta\theta = 2/\gamma$. An observer tangential to the ring will record a first photon emitted from point A. The first photon runs from A to B on the secant of the circle with the horizontal emission angle. Simultaneously, the electron moves from A to B on the corresponding arc of that circle and emits the last photon from point B. After that, both photons reach the observer on a straight line.

The duration of the electromagnetic pulse recorded by the observer is the time difference between both photons. It is identical to the difference between the time of flight of the electron and of the first photon, in both cases from A to B, on the condition that the electron runs behind the photon with $\boldsymbol{v} < c$. According to Longair [62] and Weis *et al.* [63], the pulse length caused by the photons is approximately

$$\Delta t \approx L\left(\frac{1}{\upsilon} - \frac{1 - 1/6\Delta\Theta^2}{c_0}\right) \tag{1.30}$$

where L is the secant of the circle or the distance between A and B. This quantity can be estimated to

$$\Delta t \approx \frac{4}{3}\frac{\rho_m}{c_0} \cdot \frac{1}{\gamma^3} \tag{1.31}$$

which is typically of the order of 10^{-18} s down to 10^{-20} s, that is, the photon pulses are ultrashort. After the mentioned electron, the next electron of the bunch will produce the next pulse. Altogether some 10×10^{12} (some 10 billion, Europe; and some 10 trillion, US and UK) of electrons typically belong to a bunch and participate in the emission of photons. The distance of electrons is about some pm, while the secant of the electron orbit being observed is some mm.

Usually, many bunches of electrons or positrons (up to some hundred) are grouped in a storage ring separated by just as many gaps. The bunch length is some cm, the gaps or empty parts are about 50 times longer, and the storage ring may typically have a circumference of some 100 m. The chain of bunches with the electrons rotate within the ring like a train with nearly light velocity. They rotate with the gyrofrequency $\omega = c_0/\rho_m$, also called Larmor frequency, which amounts to about 100 MHz for magnets with a bending radius of 3 m. The electrons of a bunch induce the emission of photons with a period of some 10^{-20} s, the photons recorded by an observer cause an electronic pulse of some 10^{-19} s, a bunch emits photons during some 10^{-11} s, the bunches follow each other after some 10^{-9} s. They cross the homogeneous magnetic field within some 100 ns up to 100 μs and yield a beam current of some 100 mA.

According to Wille [53], the radiation pulses have first to be corrected for effects of aberration and time retardation. Afterward, a Fourier transformation for the pulses has to be carried out in order to get the spectrum of this radiation. In accord with Heisenberg's relation, the very short pulses of some 10^{-19} s give a spectrum with a broad bandwidth, $\Delta E = h/\Delta t$, which amounts to nearly 10 keV. The spectral photon flux $(\Delta W/\Delta E)$ dependent on the energy of the emitted photons was first derived from theory by Schwinger [54] as mentioned in [53,58,62]. In order to compare the results with the spectra used in energy-dispersive X-ray spectrometry, the photon energy was chosen here as the variable quantity instead of the wavelength or the frequency. The calculation is discussed in detail (e.g., by Jackson [65] and Hoffmann [66]) leading to the expression

$$\left(\frac{\Delta W}{\Delta E}\right) = \sqrt{3} \cdot \alpha_f \cdot \gamma \cdot x \cdot \int_x^\infty K_{5/3}(\xi)\mathrm{d}\xi \tag{1.32}$$

where x is the photon energy E related to E_{crit}, which will be defined later. $K_{5/3}$ is a modified Bessel function of the fractional order 5/3; its integral goes from $x = E/E_{crit}$ up to ∞. This basic quantity $(\Delta W/\Delta E)$ is dimensionless and can be converted into the number of photons per electron. It is usually standardized to

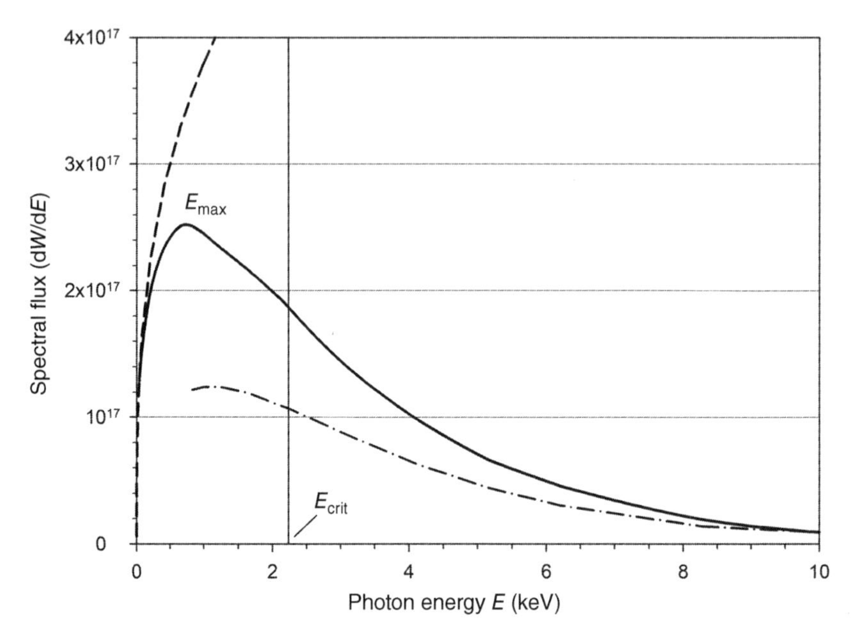

Figure 1.19. Typical continuous spectrum of synchrotron radiation. The spectral flux (dW/dE) is dependent on the energy E of the emitted photons. It is given as a dimensionless quantity in arbitrary units. The energy E_{el} of the relativistic electrons in the storage ring was chosen to be 1.5 GeV (DELTA at the TU-Dortmund) so that the critical energy E_{crit} was 2.24 keV. The area below the curve for $E \leq E_{crit}$ is equal to the area for $E \geq E_{crit}$. Two asymptotic approximations are presented: for low photon energies (dashed curve) and for high photon energies (dashed-dotted).

an electron beam current of 1 Ampere with 6.242×10^{18} electrons per second (trillions, Europe; or quintillions, US and UK) and to a spectral bandwidth of 0.1% (that means a bandwidth ΔE of 1 eV at $E = 1$ keV).

The radiation is emitted in a narrow cone and shows a bell-shaped distribution, which is dependent on the observation angle in the horizontal orbit plane and in the vertical plane. The flux shows a maximum for in-plane and on-axis and is usually given for in-plane ($\psi = 0$) or it is averaged by integration over the vertical divergence $\Delta\psi$. A typical distribution is demonstrated in Figure 1.19 for a small synchrotron facility. It shows a linear plot of the averaged spectral flux dependent on the energy of the emitted photons. The black curve looks like a broad hump with an increasing and a decreasing branch. The photon energy, which divides the area below the curve into two equal parts, is denoted as the critical (or characteristic) energy. Half the flux is emitted for $E \leq E_{crit}$, the other half presents energies $E \geq E_{crit}$. The critical energy is given by

$$E_{crit} = \frac{3}{2} \cdot \hbar \cdot \frac{c_0}{\rho_m} \cdot \gamma^3 \tag{1.33}$$

Different values for E_{crit} have already been listed in Table 1.6 provided that the bending radius is chosen to be 3.34 m (at DELTA in Dortmund).

The integral of the Bessel function in Equation 1.32 can be approximated for two special cases. For small energies ($E \ll E_{crit}$) of the emitted photons, the increasing branch can be written as

$$\left(\frac{\Delta W}{\Delta E}\right) \cong \frac{11}{9} \cdot \pi \cdot \alpha_f \cdot \gamma \cdot \left(\frac{E}{E_{crit}}\right)^{1/3} \tag{1.34}$$

For high energies ($E \gg E_{crit}$) of photons, the decreasing branch is

$$\left(\frac{\Delta W}{\Delta E}\right) \cong \frac{2}{3} \cdot \pi \cdot \alpha_f \cdot \gamma \cdot \left(\frac{E}{E_{crit}}\right)^{1/2} \cdot \exp\left(-\frac{E}{E_{crit}}\right) \tag{1.35}$$

Both asymptotes were calculated and also presented in Figure 1.19 as dimensionless quantities (Watt/Watt). It can simply be shown that Equation 1.35 has a maximum at 0.50 E_{crit} and reaches a spectral flux ($\Delta W/\Delta E$) with a value of about 0.9 $\alpha_f\,\gamma$. The maximum of the actual curve in black, however, is located at $E_{max} \approx 0.29\ E_{crit}$ [52,53] with a value of nearly 1.8 $\alpha_f\ \gamma$ according to Equation 1.32.

As stated, the basic quantity of SR is the spectral flux, which can be given by the number of photons with a certain energy emitted per unit time or second. Usually a bandwidth of 0.1% of the respective average energy is chosen while the beam current is normalized to 1 ampere and the electron energy is assumed to be 1 GeV. However, for local investigations it is decisive that the radiation source is a spot with a very small angular divergence. For that reason, the flux is related to the respective solid angle and is called angular density of the flux or spectral brightness [53]. When the photon flux is related to a small source area (cross-section of the beam) in addition, it is called brilliance [56]. These quantities are commonly used in the synchrotron literature though brilliance is also called spectral brightness in English-speaking areas (USA, UK). Brilliance is mostly related to number of photons per second, per mrad^2, per mm^2, and per 0.1% spectral bandwidth but *not* to the SI units of 1 sr and of 1 m^2 [67].

Figure 1.20 represents the brilliance derived from Figure 1.19 in a double logarithmic plot with both asymptotic approximations. It represents the *average* brilliance integrated over the small vertical divergence. For low photon energies, we find an increasing straight line with the slope 1/3;[1] for high photon energies, we have a decreasing exponential curve. A comparison with the radiation of a black body is described later.

[1] In comparison to the average brilliance with a slope of 1/3 for the increasing branch, the in-plane brilliance shows a slope of 2/3 and a maximum which is nearly doubled.

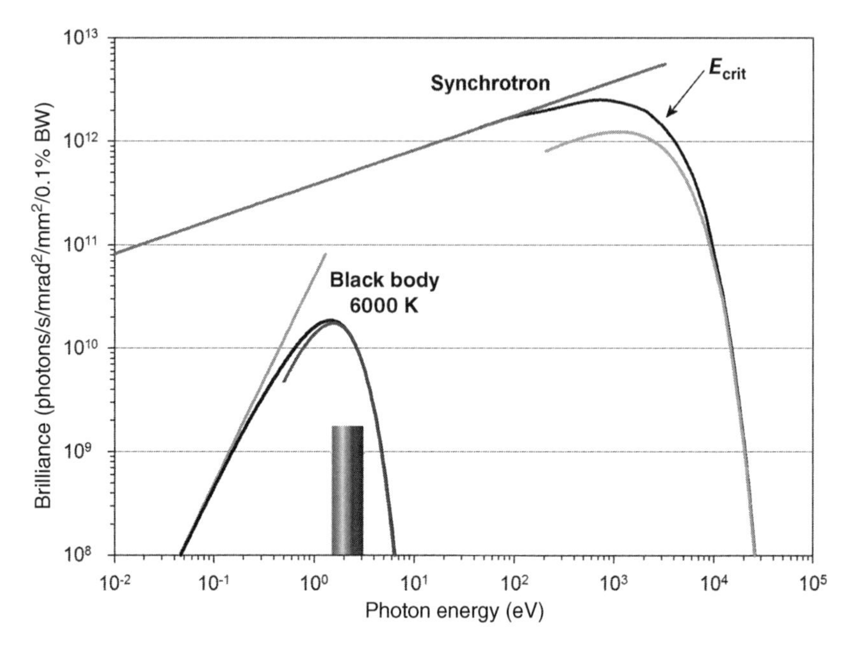

Figure 1.20. The continuous spectrum of Figure 1.19 as a double-logarithmic plot. The ordinate was converted into the brilliance or number of photons/s/mrad^2/mm^2/0.1 % bandwidth, respectively. Again, the two asymptotes are presented for low and for high photon energies (red straight line and blue curved line). The range of transition is marked in black, the maximum is at 0.29 E_{crit}. For comparison, the photon flux of a black-body is plotted according to Planck's law. At a temperature $T = 6000$ K, the black curve on the left shows a maximum at $E_{max} \approx 1.5$ eV in the visible color-coded region. (See colour plate section)

Synchrotron radiation shows a dependence on the kinetic energy, E_{el}, of the beam electrons. If the electron energy is increased synchronously with the magnetic field, the radius of the electron orbit is held constant—as already mentioned earlier. The result for the photon flux is illustrated in Figure 1.21. There is no change of the flux on the low-energy side, but with higher electron energies, the curves grow and are shifted to the right, that is, to the high-energy side. The positions E_{max} are shifted according to 0.29 E_{crit} and the maxima are increased proportional to E_{el}^2. It may be mentioned that two parts of radiation appear: one part oscillating parallel to the magnetic field and the other part perpendicular to it. It can be shown that the perpendicular part is seven times larger than the parallel one [52,58,68]. Further details of SR facilities especially with undulators, wigglers, and free-electron lasers are given in Section 3.3.2.

1.3.4.4 Comparison with Black-Body Radiation

Because of the excellent properties of SR, storage rings can even serve as radiation standards like the European calibration standard for electromagnetic radiation, MLS, in Berlin, Germany. The continuous spectrum of SR is of relativistic origin

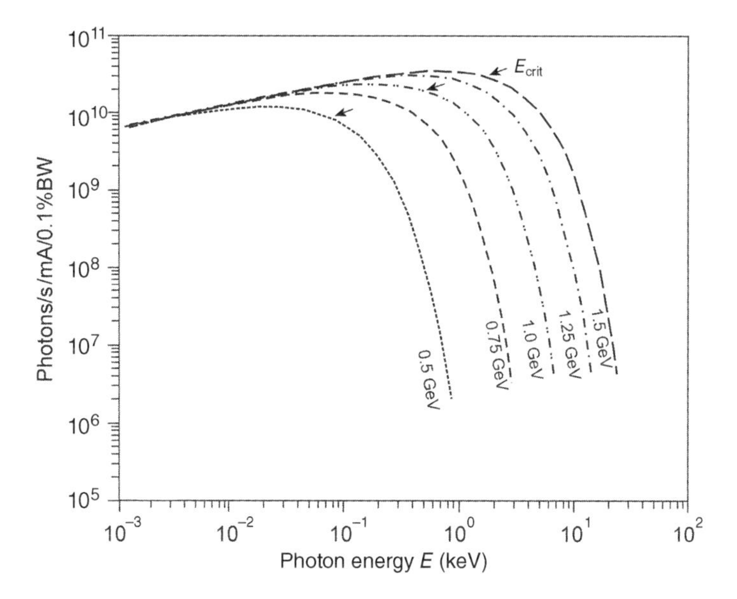

Figure 1.21. Spectral flux of synchrotron radiation depending on the photon energy as a double logarithmic plot. When the kinetic energy of the electrons or the beam energy is increased from 0.5 to 1.5 GeV in steps of 0.25 GeV, the curves rise and move to the high-energy side. The maxima occur at 0.29 E_{crit} and are shifted from 0.024 to 0.64 keV in four different steps. The curves were calculated after Equations 1.34 and 1.35, linked together and scaled for DELTA in Dortmund.

but resembles the thermal radiation of a black body (BB). Both, the continuous spectrum of SR and the "white" spectrum of BB can be calculated with sufficient accuracy in contrast to the continuous spectrum of an X-ray tube.

A black body is in thermodynamic equilibrium and characterized by its temperature T, in Kelvin. It emits electromagnetic radiation, which is homogeneous, isotropic, and unpolarized. The spectral power ΔP is given by Planck's law usually expressed in terms of frequency ν, or wavelength λ, or the reciprocal value $1/\lambda$, called wavenumber [69]. In order to compare both radiation sources, this law was adapted to a function of the energy $E = h\nu$ of emitted photons:

$$\Delta P(E, T) = 2c_0 \cdot \left(\frac{E}{hc_0}\right)^3 \cdot \frac{\Delta A\, \Delta\Omega\, \Delta E}{\exp\left(\frac{E}{kT}\right) - 1} \tag{1.36}$$

where h is Planck's constant and k is Boltzmann's constant ($hc_0 = 1.2398 \times 10^{-6}$ eV·m and $k = 8.617 \times 10^{-5}$ eV/K). The infinitesimal quantity $\Delta P(E, T)$ is determined as the power in Watt for photons with energy in eV. It is radiated normally from the surface of the BB with an area ΔA, into a solid angle of detection $\Delta\Omega$, within an energy band ΔE, centered on E. For a comparison with SR, ΔP has been related to an energy band set to 0.1% of the relevant photon energy. $(\Delta P/\Delta E)$ can be calculated in Watt per Joule or can be converted into the number of photons emitted per second in the relevant energy band. Differentiation shows that the distribution ($\Delta P/\Delta E$) peaks at E_{max} of nearly 3 kT (more exactly at 2.82 kT). This is in full agreement with Wien's displacement law $E_{max} = b \cdot T$ where $b \approx 2.431 \times 10^{-4}$ eV/K. The maximum itself amounts to about $c_0(E_{max}/2hc_0)^3\ \Delta A\ \Delta\Omega$.

Again two approximations can be deduced. For photons with low energies ($E \ll kT$), the law of Rayleigh–Jeans is valid:

$$\frac{\Delta P(E,T)}{\Delta E} = 2c_0 \cdot \left(\frac{E}{hc_0}\right)^3 \cdot \frac{kT}{E} \Delta A\ \Delta\Omega \tag{1.37}$$

For photons with high energies ($E \gg kT$), the approximation of Wien can be used:

$$\frac{\Delta P(E,T)}{\Delta E} = 2c_0 \cdot \left(\frac{E}{hc_0}\right)^3 \cdot \exp\left(-\frac{E}{kT}\right) \Delta A\ \Delta\Omega \tag{1.38}$$

Planck's law is presented in Figure 1.20 for a black body with a temperature of 6000 K in addition to the synchrotron curves. Source area and solid angle of the BB radiation were assumed to be as large as for SR with $\Delta A = 1.75 \times 10^{-2}$ mm^2, and $\Delta\Omega = 0.17$ (mrad)^2. It can be read from the curves that the continuum of BB is similar to the continuum of SR with a strong exponentially decreasing branch, but BB shows a steeper increasing branch with a slope of 2 instead of 1/3. The black body at 6000 K mainly radiates in the visible region indicated by a colored ribbon.

1.4 ATTENUATION OF X-RAYS

Different phenomena have to be considered as forming the basis of X-ray spectrometry; the attenuation of X-rays as well as their deflection and diffraction. These phenomena result from the interaction between radiation and matter and can be described partly by the wave picture and partly by the corpuscle picture as already mentioned earlier.

If an X-ray beam passes through matter, it will lose intensity due to different effects. According to Figure 1.22, the number of photons N_0 hitting upon a homogeneous sheet or layer of density ρ and thickness d is reduced to a fraction N being transmitted while the difference, $\Delta N = N_0 - N$, has been lost. Generally, the attenuation of intensity is controlled by the Lambert–Beer law. This law can be written either in the differential form

$$\frac{\Delta N}{N} = -\left(\frac{\mu}{\rho}\right)\rho\ \Delta d \tag{1.39}$$

or in the integral form

$$N(d) = N_0 \exp\left[-\left(\frac{\mu}{\rho}\right)\rho d\right] \tag{1.40}$$

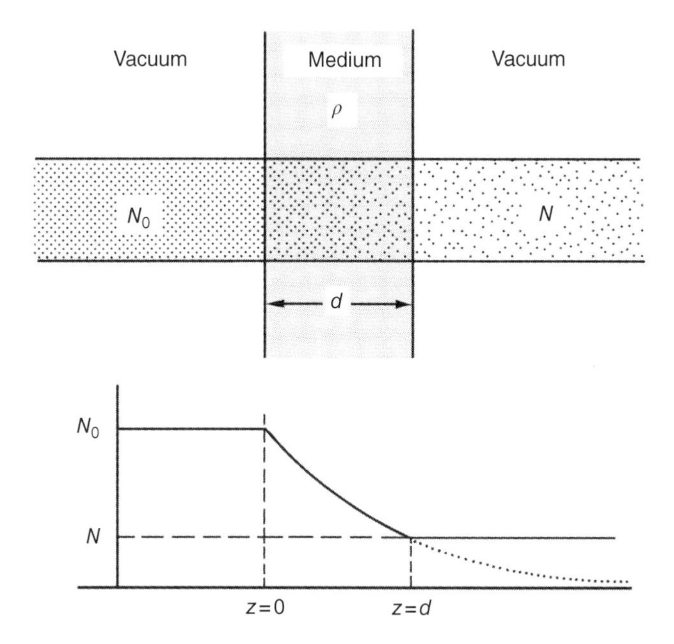

Figure 1.22. Attenuation of an X-ray beam penetrating through a homogeneous medium of density ρ and thickness d. The number of photons is reduced exponentially from N_0 to N. After Ref. [8], reproduced with permission. Copyright © 1996 by John Wiley and Sons.

where μ is the linear attenuation coefficient, and (μ/ρ) is called the mass-attenuation coefficient defined by the Lambert–Beer law in Equation 1.11. The intensity exponentially depends on the thickness of the layer and will be reduced to $1/e$ or nearly 37% if, for example, X-ray photons of about 20 keV pass through metal sheets of medium density with about 10–200 μm thickness.

The mass-attenuation coefficient expressed in cm^2/g is a quantity that depends on the element composition of the material and the energy of the X-ray photons. Since the density is incorporated, (μ/ρ) is independent of the state of aggregation. Values for a solid, liquid, or gas, whether it is a compound, solution, or mixture, will be equal if the composition of the material is equal. For the sake of clarity, this notation is preferred here. Some authors only use μ instead of the product, $(\mu/\rho)\rho$, for simplicity; others choose the symbol μ instead of (μ/ρ).

The mass-attenuation coefficient follows an additive law so that values of a compound, solution, or mixture can readily be calculated from values of the individual elements if the element composition is known:

$$\left(\frac{\mu}{\rho}\right)_{\text{total}} = \sum c_i \left(\frac{\mu}{\rho}\right)_i \tag{1.41}$$

where the values of c_i are the mass fractions of the different elements present in the total mixture. Of course, the individual coefficients, $(\mu/\rho)_i$, are

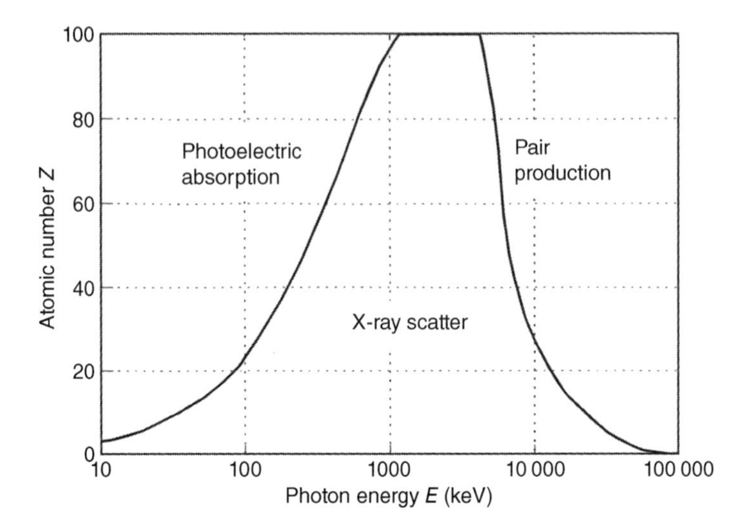

Figure 1.23 Isofrequency lines and predominant effects for the attenuation of X-rays in material of atomic number Z plotted against the photon energy E. Figure from Ref. [71], reused with permission from Springer Science + Business Media.

functions of the energy of the X-ray photons, so that the total value is determined only for X-ray photons of a certain energy, that is, for a monoenergetic X-ray beam. The individual values for each element and for several energies of X-ray photons can be taken from tables (e.g., from Bertin [48] or Tertian and Claisse [50] or the Internet [61]) or can be calculated using functions (e.g., given in Ref. [70]).

The attenuation of X-rays is caused by the interaction of their photons with the inner or the outer electrons, or even with the nuclei of atoms. It results from three competing effects, called photoelectric absorption, X-ray scatter, and pair production. As indicated in Figure 1.23 [71], the photoelectric effect predominates for E below 100 keV and is the most important in X-ray spectroscopy. Pair production does not occur for E smaller than 1 MeV, so it is insignificant for X-ray spectroscopy and will not be considered further here.

1.4.1 Photoelectric Absorption

The major component of X-ray absorption is caused by the photoelectric effect by which an electron of an inner shell of an atom is expelled by a photon of sufficient energy. The primary photon itself is completely annihilated while a secondary photon of lower energy is emitted immediately after the rearrangement of electrons. The *secondary emission* is called X-ray fluorescence (already described in Section 1.3.2.1).

Photoelectric absorption is evaluated numerically by a specific mass-absorption coefficient (τ/ρ). It can be considered as the sum of all possible expulsions of electrons from the various atomic shells K, L, M, N, O, and P, and consequently is determined by

$$(\tau/\rho) = (\tau/\rho)_K + (\tau/\rho)_L + (\tau/\rho)_M + (\tau/\rho)_N(\tau/\rho)_O + (\tau/\rho)_P \tag{1.42}$$

The different additive parts can further be split up according to the corresponding subshells. All the individual coefficients approximately follow the Bragg–Pierce law:

$$\left(\frac{\tau}{\rho}\right)_j = k_j Z^3 / E^{8/3} \tag{1.43}$$

with different constants k_j of the different subshells or levels j. In a double logarithmic plot of (τ/ρ) versus E presented in Figure 1.24, the linear segments show a negative slope of −8/3 and are mutually parallel. At the absorption edges, abrupt jumps of (τ/ρ) appear because further electrons of the next outer shell can be expelled if the photon energy exceeds the corresponding edge energy. For higher energies, the mass-absorption

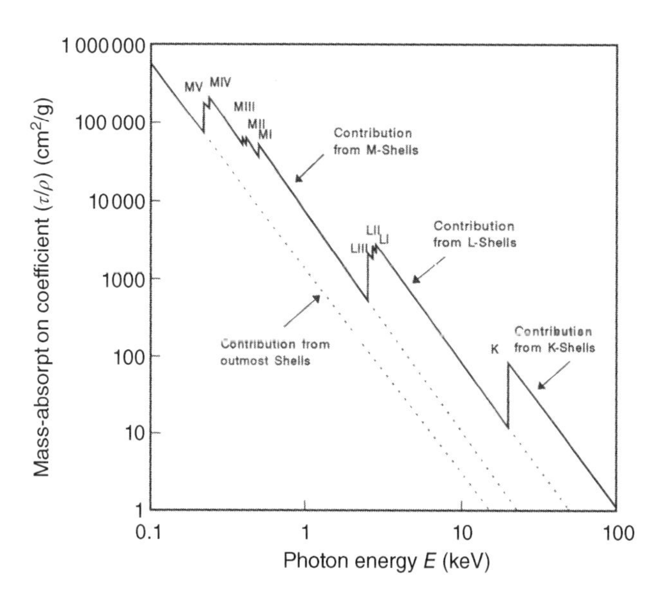

Figure 1.24. Total photoelectric mass-absorption coefficient (τ/ρ) for molybdenum vs. the photon energy E. Each discontinuity corresponds to an additional photoelectric process that occurs if the respective absorption edge K, L_I . . . L_{III}, or M_I . . . M_V is exceeded or jumped over. Data from Ref. [48], reproduced with permission from Plenum Press.

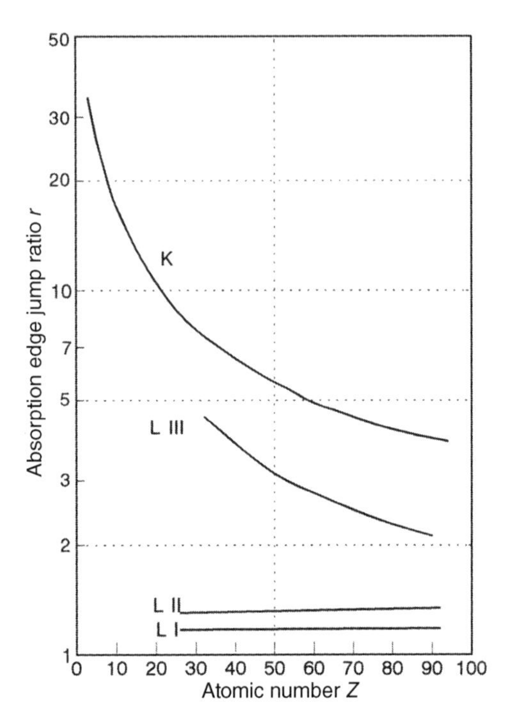

Figure 1.25 Absorption-edge jump ratio r for K, L_I, L_{II}, and L_{III} edges plotted against the atomic number Z of the material. Figure from Ref. [48], reproduced with permission from Plenum Press.

coefficient gradually falls again with the slope −8/3. The specific edge energies correspond to the binding or ionization energies of electrons in the respective shells or subshells. They follow the extended Moseley's law of Equation 1.6. The respective characteristic emission peaks always lie at somewhat lower energies (see Section 1.3.2.1).

The jump ratio at an absorption edge is defined by the quantity

$$r_j = \frac{(\tau/\rho)_{\mathrm{high}}}{(\tau/\rho)_{\mathrm{low}}} \tag{1.44}$$

where the subscripts "high" and "low" refer to the high- and low-energy side of an edge. The jump ratios of the K and L edges are represented in Figure 1.25 for various elements. From the jump ratio, another useful quantity can be derived called the absorption jump factor f_j. It is defined as the fraction $(\tau/\rho)_j$ of a certain shell or subshell j with respect to the total value (τ/ρ) according to

$$f_j = \frac{(\tau/\rho)_j}{\Sigma(\tau/\rho)_j} \tag{1.45}$$

Jump factor and jump ratio are correlated according to

$$f_j = \frac{r_j - 1}{r_j} \tag{1.46}$$

1.4.2 X-Ray Scatter

The second and generally minor component of X-ray absorption is caused by the scattering of X-ray photons. In contrast to photoelectric absorption, the primary photons do not ionize and excite an atom or molecule but are only deflected from their original direction. Three processes can be distinguished:

1. The collision of a photon with a firmly bound inner electron of an atom can lead to a change of direction of the photon without energy loss. This process is called *elastic scattering* or *Rayleigh scattering*.
2. The collision of a photon with outer electrons of molecules can also lead to a deflection and additionally, in a very few cases (1 in 10 million), to a change of its energy. The photons can win or lose energy in discrete steps or continuously in a wide band. The effect is called *inelastic Raman scattering*.
3. The collision of a photon with a loosely bound outer electron of an atom or molecule or even with a free electron can lead to a change of direction *and* a loss of energy of the photon. This process is called *inelastic scattering* or *Compton scattering*. A strict relationship exists between energy loss and angle of deflection.

Generally, the photons can be deflected in all directions. Rayleigh and Raman scattering can be coherent, that is, there is a fixed relation of phases for the incident and the scattered photons. By way of contrast Compton scattering is always incoherent. Rayleigh scattering occurs at crystal planes or multilayer interfaces (Section 2.3.5). Raman scattering is mainly concerned with the translation and rotation of molecules in gases, and spectra are observed in the visible and moreover in the infrared region.

In the corpuscle picture, the loss of *energy* a photon suffers in Compton scattering results from the conservation of total energy and total momentum at the collision of the photon and the electron. A photon with the initial energy E keeps the part E' when it is deflected by an angle ψ, while the electron takes off the residual part of energy $\mathrm{d}E = E - E'$. The fraction E'/E can be calculated according to

$$E'/E = \left[1 + (1 - \cos\psi)E/E_0\right]^{-1} \tag{1.47}$$

where E_0 is the rest energy of an electron which amounts to 511 keV. Figure 1.26 represents the distribution of E'/E for any given direction ψ in

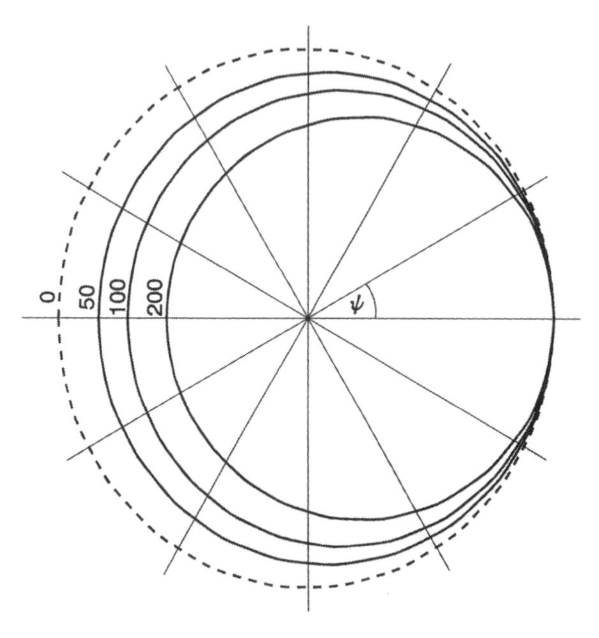

Figure 1.26. Compton-scatter of a photon with an initial energy E. On collision with a loosely bound or free electron the photon is deflected by an angle ψ. It loses energy and keeps only the fraction E'/E. This fraction is represented by an ellipse in polar coordinates for a certain energy E (50, 100, and 200 keV). It may be compared with the outer dashed circle, representing conservation of the photon energy. Figure from Ref. [8], reproduced with permission. Copyright © 1996, John Wiley and Sons.

polar coordinates. The fraction depends on the initial energy but is independent of the substance of the scatterer. It shows maxima for $\psi = 0°$ (forward scatter), reaches minima for $\psi = 180°$ (backward scatter), and decreases with the initial energy.

In the wave picture, Compton scattering is controlled by a fixed wavelength shift, $d\lambda = \lambda' - \lambda$. It amounts to

$$d\lambda = \lambda_C \cdot (1 - \cos\psi) \tag{1.48}$$

where λ_C is a small constant called Compton wavelength, which is defined by $\lambda_C = h\ c_0/E_0 = 0.002426$ nm. The wavelength shift depends only on the deflection angle ψ and is independent of the wavelength λ itself. For $\psi = 0°$, the shift is always zero, and for $\psi = 180°$, it is always $2\lambda_C$, which is maximum.

The *intensity* of the scattered radiation shows a dependence on E and ψ, as shown in Figure 1.27 [72]. Minimum intensity or scattering is achieved for a deflection around 90°–100°. For that reason, a rectangular geometric arrangement of the X-ray tube, sample, and detector is generally chosen in X-ray spectrometry in order to minimize the inelastic scatter into the detector. Nevertheless, any primary radiation of an X-ray tube is scattered by the sample and is reproduced as a blank spectrum. In particular, the characteristic peaks of

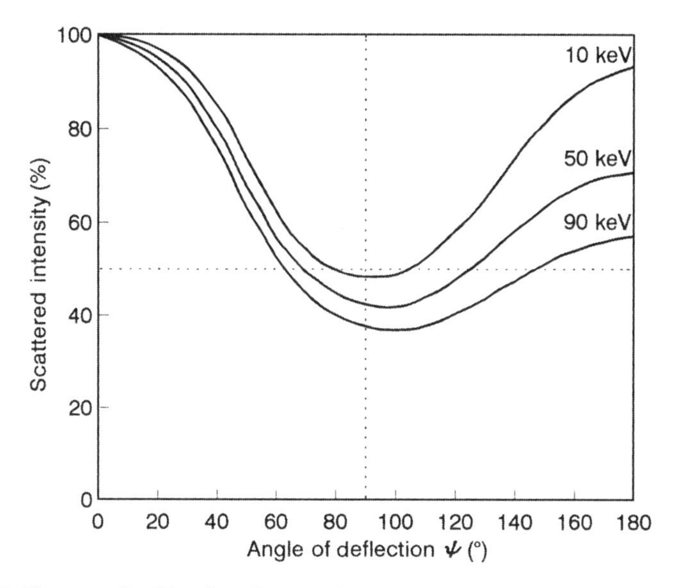

Figure 1.27. The normalized fraction of scattered *intensity*—Compton scatter—as a function of the angle ψ by which the incident photon is deflected after collision with an electron. Figure from Ref. [72], reproduced with permission from Thermo Fisher.

the tube anode give rise to the so-called Rayleigh and Compton peaks in a sample spectrum. Their corresponding *energies* can be calculated by Equation 1.47 for $\psi = 90°$, independent of the sample itself. Their *intensities*, however, depend on the photon energy and moreover on the sample substance. Rayleigh scattering will increase if the energy of X-ray photons decreases or the mean atomic number of the scattering sample increases. Compton scattering, in contrast, will decrease if the photon energy decreases or the atomic number increases.

1.4.3 Total Attenuation

Photoelectric absorption and scattering jointly lead to the attenuation of X-rays in matter. The total mass-attenuation coefficient is composed additively by the photoelectric mass-absorption coefficient (τ/ρ) and the mass-scatter coefficient (σ/ρ) according to

$$(\mu/\rho) = (\tau/\rho) + (\sigma/\rho) \tag{1.49}$$

Both fractions are shown in Figure 1.28 for the element palladium, as functions of the photon energy [57,73]. The scatter coefficient is further divided into the Rayleigh and the Compton part. In contrast to the exponential decrease of (τ/ρ) with discontinuities at the absorption edges, the function (σ/ρ) varies more slightly and steadily. It decreases for Rayleigh scattering and increases for

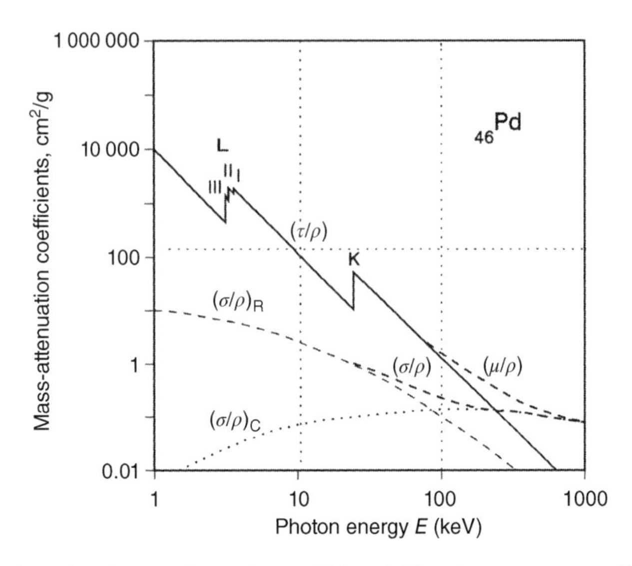

Figure 1.28. Photoelectric mass-absorption coefficient (τ/ρ) and mass-scatter coefficient (σ/ρ) as functions of the energy E of primary photons incident on palladium. The total mass-attenuation coefficient (μ/ρ) results from the sum $(\tau/\rho)+(\sigma/\rho)$, the latter term from the sum of Rayleigh and Compton scattering $(\sigma/\rho)_R + (\sigma/\rho)_C$. Figure from Ref. [57], reproduced with permission from Center for X-Ray Optics and Advanced Light Source, Lawrence Berkeley National Laboratory, http://xdb.lbl.gov.

Compton scattering in the given energy range; the Rayleigh part is predominant for energies below 90 keV, the Compton part dominates for energies above 90 keV. Both effects are relatively minor compared to the photoelectric absorption. But for energies above 200 keV, the Compton effect becomes the decisive component of total attenuation. Similar conditions are valid for elements lighter or heavier than palladium. For light elements like carbon, the point of balance between Rayleigh and Compton scattering decreases to 10 keV; for heavy elements like lead, it increases to 150 keV [57].

It should be noted that the total mass-attenuation coefficient (μ/ρ) is mainly determined and equal to the photoelectric mass-absorption coefficient (τ/ρ) for lower photon energies (< 20 keV). For most elements $(Z > 14)$ and energies between 5 and 20 keV, the quantities (μ/ρ) and (τ/ρ) differ by only about 0.01% up to 3%, relatively. In such cases, only one single set of data may be necessary for both quantities. For light elements like carbon and energies >20 keV, however, the quantities (μ/ρ) and (τ/ρ) are different and have to be distinguished.

In practice, energy-dependent attenuation is used to alter the spectrum of an X-ray beam. For that purpose, a thin metal sheet called a selective attenuation filter can be employed. It can easily be inserted into a beam path in order to reduce a particular spectral peak or an entire energy band with respect to other peaks or spectral regions.

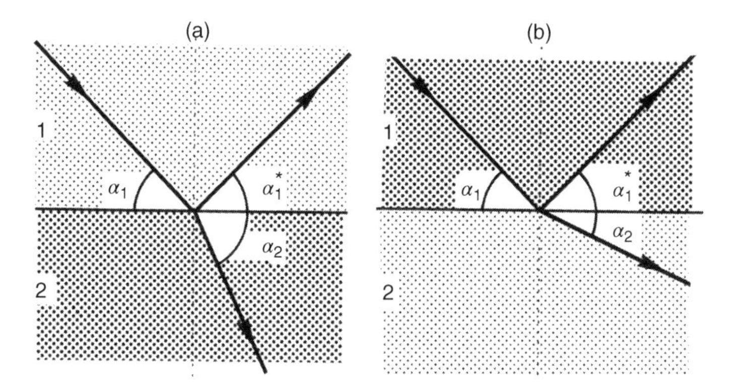

Figure 1.29. The incident, the reflected, and the refracted beam at the interface of media 1 and 2. On the left, medium 2 is optically denser than medium 1 ($n_2 > n_1$); on the right, it is vice versa ($n_1 > n_2$). The latter is usually the case for X-rays though it is exaggerated here. Figure from Ref. [8], reproduced with permission. Copyright © 1996, John Wiley and Sons.

1.5 DEFLECTION OF X-RAYS

In a homogeneous medium, the X-ray beam just behaves like a light beam and follows a straight path on which the photons travel. But if the beam hits the boundary surface of a second medium, the incident X-ray beam will be deflected from its original direction. It can even be split, that is, partly reflected into the first medium and partly refracted into the second medium.

1.5.1 Reflection and Refraction

In accord with Figure 1.29, the following rules are valid:

- The incident, the reflected, and the refracted beam span a plane that is normal to the boundary plane.
- The *glancing* angles[2] of the *incident* and the *reflected* beam are equal:

$$\alpha_1 = \alpha_1^* \tag{1.50}$$

- The *glancing* angles of the *incident* and the *refracted* beam follow Snell's law (also called Snellius' law):

$$v_2 \cos \alpha_1 = v_1 \cos \alpha_2 \tag{1.51}$$

[2] Glancing angles are considered in X-ray optics. They are complements of the angles of incidence conventionally used in light optics.

where υ_1 and υ_2 are the phase velocities of the beam in media 1 and 2, respectively. Phase velocity is the velocity at which the planes of constant phase, for example, crests or troughs, propagate within a medium. It is dependent on the wavelength, λ, and the medium itself. *In vacuo*, the phase velocity takes the value c_0 (the light velocity) independent of λ.

Division of Equation 1.51 by c_0 results in

$$n_1 \cos\alpha_1 = n_2 \cos\alpha_2 \tag{1.52}$$

where n_1 and n_2 are the absolute refractive indices of media 1 and 2, respectively, which are defined by

$$n_{1,2} = c_0/\upsilon_{1,2} \tag{1.53}$$

The refractive index is the fraction by which the phase velocity of the radiation is changed with respect to its vacuum value. For vacuum, the refractive index is 1.

The wavelength of a wave is changed by refraction and the photon energy is changed inversely to phase velocity and wavelength, though only a little bit. Division of Equation 1.51 by the frequency leads to

$$\lambda_2 \cos\alpha_1 = \lambda_1 \cos\alpha_2 \tag{1.54}$$

Two cases can be distinguished, as demonstrated in Figure 1.29. If $n_2 > n_1$, that is, if medium 2 is optically denser than medium 1, the refracted beam in medium 2 will be deflected *off* the boundary. If $n_2 < n_1$, that is, if medium 2 is optically thinner than medium 1, the refracted beam in medium 2 will be deflected *toward* the boundary.

The refractive index is the decisive quantity and can be derived from the so-called Lorentz theory assuming that the quasi-elastically bound electrons of the atoms are forced to oscillations by the primary radiation. As a result, the oscillating electrons radiate with a phase difference. By superposition of both radiations the primary one is altered in phase velocity. This alteration becomes apparent by a modified refractive index, deviating from the vacuum value $n_{vac} = 1$ by a small quantity δ.

If absorption cannot be neglected but has to be taken into account, the refractive index n has to be written as a complex quantity. Conventionally, n is defined by

$$n = 1 - \delta - i\beta \tag{1.55}$$

where i is the imaginary unit or the square root of -1.[3]

[3] The refractive index n can be also defined by the conjugate complex quantity $1 - \delta + i\beta$.

The *imaginary* component β is a measure of the attenuation already treated in Section 1.4. It can be expressed by

$$\beta = \frac{\lambda}{4\pi}\left(\frac{\mu}{\rho}\right)\rho \tag{1.56}$$

The *real* part δ, called the *decrement*, measures the deviation of the *real* component $n' = 1 - \delta$ of the refractive index from unity. n' determines the phase velocity according to $\upsilon \approx c_0/n'$, which can even be greater than the light velocity.[4]

From theory it follows that δ can be written as [74]

$$\delta = \frac{N_\mathrm{A}}{2\pi} r_\mathrm{el} \frac{1}{A}\left[f_0 + f(\lambda)\right]\rho\lambda^2 \tag{1.57}$$

where N_A is Avogadro's number $= 6.022 \times 10^{23}$ atoms/mol, r_el is the classical electron radius $= \alpha_\mathrm{f}\ \hbar\ c_0/E_0 = 2.818 \times 10^{-13}$ cm, A is the atomic mass of the respective element (in g/mol), ρ is the density (in g/cm^3), and λ is the wavelength of the primary beam. f_0 is a quantity that for X-rays is equal to the atomic number Z of the particular element and $f(\lambda)$ is a correction term [75] that is only decisive at and above the absorption edges ($E \leq E_j$ or $\lambda \geq \lambda_j$) and is generally negative. Consequently, δ includes some constants of matter and moreover strongly depends on the wavelength. This dependence is known as dispersion and demonstrated in Figure 1.30 for the elements copper and gold.

For primary X-rays—shorter in wavelength than the absorption edges—the f values disappear and Equation 1.48 can be simplified by

$$\delta = \frac{N_\mathrm{A}}{2\pi} r_\mathrm{el} \frac{Z}{A}\rho\lambda^2 \tag{1.58}$$

The first factors can be combined to one single factor, which is a material constant:

$$C_m = \frac{N_\mathrm{A}}{2\pi} r_\mathrm{el} \frac{Z}{A}\rho \tag{1.58a}$$

In addition to Z, A, and ρ, this product is listed in Table 1.7 and represented in Figure 1.31 for pure elements. The values cover the range between 1×10^{10}/cm^2 for light elements, 3.137×10^{10}/cm^2 for silicon, and 25×10^{10}/cm^2 for heavy

[4] For X-rays with positive δ values, the phase velocity υ exceeds the light velocity c_0. Only the speed of particles with a rest mass above zero, that is, the velocity of signals, has to be smaller than the upper limit c_0. The phase velocity, however, is not a velocity by which a real signal can be transmitted. Consequently, υ can exceed c_0.

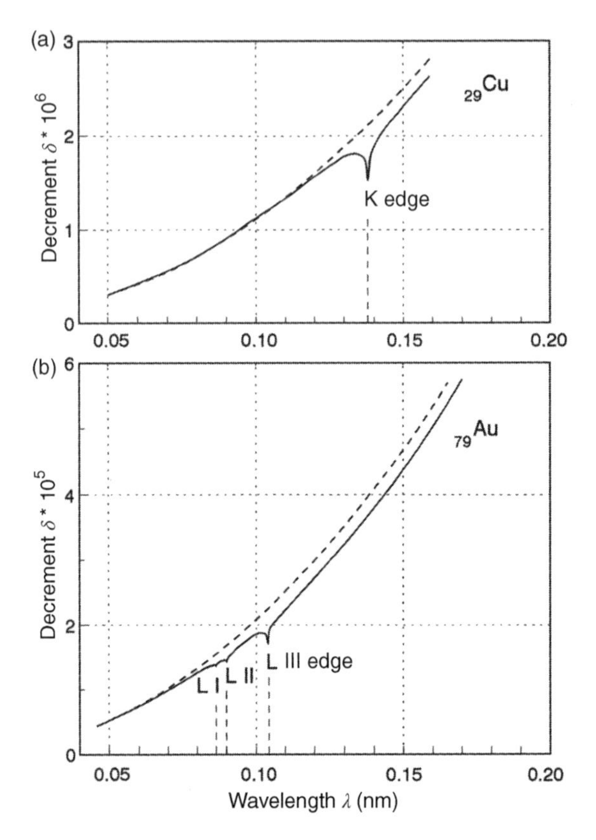

Figure 1.30. Dependence of the decrement δ on the wavelength λ for the elements (a) copper and (b) gold. The theory is based on forced oscillations of the atomic electrons—forced by the electromagnetic radiation of X-rays. At the "resonance" wavelengths or absorption edges, the decrement shows a strong variation. The asymptotic approximation for short wavelengths is represented by a dashed line. Data from Ref. [75], reproduced with permission from Forschungszentrum Jülich.

metals. The material constant contains (N_A/A), which represents the reciprocal volume of a single atom. This dependency becomes apparent in the figure.

Equation 1.58 leads to the simple formula

$$\delta = C_m \lambda^2 \tag{1.59}$$

For given λ values of the X-ray region, the δ values can easily be calculated. Because these values are quite small (between 2×10^{-7} and 5×10^{-3}), the real part $1 - \delta$ of the refractive index deviates only slightly from unity (between 0.995 and 0.9999998). The minus sign in $1 - \delta$ reflects the fact that the bound electrons follow the excitant photons only slowly, that is, with phase *opposition*.

TABLE 1.7. Atomic Number, Atomic Mass, Density, and Material Constant of Pure Elements

At. No. Z	Symbol	Atomic Mass A (g/mol)	Density ρ (g/cm^3)	Constant C_m (10^{10}/cm^2)	At. No. Z	Symbol	Atomic Mass A (g/mol)	Density ρ (g/cm^3)	Constant C_m (10^{10}/cm^2)
3	Li	6.941	0.534	0.623	48	Cd	112.411	8.65	9.976
4	Be	9.0122	1.848	2.215	49	In	114.818	7.31	8.426
5	B	10.811	2.34	2.923	50	Sn	118.71	7.31	8.316
6	C	12.0107	2.1	2.833	51	Sb	121.76	6.691	7.569
7	N	14.0067	0.808	1.091	52	Te	127.6	6.24	6.868
8	O	15.9994	1.14	1.540	53	I	126.9045	4.93	5.561
9	F	18.9984	1.5	1.919	54	Xe	131.29	3.52	3.910
10	Ne	20.1797	1.207	1.615	55	Cs	132.9055	1.873	2.093
11	Na	22.9898	0.971	1.255	56	Ba	137.327	3.62	3.987
12	Mg	24.305	1.738	2.318	57	La	138.9055	6.145	6.811
13	Al	26.9815	2.6989	3.512	58	Ce	140.116	6.77	7.569
14	Si	28.0855	2.33	3.137	59	Pr	140.9077	6.773	7.660
15	P	30.9738	1.823	2.384	60	Nd	144.24	7.008	7.873
16	S	32.0653	2.07	2.790	61	Pm	144.91	7.264	8.259
17	Cl	35.4527	1.56	2.020	62	Sm	150.36	7.52	8.375
18	Ar	39.948	1.4	1.704	63	Eu	151.964	5.244	5.872
19	K	39.0983	0.862	1.131	64	Gd	157.25	7.901	8.685
20	Ca	40.078	1.55	2.089	65	Tb	158.9253	8.23	9.091
21	Sc	44.9559	2.989	3.771	66	Dy	162.5	8.551	9.380
22	Ti	47.867	4.54	5.636	67	Ho	164.9303	8.795	9.650
23	V	50.9415	6.11	7.451	68	Er	167.26	9.066	9.955
24	Cr	51.9961	7.19	8.963	69	Tm	168.93421	9.321	10.282
25	Mn	54.938	7.32	8.997	70	Yb	173.04	6.966	7.611
26	Fe	55.845	7.874	9.901	71	Lu	174.967	9.841	10.786
27	Co	58.9332	8.9	11.013	72	Hf	178.49	13.31	14.501

(continued)

TABLE 1.7. (*Continued*)

At. No. Z	Symbol	Atomic Mass A (g/mol)	Density ρ (g/cm^3)	Constant C_m ($10^{10}/cm^2$)	At. No. Z	Symbol	Atomic Mass A (g/mol)	Density ρ (g/cm^3)	Constant C_m ($10^{10}/cm^2$)
28	Ni	58.6934	8.902	11.470	73	Ta	180.9479	16.654	18.146
29	Cu	63.546	8.96	11.044	74	W	183.84	19.3	20.982
30	Zn	65.39	7.133	8.839	75	Re	186.207	21.02	22.867
31	Ga	69.723	5.904	7.090	76	Os	190.23	22.57	24.354
32	Ge	72.61	5.323	6.336	77	Ir	192.217	22.42	24.257
33	As	74.9216	5.73	6.817	78	Pt	195.078	21.45	23.164
34	Se	78.96	4.79	5.571	79	Au	196.9665	19.31	20.918
35	Br	79.904	3.12	3.691	80	Hg	200.59	13.546	14.591
36	Kr	83.801	2.16	2.506	81	Tl	204.3833	11.85	12.684
37	Rb	85.4678	1.532	1.791	82	Pb	207.2	11.35	12.132
38	Sr	87.62	2.54	2.975	83	Bi	208.98	9.747	10.456
39	Y	88.9059	4.469	5.295	84	Po	208.98	9.32	10.118
40	Zr	91.224	6.506	7.705	85	At	209.99	—	10.149
41	Nb	92.9064	8.57	10.215	86	Rn	222.02	9.73	10.179
42	Mo	95.9108	10.22	12.087	87	Fr	223.02	—	10.704
43	Tc	97.907	11.5	13.641	88	Ra	226.03	—	11.229
44	Ru	101.07	12.41	14.592	89	Ac	227.03	—	11.754
45	Rh	102.9255	12.41	14.654	90	Th	232.038	11.72	12.278
46	Pd	106.4252	12.02	14.032	91	Pa	231.036	15.37	16.351
47	Ag	107.8682	10.49	12.345	92	U	238.029	19.16	20.001

Density of solids and liquids were determined at 20 °C and given in g/cm^3; density of gaseous elements were determined at the boiling point for the liquids.
Source: From Ref. [57]; courtesy of Center for X-Ray Optics and Advanced Light Source, Lawrence Berkeley National Laboratory.

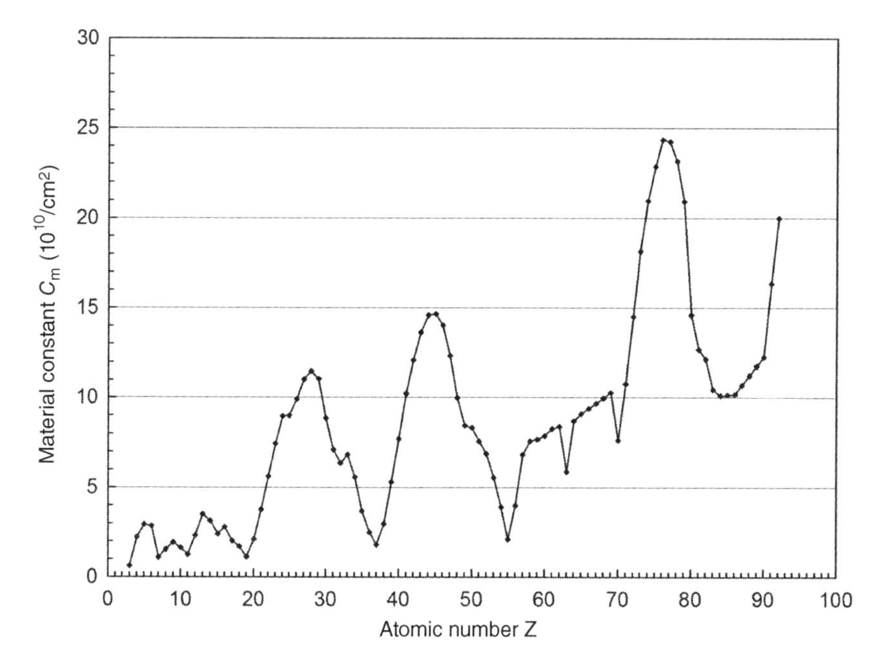

Figure 1.31. Material constant C_{m} defined by Equation 1.58a and calculated for different pure elements with atomic number Z. The maxima at $Z = 14, 28, 45$, and 77 reflect a small volume of the respective atoms while the minima at $Z = 11$, 19, 37, 55, and 85 occur for relatively large atoms (alkali metals). The quantity is important for the calculation of the refractive index and the critical angle of total reflection. Data from http://xdb.lbl.gov; reproduced with permission from Center for X-Ray Optics and Advanced Light Source, Lawrence Berkeley National Laboratory.

The small quantity of δ is due to the small amplitude of the electrons' oscillations. Because of the high photon frequencies corresponding to the short wavelengths of X-rays, only small amplitudes can occur.

The quantity β is even smaller than δ. Table 1.8 lists values of δ and β for some compounds and pure elements, calculated for Mo-Kα radiation. For compounds, solutions, or mixtures, δ and β have to be calculated according to the additive law already applied in Equation 1.41:

$$\delta_{\mathrm{total}} = \sum c_i \delta_i \tag{1.60}$$

$$\beta_{\mathrm{total}} = \sum c_i \beta_i \tag{1.61}$$

Again, the c_i terms are the different mass fractions of the individual elements i with respective values δ_i and β_i.

1.5.2 Diffraction and Bragg's Law

The phenomenon of a so-called diffraction occurs when a wave hits an obstacle. A parallel wave front hitting a small obstacle or opening deviates from its

TABLE 1.8. The Real Part δ and the Imaginary Part β of the Refractive Index n Calculated for Mo-Kα X-rays with $\lambda = 0.071$ nm[a]

Medium	ρ (g/cm^3)	δ (10^{-6})	β (10^{-8})
Plexiglas	1.16	0.9	0.055
Glassy carbon	1.41	1.0	0.049
Boron nitride	2.29	1.5	0.090
Quartz glass	2.20	1.5	0.46
Aluminum	2.70	1.8	0.79
Silicon	2.33	1.6	0.84
Cobalt	8.92	5.6	19.8
Nickel	8.91	5.8	21.9
Copper	8.94	5.6	24.1
Germanium	5.32	3.2	18.7
Gallium arsenide	5.31	3.2	18.7
Tantalum	16.6	9.1	87.5
Platinum	21.45	11.7	138.2
Gold	19.3	10.5	129.5

[a] The various media with density ρ are listed in order of increasing (mean) atomic number.
Source: From Ref. [8], reproduced with permission. Copyright © 1996, John Wiley and Sons.

original straightforward direction, bends, and spreads into the geometric shadow. Respective patterns have been observed for different waves, including visible light already in the seventeenth century. When several closely spaced obstacles or openings are encountered, the original primary wave will induce several original secondary waves after Huygens' principle. These waves will interfere with each other and show intensity maxima by constructive interference and show minimum or even zero intensity by destructive interference.

Around the turn of the nineteenth century, it was assumed that most solids are composed of crystals with a regular periodic arrangement of atoms. These atoms are fixed in a three-dimensional lattice with various two-dimensional lattice planes. Each set of planes has a spacing d_{hkl}, where (h,k,l) are called Miller indices. Values for spacing and X-ray wavelengths are of the same order of magnitude (between 0.05 and 0.5 nm), which is a prerequisite for diffraction to occur.

In 1912, Max von Laue, Friedrich, and Knipping demonstrated how X-rays were diffracted by crystals. They irradiated a single inorganic crystal (copper sulfate pentahydrate) with a fine, millimeter-wide X-ray beam and discovered several symmetrical points of deflection behind the crystal. Difficult calculations were needed to explain the diffraction by a three-dimensional array of atoms. The Braggs, however, used a narrow line-focused X-ray beam. As already demonstrated in Figure 1.2, it was diffracted at a polished plate of an inorganic crystal (e.g., NaCl, ZnS, and diamond). The X-ray beam was scattered by the atoms of the rotating crystal in a goniometer at certain angles

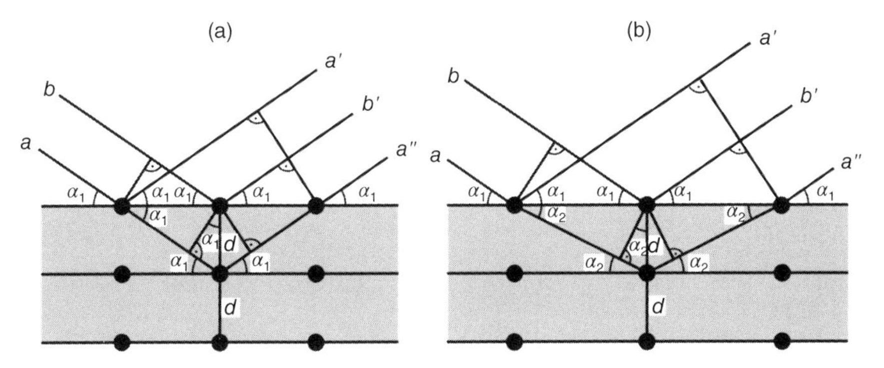

Figure 1.32. Bragg's reflection of two parallel X-ray beams *a* and *b* at parallel lattice planes of a crystal filled with scattering atoms. Neighboring planes have a constant distance *d*. (a) Without refraction of the incoming beam *a*. (b) With its refraction at the upmost lattice plane. The beams *aa*′ and *bb*′ are reflected at the upmost plane, the beam *aa*″ is refracted at the upmost plane and reflected at the next deeper plane. The geometric path difference is the length of the small sides of the kites within both figures given by $2d\sin\alpha_1$ (Figure 1.32a) or $2d\sin\alpha_2$ (Figure 1.32b). For maximum reflectivity, this difference has to be an integer multiple of the respective wavelength λ_1 of the incoming beam (Figure 1.32a) or λ_2 of the refracted and penetrating beam (Figure 1.32b).

α_1, blackening a photographic film under angles $2\alpha_1$. Calculations became easy since the diffraction could be interpreted as reflections of X-rays at crystal lattice planes followed by their constructive interference at certain angles α_1.

The derivation of Bragg's law is demonstrated in Figure 1.32. A wave between two parallel beams coming from vacuum *a* and *b* hits a crystal with parallel lattice planes under an incident angle α_1. These planes are filled with atoms, the inner electrons of which can elastically scatter the incoming wave (Rayleigh scattering). The wave *ab* is reflected at the upmost layer under the angle α_1 leading to the wave *a*′*b*′. On the left of this figure, the incoming beam *a* also penetrates the upmost plane without refraction, is reflected at the next deeper plane, and returns to vacuum under the angle α_1. On the right, beam *a* is additionally refracted and enters the crystal at a glancing angle α_2, which for X-rays is somewhat smaller than α_1. The refracted beam is reflected at the next deeper plane leading to the beam *a*″ that leaves the crystal at the glancing angle α_1. The path difference of the neighboring reflected beams *aa*′ and *bb*′ is zero, but the path difference of these two beams with the refracted and reflected beam *aa*″ is either $2d\sin\alpha_1$ (Figure 1.32a) or $2d\sin\alpha_2$ (Figure 1.32b). If this difference is an integer multiple of wavelength λ_1 (Figure 1.32a) or λ_2 (Figure 1.32b), both beams are in phase and we have constructive interference. All lattice planes cooperate in the same direction and we get maximum intensity of all reflected and refracted beams. If the difference is an odd multiple of half the wavelength we have destructive interference with minimum intensity or extinction.

Originally, Bragg's law was derived for X-rays when the refraction of the incoming beam by the crystal could be neglected (Figure 1.32a); that means $\alpha_1 = \alpha_2$ and $\lambda_1 = \lambda_2$. The condition for maximum reflectivity is given by

$$2d \sin \alpha_1 = m\lambda_1 \tag{1.62}$$

where m is a positive integer (1, 2, 3 . . .) also called order of reflection. If the refraction is taken into account we get a similar relation (Figure 1.32b):

$$2d \sin \alpha_2 = m\lambda_2 \tag{1.63}$$

where λ_2 and α_2 in the crystal medium can be related to λ_1 and α_1 in vacuum according to Snell's law of Equation 1.54. As already mentioned by Compton [7], Bragg's law in Equation 1.62 has to be corrected by a factor that depends on the decrement δ defined in Section 1.5.1:

$$2d \sin \alpha_1 \cdot \left(1 - \delta/\sin^2\alpha_1\right) = m\lambda_1 \tag{1.64}$$

For X-rays with δ values smaller than 10^{-5} and angles above 1°, the correction factor in brackets is of the order of 0.99.

The diffraction of X-rays at crystals together with Bragg's law is the basis of wavelength-dispersive X-ray spectrometry when d is known and λ_1 has to be determined (see Chapter 3.7). On the other hand, it is the prerequisite for X-ray diffractometry if—conversely—λ_1 is known and d has to be measured, for example, in crystallography.

Because of experimental difficulties, further diffraction patterns for X-rays were observed not until 1929. X-rays with a wavelength of 0.08 nm were transmitted through a single fine slit with a width of 5.5 μm and showed a pattern with several lines in the shadow region, to the left and to the right of the passing beam and parallel to it [1]. Even later, ruled gratings were developed by Siegbahn and used for diffraction of X-rays at grazing incidence. By this means, X-ray wavelengths could be determined directly and very precisely. Together with Bragg's law, the spacing of crystal lattice planes could be measured to yield a precise value of Avogadro's number that may ultimately lead to the definition of the kilogram in terms of atomic constants.

1.5.3 Total External Reflection

For X-rays, any medium is optically less dense than vacuum ($n' < n_{vac} = 1$) and any solid is optically less dense than air ($n' < n'_{air} \approx 1$), which is in contrast to visible light. This results in a refracted beam that is deflected toward the boundary plane (Figure 1.32b). For a better understanding, the refraction is exaggerated here. Since δ is very small for X-rays, the refraction is very weak.

If the respective glancing angle α_2 of the refractive beam becomes zero, the refracted beam will emerge tangentially to the boundary surface. Consequently, there is a minimum critical angle $\alpha_1 = \alpha_{crit}$ for which refraction is just possible. According to Equation 1.52, this angle of incidence is determined by

$$\cos \alpha_{crit} = n_2 \tag{1.65}$$

For angles α_1 even lower than α_{crit}, Equation 1.52 gives no real value for the refraction angle α_2 since the cosine cannot be > 1. In this case, no beam enters the second medium, but the boundary, like an ideal mirror, completely reflects the incident beam back into the first medium, that is, vacuum or air. This phenomenon is called "total *external* reflection." In contrast to X-rays, visible light can undergo "total *internal* reflection" when the light comes from a solid medium below a critical angle. It does not enter the adjacent vacuum or air as the second medium, but is totally reflected back into this first medium.

The critical angle of total reflection can easily be calculated from Equation 1.65. Since α_{crit} is small, its cosine can be approximated by

$$\cos \alpha_{crit} \approx 1 - \frac{\alpha_{crit}^2}{2} \tag{1.66}$$

The combination with Equation 1.55 leads to the simple relation

$$\alpha_{crit} \approx \sqrt{2\delta} \tag{1.67}$$

Insertion of Equation 1.58 gives the approximation

$$\alpha_{crit} \approx \frac{1.651}{E} \sqrt{\frac{Z}{A} \rho} \tag{1.68}$$

where E has to be given in keV and ρ in g/cm^3 in order to get α_{crit} in degrees. This formula may be converted into

$$\alpha_{crit} \approx \frac{10^{-5}}{E} \sqrt{C_m} \tag{1.69}$$

As already mentioned for Equation 1.58, this approximation is exactly valid for photon energies above the decisive absorption edges of the material. Table 1.9 gives values for different media and photon energies frequently used for excitation: 8.4 keV is the energy of W-Lα photons, 17.44 keV is the energy of Mo-Kα photons, and 35 keV may represent the photon energy of the hump appearing in a continuous spectrum of an X-ray tube. All values of α_{crit} lie between 0.04° and 0.6°. For all other combinations of medium and photon energy, the critical angle can simply be calculated after Equation 1.69 and by use of Table 1.7.

TABLE 1.9. Critical Angle α_{crit} of Total Reflection, Calculated for Various Media and X-rays of Different Photon Energies

	α_{crit} at Photon Energy of		
Medium	8.4 keV (degree)	17.44 keV (degree)	35 keV (degree)
Plexiglas	0.157	0.076	0.038
Glassy carbon	0.165	0.080	0.040
Boron nitride	0.21	0.10	0.050
Quartz glass	0.21	0.10	0.050
Aluminum	0.22	0.11	0.054
Silicon	0.21	0.10	0.051
Cobalt	0.40	0.19	0.095
Nickel	0.41	0.20	0.097
Copper	0.40	0.19	0.095
Germanium	0.30	0.15	0.072
Gallium arsenide	0.30	0.15	0.072
Tantalum	0.51	0.25	0.122
Platinum	0.58	0.28	0.138
Gold	0.55	0.26	0.131

Source: From Ref. [8], reproduced with permission. Copyright © 1996, John Wiley and Sons.

In the range of total reflection, calculations have to be carried out on complex refraction angles α_2 with a real and an imaginary part. Nevertheless, for our purposes, it is possible to make this task a lot easier, as simple approximations can be applied for the small glancing angles considered in X-ray optics.

In accord with Figure 1.32 b, X-rays are assumed to run through a vacuum and then to strike a medium at an angle α_1. In this case, the angle α_2 of the refracted beam has to be considered complex. Assuming Snell's law (1.52) and neglecting higher powers of small quantities we get

$$\alpha_2 \approx \sqrt{\alpha_1^2 - 2\delta - 2i\beta} \tag{1.70}$$

where δ and β belong to the complex refractive index n of the medium. The real and imaginary components of this angle, α_2' and α_2'', respectively, can be written as (see, e.g., Refs [75,76])

$$\alpha_2'^2 = \frac{1}{2}\left[\sqrt{\left(\alpha_1^2 - 2\delta\right)^2 + (2\beta)^2} + \left(\alpha_1^2 - 2\delta\right)\right] \tag{1.71}$$

$$\alpha_2''^2 = \frac{1}{2}\left[\sqrt{\left(\alpha_1^2 - 2\delta\right)^2 + (2\beta)^2} - \left(\alpha_1^2 - 2\delta\right)\right] \tag{1.72}$$

Both components are represented in Figure 1.33 for Mo-Kα X-rays striking a flat silicon substrate. The real component, α_2', is dominant in the range above the critical angle α_{crit} and is asymptotically equal to α_1 for large angles. The

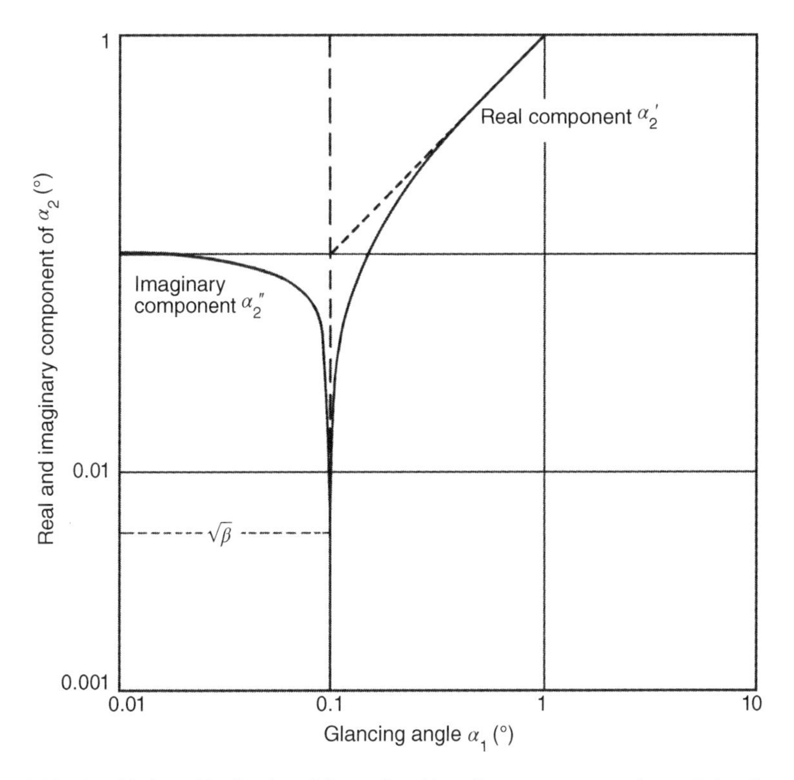

Figure 1.33. Double logarithmic plot of the real and imaginary component determining the angle α_2 of the refracted X-ray beam in dependence on the glancing angle α_1 of the incident beam. Calculation for Mo-Kα X-rays striking a flat silicon substrate above and below the critical angle of total reflection that amounts to 0.1°. Figure from Ref. [8], reproduced with permission. Copyright © 1996, John Wiley and Sons.

imaginary component, α_2'', is decisive for angles below α_{crit} and is asymptotically equal to α_{crit} for small angles. Both components become equal at the critical angle and amount to $\sqrt{\beta}$, which is extremely small. Moreover, the product of both components always equals β independent of the given glancing angle of incidence:

$$\alpha_2' \cdot \alpha_2'' = \beta \tag{1.73}$$

This relationship can easily be verified by multiplication of Equations 1.71 and 1.72 and extraction of the roots. It is important for subsequent calculations in Section 2.4.

For total reflection, two important quantities are characteristic:

- The reflectivity R, which is increased to 100% below the critical angle.
- The penetration depth z_n, which is reduced to a few nanometers in this case.

Both quantities can be calculated from the theory of a harmonic and plane electromagnetic wave, as discussed in Sections 1.5.3.1 and 1.5.3.2.

1.5.3.1 Reflectivity

The reflectivity is defined by the intensity ratio of the reflected beam and the incident beam and can be derived from the Fresnel formulas. These well-known formulas connect the vectors of the electromagnetic field of the reflected and the transmitted beam with those of the incident beam [77]. For the grazing incidence considered here, the amplitudes E_1^i, E_1^r, and E_2^t of the electric field vectors are expressed by the simple formulas

$$\frac{E_1^r}{E_1^i} = \frac{\alpha_1 - \alpha_2}{\alpha_1 + \alpha_2} \qquad \frac{E_2^t}{E_1^i} = \frac{2\alpha_1}{\alpha_1 + \alpha_2} \tag{1.74}$$

These formulas are valid independent of the polarization of the incident beam because of the assumed small angles α_1 and α_2. The reflectivity and transmissivity follow from these formulas after the absolute magnitude is squared.

The reflectivity R is given by

$$R = \left| \frac{\alpha_1 - \alpha_2}{\alpha_1 + \alpha_2} \right|^2 \tag{1.75}$$

With the help of the components α_2' and α_2'' of the complex angle α_2, the reflectivity can be calculated:

$$R = \frac{(\alpha_1 - \alpha_2')^2 + \alpha_2''^2}{(\alpha_1 - \alpha_2')^2 + \alpha_2''^2} \tag{1.76}$$

Three highly useful approximations result:

$$\begin{aligned} \alpha_1 \ll \alpha_{\text{crit}}: \quad & R \cong 1 - \sqrt{\frac{2}{\delta}\frac{\beta}{\delta}}\,\alpha_1 \\ \alpha_1 = \alpha_{\text{crit}}: \quad & R \approx \frac{\delta + \beta - \sqrt{(2\beta\delta)}}{\delta + \beta + \sqrt{(2\beta\delta)}} \\ \alpha_1 \gg \alpha_{\text{crit}}: \quad & R \cong \frac{\delta^2}{4\alpha_1^4} \end{aligned} \tag{1.77}$$

The dependence of the reflectivity on the glancing angle is demonstrated in Figure 1.34. The effect of total reflection is shown for three different elements.

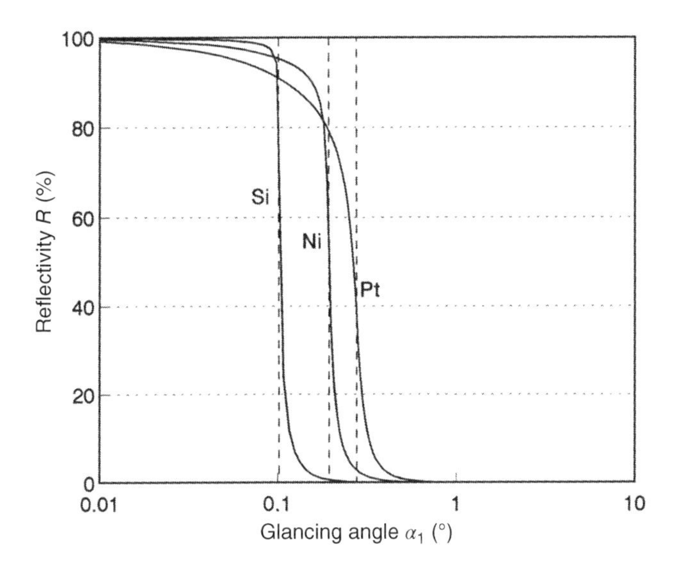

Figure 1.34. Reflectivity of three different media dependent on the glancing angle of X-rays. The curves were calculated for X-rays of Mo-Kα. Reflection below a critical angle α_{crit} is called *total reflection*. These angles are determined by the point of inflection of the curves and are marked by dashed vertical lines. Figure from Ref. [8], reproduced with permission. Copyright © 1996, John Wiley and Sons.

For glancing angles of 1° or more, the reflectivity is below 0.1%, independent of absorption, and can be neglected generally. Around the critical angle, the reflectivity rises to high values. However, the rise to 100% is not steplike but more or less gradual, dependent on the absorption or attenuation quantity β. The critical angle determines the point of inflection of the curves. For a less absorbing medium like silicon, the reflectivity shows the most distinct transition. For this reason silicon or quartz glass and even Plexiglas are used as sample carriers for TXRF.

The curves of Figure 1.34 were calculated for X-rays of Mo-Kα with photon energy of 17.44 keV. For higher energies, the α_{crit} values are decreased according to the $1.65/E$ term in Equation 1.68 and consequently the curves are shifted to the left. For lower energies, they are shifted to the right. Table 1.10 lists the corresponding reflectivity values calculated by Equation 1.77.

1.5.3.2 Penetration Depth

The penetration depth is defined by that depth of a homogeneous medium a beam can penetrate while its *intensity* is reduced to $1/e$, or 37% of its initial value. This depth z_n, which is normal to the boundary of the medium, follows the equation

$$z_n \approx \frac{\lambda}{4\pi} \frac{1}{\alpha_2''} \tag{1.78}$$

TABLE 1.10. Reflectivity R_{crit} of Various Media at the Critical Angle of Total Reflection, Calculated for X-rays of Different Photon Energies

Medium	8.4 keV (%)	R_{crit} at Photon Energy of: 17.44 keV (%)	35 keV (%)
Plexiglas	87.9	93.2	94.8
Glassy carbon	88.4	93.9	95.0
Boron nitride	87.6	93.3	94.6
Quartz glass	73.4	85.5	91.4
Aluminum	69.7	82.9	90.3
Silicon	67.3	81.5	89.5
Cobalt	37.4	59.1	75.2
Nickel	37.0	58.1	74.9
Copper	66.9	56.1	82.7
Germanium	62.3	51.2	69.7
Gallium arsenide	62.4	51.1	69.5
Tantalum	49.3	42.9	63.4
Platinum	45.3	39.4	60.2
Gold	44.8	38.7	59.5

Source: From Ref. [8], reproduced with permission. Copyright © 1996, John Wiley and Sons.

Again three approximate values can be given:

$$\begin{aligned} \alpha_1 \ll \alpha_{crit}: \quad & z_n \approx \frac{\lambda}{4\pi}\frac{1}{\sqrt{2\delta}} \\ \alpha_1 = \alpha_{crit}: \quad & z_n \approx \frac{\lambda}{4\pi}\frac{1}{\sqrt{\beta}} \\ \alpha_1 \gg \alpha_{crit}: \quad & z_n \approx \frac{\lambda}{4\pi}\frac{\alpha_1}{\beta} \end{aligned} \tag{1.79}$$

For $\alpha \geq \alpha_{crit}$, the penetration depth is dependent on β, which is proportional to the mass-absorption coefficient of the medium. For $\alpha < \alpha_{crit}$, however, δ becomes the decisive quantity, which is mainly determined by atomic constants of the medium.

Figure 1.35 shows a double-logarithmic presentation of the penetration depth dependent on the glancing angle for the three elements already considered in Figure 1.34. For angles above and down to 0.5°, the penetration depth linearly decreases with the glancing angle and the depth values are of the order of 0.1–10 μm. At the critical angle, the penetration depth drastically decreases especially for nonabsorbing media like silicon. Below this critical angle, the penetration depth reaches a constant level of only a few nanometers and the beam is called "evanescent." For silicon, the three z_n values of Equation 1.79 come to 3.2 nm, 62 nm, and 1.2 μm × (α/α_{crit}).

Of course, the effect of total reflection only appears when the medium is flat and smooth. For a rough surface, total reflection disappears. The penetration

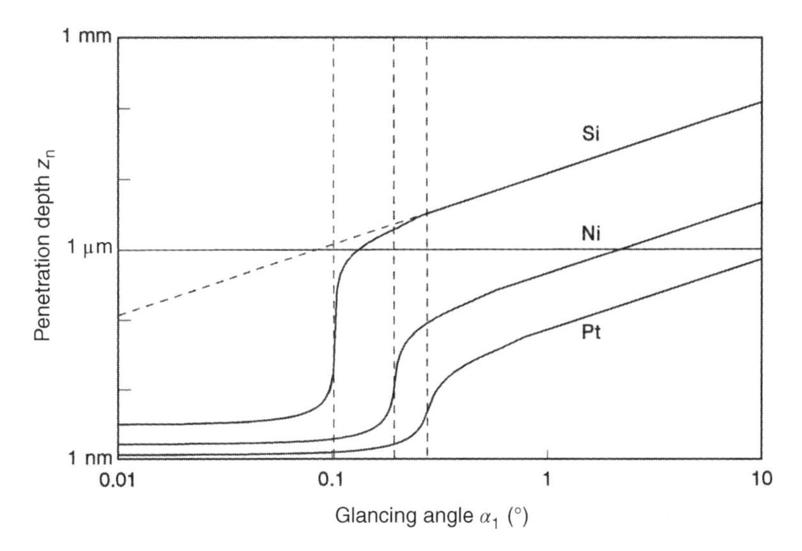

Figure 1.35. Penetration depth of X-rays hitting three different media at a variable glancing angle. The curves were calculated for X-rays of Mo-Kα with a photon energy of 17.44 keV. The critical angles are marked by dashed vertical lines. The dashed oblique straight line represents the penetration depth in a roughened silicon surface for which total reflection disappears. Figure from Ref. [8], reproduced with permission. Copyright © 1996, John Wiley and Sons.

depth linearly decreases with the glancing angle even below the critical angle, as is demonstrated in Figure 1.35 for silicon.

The curves of Figure 1.35 were calculated for the photon energy of the chosen Mo-Kα radiation. The influence of the different photon energies on the penetration depth is shown in Figure 1.36—here for silicon. The points of inflection shift to lower critical angles with increasing photon energy. Furthermore, the curves are stretched to higher depth values for normal reflection while the depth values for total reflection remain constant at

$$z_0 \cong 3.424\sqrt{\frac{A}{Z}\frac{1}{\rho}} \tag{1.80}$$

This minimum is a material constant that is only dependent on the quantity C_m but is independent of the photon energy and is listed in Table 1.11 for various media. This table also gives critical penetration depths calculated for α_{crit} and three different photon energies according to Equation 1.79.

In comparison to the penetration depth, a further quantity is important—the information depth. It is the depth of a sample from which secondary radiation emerges and reaches the detector for X-ray fluorescence analysis. This information depth is always smaller than the penetration depth. For high-energy peaks (energy of the emitted radiation is only a little smaller than the photon energy of the primary radiation used for excitation) it is of the order of the

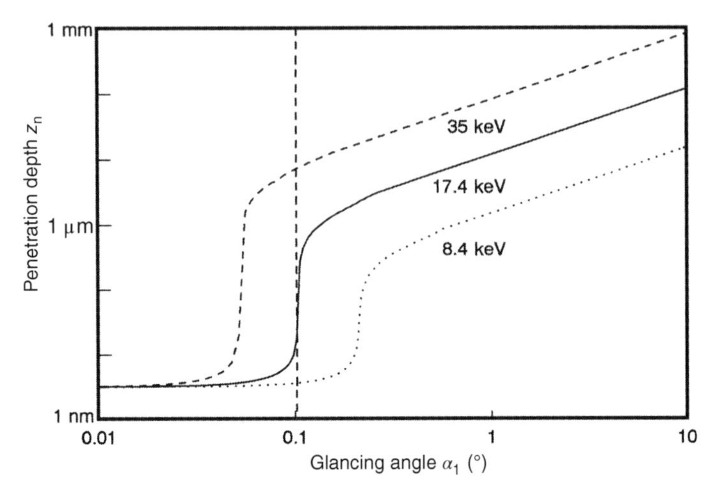

Figure 1.36. Penetration depth of X-rays striking on silicon at a variable glancing angle. The curves were calculated for three different photon energies. The dashed vertical line identifies the respective critical angle for photons with $E = 17.4$ keV. Figure from Ref. [8], reproduced with permission. Copyright © 1996, John Wiley and Sons.

TABLE 1.11. Minimum Penetration Depth z_0 and Critical Penetration Depth z_{crit} of Various Media, Calculated for X-rays of Different Photon Energies

Medium	Minimum z_0 (nm)	8.4 keV (nm)	z_{crit} at Photon Energy of: 17.44 keV (nm)	35 keV (nm)
Plexiglas	4.3	132	241	319
Glassy carbon	4.1	130	255	311
Boron nitride	3.2	97	188	238
Quartz glass	3.2	42	83	146
Aluminum	3.0	33	64	116
Silicon	3.2	32	62	115
Cobalt	1.7	6.6	12.7	24
Nickel	1.7	6.4	12.1	23
Copper	1.7	16.8	11.5	22
Germanium	2.2	18.8	13.1	25
Gallium arsenide	2.2	18.8	13.0	24
Tantalum	1.3	7.3	6.0	11.4
Platinum	1.2	5.8	4.8	9.1
Gold	1.2	6.0	5.0	9.4

Source: From Ref. [8], reproduced with permission. Copyright © 1996, John Wiley and Sons.

penetration depth. However, for low-energy peaks (energy of emitted photons is much smaller than that of primary photons) the information depth can be 10 times smaller or even less compared to the penetration depth.

1.5.4 Refraction and Dispersion

In light optics, the phenomenon of dispersion can be demonstrated by a triangular prism or by a plane parallel plate.[5] A rainbow is the most familiar example where the moisture in the atmosphere represents a lot of tiny drops. By refraction, the white polychromatic light beam splits into its component colors geometrically and leads to a "spectrum." The red color is always bent least and the violet color is bent most. The phenomenon is based on the fact that the refractive index of a substance depends on the wavelength or energy of the photons used to measure it. The refractive index is about 1.3 for water, about 1.5 for different glasses, and about 2.4 for diamond at $\lambda = 589$ nm (Na yellow).

The phenomena of refraction and dispersion also occur for X-rays with a subtle distinction; the refractive index is usually a little bit smaller than 1 (about 0.995 for 1 keV photons in platinum, up to 0.9999998 for 40 keV photons in Plexiglas). The effect is demonstrated by a polychromatic X-ray beam coming from vacuum or air and refracted at a plane-parallel plate, for example, a rectangular piece of a wafer, as illustrated in Figure 1.37. This incident beam may hit the plate at an incident glancing angle α_1 is refracted at the upper plane, and split into beams of particular wavelengths or photon energies at different exit angles α_2. All these X-ray beams are deflected in the direction of the plate surface (in contrast to visible light). The larger the wavelength of the X-radiation, that is, the smaller the photon energy E_2 the smaller the corresponding refraction angle. All the refracted beams are refracted a second time at the bottom plane and leave the plate in parallel to each other but with a different distance in the x direction. The exit beam in total is parallel to the incident primary beam but split into parts of different photon energies and shifted to the right.

Investigations go back to an early paper of Yoneda in 1963 [79] who investigated the refraction and reflection of an X-ray beam of a Mo tube at a plane-parallel plate of silicon. He observed an anomalous surface reflection (ASR) near the critical angle of total reflection (see also Ref. [80]). A photograph shows several lines: the totally reflected line, an ASR band, an artificial center line between totally reflected and directly passing beam, the Kα and Kβ lines of Mo, and the directly passing beam (without plate).

For a given glancing angle of the incident beam, the glancing angle of the exit beam is determined by Snell's law according to Equation 1.52. For glancing angles above the critical angle of total reflection, the wavelength of the refracted beam increases only slightly and α_2 is somewhat smaller than α_1 as

[5] A plane-parallel plate can be defined as a rectangular parallelepiped or a rectangular prism.

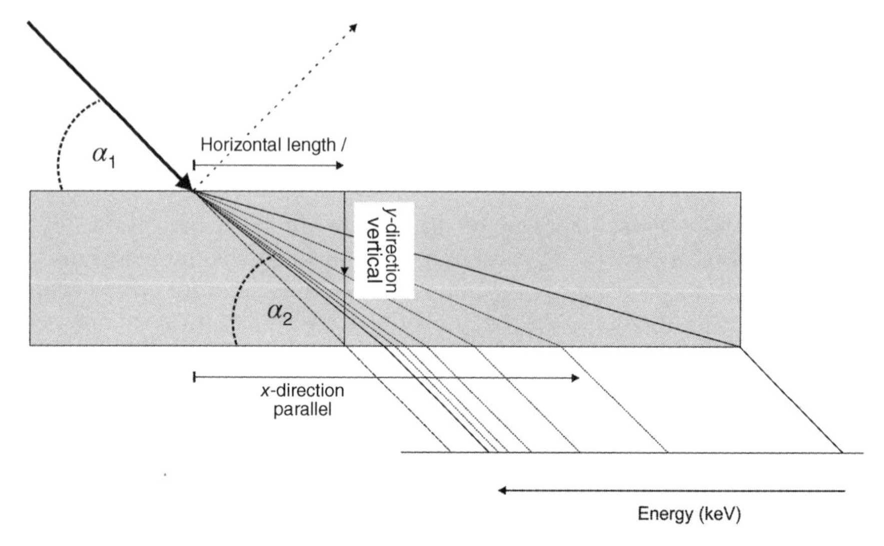

Figure 1.37. Dispersion of X-rays by refraction at a plane-parallel plate or a "rectangular prism," respectively. The incident polychromatic beam with several different photon energies strikes a silicon wafer plate at a fixed glancing angle α_1. The beam is reflected at the same angle with a small intensity of only about 1% in the range of 0.1° to 0.2° as is indicated by a dashed line. In addition, the beam is refracted under exit angles α_2 dependent on respective photon energies, E_2, with an intensity of about 99% (after Ref. [78]). In a horizontal distance x dependent on α_2 and E_2, all the refracted beams leave the silicon plate at the original α_1. The exit beam in total is parallel to the incident primary beam but is shifted to the right in dependency of the energy E_2. If the wafer has a thickness of 0.5 mm the parallel shift is of the order of several 10 cm. A vertical shift of about some 10 μm can be observed if the wafer is cut in a distance of 10 mm.

demonstrated in Figure 1.32b.[6] For glancing angles below the critical angle of total reflection $\alpha_1 < \alpha_{crit}$, there is no refracted beam but only the reflected beam with an intensity of nearly 100%.

The glancing angle of the refracted exit beam α_2 for a given incident glancing angle α_1 can be calculated in accord with Snell's law. We find the approximation for the refractive index of a substance [78,81,82]:

$$n_2 = 1 - \left(\frac{\alpha_1^2}{2} - \frac{\alpha_2^2}{2}\right) \tag{1.81}$$

Comparison with Equation 1.67 leads to

$$\lambda_2 = \lambda_{cut} \frac{\sqrt{\alpha_1^2 - \alpha_2^2}}{\alpha_{crit}} \tag{1.82}$$

[6] The wavelength of the refracted beam is increased negligibly by $1/n_2$; that is, by a factor smaller than 1.0005 for silicon or quartz glass.

where λ_{cut} is the longest wavelength of an X-ray beam that can be refracted at the given incident angle. A beam with a wavelength that is still longer than λ_{cut} will not be refracted but totally reflected. For X-rays, α_2 is somewhat smaller than α_1 and consequently the radicand is always positive.

Equation 1.82 can be transformed for photon energies of the refracted beam:

$$E_2 = E_{cut} \frac{\alpha_{crit}}{\sqrt{\alpha_1^2 - \alpha_2^2}} \tag{1.83}$$

where E_{cut} is the cutoff energy, which is the smallest possible photon energy of a beam that can be refracted; a beam with an even smaller photon energy is totally reflected. The product of E_{cut} and α_{crit} is constant for a specific material. It is equal to $hc_0 \cdot \sqrt{2C_m}$ and amounts to 1.778 for a silicon wafer if the energy is given in keV and the angle is entered in degrees. The photon energy of the exit beam is demonstrated in Figure 1.38 in dependence of the exit angle and four fixed incident angles. The range of possible exit angles is between 0° and α_1 and the respective range of photon energies is between E_{crit} and ∞. In all cases, the photon energy of the incident beam E_1 is only a little bit higher than that of the exit beam E_2.

Hayashi *et al.* investigated the refraction of X-rays in silicon at grazing incidence [78]. The authors turned a polychromatic beam of a Mo tube on a Si

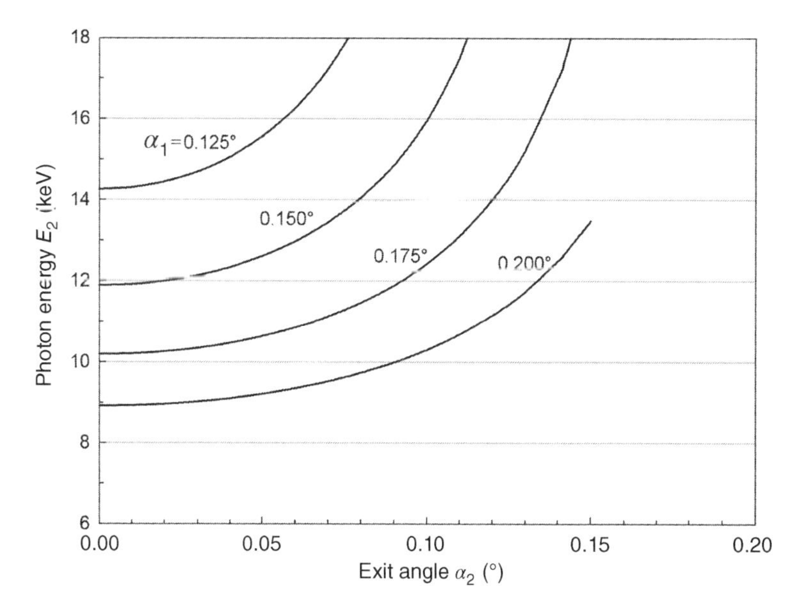

Figure 1.38. Energy of photons *in* the refracted beam dependent on the respective exit angle. It is provided that the incident beam strikes a silicon wafer at a fixed glancing angle of 0.125°, 0.150°, 0.175°, and 0.200°.

wafer under a small angle $\alpha_1 = 0.124°$ and verified three single refracted peaks at angles of 0.02°, 0.04°, and 0.06° with an energy of 14.5, 15.1, and 16.4 keV, respectively, in accord with Equation 1.83. The width of these peaks was about 1 keV. Furthermore, a thin organic film of n-$C_{33}H_{68}$ was deposited on the Si carrier and the experiment repeated with $\alpha_1 = 0.095°$. Again a refracted peak was found depending on α_2 with energies between 18.4–20.6 keV, but moreover a strong peak at 16.2 keV was recognized independent of α_2. The relationships were proved true by Ebel *et al.* [81–83] in accord with Equation 1.83. Moreover, the additional peak could be explained by the organic overlay acting as a waveguide. Obviously, a standing wave is built by superposition of two plane waves, propagating as a monochromatic beam of 16.2 keV within the film of 15.4 nm thickness. The authors detected a refracted beam at 10.2 keV for an incident angle of 0.18° and an exit angle of 0.0433° in accord with Equation 1.83. The width of this peak was about 0.6 keV for a beam with a divergence of 0.1°. The authors suggested the use of a silicon wafer as a refractive monochromator and the application of cutoff and refracted beams for angular calibration of the experimental setup [82].

REFERENCES

1. Franks, A. (1997). The First Hundred Years. In: Michette, A. and Pfauntsch, S. (editors) *X-Rays*, John Wiley & Sons: New York, pp. 1–19.
2. Friedrich, W., Knipping, P., and von Laue, M. (1912). *Interferenzerscheinungen bei Röntgenstrahlen*, Bayerische Akademie der Wissenschaften zu München, Sitzungsberichte math.-phys. Kl., p. 303–322. Reprinted (1913): Annalen der Physik **41**, 971–988.
3. Moseley, H.G.J. (1913). The high frequency spectra of the elements. *Phil. Mag.*, **26**, 1024–1034; and Moseley, H.G.J. (1914). The high frequency spectra of the elements. Part II. *Phil. Mag.*, **27**, 703–713.
4. Long, J. (1997). X-ray microanalysis. In: Michette, A. and Pfauntsch, S. (editors) *X-Rays*, John Wiley & Sons: New York, pp. 21–40.
5. Barkla, C.G. (1905). Polarised Röntgen radiation. *Philos. Trans. R. Soc. London*, **A204**, 467–480.
6. Barkla, C.G. (1911). The spectra of the fluorescent Röntgen radiations. *Phil. Mag.*, **22**, 396–412.
7. Compton, A.H. (1923). The total reflection of X-rays. *Phil. Mag.*, **45**, 1121–1131.
8. Klockenkämper, R. (1997), *Total-Reflection X-Ray Fluorescence Analysis*, 1st ed., John Wiley & Sons, Inc., New York.
9. Hennig, U. (1989). *Deutsches Röntgen-Museum Remscheid-Lennep*, G. Westermann Verlag: Braunschweig, pp. 3–128.
10. Yoneda, Y. and Horiuchi, T. (1971). Optical flats for use in X-ray spectro-chemical microanalysis *Rev. Sci. Instrum.*, **42**, 1069–1070.
11. Wobrauschek, P. (1975). *Totalreflexions-Röntgenfluoreszenzanalyse*, PhD thesis. Available from: Atominstitut der Österreichischen Universitäten, Technical University of Vienna, Austria.

12. Aiginger, H. and Wobrauschek, P. (1974). A method for quantitative X-ray fluorescence analysis in the nanogram region. *Nucl. Instr. Meth.*, **114**, 157–158.
13. Aiginger, H. and Wobrauschek, P. (1975). Total-reflection X-ray fluorescence spectrometric determination of elements in nanogram amounts. *Anal. Chem.*, **47**, 852–855.
14. Knoth, J. and Schwenke, H. (1978). An X-ray fluorescence spectrometer with totally reflecting sample support for trace analysis at the ppb level. *Fresenius Z. Anal. Chem.*, **291**, 200–204.
15. Schwenke, H. and Knoth, J. (1982). A highly sensitive energy-dispersive X-ray spectrometer with multiple total reflection of the exciting beam. *Nucl. Instr. Meth.*, **193**, 239–243.
16. Boumans, P.W.J.M. and Klockenkämper, R. editors (1989). Total reflection X-ray fluorescence spectrometry. *Spectrochim. Acta*, **44B**, 433–549.
17. Boumans, P.W.J.M., Wobrauschek, P., and Aiginger, H. editors (1991). Total reflection X-ray fluorescence spectrometry: Proceedings *Spectrochim. Acta* **46B**, 1313–1436.
18. Boumans, P.W.J.M. and Prange, A. editors (1993). Total reflection X-ray fluorescence spectrometry: Proceedings *Spectrochim. Acta*, **48B**, 107–299.
19. Taniguchi K. editor (1995). *Adv. X-Ray Chem. Anal. Jpn.*, **26s**, 1–206.
20. de Boer, D.K.G. and Klockenkämper, R. editors (1997). Total reflection X-ray fluorescence analysis: 6th conference on total reflection X-ray fluorescence analysis and related methods. *Spectrochim. Acta*, **52B**, 795–1072.
21. Wobrauschek, P. editor (1999). *Spectrochim. Acta*, **54B**, 1383–1544.
22. Streli, C. and Wobrauschek, P. editors (2001). *Spectrochim. Acta*, **56B**, 2003–2336.
23. de Carvalho, M.L. editor (2003). *Spectrochim. Acta*, **58B**, 2023–2260.
24. Gohshi, Y., Kawai, J., and Taniguchi, K. editors (2004). *Spectrochim. Acta*, **59B**, 1047–1334.
25. Zaray, Gy. and Ovari, M. editors (2006). *Spectrochim. Acta*, **61B**, 1081–1239.
26. Pepponi, G. editor (2008). *Spectrochim. Acta*, **63B**, 1349–1510.
27. Boman, J. editor (2010). *Spectrochim. Acta*, **65B**, 427–508.
28. Becker, R.S., Golovchenko, J.A., and Patel, J.R. (1983). X-ray evanescent-wave absorption and emission. *Phys. Rev. Lett.*, **50**, 153–156.
29. Iida, A., Yoshinaga, A., Sakurai, K., and Gohshi, Y. (1986). Synchrotron radiation excited X-ray fluorescence analysis using total reflection of X-rays. *Anal. Chem.*, **58**, 394–397.
30. Klockenkämper, R., Knoth, J., Prange, A., and Schwenke, H. (1992). Total-reflection X-ray fluorescence spectroscopy. *Anal. Chem.*, **64**, 1115A–1123A.
31. Klockenkämper, R. (1991). Totalreflexions-Röntgenfluoreszenzanalyse. In: Günzler, H. et al., (editors) *Analytiker Taschenbuch*, Vol. 10, Springer-Verlag: Berlin; Band 111–152.
32. von Bohlen, A. (2009). Total reflection X-ray fluorescence and grazing incidence X-ray spectrometry – Tools for micro- and surface analysis: a review. *Spectrochim. Acta*, **64B**, 821–832.
33. Alov, N.V. (2011). Total reflection X-ray fluorescence analysis: physical foundations and analytical application (A review). *Inorganic Materials*, **47**, 1487–1499.

34. Wobrauschek, P. (2007). Total reflection X-ray fluorescence analysis: a review. *X-Ray Spectrom.*, **36**, 289–300.
35. Kregsamer, P., Streli, C., and Wobrauschek, P. (2002). Total reflection X-ray fluorescence. In: Van Grieken, R. and Markowicz, A. (editors) *Handbook of X-Ray Spectrometry*, 2nd ed., Marcel Dekker, pp. 559–602.
36. Fabry, L., Pahlke, S., and Beckhoff, B. (2011). Total-reflection X-ray fluorescence (TXRF) analysis. In: Friedbacher, G. and Bubert, H. (editors) *Surface and Thin Film Analysis*, 2nd ed., Wiley-VCH, pp. 267–292.
37. Mori, Y. (2004). Total-reflection X-ray fluorescence for semiconductors and thin films. In: Tsuji, K., Injuk, J., and Van Grieken, R. (editors) *X-Ray Spectrometry: Recent Technological Advances*, John Wiley & Sons: New York, pp. 517–533.
38. Meirer, F., Singh, A., Pepponi, G., Streli, C., Homma, T., and Pianetta, P. (2010). Synchrotron radiation-induced total reflection X-ray fluorescence analysis. *Trends Anal. Chem.*, **29**, 479–496.
39. Streli, C., Wobrauschek, P., Meirer, F., and Pepponi, G. (2008). Synchrotron radiation induced TXRF. *J. Anal. At. Spectrom.*, **23**, 792–798.
40. von Bohlen, A., Krämer, M., Sternemann, C., and Paulus, M. (2009). The influence of X-ray coherence length on TXRF and XSW and the characterization of nanoparticles observed under grazing incidence of X-rays. *J. Anal. At. Spectrom.*, **24**, 792–800.
41. von Bohlen, A., Brücher, M., Holland, B., Wagner, R., and Hergenröder, R. (2010). X-ray standing waves and scanning electron microscopy – Energy dispersive X-ray emission spectroscopy study of gold nanoparticles. *Spectrochim. Acta*, **65B**, 409–414.
42. Szoboszlai, N., Polgari, Z., Mihucz, V., and Zaray, G. (2009). Recent trends in total reflection X-ray fluorescence spectrometry for biological applications. *Analytica Chimica Acta*, **633**, 1–18.
43. Schmeling, M. and Van Grieken, R. (2002). Sample preparation for X-ray fluorescence. In: Van Grieken, R. and Markowicz, A. (editors) *Handbook of X-Ray Spectrometry*, 2nd ed., Marcel Dekker, pp. 933–976.
44. Kunimura, S., Watanabe, D., and Kawai, J. (2009). Optimization of a glancing angle for simultaneous trace element analysis by using a portable total reflection X-ray fluorescence spectrometer. *Spectrochim. Acta*, **64B**, 288–290.
45. Kiessig, H. (1931). Untersuchungen zur Totalreflexion von Röntgenstrahlen. *Ann. Phys.*, **10**, 715–768.
46. Du Mond, J. and Youtz, J.P. (1940). An X-ray method of determining rates of diffusion in the solid state. *J. Appl. Phys.*, **11**, 357–365.
47. National Institute of Standards and Technology (2010). The NIST Reference on Constants, Units, and Uncertainty. http://physics.nist.gov/cuu/Constants/
48. Bertin, E.P. (1975). *Principles and Practice of Quantitative X-ray Fluorescence Analysis*, 2nd ed., Plenum Press: New York.
49. Jenkins, R., Manne, R., Robin, R., and Senemaud, C. (1991). IUPAC - Nomenclature system for X-ray spectroscopy. *X-Ray Spectrom.*, **20**, 149–155.
50. Tertian, R. and Claisse, F. (1982). *Principles of Quantitative X-ray Fluorescence Analysis*, Heyden: *London*.
51. Elder, F.R., Gurewitsch, A.M., Langmuir, R.V., and Pollock, H.C. (1947). Radiation from Electrons in a Synchrotron. *Phys. Rev.*, **71**, 829–830.

52. http://asd.gsfc.nasa.gov/Volker.Beckmann/school/download/Longair_Radiation2.pdf
53. Wille, K. (1992). *Physik der Teilchenbeschleuniger und Synchrotronstrahlungsquellen*, Teubner Studienbücher, Physik: Stuttgart.
54. Schwinger, J. (1949). On the classical radiation of accelerated electrons. *Phys. Rev.*, **75**, 1912–1925.
55. Munro, I. (1997). The First Hundred Years. In: Michette, A. and Pfauntsch, S. (editors) *X-Rays*, John Wiley & Sons: New York, pp. 131–154.
56. Als-Nielson, J. and Mc Morrow, D. (2001). *Elements of Modern X-Ray Physics*, 4th reprint 2010, John Wiley & Sons, Ltd.: New York.
57. Author collective (1986). Vaughan, D. (editor) *X-ray Data Booklet*. 1st ed., Lawrence Berkeley Laboratory: Berkeley; third edition (2009), editor A.C. Thompson.
58. http://en.wikipedia.org/wiki/Synchrotron_radiation; August 2014, Wikimedia Foundation, Inc., USA, October 2014
59. http://www.ira.inaf.it/~ddallaca/P-Rad_3.pdf; 2014, Daniele Dallacasa, INAF - ISTITUTO DI RADIOASTRONOMIA, Bologna, Italy, October 2014
60. http://www.lightsources.org, Management Board Lightsources.org, October 2014
61. http://www.cxro.lbl.gov/A U.S. Department of Energy National Laboratory Operated by the University of California 2014, X-ray Data Base, 2014.
62. Longair, M.S. (1981). *High Energy Astrophysics*, 3rd ed. 2011, Cambridge University Press: Cambridge, *440 pages*.
63. Weis, T., Bergers, U., Friedl, J., Hartmann, P., Heine, R., Huck, H., Kettler, J. Kopitetzki, O., Schirmer, D., Schmidt, G., and Wille, K. (2006). *Status of the 1.5 GeV Synchrotron Light Source DELTA and related Accelerator Physics Activities.* In: *RuPAC*, Novosibirsk.
64. http://www.delta.tu-dortmund.de; 2014, Technische Universität Dortmund, 44227 Dortmund, Germany, October 2014
65. Jackson, J.D. (1982). *Klassische Elektrodynamik*, 2nd ed., de Gruyter: Berlin, New York.
66. Hoffmann, A. (1986) *Theory of Synchrotron Radiation*, SSRL (ACD-note 38).
67. Mills, D.M., Helliwell, J.R., Kvick, A., Ohta, T., Robinson, I.A., and Authier, A. (2005). Report on the Working Group on Synchrotron Radiation Nomenclature – brightness, spectral brightness or brilliance. *J. Synchrotron Rad.*, **12**, 385.
68. Schirmer, D. (2005). *Synchrotron Radiation Sources at DELTA*, University of Dortmund, DELTA: Int. Rep. 001-05.
69. en.wikipedia.org/wiki/Planck's_law
70. Williams, K.L. (1987). *An Introduction to X-Ray Spectrometry*, Allen & Unwin: London.
71. Krieger, H. and Petzold, W. (1989). *Strahlenphysik, Dosimetrie und Strahlenschutz*, Bd. 2, Teubner: Stuttgart.
72. Woldseth, R. (1973). *All you ever wanted to know about X-Ray Energy Spectrometry*, Kevex Corp.: Burlingame, California.
73. Veigele, W.J. (1973). *Atomic Data Tables*, **5**, 51–111.
74. James, R.W. (1967). *The Optical Principles of the Diffraction of X-rays*, Cornell University Press: Ithaca, NY.

75. Stanglmeier, F. (1990). *Bestimmung der dispersiven Korrektur f'(E) zum Atomformfaktor aus der Totalreflexion von Röntgenstrahlen*, Forschungszentrum Jülich: Berichte Nr. 2346. (Dissertation TH Aachen).

76. Blochin, M.A. (1957). *Physik der Röntgenstrahlen*, VEB Verlag Technik: Berlin.

77. Born, M. and Wolf, E. (1959). *Principles of Optics*, 6th ed., Pergamon Press: London, p. 36–51.

78. Hayashi, K., Kawai, J., Moriyama, Y., Horiuchi, T., and Matsushige, K. (1999). Refracted X-rays propagating near the surface under grazing incidence condition. *Spectrochim. Acta*, **B54**, 227–230.

79. Yoneda, Y. (1963). Anomalous surface reflection of X rays. *Phys. Rev.*, **131**, 2010–2013.

80. Nigam, A.N. (1965). Origin of anomalous surface reflection of X rays. *Phys. Rev.*, **138**, A1189–A1191.

81. Ebel, H., Svagera, R., and Ebel, M.F. (2001). X-ray waveguide phenomenon in thin layers under grazing incidence conditions. *X-Ray Spectrom.*, **30**, 180–185.

82. Ebel, H., Streli, C., Pepponi, G., and Wobrauschek, P. (2001). Energy dispersion of X-ray continua in the energy range 8 keV to 16 keV by refraction on Si wafers. *Spectrochim. Acta*, **B56**, 2045–2048.

83. Ebel, H. (2002). Quantification of the monochromatic photon flux of refractive X-ray monochromators in the energy range from 10 to 30 keV. *X-Ray Spectrom.*, **31**, 368–372.

CHAPTER

2

PRINCIPLES OF TOTAL REFLECTION XRF

Our consideration of the fundamentals of X-ray fluorescence so far has treated the nature of X-rays and their production, attenuation, and deflection. These fundamentals are sufficient as a basis of conventional XRF spectrometry. However, some additional premises are required to elucidate the peculiarities of total reflection X-ray fluorescence (TXRF) and related methods. These premises especially concern the phenomena of interference and standing waves. While conventional excitation takes place in a uniform field of X-radiation, TXRF occurs in the inhomogeneous field of standing waves in front of optically flat substrates or within thin layers on such substrates. The occurrence of such a field consequently has to be considered in some detail.

Total-Reflection X-ray Fluorescence Analysis and Related Methods, Second Edition.
Reinhold Klockenkämper and Alex von Bohlen.

2.1 INTERFERENCE OF X-RAYS

The phenomena of interference result from the superposition of two beams (double-beam interference) or even more than two beams (multiple-beam interference). They are generally explained in the wave picture. In the region of superposition, the resulting wave field can show a pattern with maxima and minima. These fluctuations will be highly distinct if two superimposing waves are monochromatic and coherent, that is, if they have the same wavelength and a fixed phase difference. If the phase difference is an odd multiple of π, the amplitudes are subtracted to a minimum. This kind of interference is called destructive, and points with minima are called nodes. On the other hand, interference is called constructive if the phase difference is an even multiple of π and the amplitudes sum up to a maximum. The corresponding points are called antinodes. Nodes and antinodes can be extended to nodal and antinodal lines or planes, respectively.

A most simple way to produce interference is the superposition of two beams propagating on the same straight line. This two-beam interference can be affected by reflection of X-rays at the upper and lower boundaries of a thin layer deposited on a thick substrate. It can become multiple-beam interference if a multilayer is chosen with a reflection at various boundaries. Furthermore, the scattering of X-rays at different atoms of a crystal can lead to multiple-beam interference. The specific behavior can be explained by reflection at the various lattice planes, which results in the well-known Bragg's law. This interference of coherently scattered X-rays is usually called diffraction.

2.1.1 Double-Beam Interference

A thin homogeneous layer may be deposited on a thick and flat substrate. If a monochromatic X-ray beam hits this layer at grazing incidence, double-beam interference can be observed. Such experiments require a glancing angle of about 0.1° and a layer thickness of 1 nm to 1 μm. Furthermore, a beam width of ~10 μm and a layer size of ~10 mm are required for a wide superposition.

Figure 2.1 depicts the paths of X-rays within the three media denoted by subscripts 0, 1, and s: the first medium, from which the X-rays are coming, is assumed to be vacuum or air, the second medium is a thin plane-parallel layer of thickness d, and the third medium is a thick substrate. Two cases A and B can be distinguished. In case A, the layer is optically denser than the substrate ($n'_1 > n'_s$ where n' is the real part of the refractive index n) (see Section 1.5.1). In case B, the layer is optically thinner than the substrate ($n'_1 < n'_s$).

In case A, the glancing angle α_1 is smaller than α_0 and α_s is even smaller than α_1, so total reflection is possible at *both* interfaces: air–layer and layer–substrate. The first happens at or below a small angle determined by $\alpha_{01} = \sqrt{2\delta_1}$ when α_1 becomes zero or imaginary (see Section 1.5.3). The last happens below the greater angle $\alpha_{0s} = \sqrt{2\delta_s}$ when α_1 becomes $\sqrt{2(\delta_s - \delta_1)}$ and α_s becomes zero

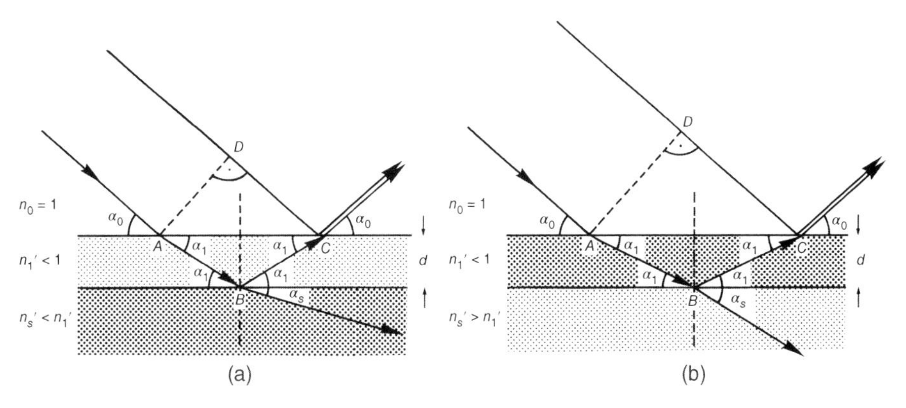

Figure 2.1. Paths of X-rays striking a thin layer of thickness d deposited on a thick substrate. (a) Case A: the layer is optically denser than the substrate ($n_0 = 1 > n_1' > n_s'$). (b) Case B: the layer is optically thinner than the substrate ($n_0 = 1 > n_1' < n_s'$). Figure from Ref. [1], reproduced with permission. Copyright © 1996, John Wiley and Sons.

or imaginary. In case B, α_1 is likewise smaller than α_0 but α_s is generally greater than α_1 so total reflection is possible only at the upper, air–layer interface but not at the lower layer–substrate interface. The first again happens below the angle $\alpha_{01} = \sqrt{2\delta_1}$ when α_1 becomes zero or imaginary.

Of course, the actual glancing angles for grazing incidence of X-rays are much smaller than those illustrated in the figure. Moreover, multiple zigzag reflections not considered here are generally possible. Finally, the two boundaries assumed to be perfectly smooth usually have a certain roughness. Nevertheless, the idealized model facilitates an understanding of the approximate description of the interference phenomena.

To further understand interference, the phase difference of the two reflected beams coming from the lower and upper boundaries has to be determined. These two beams have a path difference Δ given by $|\mathrm{AB}| + |\mathrm{BC}| - |\mathrm{CD}|$:

$$\Delta = 2\,d \sin \alpha_1 \tag{2.1}$$

The path difference can be approximated using Equation 1.70:

$$\Delta \approx 2d\sqrt{\alpha_0^2 - 2\delta_1} \tag{2.2}$$

where α_0 is the glancing angle of the incoming beam and δ_1 is the decrement of the layer. The phase difference φ of the two reflected beams is determined by $(2\pi/\lambda)\cdot\Delta$, at least in case A. In case B, the amount of π has still to be added due to a phase jump at the optically denser substrate.

As already mentioned, a constructive or destructive interference occurs if the phase difference φ is an even or odd multiple of π, respectively. If k

represents this even or odd integer, the condition can be written as

$$\frac{4\pi}{\lambda} d \sqrt{\alpha_0^2 - 2\delta_1} \approx k\pi \tag{2.3}$$

This relation is fulfilled for two sets of glancing angles $\alpha_0 = \alpha_k$. The maxima occur for

$$\alpha_k^2 \approx \alpha_{01}^2 + \left(k \frac{\lambda}{2d}\right)^2 \tag{2.4a}$$

and the minima occur for

$$\alpha_k^2 \approx \alpha_{01}^2 + \left(\frac{2k+1}{2} \frac{\lambda}{2d}\right)^2 \tag{2.4b}$$

where α_{01} is the critical angle of total reflection for the layer material.

The angles α_k determine the directions in which the two reflected beams are reinforced or annihilated. They lead to maxima and minima for the reflectivity of the layered substrate. Case A and case B differ by exchange of maxima and minima at the same k value because of the phase jump π.

The reflectivity can be determined by measuring the intensity of the reflected beams at an angle of $\alpha_0^* = \alpha_0$. The ratio of this value to the intensity of the incoming beam gives the reflectivity R. This quantity can also be calculated theoretically. It depends on the two values of reflectivity—that of the layer R_{01} and that of the substrate R_{1s}. These two values result from Equation 1.41 and finally lead to the reflectivity of the layered substrate [2,3]:

$$R = \frac{R_{01} + 2\sqrt{(R_{01}R_{1s})}\exp(-\varphi'')\cos\varphi' + R_{1s}\exp(-2\varphi'')}{1 + 2\sqrt{(R_{01}R_{1s})}\exp(-\varphi'')\cos\varphi' + R_{01}R_{1s}\exp(-2\varphi'')} \tag{2.5}$$

where φ' and φ'' are the real and imaginary components, respectively, of the complex phase difference φ between the two reflected beams. φ is determined by

$$\varphi = \frac{4\pi}{\lambda} d\, \alpha_1 \tag{2.6}$$

where α_1 is the complex glancing angle of the refracted beam given by Equation 1.70. As already mentioned, the value π has to be added if case B takes place instead of case A.

The reflectivity, R, ultimately is a function of the glancing angle α_0 of the incoming beam. This function is demonstrated in Figure 2.2 (after Ref. [4]) for an example of case A and in Figure 2.3 (after Ref. [5]) for a chosen case B. The

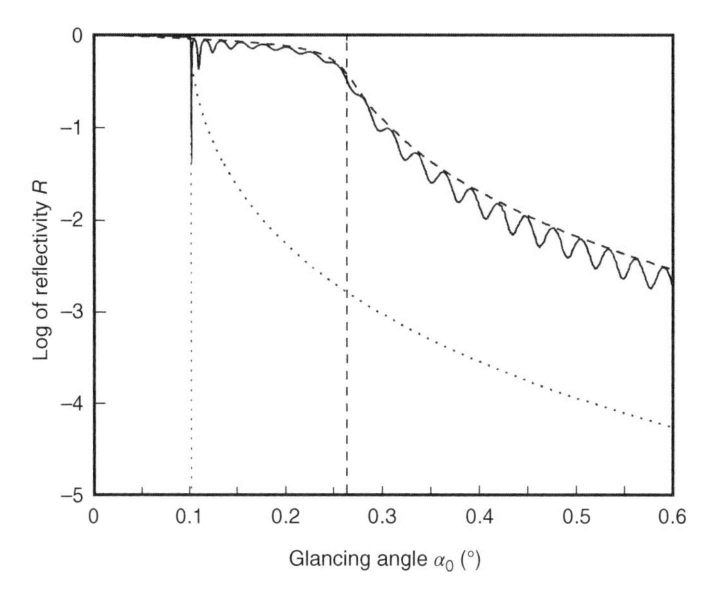

Figure 2.2. Reflectivity R of a thin layer deposited on a thick substrate in dependence on the glancing angle α_0. Case A: a 70 nm silicon layer on a gold substrate [4]. Curves were calculated for Mo-Kα radiation with $\lambda = 0.071$ nm. The dotted curve indicates the reflectivity of the layer material Si; the dashed curve indicates that of the substrate material Au. The dotted and dashed vertical lines represent the critical angles for the layer Si and the substrate Au, respectively. Figure from Ref. [1], reproduced with permission. Copyright © 1996, John Wiley and Sons.

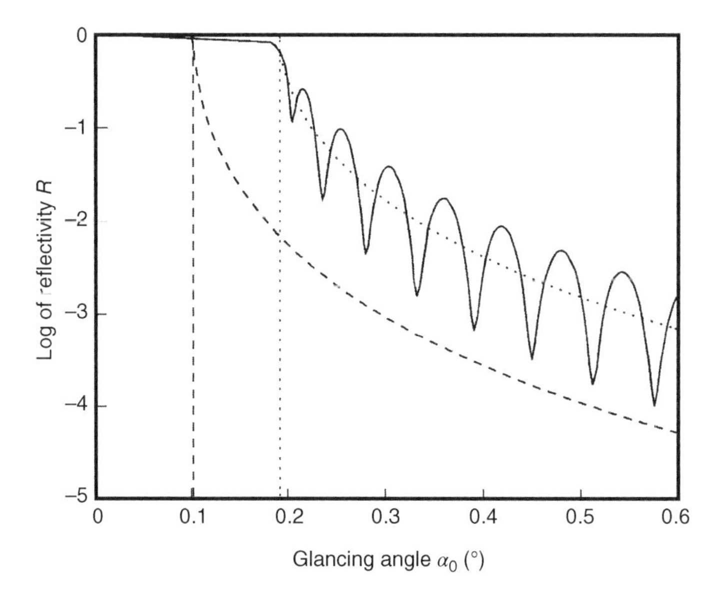

Figure 2.3. Reflectivity R of a thin layer deposited on a thick substrate as a function of the glancing angle α_0. Case B: a 30 nm cobalt layer on a silicon substrate. Curves were calculated for Mo-Kα radiation with $\lambda = 0.071$ nm. The dotted curves indicate the reflectivity of the layer material cobalt; the dashed curve, that of the substrate material silicon. The dotted and dashed vertical lines represent the critical angles for the layer Co and the substrate Si, respectively. Figure from Ref. [5], reproduced with permission. Copyright © 1991, ICDD.

semilogarithmic plot spans five orders of magnitude for the reflectivity. The value of R is roughly approximated by the greater values of R_{01} and R_{1s}. Maxima and minima appear just above the critical angle α_{01} of the respective layer at the angles α_k in accord to Equation 2.4. They are called Kiessig maxima and minima because Kiessig was the first to observe equivalent fringes for the reflectivity of a layer [6]. The first minimum is extremely marked in case A, which is discussed in Section 2.2.2.

The distances between the maxima or minima asymptotically follow with an angular period:

$$\Delta\alpha = \lambda/2\,d \tag{2.7}$$

Consequently, the extrema of a thinner layer show a greater period, for example, for case B compared to case A. Reflectivity measurements can be carried out in order to determine the thickness of a layer. The thickness can be estimated by comparison of the measured and calculated curves of the reflectivity. This method, however, cannot determine the chemical composition of an unknown layer or substrate.

2.1.2 Multiple-Beam Interference

In addition to the great interest in single thin layers, there is much interest in double layers (or bilayers) and even in more complicated multilayers.

An ideal multilayer consists of different layers, each being isotropic and homogeneous. Their boundaries are plane-parallel to each other and to the surface of a substrate on which the layers are deposited. The properties of each layer are considered constant throughout each plane, parallel to the surface. Such a system is generally called a *stratified medium* or a medium with a *stratified structure*. The most important structures are periodic, composed of a stack of N bilayers. Figure 2.4 shows a succession of bilayers consisting of alternately lower and higher refractive indices n_1 and n_2, respectively, and with thickness d_1 and d_2. The sum $d = d_1 + d_2$ is the thickness of each bilayer. With respect to their specific function, the first layer is called the *reflector*; the second, the *spacer*. Details on available multilayers are given in Section 3.4.2.

A special kind of periodic multilayers, called Langmuir–Blodgett layers, consists of multiple smectic soap films with a pseudocrystalline structure. Even flat plates of true inorganic or organic crystals can be regarded as periodic multilayers. In this case, the atoms of a crystal lattice plane build the one "layer," and the intermediate space between two successive crystal planes build the other "layer," all the layers being parallel to the surface plane. The interplanar spacing d between two neighboring crystal planes corresponds to the aforementioned sum $d_1 + d_2$.

A special feature of all these periodic multilayers is the so-called *Bragg reflection*. Figure 2.4 shows X-ray beams incident under a glancing angle α_0 and

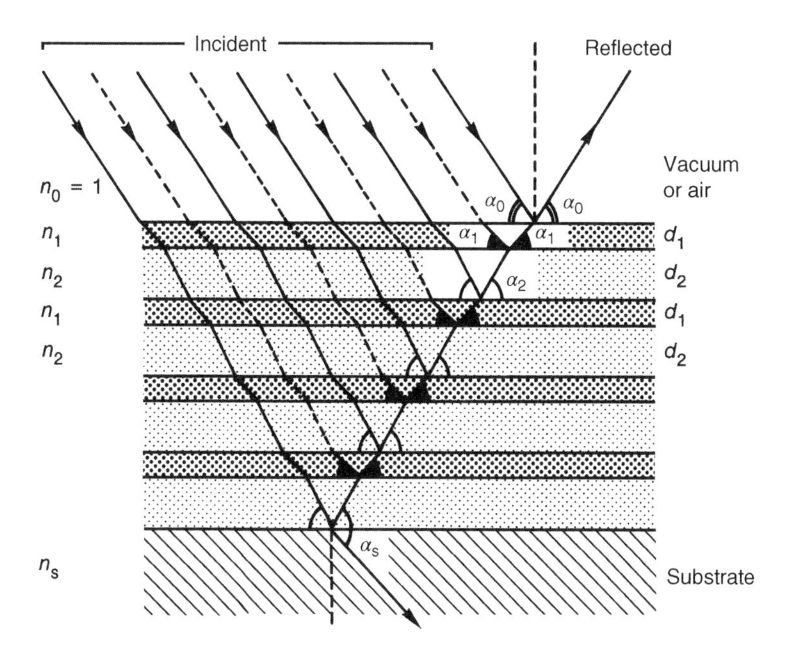

Figure 2.4. A periodic multilayer consisting of N bilayers with a period $d = d_1 + d_2$ and a total thickness $D = Nd$, deposited on a thick flat substrate. The layers with index 1 are the reflectors; the layers with index 2 are the spacers. After coherent scattering, several reflected X-ray beams can interfere with each other. Figure from Ref. [1], reproduced with permission. Copyright © 1996, John Wiley and Sons.

following polygonal courses within a multilayer. At any interface, the beams are refracted and reflected under an angle α_1 or α_2, respectively. In total, reflection results from X-rays being scattered coherently from crystal planes or layers and interfering with each other. Several beams can leave the multilayer together under the glancing angle α_0 of reflection. X-rays of certain wavelengths, λ, are reinforced in intensity by constructive interference, while other X-rays are more or less annihilated by destructive interference. Constructive interference will happen if the path difference of X-rays reflected from two neighboring planes or reflectors is an integer of λ. Neglecting absorption and refraction, Bragg's law can be written as a simple equation:

$$2\,d \sin \alpha_m \approx m\lambda \tag{2.8}$$

where m is an arbitrary integer, λ the wavelength of X-rays, and α_m the glancing angles of incidence and reflection of these X-rays. The law implies that a strong reflection for a given wavelength is possible for certain *angles* α_m or in certain *directions* α_m as is demonstrated in Figure 2.5. The different possible reflections are called reflections of the mth order.

A more stringent equation for multilayers [7] accounts for the refraction and absorption of the incoming beam within the reflector, 1, and the spacer, 2:

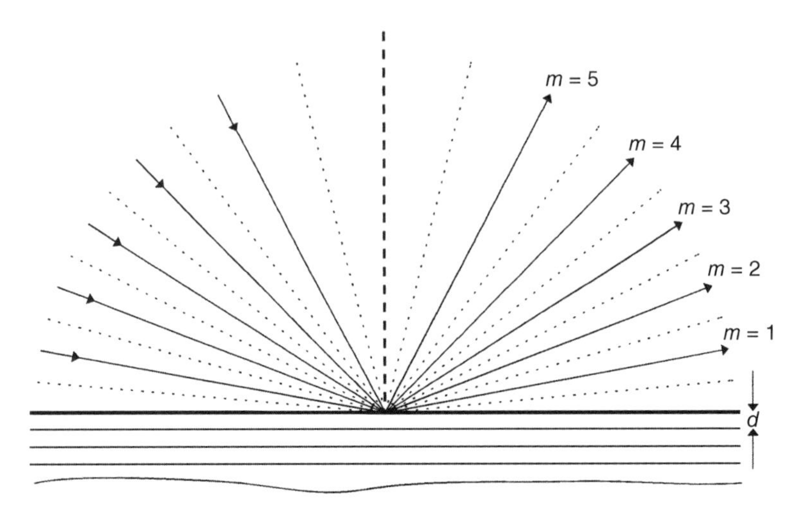

Figure 2.5. Bragg reflection at a periodic multilayer or crystal with a period d. Maxima of reflection appear for certain directions (——), minima are in between (⋯ ⋯). This illustration is valid for X-rays of Mo-Kα reflected at a LiF crystal with $d = 0.20$ nm. Figure from Ref. [1], reproduced with permission. Copyright © 1996, John Wiley and Sons.

$$2d_1 \sin \alpha_1 + 2d_2 \sin \alpha_2 = m\lambda \tag{2.9}$$

where d_1 and d_2 are the thicknesses of the reflector and spacer, respectively; α_1 and α_2 are the complex refraction angles determined by Equation 1.70. For X-rays, the original Bragg Equation 2.8 is a good approximation of the stringent relation (see Section 1.5.2).

Natural crystals and pseudocrystals have long been used as Bragg monochromators. They constitute the cornerstone of wavelength-dispersive spectrometry. In recent years, periodic multilayers have been used as well. They are especially suitable for the long wavelength region and consequently for the determination of light elements with low atomic numbers, Z. In TXRF instruments, multilayers are employed as monochromators for short wavelengths (see Section 3.4.2). Furthermore, such multilayers can serve as test samples to be examined by TXRF or related methods. In this case, the decisive glancing angles are small and connected with the range of total reflection. Bragg's law in Equation 2.8 can be rearranged after a simple approximation

$$\alpha_m^2 \approx \alpha_{01}^2 + \left(m\frac{\lambda}{2d}\right)^2 \tag{2.10}$$

where α_m values are the Bragg angles for maximum reflectivity of the mth order; α_{01} is a critical angle for total reflection of the multilayer.

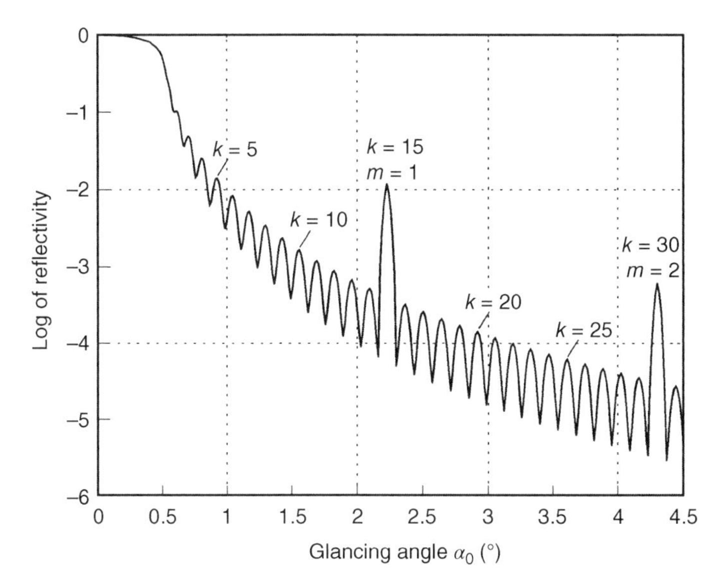

Figure 2.6. Reflectivity of a multilayer dependent on the glancing angle α_0. The multilayer was assumed to consist of 15 bilayers of platinum and cobalt, each of them 1.9 and 0.2 nm thick, respectively [8]. Reflection is observed for the Cu-Kα peak with $E = 8.047$ keV. The lower Kiessig maxima can easily be distinguished from the higher Bragg maxima with $m = 1$ and $m = 2$. Figure from Ref. [1], reproduced with permission. Copyright © 1996, John Wiley and Sons.

As was already mentioned for a single layer, the reflectivity can be measured or even calculated. A suitable algorithm for a multilayer is given later (in Section 2.4.3). The results are somewhat complicated, but their depiction, for example, in Figure 2.6, is not. The reflectivity curve is shown as a semilogarithmic plot dependent on the glancing angle (after Ref. [8]). The range of total reflection with $R \approx 1$ and the first and second Bragg maxima can clearly be assigned to the chosen periodic multilayer. Besides, there are several smaller oscillations that may be characterized as Kiessig maxima and minima.

The multilayer can obviously be considered as a single layer of thickness $D = N\,d$. The Kiessig maxima occur for angles α_k in accord with Equation 2.4a where d is to be replaced by D. The comparison with Equation 2.10 reveals the first Bragg reflection as the Nth Kiessig maximum. In general, Bragg reflections are identical with certain Kiessig maxima. They appear if k equals the number N of bilayers or an m-fold of this value ($k = Nm$). The Bragg maxima significantly exceed the "normal" Kiessig maxima by about one order of magnitude. A higher degree of resonance or symmetry is obviously reached in these cases (considered further in Section 2.2.3).

The Bragg maxima usually decrease as the order of reflection becomes higher. There is, however, one important exception. If the thickness ratio d_2/d_1 is an integer i, the order $(i+1)$ of reflection will vanish [7].

2.2 X-RAY STANDING WAVE FIELDS

The interference as treated in Section 2.1 results from the superposition of two or more beams propagating in the same direction. Interference, however, is even possible for beams of different directions. If the region of superposition is extensive, a total wave field will be observed. The resulting wave pattern can propagate with a certain velocity in a certain direction. However, it can also be stationary in a direction where it does not move at all. This phenomenon is called standing wave or stationary wave. It can be described as simple oscillations with locally dependent amplitudes. The locally fixed minima are called nodes and the maxima are called antinodes, as mentioned previously.

A very simple way to produce a standing wave is the superposition of a wide incoming and a wide reflected wave. It can be accomplished in front of a totally reflecting surface of a thick substrate, or within a thin layer on such a substrate, or even within a multilayer. In any case, standing waves are formed that have a fundamental importance for TXRF, where excitation is performed either in front of totally reflecting flat substrates or within highly reflecting layers.

Real substrates or layers unfortunately are not ideally flat and smooth but show at least a certain microroughness. This should and can be taken into account and included in the calculations. For simplicity and clarity, however, this phenomenon will be ignored in this section. Substrates or layers are assumed to be optically flat in the nanometer range. For the practical examples of Chapter 4, however, even a rough surface is taken into consideration.

2.2.1 Standing Waves in Front of a Thick Substrate

When a plane monochromatic wave hits a flat reflecting surface of a thick substrate, interference occurs in the triangular section above the surface where the two waves with the same wavelength cross each other. Figure 2.7 shows an

Figure 2.7. Interference of the incoming and the reflected X-ray waves in the triangular region above a flat and thick reflecting substrate. The strength of the electromagnetic field is represented in a gray scale with instantaneous crests (white) and troughs (dark or black). In the course of time, the pattern moves from the left to the right. Figure from Ref. [1], reproduced with permission. Copyright © 1996, John Wiley and Sons.

instantaneous picture of the interference pattern with a graded scale for the strength E_{int} of the electromagnetic field. This pattern moves from the left to the right parallel to the surface. In all directions normal to the surface, however, standing waves can be observed with nodes (zero field strength) and antinodes (maximum field strength). Parallel to the surface, there are stationary nodal lines with zero amplitude (gray) that follow one another at a constant distance. The stationary antinodal lines with extreme amplitudes (crests are white and troughs are black) are exactly between the nodal lines.

The pattern of course depends on the angle at which the primary beam is incident. Four examples are demonstrated in Figure 2.8. For $\alpha = 45°$, the nodal lines are the same distance apart as crests and troughs are, so that a checkered pattern arises (Figure 2.8a). For glancing angles $\alpha < 45°$, the nodal lines are pulled apart while the crests and troughs of the antinodal lines move closer together. This pattern is formed in X-ray optics because of the small angles at total reflection (Figure 2.8b). For angles $\alpha > 45°$, the nodal lines are compressed while the crests and troughs move further apart (Figure 2.8c). At normal incidence, that is, $\alpha = 90°$, the crests and troughs are horizontally fixed and the nodal lines follow one another at a distance of half a wavelength. This pattern represents the most simple and familiar picture of a standing wave as is known from light optics (Figure 2.8d).

For a mathematical description of the pattern some assumptions have to be made. The substrate is first regarded as a totally reflecting mirror with a reflectivity $R = 100\%$. The xy plane is taken to be the surface plane; the z-axis, a normal of that plane; and α, the glancing angle of an X-ray beam incident from the vacuum and reflected at the substrate. The incident and reflected beam represent a monochromatic plane wave with wavelength λ, velocity c_0, and maximum amplitude E_0 of the electric field strength. In front of the substrate, both beams cross at an angle 2α. Based on these conditions, the interference can be described by the electric field strength E_{int} according to

$$E_{int} = 2E_0 \sin(k_0 z \sin\alpha + \varphi)\cos(k_0 c\, t - k_0 x \cos\alpha) \qquad (2.11)$$

where $k_0 = 2\pi/\lambda$ is the wave number; t, the time; φ, a fixed phase difference between incoming and reflected waves; and z, the height above the xy plane [2]. The formula, which is independent of y, represents a standing wave for any fixed x value but a propagating wave for any fixed z value. Its amplitude is $2\,E_0 \sin(k_0 z \sin\alpha + \varphi)$, which periodically varies with height z due to the sine factor. There are nodal and antinodal planes parallel to the surface where the sine factor becomes zero or unity, respectively. These planes follow one another at a distance or period $a_{vertical}$, normal to the surface plane:

$$a_{vertical} = \frac{\lambda}{2\sin\alpha} = \frac{hc_0}{E}\frac{1}{2\sin\alpha} \qquad (2.12a)$$

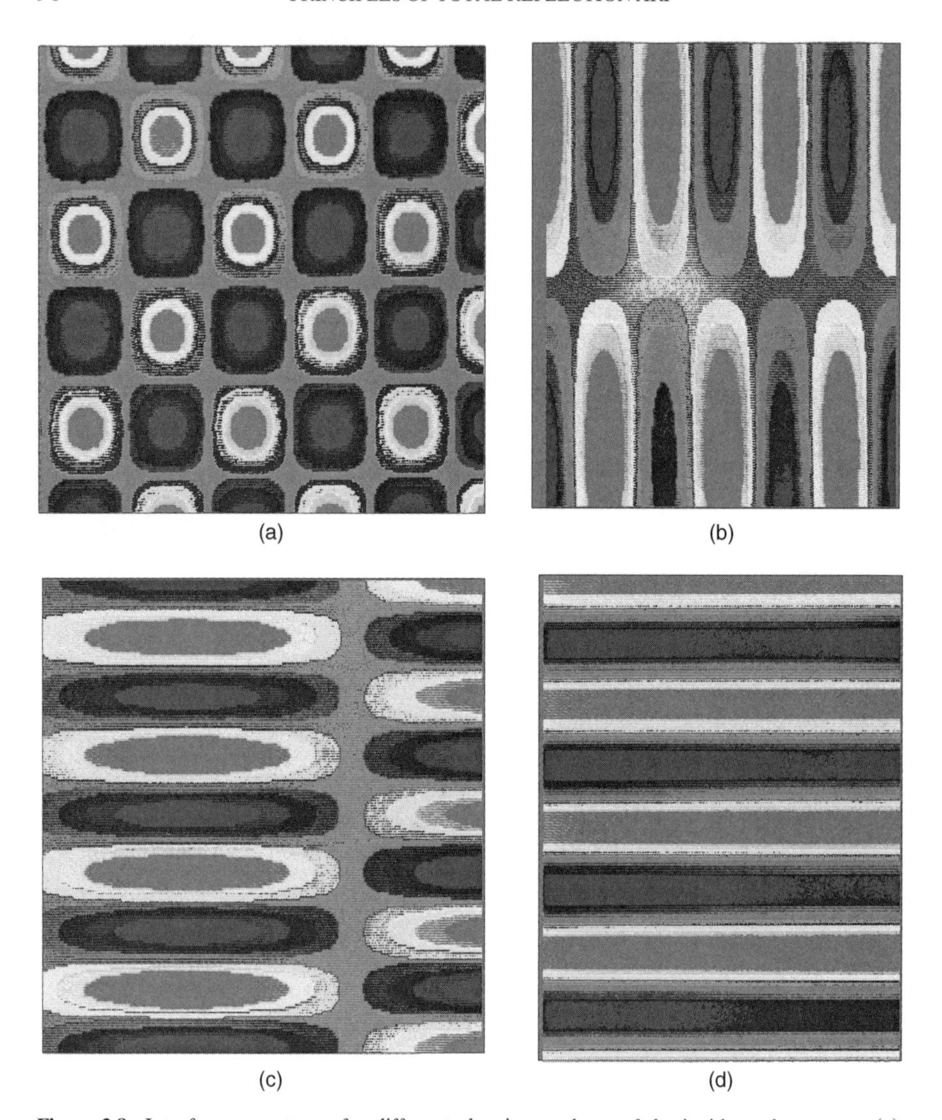

Figure 2.8. Interference patterns for different glancing angles α of the incident plane wave: (a) $\alpha = 45°$, (b) $\alpha = 10°$, (c) $\alpha = 80°$, (d) $\alpha = 90°$. Figure from Ref. [1], reproduced with permission. Copyright © 1996, John Wiley and Sons. (See colour plate section)

where λ is the wavelength of the incoming X-ray beam, E the respective energy of its photons, and $hc_0 = 1239.8$ nm keV. In the case of a metallic plate, the first nodal plane of the electric field is the reflecting surface (corresponds with the fixed end of a string or cord). For a nonconducting plate, the surface is the first antinodal plane (corresponds with the free end of a string or cord). In Figure 2.7 we have an antinodal plane at the surface, which corresponds to Equation 2.14 with fourfold intensity (see also Ref. [2]). The electric and the magnetic field vector are always vertical to each other and the magnetic field is moved by a phase shift of $\pi/2$ or 90° in comparison to the electric field. Consequently, nodes of the electric field coincide with antinodes of the magnetic field, and vice versa. Several physical and chemical effects of electromagnetic radiation are correlated with the antinodes of the electric field [2].

Crests and troughs of the propagating wave travel with a velocity $c/\cos\alpha$ in the $+x$-direction parallel to the surface plane. They have a distance

$$a_{\text{parallel}} = \frac{\lambda}{2\cos\alpha} = \frac{hc_0}{E}\frac{1}{2\cos\alpha} \tag{2.12b}$$

At normal incidence, Equation 2.11 is reduced to

$$E_{\text{int}} = 2E_0 \sin(k_0 z + \varphi)\cos(k_0 ct) \tag{2.13}$$

This equation describes the most familiar standing wave with a period a_{vertical} of $\lambda/2$.

The radiation intensity is dependent on the Poynting vector, $\boldsymbol{S}$, which is proportional to the cross-product of the electric and magnetic field strength. In order to determine this quantity within the electromagnetic field of a plane wave, the electric field strength has to be squared at any point in space and temporally averaged. A general formula for the intensity can be derived [9] by assuming a reflectivity $R < 100\%$ for the substrate in accord with Equation 1.76. In front of a flat substrate, the intensity I_{int} is given by

$$I_{\text{int}}(\alpha, z) = I_0\left[1 + R(\alpha) + 2\sqrt{R(\alpha)}\cos(2\pi z/a_{\text{vertical}} - \varphi(\alpha))\right] \tag{2.14}$$

where I_0 is a measure for the intensity of the *primary* beam, which is supposed to be constant in time and space. The argument of the cosine is the phase difference of the incoming and reflected waves, including two components: a travel distance $2\pi z/a_{\text{vertical}}$ and a phase shift φ depending on α. This phase shift [9] only occurs in the region of total reflection and is determined by

$$\varphi(\alpha) = \arccos\left[2\left(\frac{\alpha}{\alpha_{\text{crit}}}\right)^2 - 1\right] \tag{2.15}$$

It falls from π to 0 if α is changed from 0 to α_{crit} but is continuously zero for $\alpha > \alpha_{\text{crit}}$.

The intensity, I_{int}, given by Equation 2.14 is dependent on α and z but independent of x and y. Nodes and antinodes can be characterized by a minimum or by a maximum of the intensity, respectively:

$$I_{\text{min,max}} = I_0\left[1 + R(\alpha) \mp 2\sqrt{R(\alpha)}\right] \tag{2.16}$$

The highest contrast between minimum and maximum is reached for $R = 100\%$, leading to $I_{min} = 0$ and $I_{max} = 4$. This fact can be explained in the following simple way. In case of total reflection, the amplitude of the incoming wave and reflected wave are equal and can be subtracted so as to go down to zero or added so as to double in value. Squaring of these extreme values gives the aforementioned results of 0 and 4, respectively, for the intensities.

Within the substrate, the intensity is exponentially decreasing with the depth z according to

$$I_{int}(\alpha, z) = I_0 \left[1 + R(\alpha) + 2\sqrt{R(\alpha)} \cos\varphi\right] \exp\left(-\frac{z}{z_n}\right) \tag{2.17}$$

where z_n is the penetration depth defined by Equation 1.78. Both functions of intensity 2.14 and 2.17 coincide at the surface of the substrate, that is, for $z = 0$. This continuity is of course a must.

The dependence of the primary intensity on the glancing angle and the height in accord with Equations 2.14 and 2.17 can be demonstrated as a three-dimensional graph $I_{int}(\alpha,z)/I_0$ after Krämer *et al.* [10–12]. A flat silicon plate was chosen as substrate and Mo-Kα radiation as the incident monochromatic X-ray beam. Figure 2.9 is plotted for the height z above the surface. It consists of multiple cosine functions and looks like a range of fold mountains. If α is below the critical angle of total reflection (<0.10°), antinodes (mountains) show a nearly fourfold height and nodes (valleys) are at a zero level. For larger angles, nodes and antinodes recede. Figure 2.10 is relevant for the depth z below the surface and looks like one-half of a single sloping ridge. Below the critical angle, the intensity is evanescent while the penetration depth z_n is suddenly decreasing from about 62 to 3.2 nm. The intensity at the surface ($z = 0$) increases with α from zero to a maximum of 3.6 for $\alpha_{crit} = 0.1°$. For $\alpha > \alpha_{crit}$, the intensity approximates a value of $\exp(-z/z_n)$ with a penetration depth z_n in the micrometer range according to Equation 1.79.

Three cross-sections of Figure 2.9 and Figure 2.10 are repeated in Figure 2.11. The dependence of the intensity ratio I_{int}/I_0 on the height above and the depth below the silicon surface is demonstrated here for three different fixed angles α as "angle cuts." (i) For $\alpha_{crit} = 0.1°$, nodes and antinodes follow with a distance $\alpha_{crit} = 20$ nm and the first antinode coincides with the surface (at $z = 0$). With $R_{crit} = 81.5\%$, antinodes have a 3.6-fold intensity of the primary beam and the nodes have a nearly zero intensity (about 0.01-fold). Below the surface, the intensity is exponentially damped within a depth z_n of about 62 nm. (ii) For decreasing angles $\alpha < \alpha_{crit}$, nodes and antinodes are stretched while the first antinode is moving away from the surface. The intensity of the antinodes is increased to a nearly fourfold value. In the substrate, the intensity is evanescent within only some nm. (iii) For increasing angles $\alpha > \alpha_{crit}$, nodes and antinodes are compressed toward the surface. The oscillations of intensity vanish, and the intensity approaches unity. Below the surface, the intensity reaches a depth of several micrometers instead of only some nanometers.

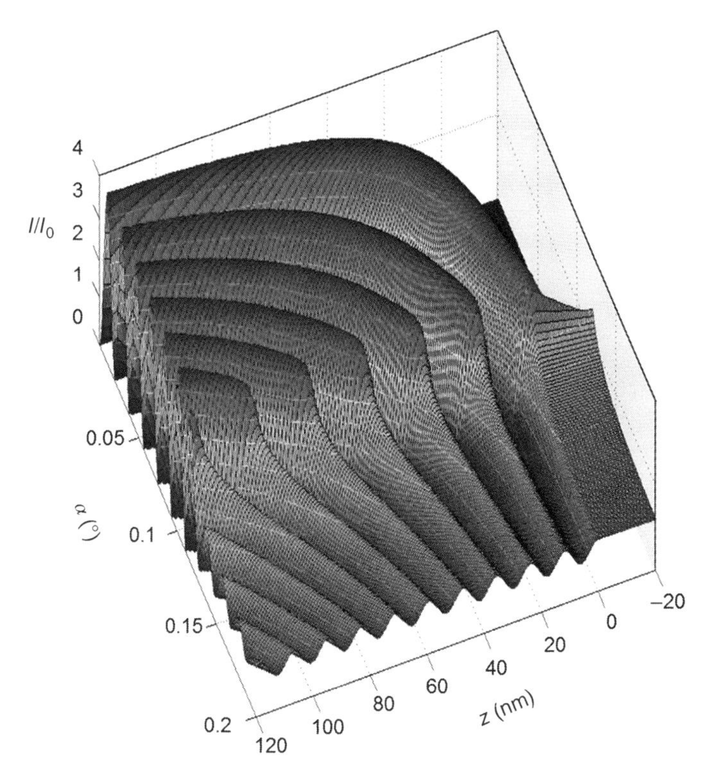

Figure 2.9. The normalized intensity I_{int}/I_0 is presented here as a three-dimensional graph depending on the glancing angle α and on the height z. An X-ray beam of Mo-Kα is assumed striking a thick Si flat. In this case, the critical angle of total reflection is about 0.1°. The region above the wafer surface is represented by positive z values up to 120 nm, the region below the surface has negative z values down to −20 nm. Figure from Ref. [10], reproduced with permission from the author. (See colour plate section)

Three further cross-sections of Figure 2.9 are presented in Figure 2.12. The intensity ratio I_{int}/I_0 is given here for three fixed values z above the surface depending on the glancing angle as "angle scans." (i) For $z = 20\,\text{nm} = a_{crit}$, only one single node and antinode appear below α_{crit}. (ii) For $z = 40\,\text{nm} = 2a_{crit}$, antinodes with fourfold intensity appear at two certain angles below α_{crit} and additionally at α_{crit}. Nodes with zero intensity are found between them. Even for angles $\alpha > \alpha_{crit}$, further nodes and antinodes can be found. They oscillate around 1.0 with a declining contrast. (iii) For $z = 0$ nm, only one antinode of 3.6 occurs at α_{crit}. For $\alpha > \alpha_{crit}$, the intensity asymptotically approximates the value 1.

At the critical angle α_{crit}, the period of standing waves is determined by

$$a_{crit} = \frac{\lambda}{2}\frac{1}{\sqrt{2\delta}} \tag{2.18}$$

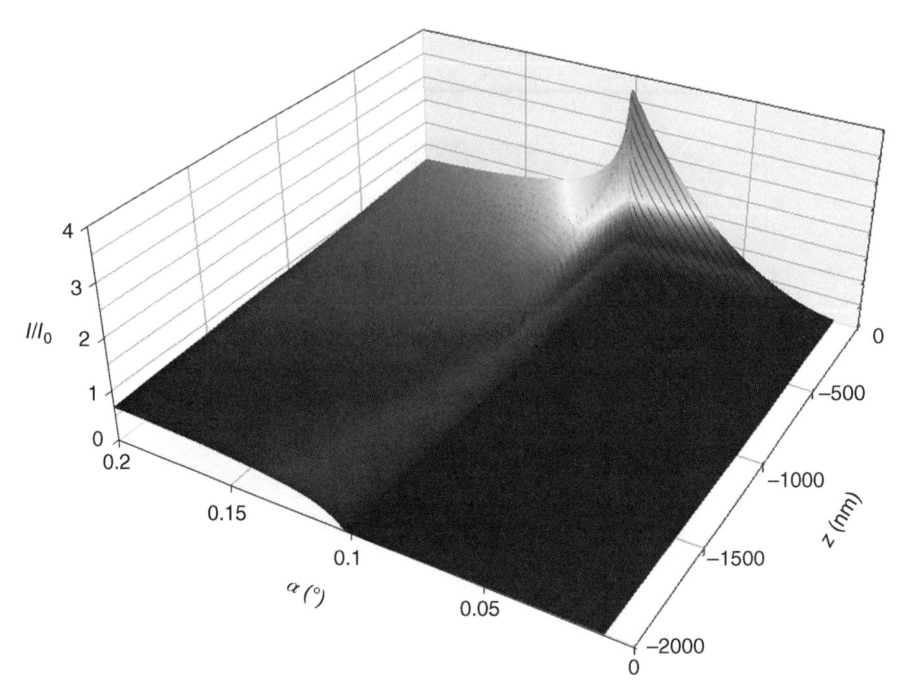

Figure 2.10. The normalized intensity I_{int}/I_0 depending on glancing angle α and depth z. The negative z values characterize the region below the surface inside the Si wafer. There is one single antinode at $z=0$ for $\alpha_{crit}=0.1°$. For $\alpha>\alpha_{crit}$, the intensity exponentially decreases with the penetration depth z_n, which is about 62 nm. For $\alpha<\alpha_{crit}$, the intensity is evanescent within a few nanometers. Figure from Ref. [10], reprinted with permission from the author. (See colour plate section)

This quantity may be of special interest. It is correlated to the minimum penetration depth z_0 defined by Equation 1.79 for $\alpha << \alpha_{crit}$. Both quantities a_{crit} and z_0 differ by the factor 2π. Like the minimum depth z_0, the period a_{crit} is also a material constant that is independent of the wavelength or photon energy of the primary beam. Table 2.1 summarizes a_{crit} and z_0 values calculated according to Equations 2.18 and 1.80, respectively.

2.2.2 Standing Wave Fields Within a Thin Layer

Next to an infinitely thick substrate, a thin homogeneous layer on such a substrate is of particular interest. The paths of X-rays for the two different cases, A and B, have already been shown in Figure 2.1. Constructive or destructive interference was demonstrated for the two reflected beams propagating in the same direction. They result in Kiessig maxima and minima of the reflectivity appearing for certain angles or *directions*, α_k.

Both reflected beams not only interfere with each other, they also jointly interfere with the incoming beam though propagating in a different direction.

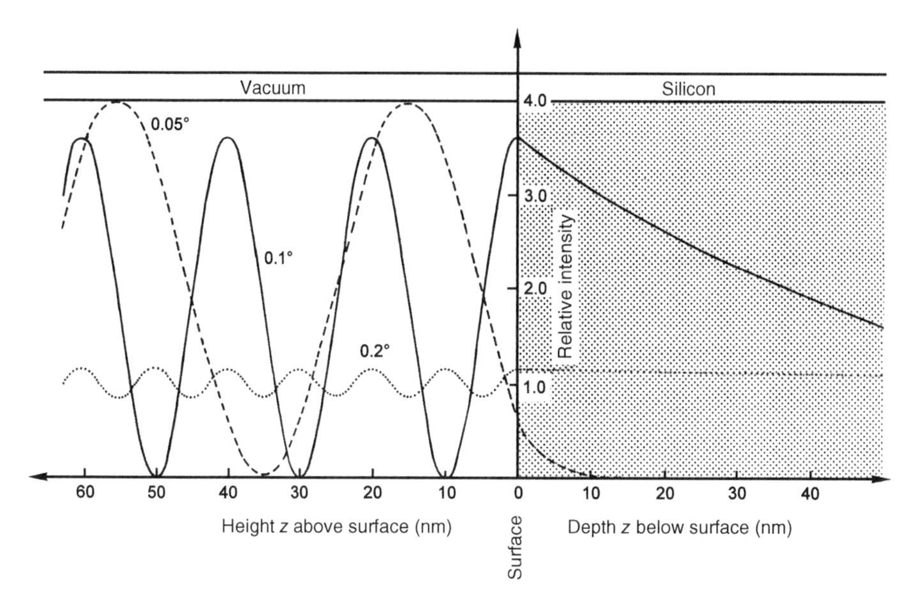

Figure 2.11. Cross-sections of the standing wave fields represented in Figures 2.9 and 2.10, called "angle-cuts" [10], at three fixed glancing angles of 0.05° (------), 0.1° (——), and 0.2° (·····). The normalized primary intensity is dependent on z (from left to right). Standing-wave-fields distinctly arise above the surface for glancing angles at and below the critical angle of total reflection, which is 0.1°. Figure from Ref. [1], reproduced with permission. Copyright © 1996, John Wiley and Sons.

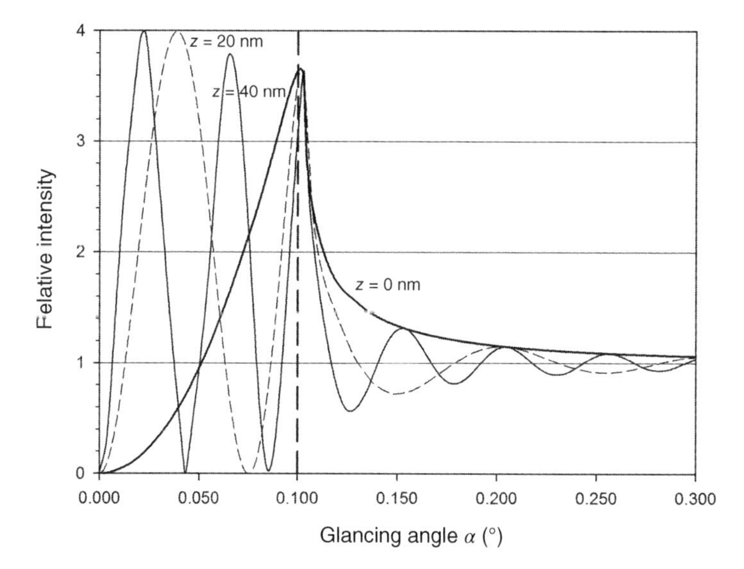

Figure 2.12. Cross-sections of the standing wave fields represented in Figure 2.9, called "angle-scans," at three fixed heights above the surface, that is, for $z = 40$ nm (——), 20 nm (------), and 0 nm (——).The normalized primary intensity is dependent on α (from left to right). Standing wave fields arise for positive z values chosen here. For negative z values, that is, below the surface, the curves are similar to the curve with $z = 0$ nm (not shown here) with decreasing intensity. Figure from Ref. [10], reproduced with permission from the author.

TABLE 2.1. Distance α_{crit} of Nodal Planes in Front of Various Substrates and Minimum Penetration Depth z_0

Substrate	α_{crit} (nm)	z_0 (nm)
Plexiglas	26.9	4.3
Glassy carbon	25.6	4.1
Boron nitride	20.3	3.2
Quartz glass	20.3	3.2
Aluminum	18.9	3.0
Silicon	20.0	3.2
Cobalt	10.7	1.7
Nickel	10.4	1.7
Copper	10.7	1.7
Germanium	14.0	2.2
Gallium arsenide	14.0	2.2
Tantalum	8.3	1.3
Platinum	7.3	1.2
Gold	7.7	1.2

Source: Ref. [1], reproduced with permission.

As demonstrated in Figure 2.13, the beams cross over in a triangular region, I, in front of the layer and in a trapezoidal region, II, within the layer. They overlap under an angle $2\alpha_0$ in front of the layer and under an angle $2\alpha_1$ within the layer. This superposition gives rise to standing waves as was already shown in the foregoing subsection concerning a thick substrate. The standing wave pattern in both regions essentially looks like that of Figure 2.8b. Nodes and antinodes run parallel to the surface and to the interface. In region III of the underlying substrate, a simple propagating wave appears with amplitudes decreasing with depth. A propagating wave also arises in region IV, where only the two reflected waves overlap. Its amplitude strictly depends on the angle α_0, showing maxima and minima as was demonstrated in Section 2.1.1.

As just noted, standing waves are expected to arise in a triangular and trapezoidal region in front of and within the layer, respectively. Of course, the incoming wave has to be monochromatic and plane. In that case, the intensity of the radiation field can be expected to depend only on the height or depth normal to the layer and not on the wavelength.

The intensity of the X-radiation field again results from squaring the amplitudes of the electromagnetic wave at any point in the field. This intensity normalized to the intensity of the primary incoming beam is demonstrated in Figure 2.14 for case A and in Figure 2.15 for case B. These two cases chosen from Refs [4,5], respectively, have already been considered in Section 2.1.1. The interfaces of the layer are indicated by vertical lines. The intensity curves are calculated for five different glancing angles of an incoming Mo-Kα beam.

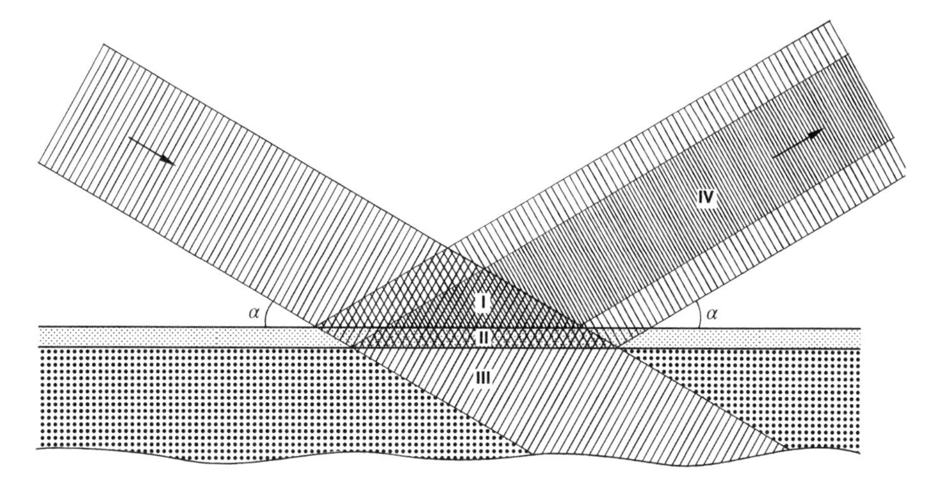

Figure 2.13. Regions of interference in front of and within a thin layer (small dots) deposited on a thick substrate (big dots). Standing waves appear in the triangular region I and in the trapezoidal region II. Regions III and IV show simple propagating waves with a depth-dependent amplitude (region III) or an angle-dependent amplitude (region IV). It is not demonstrated in this figure that the incident beam is shifted some 10–100 nm parallel to the surface according to the Goos–Hänchen effect. Figure from Ref. [1], reproduced with permission. Copyright © 1996, John Wiley and Sons.

In both cases A and B, nodes and antinodes can be realized in front of and within the thin layer. *In front of the layer*, they are extremely marked at angles below the critical angle of the layer (dotted lines in Figures 2.14a,b and 2.15a,b). Nodal lines with zero intensity follow in height with a periodicity of $\lambda/2\alpha$ according to Equation 2.12a. The antinodes in their midst have a fourfold intensity as already observed for a thick substrate. With increasing angle α, the standing wave is generally more and more lacking in contrast. *Within the layer*, the incoming wave is evanescent for angles below the critical angle of the layer (dotted lines in Figures 2.14 and 2.15). Just above this angle, however, standing waves arise with distinct nodes and antinodes. With increasing angle α, more and more periods fit between the two interfaces according to Equation 2.12a. In the upper parts of both figures, standing waves are shown for angles of the first and second Kiessig minima of reflectivity (full and dashed line, respectively). The lower parts of both figures represent standing waves for the first and second Kiessig maxima of reflectivity determined by Equation 2.4. From these figures, it can be observed that an integral number of half periods of a standing wave fit between the interfaces in case of a Kiessig minimum or maximum. The number is even for a minimum and odd for a maximum. The first case may be called *resonance*; the second, *antiresonance*. Resonances within the layer obviously correspond to minima of reflectivity; antiresonances correspond to maxima.

The resonances lead to especially high antinodes lying symmetrically in the layer. In case A the intensity is still higher (27-fold) than in case B (8-fold). The

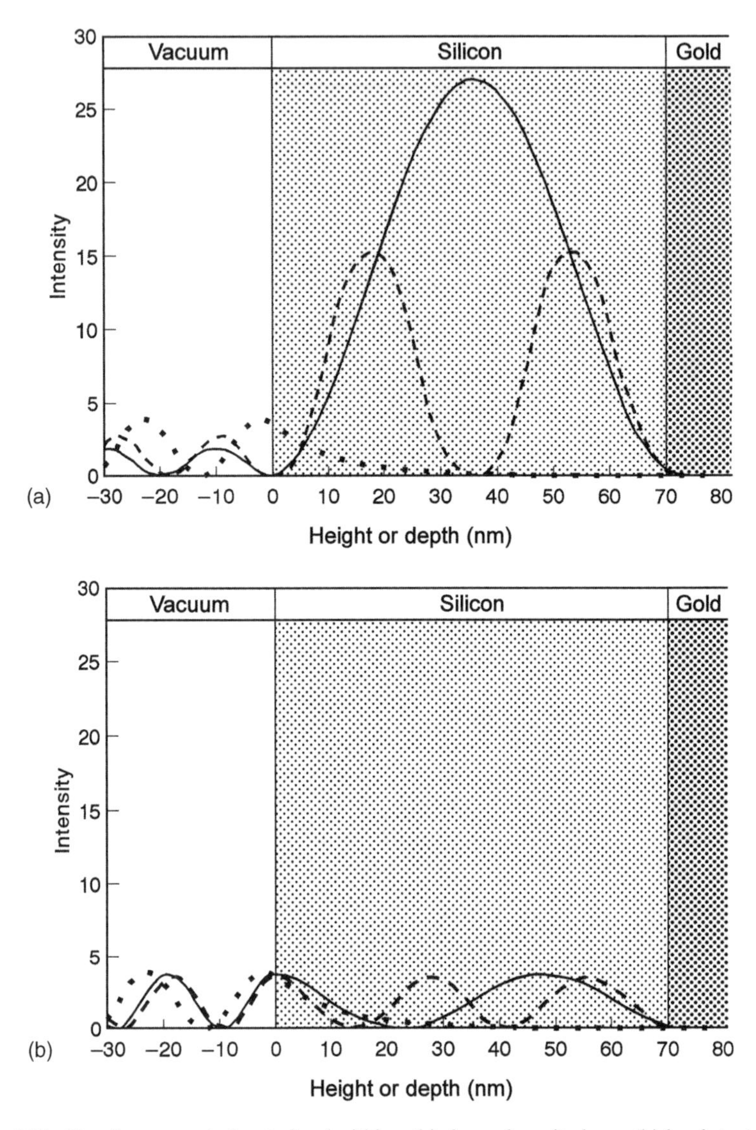

Figure 2.14. Standing waves in front of and within a thin layer deposited on a thick substrate. Case A: 70 nm silicon on gold. The normalized intensity is plotted vs. the depth normal to the Si layer at various glancing angles: (a) $\alpha = 0.097°$ dotted, 0.103° solid, and 0.111° dashed. (b) $\alpha = 0.097°$ dotted, 0.106° solid, and 0.117° dashed. Figure from Ref. [4], reproduced with permission from APS. Copyright © 1991, American Physical Society

excessive intensity can be explained by the layer acting as a wave guide [4]. Radiation is swapped back and forth in this waveguide and consequently is reinforced. The extremely high antinode of Figure 2.14 corresponds to the first extremely deep break in the reflectivity curve of Figure 2.2. X-ray waveguide effects are specifically dealt with and described by Zheludeva *et al.* [13].

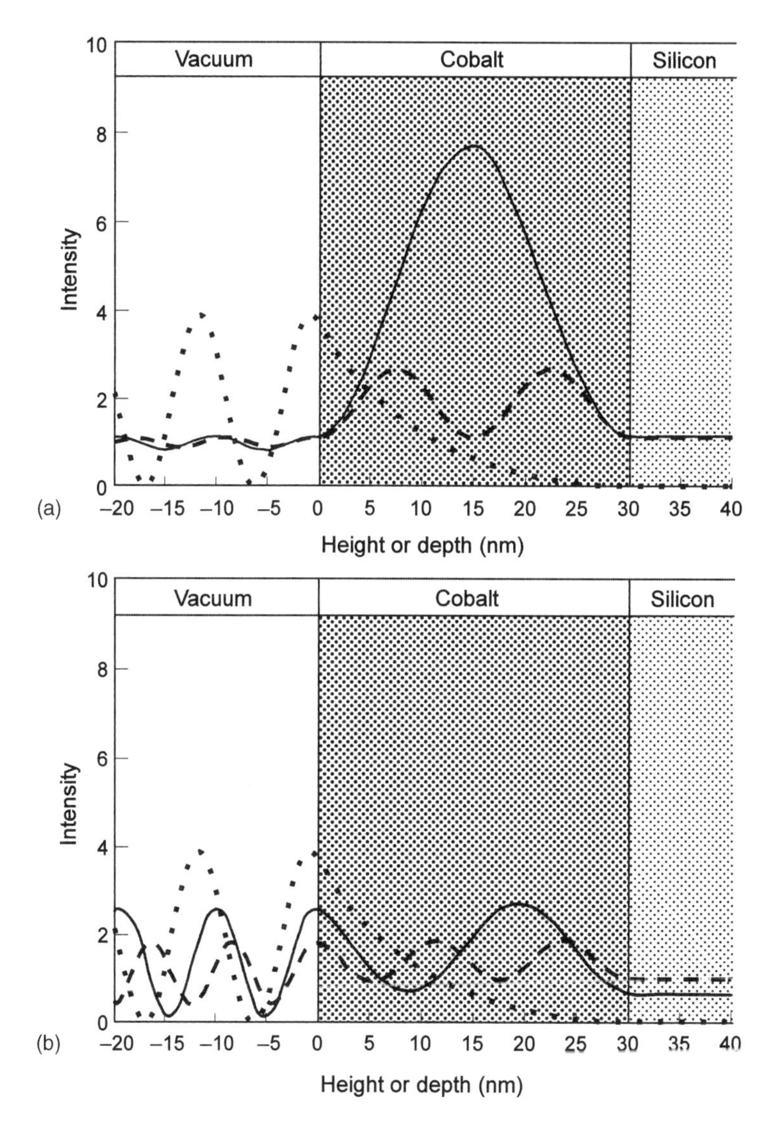

Figure 2.15. Standing waves in front of and within a thin layer deposited on a thick substrate. Case B: 30 nm cobalt on silicon. The normalized intensity is plotted vs. the depth normal to the surface of the Co layer at various glancing angles: (a) $\alpha = 0.189°$ dotted, 0.203° solid, 0.235° dashed. (b) $\alpha = 0.189°$ dotted, 0.214° solid, and 0.253° dashed. Figure from Ref. [1], reproduced with permission. Copyright © 1996, John Wiley and Sons.

Cases A and B mainly differ in the contrast of nodes and antinodes within the layer. This contrast is more distinct in case A (Figure 2.14) than in case B (Figure 2.15). A further difference can be recognized for the underlying *thick substrate*. In case A the X-ray beam is evanescent in the substrate gold, whereas in case B it deeply penetrates into the substrate silicon.

2.2.3 Standing Waves Within a Multilayer or Crystal

Standing wave fields also arise in multilayers and even in crystals. They can attain a high degree of symmetry especially in crystals but also in *periodic* multilayers. In Figure 2.4, a periodic multilayer has been demonstrated with the geometric pathways of X-rays. Transmitted and reflected beams overlap under an angle $2\alpha_1$ in all reflectors and under $2\alpha_2$ in all spacers. In accord with the aforementioned considerations, standing waves are formed in each layer with nodes and antinodes parallel to the interfaces [14]. For perfect crystals, this phenomenon has long been known and is described by the *dynamical theory of diffraction* [15,16]. It is based on the coherent scattering of X-rays at the individual atoms of the different crystal lattice planes.

In a rough approximation, such a multilayer can be considered a single layer of thickness $D = N\,d$ to which the relations of Section 2.1.1 are applicable to a certain degree. If the glancing angle α just exceeds the critical angle α_{01} of the upmost reflector layer, an incoming monochromatic beam will deeply penetrate into the multilayer and produce a standing wave field. With increasing angle, more and more nodal and antinodal planes will be formed in parallel to its surface. At distinct angles α_k, they will exactly fit in the multilayer and give rise to relative maxima or minima of the reflectivity introduced earlier as Kiessig maxima or minima. The corresponding glancing angles α_k follow Equation 2.4.

The exceptional case of the Bragg reflection becomes apparent from Figure 2.16. If the antinodes or nodes coincide with all reflectors (or crystal planes), the greatest degree of symmetry will be reached. The periodicity of the standing wave is then equal to that of the multilayer. The antinodes can be situated either in the spacers (Figure 2.16a) or in the reflectors (Figure 2.16c). This shift happens due to the phase jump of π occurring if reflection changes from the optically denser spacer to the optically thinner reflector [4,14]. Figure 2.16b illustrates Bragg reflection when the antinodes lie on top of the reflectors and the nodes on top of the spacers. Bragg reflection appears for the Nth Kiessig maximum when k corresponds to N, the number of bilayers, or an m-fold of this value ($k = Nm$). In this case, Bragg's law is valid for glancing angles α_m determined by Equations 2.8 or 2.9. These angles are distinctly above the critical angle α_{01} of total reflection but small enough for the applied sine approximation. These α_m values represent the Bragg maxima of reflectivity for a periodic stratified medium.

2.3 INTENSITY OF FLUORESCENCE SIGNALS

As was already mentioned in Section 1.4.1, an X-ray photon with sufficient energy can be absorbed by the photoelectric process. An inner electron of the absorbing atom is ejected during this process. The instantaneous rearrangement of electrons leads to the emission of X-ray photons called X-ray fluorescence.

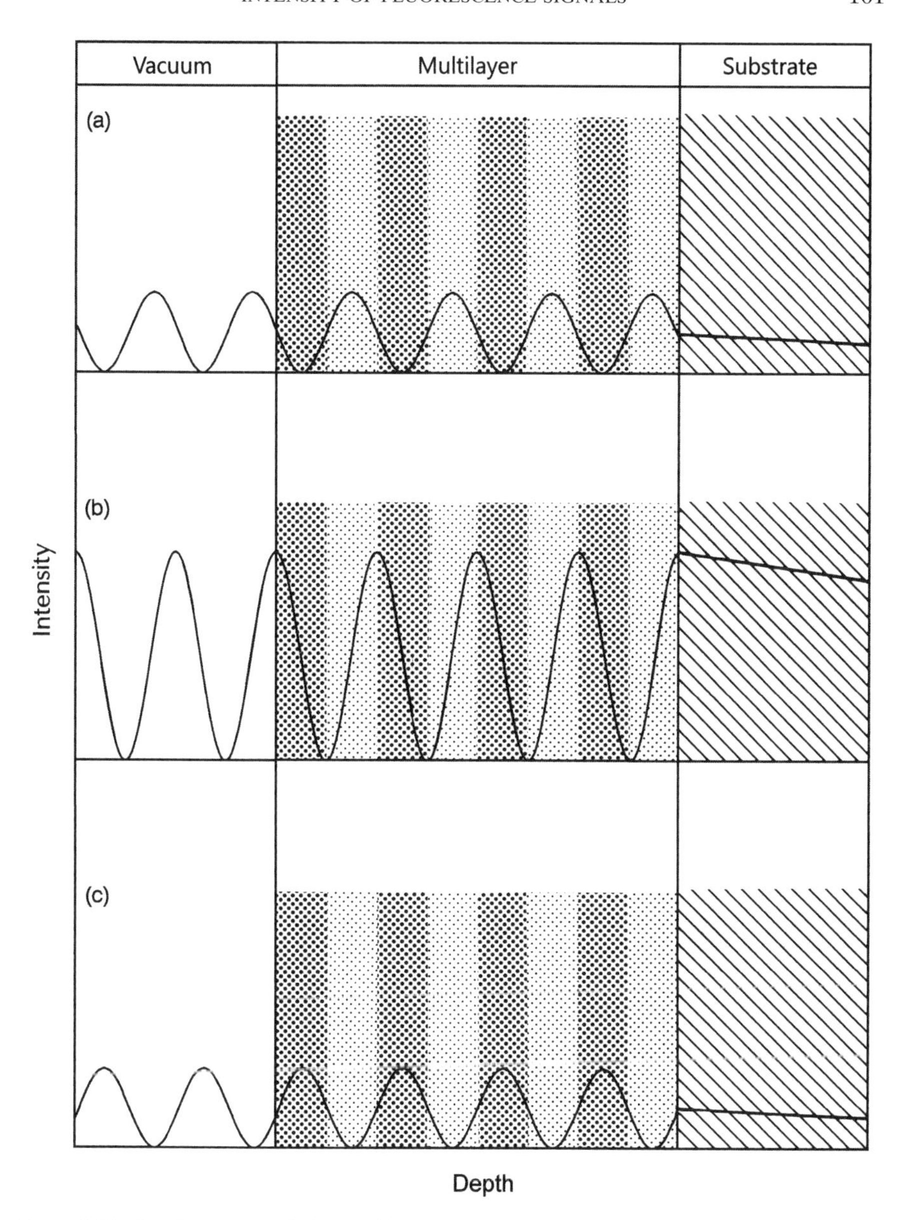

Figure 2.16. Standing waves within a multilayer consisting of a stack of four bilayers for the case of Bragg reflection. The reflectors are dotted darker, the spacers, dotted lighter. Part (a) represents an angle just below the Bragg angle, which is exactly reached in part (b), while part (c) represents an angle just above the Bragg angle. In all cases, the wave-pattern has the periodicity of the multilayer itself, but the antinodes jump from the spacers in part (a) to the reflectors in part (c). In part (b) the Bragg angle with highest reflectivity is reached. Figure from Ref. [1], reproduced with permission. Copyright © 1996, John Wiley and Sons.

The emitted photons give rise to the "characteristic" spectrum representing a fingerprint of atoms in a corresponding sample. The element composition can be estimated from the relative intensities of the individual characteristic peaks or lines. This is the basis of X-ray fluorescence analysis.

Excitation to fluorescence is conventionally done by a centimeter-wide primary X-ray beam representing a propagating wave. The intensity of the wave field is assumed to be locally constant *in vacuo* and to be exponentially decreasing in solids of micrometer of millimeter thickness. In TXRF, however, the primary beam appears as a standing wave field with locally dependent oscillations or as an evanescent wave field. The atoms in these fields are excited to fluorescence with a probability directly proportional to the wave field intensity. Consequently, the fluorescence signal reflects the intensity of the standing or evanescent wave field in the sample and additionally indicates the elemental composition. On the one hand, an internal standard can be used to compensate for inhomogeneity of the primary field. On the other hand, the fluorescence signal can indicate the varying field intensity, and its dependence on the glancing angle can even be used for depth-profiling of layered structures.

In Sections 2.3.1–2.3.5, the fluorescence signals of various systems will be calculated and discussed: signals of a thick and flat substrate, of a residue left on a substrate, and of near-surface layers deposited *on* or buried *in* a substrate. The primary beam shall be monoenergetic, with an intensity I_0 hitting the substrate under a glancing angle α.

2.3.1 Infinitely Thick and Flat Substrates

The simplest case is that of an infinitely thick homogeneous substrate. It is presupposed to be optically flat in order to ensure total reflection of X-rays. The primary intensity I_{int} within such a substrate is already known from Equation 2.17. At a depth z of the substrate, the primary beam may induce a fluorescence signal of a constituent element. Naturally, its intensity is proportional to the primary intensity at this depth. The sum of all signals emitted from the entire thick substrate is obtained by integrating Equation 2.17 from zero to infinity. If self-absorption is neglected, the total intensity arising from the substrate will be

$$I_{\text{B}}(\alpha) = I_{\text{n}}\left[1 + R(\alpha) + 2\sqrt{R(\alpha)}\cos\varphi\right] z_{\text{n}}(\alpha) \tag{2.19}$$

where $R(\alpha)$ is the reflectivity given by Equation 1.75, $z_{\text{n}}(\alpha)$ is the penetration depth determined by Equation 1.78, and φ is the phase shift defined by Equation 2.15. I_{n} is a norm or reference for the fluorescence intensity registered by a detector, which is proportional to the primary beam intensity, I_0.

Equation 2.19 is not very transparent but it can be transformed into a more comprehensible approximation as shown in Section 2.4.1:

$$I_B(\alpha) = I_n C[1 - R(\alpha)]\alpha \tag{2.20}$$

where C is a quantity of the substrate mainly determined by $1/\beta$ or $1/[(\mu/\rho)\cdot\rho]$, respectively. This formula can easily be interpreted as follows [17,18].

If the glancing angle α is far above the critical angle α_{crit} of total reflection, the primary X-ray beam deeply penetrates into the substrate. A thick layer is passed through with a thickness proportional to α or to the sine of α if larger glancing angles are considered. The fluorescence signal comes from this layer. Because the primary beam is nearly not reflected ($R \approx 0$), the signal is only dependent on α, that is, directly proportional to α. In the region of total reflection, however, the primary beam is evanescent in the substrate, as was demonstrated in Figure 2.11. A major part of the primary beam is reflected ($R \gg 0$) and only the remainder ($1 - R$) is decisive for fluorescence. The quantity $(1 - R)\alpha$ in Equation 2.20 is called energy transfer and defines that portion of the impinging energy that penetrates into the substrate.

Equation 2.20 is demonstrated by Figure 2.17 for a Mo-Kα beam striking a thick and flat Si substrate. In general, the signal intensity linearly decreases as glancing angles α decrease. At the critical angle of total reflection, however, a

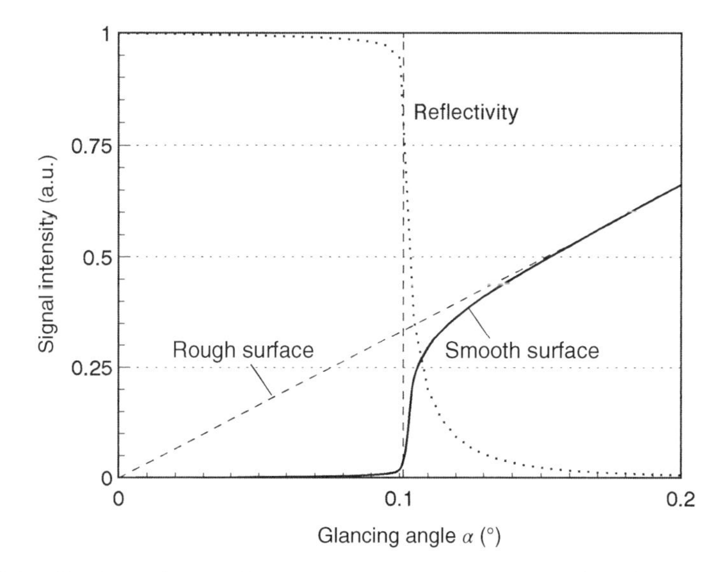

Figure 2.17. Signal intensity of a thick, flat, and smooth silicon substrate (——), calculated for an impinging Mo-Kα beam. In addition, the reflectivity R (·······) is shown, dependent on the glancing angle α. Below $\alpha_{crit} = 0.102°$, total reflection occurs with a steplike increase in reflectivity and a steplike decrease of the signal intensity. The oblique dashed line represents the intensity from a rough Si substrate. Figure from Ref. [1], reproduced with permission. Copyright © 1996, John Wiley and Sons.

steplike decrease of the signal intensity can be noticed. This decrease is on the order of 10^{-4}.

Infinitely thick and flat substrates are usually applied as sample carriers for TXRF analyses. Polished plates of silicon or quartz glass, for example, are highly suitable. They give rise to a very low blank or background spectrum consisting of fluorescence peaks, for example, Si-Kα, and of a scatter part with tube peaks and a continuum. The intensity of the *fluorescence* and of the *scatter* signals are equally dependent on the energy transfer as described by the intensity I_B [17,18]. The index B was already chosen in order to point to blank or background. At total reflection, the angle α is small, $R(\alpha)$ is nearly 1 and the intensity I_B becomes extremely small. For light substrates like silicon, quartz glass, and also Plexiglas, the blank or spectral background of TXRF is six orders of magnitude smaller than that of conventional XRF. This is the reason for the very low detection limits of TXRF.

2.3.2 Granular Residues on a Substrate

In addition to an infinitely thick and flat substrate, an analyte sample is of particular interest. It may be deposited on a substrate used as a sample carrier and may be a mineral residue of a solution or suspension after drying. But it may also be a thin section of an organic tissue, or a thin-film-like residue of a metallic smear [19]. For analysis, the analyte is positioned within the standing wave field in front of the substrate. It should not disturb this primary wave field significantly. For example, it should be so rough that a total reflection of the primary beam does not happen at the analyte itself but only at the substrate. As already mentioned, the analyte is excited to fluorescence and the emitted fluorescence signal is proportional to the primary-field intensity. The intensity of this field with nodes and antinodes is given by Equation 2.14 and illustrated by Figure 2.11.

The analyte may consist of one or several particles of the same grain size s deposited side by side, or may be a thin section or film with thickness s. In this case, the totally emitted intensity of the analyte is obtained by integrating Equation 2.14 between $z = 0$ and $z = s$. The corresponding result is shown in Figure 2.18 for different grain size or thickness but equal mass of the analyte. The fluorescence intensity is independent of the glancing angle α if this angle is far beyond the critical angle of total reflection. In the angular range of total reflection, however, strong oscillations occur for granular or thin-film like residues smaller than about 100 nm. These oscillations develop because only a few antinodes and nodes penetrate the particles, thin films, or sections. They diminish with increasing grain size or thickness and finally approach a constant value, which results from the approximation [18,20]

$$I_x(\alpha) = I_n\, m_x[1 + R(\alpha)] \tag{2.21}$$

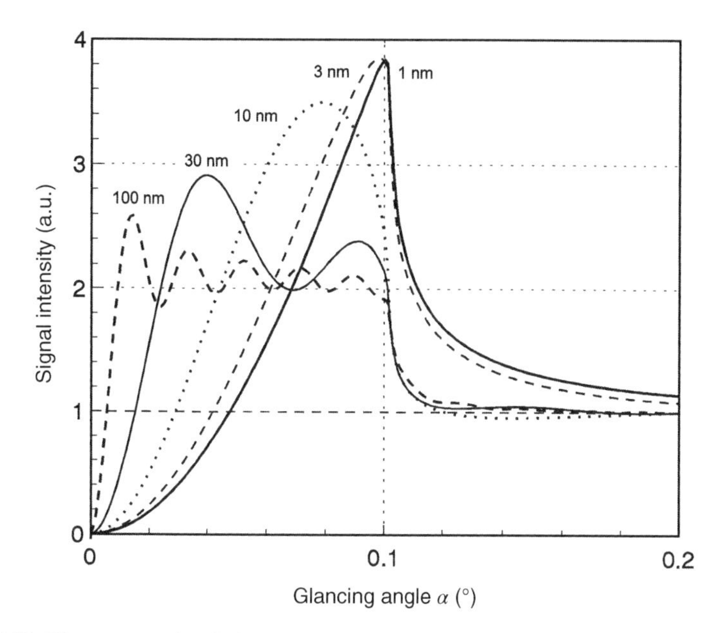

Figure 2.18. Fluorescence signal of small particles or thin films deposited on a silicon substrate used as sample carrier. The intensity was calculated for particles, thin films or sections of different thickness but equal mass of the analyte, and plotted against the glancing angle α. A Mo-Kα beam was assumed for excitation. Particles or films more than 100 nm thick show double intensity below the critical angle of 0.1°. Figure from Ref. [1], reproduced with permission. Copyright © 1996, John Wiley and Sons.

where m_x is the mass of the analyte element x on top of the substrate. The constant value of the intensity is $(1 + R)$-fold, which is asymptotically approximated. It is nearly double the amount observed for angles beyond the total reflection zone. The doubling can easily be explained by the fact that particles are excited equally by the incoming *and* the reflected beam. Thus, if the glancing angle falls short of the critical angle, the fluorescence intensity will step up to the double value.

For quantitative analyses, it is important to get an intensity that is independent of particle size or film thickness and also of glancing angle. This ideal behavior is realized for granular or thin-film-like samples of about 100 nm. However, several particles of various thicknesses may also be applied, as frequently happens in practice. A broad size distribution can average over the oscillations [21]. On the other hand, the particles have to be small enough in order to avoid absorption effects [19]. In any case, the need for internal standardization is obvious. A standard may be added to the particles and distributed homogeneously over the height of the particles. In particular, the standard must not be enriched in the nodes or antinodes of the primary field but should be uniformly mixed with the sample. In the case of a fairly homogeneous distribution, the intensity ratio of analyte and internal standard is constant, that is, independent of the particle size or thickness and also of the glancing angle α.

2.3.3 Buried Layers in a Substrate

After our consideration of an analyte placed *on* a substrate, we shall now consider an analyte that is included *in* a substrate. This analyte can be a layer containing either impurities or minor, but essential, constituents within the substrate. The layer can be produced by vaporization and diffusion or implantation. Such layers in wafers are called buried layers. If they are localized directly below the surface of the substrate, they may be called near-surface layers. However, they can also represent a deeper lying interface. In any case, they are assumed to contain the decisive constituents in a low concentration. Consequently, the refractive index of the substrate can be considered unchanged and the layer itself can be assumed to be nonreflecting. Furthermore, the concentration is assumed to be constant within the layer, representing a rectangular profile with a thickness d.

The fluorescence intensity emitted from this layer can be calculated by integration of Equation 2.17 between the boundaries of the layer. If the layer is between the depth z and $z+d$ and if absorption can be neglected, the result will be given by

$$I_{\mathrm{BL}}(\alpha, z) = I_{\mathrm{n}} c_{\mathrm{A}}\, C[1 - R(\alpha)] \frac{\alpha}{d} \exp\left(-\frac{z}{z_{\mathrm{n}}}\right) \cdot \left[1 - \exp\left(-\frac{d}{z_{\mathrm{n}}}\right)\right] \tag{2.22}$$

in accord with Schwenke *et al.* [17,18]. The index BL signifies the buried layer; c_{A} is the area-related mass (g/cm^2) or area density (atoms or ions/cm^2) of the given element; C is the quantity defined in Equation 2.20; R is the reflectivity of the substrate; and z_{n} is the penetration depth for the primary X-ray beam normal to the surface. For ultrathin layers directly below the surface, that is, for $z=0$ and $d=0$, Equation 2.22 leads to

$$I_{\mathrm{BL}}(\alpha) \cong I_{\mathrm{n}} c_{\mathrm{A}} C[1 - R(\alpha)] \frac{\alpha}{z_{\mathrm{n}}} \tag{2.23}$$

The dependence of the fluorescence intensity on the glancing angle α is demonstrated in Figure 2.19 for buried layers of different thickness, d, but equal area density. For the extremely thin layer of 1 nm, the curve is identical to that of Figure 2.18 for the same thickness. This agreement is caused by a continuous transition of the primary intensity into the substrate even at total reflection.

Thus, a single function can be expected for ultrathin layers *above* or *below* the surface of a substrate. It can be derived from Equation 2.23 by use of

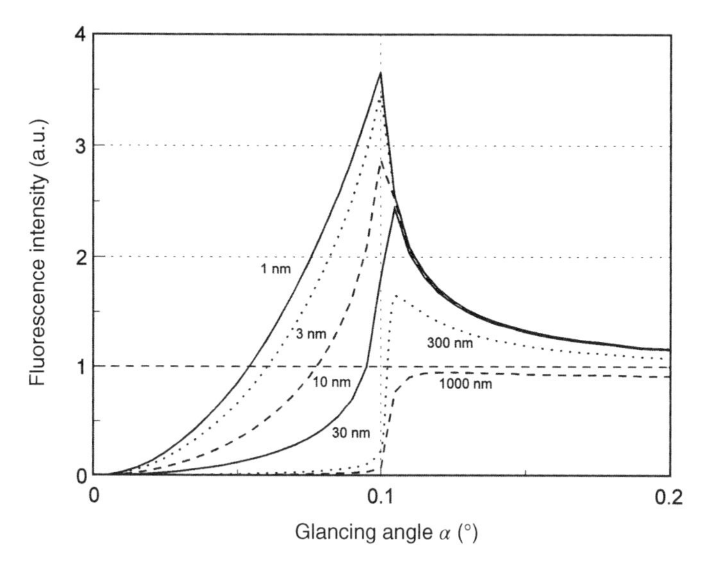

Figure 2.19. Fluorescence intensity from layers buried in a thick substrate. The intensity dependent on the glancing angle was calculated for layers of different thickness but with a constant area density of the analyte. Silicon was assumed as the substrate, Mo-Kα X-rays, as the primary beam. Total reflection occurs in the region below 0.1°. Without total reflection, the dashed horizontal line would be valid in all. Figure from Ref. [1], reproduced with permission. Copyright © 1996, John Wiley and Sons.

Equations 2.15, 2.19, and 2.20, and can be approximated by

$$I_{\mathrm{BL}}(\alpha) \cong I_{\mathrm{n}} c_{\mathrm{A}} 4\sqrt{R(\alpha)}\left(\frac{\alpha}{\alpha_{\mathrm{crit}}}\right)^2 \tag{2.24}$$

in the region of total reflection $\alpha \leq \alpha_{\mathrm{crit}}$, and by

$$I_{\mathrm{BL}}(\alpha) \cong I_{\mathrm{n}}\, c_{\mathrm{A}}\left[1 + \sqrt{R(\alpha)}\right]^2 \tag{2.25}$$

in the region beyond total reflection. The first approximation represents a parabola with a maximum intensity of nearly $4I_{\mathrm{n}}c_{\mathrm{A}}$ at the critical angle. The second equation describes an asymptotic decrease of the curve to a constant value $I_{\mathrm{n}}c_{\mathrm{A}}$ at larger angles as illustrated by Figure 2.19.

For layers of several 100 nm, the thickness d far exceeds the penetration depth z_{n}. In this case the peaks at the critical angle are reduced and the curves of Figure 2.19 get an S-like shape. Equation 2.22 can be approximated by

$$I_{\mathrm{BL}}(\alpha) \cong I_{\mathrm{n}}\, c_{\mathrm{A}}\, C[1 - R(\alpha)]\frac{\alpha}{d} \tag{2.26}$$

with symbols as defined earlier.

Finally, an infinitely thick layer may be considered. This layer may be thought of as a substrate in which the respective element is homogeneously distributed. Equation 2.26 leads to the asymptotic expression

$$I_{\mathrm{BL}}(\alpha) \cong I_{\mathrm{n}}\, c_{\mathrm{v}}\, C[1 - R(\alpha)]\alpha \tag{2.27}$$

where c_v is the concentration of the analyte in the total volume of the substrate given by the limit of c_A/d. Expression 2.27 is already known from Equation 2.20, but it additionally contains the quantity c_v as volume concentration.

2.3.4 Reflecting Layers on Substrates

The granular residues and buried layers treated so far were both assumed to be nonreflecting. Their fluorescence signal was calculated by the rather simple and transparent Equations 2.21 to 2.27. By way of contrast, reflecting layers shall now be considered. They may be plated on top of thick substrates in order to be used as conductive or nonconductive layers in wafer technology, as protective or decorative coatings, and so on. First of all, let us consider single reflecting layers. They are assumed to be flat, smooth, and equally thick over the entire surface of a substrate. The thickness may range from about 0.2 nm for a monoatomic layer up to some 100 nm.

The X-ray induced fluorescence of such layers can be calculated by means of fundamental parameters. In a first step, the primary intensity within the layer has to be calculated in dependence on the depth, as demonstrated for example in Figure 2.14 or Figure 2.15. In a second step, the excitation to fluorescence, the absorption and enhancement must be incorporated into the calculation. Finally, the fluorescence intensity values have to be integrated over the layer thickness. The mathematics is somewhat complex and will be presented later on in Section 2.4. To start with, the fluorescence signal shall be shown only for a few examples. In particular, the angular dependence of this signal shall be represented, as has already been done for nonreflecting layers.

In Figures 2.20 and 2.21, the fluorescence intensity emitted from a single layer is shown for cases A and B that were already considered in Sections 2.1.1 and 2.2.2. Case A represents a 70 nm Si layer deposited on a Au substrate, case B a 30 nm Co layer on a Si substrate. A Mo tube was chosen for excitation. The measured signal of the Si-Kα or Co-Kα radiation is demonstrated in angular dependence [4,5]. Below the critical angle of the single layer, the intensity is low but increases strongly at this angle, especially in case A. Beyond the critical angle of the layer, the intensity shows oscillations with maxima and minima but then approaches a constant value. As Figures 2.20 and 2.21 demonstrate, the maxima and minima of the fluorescence intensity correspond to the minima and maxima of the reflectivity of the layered substrates. These extrema are named Kiessig minima and maxima, as already mentioned in Section 2.1.1. A low reflectivity of course means a high energy flow into the layer and

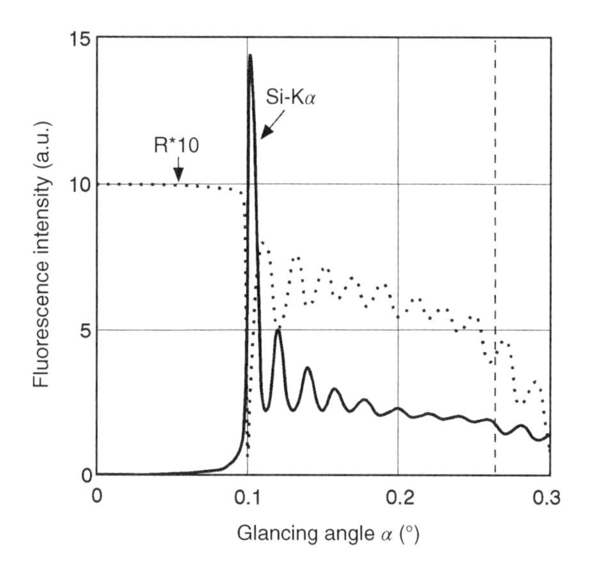

Figure 2.20. Fluorescence intensity from a thin layer deposited on a thick substrate. Case A of Section 2.1.1 was assumed for calculation: a 70 nm Si layer on a Au substrate irradiated by a Mo-Kα beam. The intensity of Si-Kα (——) was determined in dependence on the glancing angle α. From Figure 2.2, the reflectivity of the layered substrate is shown again (• • • • • •). It can be verified that Kiessig maxima of reflectivity correspond to minima of fluorescence, and vice versa. Figure from Ref. [4], reproduced with permission from APS.

consequently a strong excitation to fluorescence. For high reflectivity, the relationship is reversed. The steep Si-Kα peak in Figure 2.20 corresponds to the abrupt drop in reflectivity of Figure 2.2. It also corresponds to the excessive primary intensity demonstrated in Figure 2.14 as an extreme case of resonance.

The relationships of course depend on the thickness of the deposited layers. For the foregoing example of case B, the influence of the layer thickness is demonstrated in Figure 2.22. The thickness of Co layers varies between 1 and 200 nm, leading to a broad scale of intensity values for the Co-Kα radiation [20]. Consequently, a semilogarithmic plot is chosen to obtain a better general view.

For ultrathin layers, the curves are similar to those for nonreflecting layers. Because the radiation penetrates the nanometer-thin Co layers, total reflection does not occur at the Co layer but only at the Si substrate. For that reason, the intensity maximum appears at the critical angle of Si. If the layer exceeds a thickness of 10 nm, the maximum, however, shifts to the critical angle of cobalt. The range of intermediate thickness between 10 and 100 nm is characterized by oscillations of intensity beyond the Co critical angle. They are already known from Figures 2.20 and 2.21 and defined as Kiessig maxima and minima. Their angular period as well as their contrast decreases with thickness. Finally, the curves smooth out and show the shape of an infinitely thick substrate. On that account, the 200 nm thick layer gives a curve like that shown in Figure 2.19.

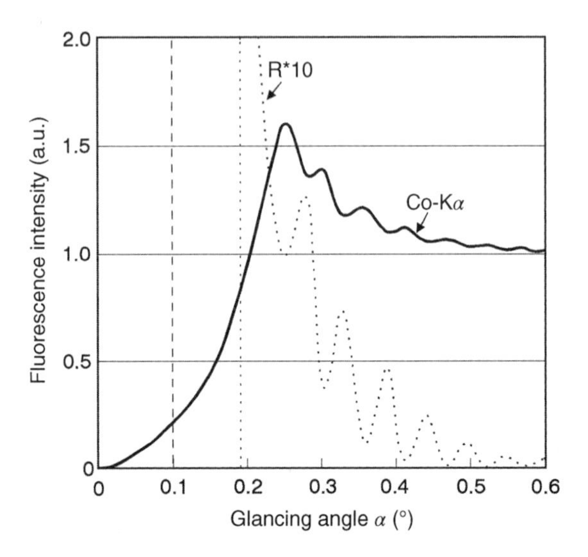

Figure 2.21. Fluorescence intensity of a thin reflecting layer deposited on a thick substrate. Case B of Section 2.1.1 was considered: a 30 nm Co layer on a silicon flat, irradiated by Mo-Kα X-rays. The Co-Kα intensity (——) is dependent on the glancing angle α and connected with the reflectivity (·······) of the layered sample. Figure from Ref. [5], reproduced with permission. Copyright © 1991 by ICDD.

The last example pertains to case B. The first maximum shifts from the critical angle of the substrate to that of the layer if its thickness exceeds 10 nm or so. In particular, the maximum shifts to a greater angle. Such behavior also occurs for examples of case A but the maximum shifts to a smaller angle [21].

2.3.5 Periodic Multilayers and Crystals

Single reflecting layers will present the characteristic features of TXRF if the glancing angle of the primary beam is varied at grazing incidence. In principle, multilayers show a similar angular dependence of the fluorescence intensity with typical Kiessig maxima and minima. The intensity can be calculated by integration over the standing wave field within the layers. The calculation is carried out by computers using a fundamental parameter approach, as will be described in Section 2.4. It can be simplified due to the small glancing angles considered.

A further phenomenon will occur if a *periodic* multilayer is taken into consideration [14]. The same effect happens for natural or synthetic crystals, as they are akin to periodic multilayers as already considered in Section 2.1.2. Figure 2.23 [22] demonstrates this effect for a multilayer already chosen in Figure 2.6. A special multilayer may consist of 15 bilayers of 1.9 nm platinum and 0.2 nm cobalt, with a period of about 2.1 nm. The fluorescence intensity in Figure 2.23 strongly increases around the critical angle of platinum at 0.6°. It further shows maxima at angles of minimum reflectivity and minima at angles

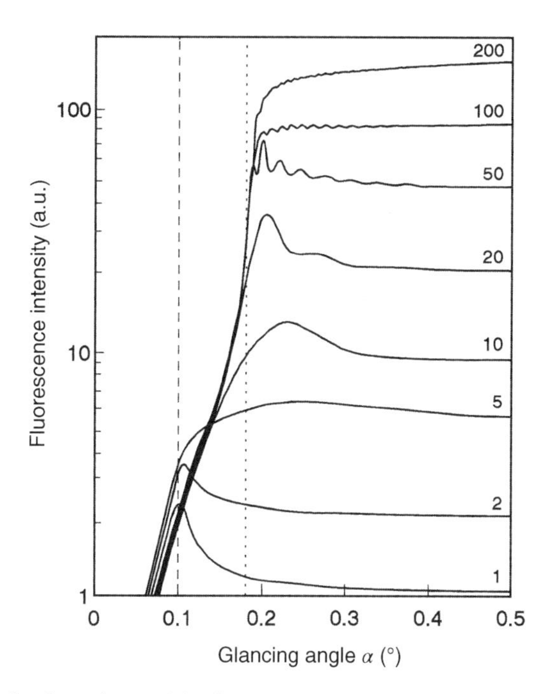

Figure 2.22. Angular dependence of the fluorescence radiation emitted from a Co-layered silicon substrate. The Co-Kα intensity is plotted semilogarithmically for layers of different thicknesses (in nm). The maxima for the ultrathin Co layers are localized at the critical angle of Si (dashed vertical line). They are shifted to the critical angle of Co (dotted vertical line) if the layer is more than 10 nm thick. Figure from Ref. [20], reproduced with permission. Copyright © 1991, John Wiley and Sons.

of maximum reflectivity. These Kiessig oscillations, however, are not very distinct in comparison to the following Bragg maxima and minima of intensity.

The first Bragg maximum of reflectivity ($m = 1$) occurs at a glancing angle of about 2.2°. In the upgoing flank of this peak, a maximum of the Co-Kα intensity appears whereas a minimum of the Pt Lα intensity occurs. Recall from Figure 2.16 that the antinodes are in the Co layers while the nodes are in the Pt layers. If the glancing angle just exceeds the Bragg angle, the antinodes are switched into the Pt layers and the nodes into the Co layers due to the phase shift of π. Consequently, the Pt-Lα intensity shows a maximum in the descending flank of the reflection peak whereas the Co-Kα intensity shows a minimum. The effect is quite distinct for the first order and will recur with smaller contrast at higher orders m. In any case, the intensity maxima of the lighter spacer appear at angles just below those of the heavier reflector.

The phenomenon is also known for crystal lattices, with the following two distinctions: (i) Because there is no spacer element but possibly more than one reflector element, for example, for NaCl, the intensity minima of these elements always lie on the upgoing flank of the reflection peak, the maxima on the downgoing flank. (ii) Because the lattice distance of inorganic crystals is

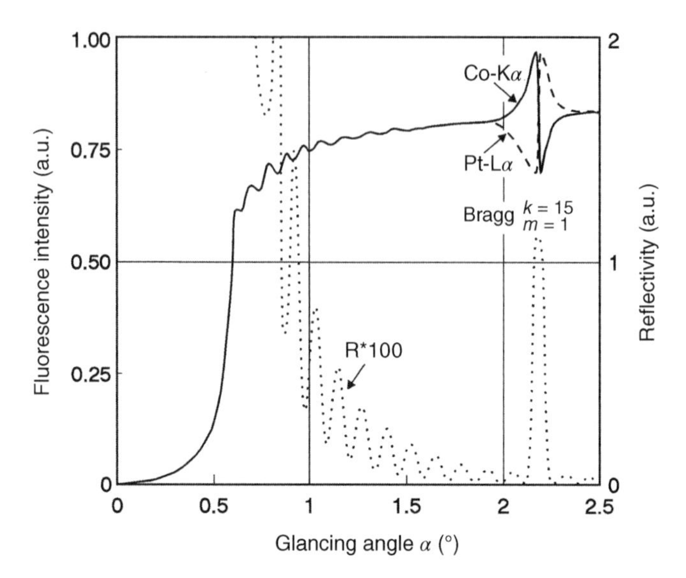

Figure 2.23. Fluorescence intensity of a periodic multilayer plotted against the glancing angle α. The multilayer of Section 2.1.2 was used for calculation, consisting of 15 bilayers of platinum and cobalt ($d = 2.1$ nm). A Cu-Kα beam was assumed for excitation. The reflectivity of the multilayer is represented by the dotted curve. Figure from Ref. [22], reproduced with permission. Copyright © 1991 by Elsevier.

mostly between 0.15 and 0.3 nm, the Bragg angle is on the order of 10° for wavelengths of about 0.1 nm. For that reason, Bragg reflection does not appear under grazing but under steeper incidence.

2.4 FORMALISM FOR INTENSITY CALCULATIONS

The phenomena qualitatively described in Section 2.3 can rigorously be calculated from theory. The *primary* beam can be described by a plane incoming and reflected wave leading to a standing wave field. Its intensity can be calculated from the optical theory of wave propagation in layers with flat interfaces, especially by the Fresnel relations. The *primary intensity* is a function of the glancing angle α_0 at grazing incidence and is further dependent on the depth z normal to the layer surface. Besides the primary intensity, the intensity ratio of the reflected and the incoming wave can be calculated. This ratio is defined as *reflectivity* and can directly be checked by reflectivity measurements. All calculations can be based either on a recursive formalism described by Parratt [23] or on an equivalent matrix formalism first proposed by Abelès [24] and extensively described by Born and Wolf [2] and Król *et al.* [25]. The latter is the more elegant method and is preferred here.

The *fluorescence intensity* can be determined by a fundamental parameter approach of XRF. This intensity can be calculated as a function of the glancing angle and be checked by fluorescence measurements. The theory has been described and applied by several authors [4,22,26–31] and has been proven valid by many experiments [5,8,26,32–35]. Even the surface and interface roughness has been taken into account [30,36–38].

In Sections 2.4.1–2.4.3, the reflectivity, the primary, as well as the fluorescence intensities are calculated for three different media already distinguished earlier: (i) a thick and flat substrate, (ii) a thin homogeneous layer on a substrate, and (iii) a stratified medium of several layers. The derivations are carried out mainly following papers by de Boer and van den Hoogenhof [22], Gutschke [26], and Weisbrod *et al.* [32].

2.4.1 A Thick and Flat Substrate

An infinitely thick and flat substrate may be present, and the z-direction shall be assumed normal to the surface (as shown in Figure 2.24). A plane wave may hit the surface under a glancing angle α_0 be reflected under an equal angle, and be transmitted under a refraction angle α_s which is given by

$$\alpha_s = \sqrt{\alpha_0^2 - 2\,\delta_s - 2\,i\beta_s} \tag{2.28}$$

where δ_s and β_s are the real and complex parts of the refractive index of the substrate, respectively. The real and imaginary components of α_s can be derived in accord with Equations 1.71 and 1.72, respectively.

The radiation intensity is determined by the amplitude of the electric field vector. At steeper incidence, a π- and σ-polarization has to be distinguished

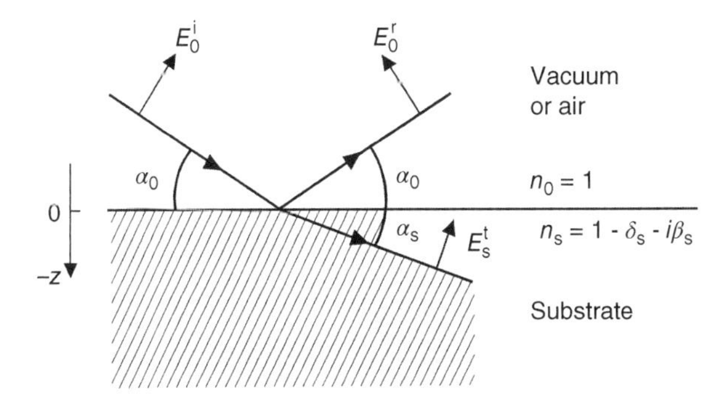

Figure 2.24. Incoming, reflected, and refracted beams at the interface between vacuum or air and a thick substrate. n_0 and n_S are the refractive indices, z is the direction normal to the surface, and $\boldsymbol{E}$ is the electric-field vector representing a π-polarization. Figure from Ref. [1], reproduced with permission. Copyright © 1996, John Wiley and Sons.

(the field vector is within the plane of incidence or perpendicular to this plane, respectively). At grazing incidence, however, the polarization does not have to be considered. Only the length or amplitude of the electric field vector is relevant for fluorescence. Consequently, only this component has to be taken into account, but an averaging over time has to be carried out.

There are different waves with different amplitudes. The incident wave amplitude E_0^{i} may be normalized to 1. The reflected wave amplitude may be denoted by E_0^{r}. The refracted wave amplitude, which is the incident wave amplitude in the substrate, is called $E_{\mathrm{s}}^{\mathrm{t}}$. Within the substrate, there is no reflected wave, so that a hypothetical amplitude $E_{\mathrm{s}}^{\mathrm{r}}$ can be set to zero. These four amplitudes are connected by the so-called Born or transfer matrix according to

$$\begin{bmatrix} 1 \\ E_0^{\mathrm{r}} \end{bmatrix} = \begin{bmatrix} m_1 & m_2 \\ m_3 & m_4 \end{bmatrix} \cdot \begin{bmatrix} E_{\mathrm{s}}^{\mathrm{t}} \\ 0 \end{bmatrix} \tag{2.29}$$

The individual components of this 2×2 matrix are given by

$$m_1 = m_4 = \frac{\alpha_0 + \alpha_{\mathrm{s}}}{2\alpha_0} \tag{2.29a}$$

$$m_2 = m_3 = \frac{\alpha_0 - \alpha_{\mathrm{s}}}{2\alpha_0} \tag{2.29b}$$

The primary intensity I_{int} is defined as the square of the modulus of the relevant amplitude. Because incoming and reflected beams interfere above the surface, the intensity I_{int} at height z for a glancing angle α_0 can be written as a sum of the downgoing and upgoing plane waves:

$$I_{\mathrm{int}}(\alpha_0, z) = I_0 \left| \exp(-i\, k_0\, \alpha_0\, z) + E_0^{\mathrm{r}} \exp(i\, k_0\, \alpha_0\, z - i\varphi) \right|^2 \tag{2.30}$$

where k_0 is $2\pi/\lambda$, φ is a phase shift already defined by Equation 2.15, and I_0 is a measure for the incoming beam intensity. Squaring leads to

$$I_{\mathrm{int}}(\alpha_0, z) = I_0 \left[1 + \left| E_0^{\mathrm{r}} \right|^2 + 2 \left| E_0^{\mathrm{r}} \right| \cos(2\, k_0 \alpha_0 z - \varphi) \right] \tag{2.31}$$

Because $\left| E_0^{\mathrm{r}} \right|^2$ is defined as reflectivity R, this equation is equivalent to expression (2.14).

The amplitude E_0^{r} is obtained from Equation 2.29 as the ratio of m_3 and m_1. The squared value of the modulus yields the reflectivity itself:

$$R(\alpha_0) = \left| \frac{\alpha_0 - \alpha_{\mathrm{s}}}{\alpha_0 + \alpha_{\mathrm{s}}} \right|^2 \tag{2.32}$$

which corresponds to Equation 1.75.

Within the substrate, the primary intensity is only determined by the refracted beam with the amplitude E_s^t:

$$I_{int}(\alpha_0, z) = I_0 \left| E_s^t \cdot \exp(i\, k_0\, \alpha_s\, z) \right|^2 \tag{2.33}$$

The amplitude E_s^t obtained from Equation 2.29 is given by $1/m_1$. Inserting the respective value of m_1 leads to

$$I_{int}(\alpha_0, z) = I_0 \left| \frac{2\,\alpha_0}{\alpha_0 + \alpha_s} \right|^2 \cdot \exp\left(-\frac{z}{z_n} \right) \tag{2.34}$$

where z_n means the penetration depth defined by Equation 1.78.

The quadratic term is the transmissivity. By means of the reflectivity defined in Equation 1.75, this quantity can easily be shown to equal $(1 - R)\ \alpha_0/\alpha_s'$. Replacing α_s' by α_s'' in accord with Equation 1.73 and inserting z_n defined by Equation 1.78 leads to

$$I_{int}(\alpha_0, z) = I_0\, C(1 - R) \frac{\alpha_0}{z_n} \exp\left(-\frac{z}{z_n} \right) \tag{2.35}$$

where C is the quantity already used in Equation 2.20 and which is equal to $(\lambda/4\pi)/\beta_s$. Equation 2.35 is equivalent to Equation 2.17 in Section 2.2.1. It is the basis for the fluorescence intensities of different buried layers that were calculated by integration and finally led to the approximate Equations 2.22–2.27.

A more detailed calculation of the X-ray fluorescence intensity can be based on a fundamental parameter approach. It may be provided that the takeoff angle for the detector is 90° and the detector area determines the area of observation. If an element named x with a mass fraction c_x is homogeneously distributed in a thick and flat substrate, the fluorescence intensity can be approximated by

$$I_x(\alpha_0) \cong I_n c_x S_{x,E_0} \varepsilon_{det}\, T_{air} \frac{1 - R(\alpha_0)}{(\mu/\rho)_{s,E_0}/\alpha_0 + (\mu/\rho)_{s,E}} \tag{2.36}$$

This equation replaces Equation 2.27. I_n is the reference intensity induced by the incident plane wave or primary beam; S_{x,E_0} is a sensitivity value of element x at photon energy E_0 of the monochromatic primary beam; $(\mu/\rho)_{s,E_0}$ and $(\mu/\rho)_{s,E}$ are the mass attenuation coefficients of the substrate at photon energy E_0 of the primary beam and at photon energy E of the detected element peak, respectively; α_0 is the glancing angle of incidence, which is assumed to be small (<5°), so that the actual sine dependence can be ignored; ε_{det} is the efficiency of the detector for X-ray photons of energy E, and T_{air} is their transmission by air on the beam path between the sample and the detector. The sensitivity itself can be further deduced from

$$S_{x,E_0} = g_x \omega_x f_x (\tau/\rho)_{x,E_0} \tag{2.37}$$

where g_x is the relative emission rate of the element peak in its series; ω_x is the fluorescence yield of the element x, f_x is its jump factor at the relevant absorption edge and $(\tau/\rho)_{x,E_0}$ is the photoelectric mass-absorption coefficient at the primary photon energy E_0. This product of fundamental parameters is independent of instrumentation or sample matrix and can be calculated for each element x. Different sets of tables can be used for this calculation [39–41].

The foregoing derivation includes the absorption of the primary radiation as well as the fluorescence radiation. Secondary fluorescence or enhancement is not taken into account because it is a second-order process. It is rather complex and may be neglected in the case of grazing incidence (but ought to be considered in principle [4,32]). The angular divergence of the instrument, however, can easily be taken into consideration. For that purpose, Equation 2.34 has to be convoluted by a triangle function of α_0 with the width of the aperture.

The intensity dependence on the energy transfer $(1 - R)\ \alpha_0$ was derived earlier in simple terms in Section 2.3.1. The matrix and the fundamental parameter approaches have been introduced in the present section in order to provide a more detailed description. Both formalisms can be expanded and applied to single- and multiple-layer systems.

2.4.2 A Thin Homogeneous Layer on a Substrate

A single layer with a thickness d_1 may be deposited on a thick and flat substrate, as demonstrated in Figure 2.25. Here n_0, n_1, and n_s shall denote the complex

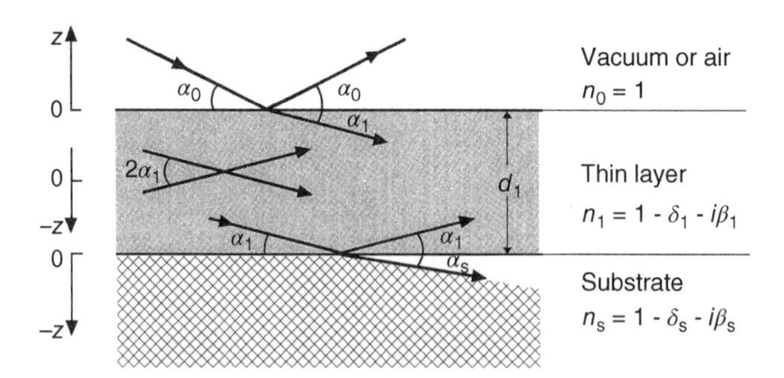

Figure 2.25. Incoming, reflected, and refracted beams above, within, and below a thin layer, respectively, deposited on a thick substrate. n_0, n_1, and n_s are the refractive indices. At any point above and within the layer, there are two beams interfering with one another at an angle $2\alpha_0$ or $2\alpha_1$, respectively. In the substrate, there is only one refracted beam penetrating at an angle α_s. Figure from Ref. [1], reproduced with permission. Copyright © 1996, John Wiley and Sons.

refractive indices of the vacuum, the layer, and the substrate, respectively, dependent on the relevant quantities δ and β. A wide X-ray beam represented by a plane wave may be directed to this simple stratified structure at a glancing angle, α_0. In this case, two beams are imaginable at any point above and within the layer: one transmitted and one reflected beam. *In vacuo*, they interfere at an angle $2\alpha_0$ and in the layer they overlap at an angle $2\alpha_1$. This is demonstrated for one point at the vacuum–layer interface, for one point at the layer–substrate interface, and for one point in the middle of the layer. In the substrate, however, there are only refracted beams at any point, penetrating at an angle α_s. The angles are determined by

$$\alpha_1 = \sqrt{\alpha_0^2 - 2\delta_1 - 2i\beta_1} \tag{2.38a}$$

and

$$\alpha_s = \sqrt{\alpha_0^2 - 2\delta_s - 2\,i\,\beta_s} \tag{2.38b}$$

respectively.

The radiation intensity of the wave field called primary intensity can easily be calculated by the matrix formalism. For simplification, the origin of the vertical z-axis is defined for the three media separately. For the vacuum and the substrate it is placed at the interfaces; for the layer it is positioned right in the middle of it. Under this condition, the amplitudes of four different electric field vectors E_0^r, E_1^t, E_1^r, and E_s^t are connected by two transfer matrices $M_{0,1}$ and $M_{1,s}$ according to

$$\begin{bmatrix} 1 \\ E_0^r \end{bmatrix} = M_{0,1} \begin{bmatrix} E_1^t \\ E_1^r \end{bmatrix} \tag{2.39a}$$

and

$$\begin{bmatrix} E_1^t \\ E_1^r \end{bmatrix} = M_{1,s} \begin{bmatrix} E_s^t \\ 0 \end{bmatrix} \tag{2.39b}$$

The individual components of the first matrix $M_{0,1}$ are given by

$$\begin{aligned} m_{1,1} &= \frac{\alpha_0 + \alpha_1}{2\alpha_0} \exp\left(-i\frac{k_0}{2}\alpha_1 d_1\right) \\ m_{2,1} &= \frac{\alpha_0 - \alpha_1}{2\alpha_0} \exp\left(+i\frac{k_0}{2}\alpha_1 d_1\right) \\ m_{3,1} &= \frac{\alpha_0 - \alpha_1}{2\alpha_0} \exp\left(-i\frac{k_0}{2}\alpha_1 d_1\right) \\ m_{4,1} &= \frac{\alpha_0 + \alpha_1}{2\alpha_0} \exp\left(+i\frac{k_0}{2}\alpha_1 d_1\right) \end{aligned} \tag{2.40a}$$

The components of the second matrix $M_{1,s}$ are determined by

$$\begin{aligned} m_{1,s} &= \frac{\alpha_1 + \alpha_s}{2\alpha_1} \exp\left(-i\frac{k_0}{2}\alpha_1 d_1\right) \\ m_{2,s} &= \frac{\alpha_1 - \alpha_s}{2\alpha_1} \exp\left(-i\frac{k_0}{2}\alpha_1 d_1\right) \\ m_{3,s} &= \frac{\alpha_1 - \alpha_s}{2\alpha_1} \exp\left(+i\frac{k_0}{2}\alpha_1 d_1\right) \\ m_{4,s} &= \frac{\alpha_1 + \alpha_s}{2\alpha_1} \exp\left(+i\frac{k_0}{2}\alpha_1 d_1\right) \end{aligned} \tag{2.40b}$$

The four different amplitudes E_0^r, E_1^t, E_1^r, and E_s^t have to be determined by two matrix equations. By combination of both Equations 2.39a and 2.39b, the amplitudes E_0^r and E_s^t can first be determined from the relation

$$\begin{bmatrix} 1 \\ E_0^r \end{bmatrix} = M_{0,1} \cdot M_{1,s} \cdot \begin{bmatrix} E_s^t \\ 0 \end{bmatrix} \tag{2.41}$$

The amplitude E_s^t can simply be written as

$$E_s^t = 1/(m_{1,1}m_{1,s} + m_{2,1}m_{3,s}) \tag{2.42}$$

The amplitude E_0^r is not needed here but it may be mentioned that the squared modulus of E_0^r results in the reflectivity of the layered substrate already given by Equation 2.5.

Afterward, the two other amplitudes of the electric field within the layer E_1^t and E_1^r can be calculated from Equation 2.39b.

$$E_1^t = m_{1,s}/(m_{1,1}m_{1,s} + m_{2,1}m_{3,s}) \tag{2.43}$$

$$E_1^r = m_{3,s}/(m_{1,1}m_{1,s} + m_{2,1}m_{3,s}) \tag{2.44}$$

The primary intensity can now be determined for the three different media, for a given glancing angle α_0 in dependence of the depth. *In vacuo*, the primary intensity results from

$$I_{\text{prim}}(\alpha_0, z) \cong I_n \cdot \left|\exp(i\,k_0\,\alpha_0\,z) + E_0^r \exp(-i\,k_0\alpha_0 z + i\varphi_0)\right|^2 \tag{2.45a}$$

φ_0 considers the phase shift of total reflection at the uppermost surface of the layer.

In the layer the primary intensity is given by

$$I_{\text{prim}}(\alpha_0, z) \cong I_n \cdot \left|E_1^t \exp(i\, k_0 \alpha_1\, z) + E_1^r \exp(-i\, k_0 \alpha_1 z + i\varphi_1)\right|^2 \tag{2.45b}$$

φ_1 considers the phase shift of total reflection at the interface between layer and substrate while z is in the range $(-d_1/2, +d_1/2)$. In the substrate, the primary intensity follows from

$$I_{\text{prim}}(\alpha_0, z) \cong I_0 \cdot \left|E_s^t \exp(i\, k_0\, \alpha_s\, z)\right|^2 \tag{2.45c}$$

Next, the fluorescence intensity has to be derived for an element named x that is present in the layer with a mass fraction c_x. It is assumed again that the detector is positioned perpendicular to the surface with a short gap in between them. The solution follows from the fundamental parameter approach after integration over the thickness of the layer [22]. For an easier presentation three auxiliary quantities may be defined, which appear in denominators and exponents later on:

$$Q_t = \left[-\frac{(\mu/\rho)_{1,E_0}}{\alpha_1} + (\mu/\rho)_{1,E}\right] \tag{2.46a}$$

$$Q_r = \left[\frac{(\mu/\rho)_{1,E_0}}{\alpha_1} + (\mu/\rho)_{1,E}\right] \tag{2.46b}$$

$$Q_{rt} = \left[-\frac{2ik_0\alpha_1}{\rho_1} + (\mu/\rho)_{1,E}\right] \tag{2.46c}$$

The different parameters are defined in correspondence to those of Equation 2.36. $(\mu/\rho)_{1,E_0}$ is the mass-attenuation coefficient of the layer at photon energy E_0; $(\mu/\rho)_{1,E}$ is that coefficient at photon energy E corresponding to the element peak; and ρ_1 is the density of the layer.

The fluorescence intensity of the thin monolayer can be derived now in dependence on the glancing angle α_0. It results from integrating Equation 2.45b. Secondary fluorescence effects are neglected and also a footprint effect caused by a limited sample and detector window is not considered here. The fluorescence intensity is represented by the trinomial sum:

$$I_{\text{fluo}}(\alpha_0) \cong I_n\, c_x\, S_{x,E_0}\, \varepsilon_{\text{det}}\, T_{\text{air}} \cdot [X_t(\alpha_1, d_1) + X_r(\alpha_1, d_1) + 2X_{rt}(\alpha_1, d_1)] \tag{2.47}$$

The three addenda are given by similar expressions:

$$X_t(\alpha_1, d_1) = \left|E_1^t\right|^2 \cdot \left|\frac{1 - \exp(-Q_t\rho_1 d_1)}{Q_t}\right| \tag{2.47a}$$

$$X_{\mathrm{r}}(\alpha_1, d_1) = \left|E_1^{\mathrm{r}}\right|^2 \cdot \left|\frac{1 - \exp\left(-Q_{\mathrm{r}}\rho_1 d_1\right)}{Q_{\mathrm{r}}}\right| \tag{2.47b}$$

$$X_{\mathrm{rt}}(\alpha_1, d_1) = \mathrm{Re}\left[E_1^{\mathrm{t}*} E_1^{\mathrm{r}} \cdot \frac{1 - \exp\left(-Q_{\mathrm{rt}}\rho_1 d_1\right)}{Q_{\mathrm{rt}}} \cos\varphi_1\right] \tag{2.47c}$$

Re means the real part of the complex number in brackets, the star means the complex conjugate. As mentioned in Section 2.4.1, the angular divergence can be taken into account by a convolution.

Two special cases may be of interest. For a very thin layer, d_1 approximates zero and the three fractions of Equations 2.47a–2.47c take on the limiting value $\rho_1\, d_1$. The fluorescence intensity simply becomes

$$I_{\mathrm{fluo}}(\alpha_0) \cong I_{\mathrm{n}}\, c_{\mathrm{A}}\, S_{x,E_0}\, \varepsilon_{\mathrm{det}}\, T_{\mathrm{air}} \cdot \left|E_1^{\mathrm{t}} + E_1^{\mathrm{r}}\right|^2 \tag{2.48}$$

where c_{A} is the area-related mass given by the product $c_x\, \rho_1\, d_1$. The second special case is a single thin layer for which vacuum or air may be regarded as a substrate. In this case, the amplitude E_1^{r} nearly vanishes because the reflection at the bottom of the layer can be neglected. Equation 2.47 is then reduced to the first term of the sum.

2.4.3 A Stratified Medium of Several Layers

To be consistent, we now consider a stratified structure of several layers. It may be composed of N layers with a thickness d_ν ($\nu = 1, \ldots, N$) that are each homogeneous and plane parallel. The refractive indices may be n_ν and the densities ρ_ν. The first layer adjoins to the vacuum or air ($\nu = 0$); the last layer, to the substrate ($\nu = N+1$). The origin of each layer is shifted right in the middle of the respective layer.

In any layer, there is one definite glancing angle of incidence α_ν. It is determined by the outer glancing angle α_0, and by δ_ν and β_ν of the layer, according to

$$\alpha_\nu = \sqrt{\alpha_0^2 - 2\delta_\nu - 2i\beta_\nu} \tag{2.49}$$

One incoming and one reflected beam interfere with each other at any point of the νth layer, as shown by Figure 2.26. They overlap under the angle $2\alpha_\nu$. The amplitudes of the electric fields can be denoted by E_ν^{t} and E_ν^{r} for the transmitted and reflected waves, respectively, in the middle of layer ν. As in Section 2.4.2, the amplitudes of two adjacent layers can be connected by a transfer matrix

$$\begin{bmatrix} E_\nu^{\mathrm{t}} \\ E_\nu^{\mathrm{r}} \end{bmatrix} = M_{\nu,\nu+1} \begin{bmatrix} E_{\nu+1}^{\mathrm{t}} \\ E_{\nu+1}^{\mathrm{r}} \end{bmatrix} \tag{2.50}$$

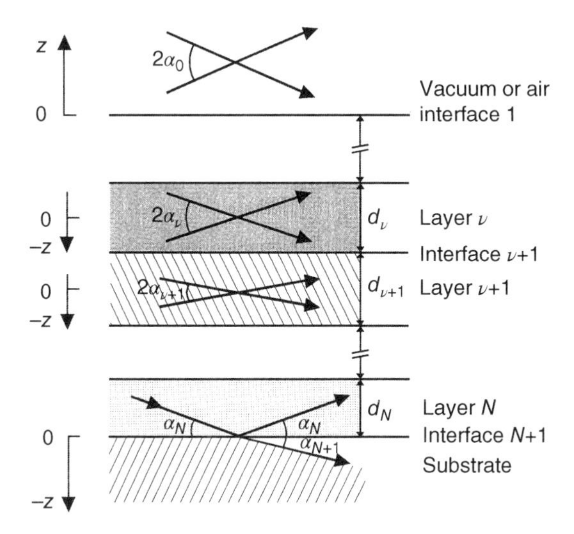

Figure 2.26. Incoming, reflected and refracted beams in a stratified structure with N layers of a finite thickness d_ν ($\nu = 1 \ldots N$) deposited on a thick substrate. At any point within the individual layers, one incident and one reflected beam interfere with each other at an angle $2\alpha_\nu$. Only in the substrate is there a single incident beam refracted at an angle α_{N+1}. For the vacuum or air and for the substrate, the origin of the z-axis is placed on the interface. However, for all layers, a separate origin is placed in the middle of the respective layer. Figure from Ref. [1], reproduced with permission. Copyright © 1996, John Wiley and Sons.

with the components

$$
\begin{aligned}
m_{\nu,\nu+1} &= \frac{\alpha_\nu + \alpha_{\nu+1}}{2\alpha_\nu} \exp\left[-i\frac{k_0}{2}(\alpha_\nu\, d_\nu + \alpha_{\nu+1}\, d_{\nu+1})\right] \\
m_{\nu,\nu+1} &= \frac{\alpha_\nu - \alpha_{\nu+1}}{2\alpha_\nu} \exp\left[-i\frac{k_0}{2}(\alpha_\nu\, d_\nu - \alpha_{\nu+1} d_{\nu+1})\right] \\
m_{\nu,\nu+1} &= \frac{\alpha_\nu - \alpha_{\nu+1}}{2\alpha_\nu} \exp\left[i\frac{k_0}{2}(\alpha_\nu\, d_\nu - \alpha_{\nu+1}\, d_{\nu+1})\right] \\
m_{\nu,\nu+1} &= \frac{\alpha_\nu + \alpha_{\nu+1}}{2\alpha_\nu} \exp\left[i\frac{k_0}{2}(\alpha_\nu\, d_\nu + \alpha_{\nu+1}\, d_{\nu+1})\right]
\end{aligned}
\tag{2.51}
$$

For the vacuum ($\nu = 0$) and for the substrate ($\nu = N+1$), the hypothetical thickness, d_0, and d_{N+1}, respectively, have to be set to zero. Due to this agreement, the Equations 2.51 are even valid for the components of the first matrix, $M_{0,1}$, and the last matrix, $M_{N,N+1}$, and also correspond with Equations 2.40a and 2.40b.

A solution for the $2(N+1)$ unknown amplitudes E_ν^{t} and E_ν^{r} can be found from the $2(N+1)$ equations included in the matrix expression 2.50. First, the two amplitudes, E^{t}_{N+1} for the substrate and E^{r}_{0} for the vacuum, have to be determined after connecting them by a matrix multiplication:

$$\begin{bmatrix} 1 \\ E^{\mathrm{r}}_{0} \end{bmatrix} = \prod_{\nu=0}^{N} M_{\nu,\nu+1} \begin{bmatrix} E^{\mathrm{t}}_{N+1} \\ 0 \end{bmatrix} \tag{2.52}$$

With the known amplitude E^{t}_{N+1} for the substrate, the remaining amplitudes for all layers can then be calculated in accord with expression 2.50 in steps backward. Beginning with the known amplitude E^{r}_{0} for the vacuum, the remaining amplitudes can be calculated also in steps forward.

The primary intensity for a given outer glancing angle α_0 in dependence on the depth of the νth layer is then given by

$$I_{\mathrm{prim},\nu}(\alpha_0, z) \cong I_{\mathrm{n}} \cdot \left| E^{\mathrm{t}}_{\nu} \exp(i\, k_0\, \alpha_\nu\, z) + E^{\mathrm{r}}_{\nu} \exp(-i\, k_0\, \alpha_\nu\, z + i\varphi_\nu) \right|^2 \tag{2.53}$$

for z values between $-d_\nu/2$ and $+d_\nu/2$. The phase shift φ_ν is taken into account because of total reflection at the different interfaces.

By means of the primary intensity, the fluorescence intensity of an element named x in the layer ν can now be determined. The element may be present in the νth layer with a mass fraction $c_{x,\nu}$. Its atoms may be excited to fluorescence in the primary wave field, $I_{\mathrm{prim},\nu}$, by a monochromatic radiation with photons of energy E_0. At these conditions, the fluorescence intensity of the layer can be derived from Equation 2.47 by replacing the fundamental parameters of the first layer by those of the νth layer. This original fluorescence radiation with a photon energy $E < E_0$ will further be absorbed on its way to the detector, which again should be mounted perpendicular to the stratified medium and close to it. Absorption takes place in all layers *above* the relevant νth layer, that is, in the $(\nu - 1)$ layers with an index $j = 1, \ldots, \nu - 1$. Consequently, the detector will measure an intensity that results from Equation 2.47 by multiplication with an absorption factor

$$A_\nu = \exp\left[-\sum_{j=1}^{\nu-1} (\mu/\rho)_{j,E}\, \rho_j\, d_j \right] \tag{2.54}$$

where $(\mu/\rho)_{j,E}$ is the mass-attenuation coefficient of the jth layer for photons of the respective energy E.

Again, three auxiliary quantities have been defined:

$$Q_{\mathrm{t},\nu} = \left[-\frac{(\mu/\rho)_{\nu,E_0}}{\alpha_\nu} + (\mu/\rho)_{\nu,E} \right] \tag{2.55a}$$

$$Q_{\mathrm{r},\nu} = \left[\frac{(\mu/\rho)_{\nu,E_0}}{\alpha_\nu} + (\mu/\rho)_{\nu,E} \right] \tag{2.55b}$$

$$Q_{\mathrm{rt},\nu} = \left[-\frac{2ik_0\alpha_\nu}{\rho_\nu} + (\mu/\rho)_{\nu,E}\right] \tag{2.55c}$$

By means of these quantities three addenda have to be built similar to Equations 2.47a, 2.47b, and 2.47c, which may be called $X_{\mathrm{t},\nu}$, $X_{\mathrm{r},\nu}$, and $X_{\mathrm{rt},\nu}$. The fluorescence intensity to be measured by the detector can then be calculated by

$$I_{\mathrm{fluo},\nu}(\alpha_0) \cong I_{\mathrm{n}}\, c_{x,\nu}\, S_{x,E_0}\, \varepsilon_{\mathrm{det}}\, T_{\mathrm{air}}\, A_\nu[X_{\mathrm{t},\nu}(\alpha_\nu, d_\nu) + X_{\mathrm{r},\nu}(\alpha_\nu, d_\nu) + 2X_{\mathrm{rt},\nu}(\alpha_\nu, d_\nu)] \tag{2.56}$$

This rather complex expression has to be summed up for an element named x, which is present in different layers. It is evident that a polychromatic instead of a monochromatic excitation would further complicate this expression. It may also be emphasized that the equation cannot be solved directly for any of the possibly unknown parameters $c_{x,\nu}$, ρ_ν, or d_ν. The mass-fractions $c_{x,\nu}$ are even concealed in the mass-attenuation coefficients $(\mu/\rho)_{\nu,E_0}$ and $(\mu/\rho)_{\nu,E}$ and not only present as single factors.

One special and simplified case should be mentioned, namely, that of a periodic multilayer. For this, only two different layers and consequently only two different transfer matrices are relevant. Any period of a double layer is characterized by one and the same characteristic matrix. On account of the periodicity, the total multilayer with N periods is characterized by the Nth power of this matrix. All further calculations are considerably shortened under this condition.

REFERENCES

1. Klockenkämper, R. (1997) *Total-Reflection X-Ray Fluorescence Analysis*, 1st ed., John Wiley & Sons, Inc., New York.
2. Born, M. and Wolf, E. (1980). *Principles of Optics*, 6th ed., (reprinted in 1993), Pergamon Press: Oxford; 808 pp.
3. Röseler, A. (1990). *Infrared Spectroscopic Ellipsometry*, Akademie Verlag: Berlin.
4. de Boer, D.K.G. (1991). Glancing incidence X-ray fluorescence of layered materials. *Phys. Rev.*, **B44**, 498–511.
5. de Boer, D.K.G. and van den Hoogenhof, W.W. (1991). Total-reflection X-ray fluorescence of thin layers on and in solids. *Adv. X-Ray Anal.*, **34**, 35–40.
6. Kiessig, H. (1931). Interferenz von Röntgenstrahlen an dünnen Schichten. *Ann. Phys. (Leipzig)*, **10**, 769–788.
7. de Boer, D.K.G., Leenaers, A.J.G., and van den Hoogenhof, W.W. (1995). Glancing-incidence X-ray analysis of thin-layered materials: a review. *X-Ray Spectrom.*, **24**, 91–102.
8. Huang, T.C. and Parrish, W. (1992). *Adv. X-Ray Anal.*, **35**, 137–142.
9. Bedzyk, M.J., Bommarito, G.M., and Schildkraut, J.S. (1989). X-ray standing waves at a reflecting mirror surface. *Phys. Rev. Lett.*, **62**, 1376–1379.

10. Krämer, M. (2007). Potentials of synchrotron radiation induced X-ray standing waves and X-ray reflectivity measurements in material analysis. PhD thesis, University of Dortmund, 128 pages.
11. Krämer, M., von Bohlen, A., Sternemann, C., Paulus, M., and Hergenröder, R. (2006). X-ray standing waves: a method for thin-layered systems. *J. Anal. At. Spectrom.*, **21**, 1136–1142.
12. Krämer, M., von Bohlen, A., Sternemann, C., Paulus, M., and Hergenröder, R. (2007). Synchrotron radiation induced X-ray standing waves analysis of layered structures. *Appl. Surf. Sci.*, **253**, 3533–3542.
13. Zheludeva, S.I., Kovalchuk, M.V., Novikova, N.N., and Sosphenov, A.N. (1995). The role of film thickness in the realization of X-ray waveguide effects at total reflection. *Adv. X-Ray Chem. Anal. Jpn.*, **26s**, 181–186.
14. Barbee Jr., T.W. and Warburton, W.K. (1984). X-ray evanescent- and standing-wave fluorescence studies using a layered synthetic microstructure. *Mater. Lett.*, **3**, 17–23.
15. Batterman, B.W. (1963). *Phys. Rev.*, **133**, A759.
16. Batterman, B.W. and Cole, H. (1964). Dynamical diffraction of X-rays by perfect crystals. *Rev. Mod. Phys.*, **36**, 681–717.
17. Schwenke, H., Berneike, W., Knoth, J., and Weisbrod, U. (1989). How to use the features of total reflection of X-rays for energy dispersive XRF. *Adv. X-Ray Anal.*, **32**, 105–114.
18. Schwenke, H. and Knoth, J. (1993). Total reflection XRF, In: van Grieken, R. and Markowicz, A. (editors) *Handbook on X-Ray Spectrometry*, Vol. **14**, Practical Spectroscopy Series, Dekker: New York, 453 pp.
19. Klockenkämper, R. and von Bohlen, A. (1989). Determination of the critical thickness and the sensitivity of thin-film analysis by total reflection X-ray fluorescence spectrometry. *Spectrochim. Acta*, **44B**, 461–470.
20. Schwenke, H., Knoth, J., and Weisbrod, U. (1991). Current work on total reflection X-ray fluorescence spectrometry at the GKSS Research Centre. *X-Ray Spectrom.*, **20**, 277–281.
21. de Boer, D.K.G. (1991). X-ray standing waves and the critical sample thickness for total-reflection X-ray fluorescence analysis. *Spectrochim. Acta*, **46B**, 1433–1436.
22. de Boer, D.K.G. and van den Hoogenhof, W.W. (1991). Total reflection X-ray fluorescence of single and multiple thin-layer samples. *Spectrochim. Acta*, **46B**, 1323–1331.
23. Parratt, L.G. (1954). Surface studies of solids by total reflection of X-rays. *Phys. Rev.*, **95**, 359–369.
24. Abelès, F. (1950). Recherches sur la propagation des ondes electromagnetiques sinusoidales dans les milieux stratifies. Application aux couches minces. *Ann. Phys.*, **5** 596–640 (part I) and 706–784 (part II).
25. Król, A., Sher, C.J., and Kao, Y.H. (1988). X-ray fluorescence of layered synthetic materials with interfacial roughness. *Phys. Rev.*, **B38**, 8579–8592.
26. Gutschke, R. (1991). Diploma thesis, University of Hamburg.
27. Iida, A. (1992). *Adv. X-Ray Anal.*, **35**, 795.
28. Sakurai, K. and Iida, A. (1992). *Adv. X-Ray Anal.*, **35**, 813.

29. Kregsamer, P. (1991). Fundamentals of total reflection X-ray fluorescence. *Spectrochim. Acta*, **46B**, 1333–1340.
30. Hüppauf, M. (1993). Charakterisierung von dünnen Schichten und von Gläsern mit Röntgenreflexion und Röntgenfluoreszenzanalyse bei streifendem Einfall. Doctoral thesis, RWTH Aachen, and JÜL-report JÜL-2730, ISSN 0366-0885.
31. Holz, Th. (1992). Diploma thesis, Technical University Dresden.
32. Weisbrod, U., Gutschke, R., Knoth, J., and Schwenke, H. (1991). Total reflection X-ray fluorescence spectrometry for quantitative surface and layer analysis. *Appl. Phys.*, **A53**, 449–456.
33. Weisbrod, U., Gutschke, R., Knoth, J., and Schwenke, H. (1991). X-ray induced fluorescence at grazing incidence for quantitative surface analysis. *Fresenius J. Anal. Chem.*, **341**, 83–86.
34. Schwenke, H., Gutschke, R., and Knoth, J. (1992). Characterization of near surface layer by means of total reflection X-ray fluorescence spectrometry. *Adv. X-Ray Anal.*, **35**, 941–946.
35. Lengeler, B. (1992). X-ray reflection, a new tool for investigating layered structures and interfaces, *Adv. X-Ray Anal.*, **35**, 127.
36. Schwenke, H., Gutschke, R., Knoth, J., and Kock, M. (1992). Treatment of roughness and concentration gradients in total reflection X-ray fluorescence analysis of surfaces. *Appl. Phys.*, **A54**, 460–465.
37. de Boer, D.K.G. and Leenaers, A.J.G. (1995). X-ray scattering from samples with rough interfaces. *Adv. X-Ray Chem. Anal. Jpn.*, **26s**, 119–124.
38. Kawamura, T. and Takenaka, H. (1994). Interface roughness characterization using X-ray standing waves. *J. Appl. Phys.*, **75**, 3806–3809.
39. Veigele, W.J. (1973). Atomic Data Tables 5.
40. Bertin, E.P. (1975). *Principles and Practice of X-Ray Spectrometric Analysis*, Plenum Press: New York.
41. Hubbel, J.H., Veigele, W.J., Briggs, E.A., Brown, R.T., Cromer, D.T., and Howerton, R.J. (1975). Atomic form factors, incoherent scattering functions, and photon scattering cross sections. *J. Phys. Chem. Ref. Data*, **4**, 471–538.

CHAPTER

3

INSTRUMENTATION FOR TXRF AND GI-XRF

Most X-ray fluorescence spectrometers today operate in the wavelength-dispersive mode, but the number of energy-dispersive spectrometers is rapidly increasing. There are an estimated 25 000 wavelength-dispersive instruments worldwide, whereas there are some 5000 energy-dispersive instruments in use.

Total-Reflection X-ray Fluorescence Analysis and Related Methods, Second Edition.
Reinhold Klockenkämper and Alex von Bohlen.

Total reflection devices are more efficient and available in the energy-dispersive version, and there may be about 600 presently in operation.

The first commercially available total reflection X-ray fluorescence (TXRF) instrument was built in 1980 (EXTRA by Rich. Seifert & Co, Ahrensburg, Germany). It was protected by a patent of Marten, Rosomm, and Schwenke [1]. The successive improved model, EXTRA IIa, was distributed by Atomika Instruments (Oberschleißheim, Germany). This company also sold model TXRF 8010, especially suitable for the examination of wafers. Since 1988, two Japanese companies have put instruments on the market: the TREX-series was built by Technos, Osaka (represented in Europe by Philips, Eindhoven, The Netherlands) and the 3700-series was constructed by Rigaku, Osaka. These instruments were compact, self-contained units consisting of a power supply, an X-ray tube, a special filter or even a monochromator, a sample chamber, a detector, and a multichannel analyzer. The optical path of X-rays is determined by apertures and is safeguarded by lead-shielding. The compact instruments were licensed only as highly protective units that switch off the X-ray tube if the protective shielding is removed.

Modular equipment of great versatility has also been constructed consisting of a collimator system, a special filter or monochromator, a sample holder device, and connection flanges for an X-ray tube and a detector. This module is distributed by the IAEA (International Atomic Energy Agency, represented by P. Wobrauschek, Vienna). The generator, X-ray tube, detector, and multichannel analyzer have to be supplied elsewhere and coupled to this module.

In the new century, old manufacturers (Atomica, Seifert, and Technos) disappeared from the market and new companies (Bruker, G.N.R., formerly Italstructures, and Ourstex) developed and produced new instruments suitable for TXRF and related methods. An overview of today's manufacturers and commercially available equipment is given in Section 6.3.1.

Maintenance and operating costs of a TXRF device are relatively low. The generator needs about 3 kW electric power, the X-ray tube has to be cooled by tap water (5 l/min), and the detector must be cooled by liquid nitrogen (8 l/week). If an indoor closed water-cooling cycle is available, water consumption can be greatly limited. Newly developed X-ray tubes are even air-cooled and modern silicon drift detectors are cooled thermo-electrically. Furthermore, no vacuum is necessary and no gases are needed in principle. The consumption of sample carriers is small. Quartz glass carriers, for example, can easily be cleaned and reused. Plexiglas carriers are disposable, each costing only pennies.

In general, all compact and self-contained units are easy to operate. The menu-driven setting of the power supply, as well as automatic changing of samples and recording of the spectra, all guarantee a user-friendly mode of operation. If two X-ray tubes are provided, a simple switchover of the power supply avoids the need for troublesome tube changing. Sophisticated software is available to facilitate calibration, evaluation, and storage of data.

3.1 BASIC INSTRUMENTAL SETUP

In Section 2.2, it was pointed out in detail that standing waves appear when X-rays interfere at external total reflection. The standing waves may arise in front of a thick and flat substrate and/or within a layered structure on top of such a substrate. Sample material placed in the field of standing waves can be excited to X-ray fluorescence. Two cases may be distinguished in general: (1) For a granular residue, the fluorescence intensity will be constant, that is, independent of the glancing angle if this angle is reduced below the critical angle of total reflection. The spectral background arising from Rayleigh and Compton scattering of the primary beam is likewise constant. (2) For thin layers, however, the fluorescence intensity will be angle dependent—the layers may be buried in or deposited on a thick substrate and may be self-reflecting or not.

In order to take such measurements, great demands are made on instrumentation:

- The glancing angle of incidence for the primary beam must be rather small in order to ensure external total reflection. The critical angle is on the order of 0.1° for primary X-rays of some 10 keV as normally applied (see Table 1.9).
- The primary beam should be shaped like a strip of paper, realized by an X-ray tube with a line-focus. Apertures have to restrict the beam to some 10 μm in height and about 10 mm in width because the detector window is usually smaller than 1 cm in diameter.

For the examination of a granular residue, case 1, only one single measurement at a fixed angle setting has to be carried out (an "angle cut" by TXRF). Two conditions have to be met:

- The glancing angle must be set to about 70% of the critical angle of total reflection. Depending on excitation energy and carrier material, it should be fixed to maybe 0.07°. The divergence of the primary beam should be restricted to only 0.01°. For a stable setting of such small angles, the instrumental arrangement has to be very solid and compact.
- Intensive spectral peaks or a broad band of the primary brems-continuum should first be selected for excitation. The high-energy part of the primary spectrum must be eliminated by a filter so that total reflection can occur at a small but not too small angle according to Equation 1.68. This high-energy part of the spectrum would not be totally reflected under the larger glancing angle but would lead to an increased background. A prior low-pass or bandpass filtration prevents this detrimental effect.

For the examination of thin layers, case 2, not only a single measurement at a fixed angle is needed but an angle-dependent intensity profile has to be recorded (an "angle scan" by grazing-incidence X-ray fluorescence or GI-XRF). Instead of the aforementioned conditions, the following requirements have to be met:

- The angle of incidence must be capable of being varied between 0° and 2° in steps of 0.01°. This can be realized by tilting the sample around an axis located on the surface. Great accuracy of the fine-angle control is necessary for quantification and should be better than 0.005° absolutely.
- A strong spectral peak of the primary spectrum should be selected by a suitable monochromator. Only a monochromatic incident beam produces the angle-dependent intensity profiles that are distinctly determined by the layered samples. Additional spectral parts, especially strong peaks, would blur the distinct correlation.

The basic design of TXRF instruments based on these conditions is demonstrated in Figure 3.1. The primary beam is generated by a high-power X-ray tube with a line focus. It may either be a fixed or a rotating anode tube. In order to increase certain peaks in relation to the spectral continuum, thin metal foils are easily placed in front of the X-ray tube. They work as filters, as mentioned in Section 1.4.3. By means of a pair of precisely aligned diaphragms or slits, the beam will be shaped like a strip of paper.

Since the mid-1980s, synchrotron beamlines have been used for excitation [3–5]. This X-ray source certainly provides an ideal primary beam with natural vertical collimation, polarization, and high brilliance. Of course, it needs a large-scale machine, the synchrotron, which is now available worldwide for research work and even for routine analyses (see Sections 1.3.4 and 3.3).

The polychromatic beam of conventional X-ray tubes is deflected by what is called the *first reflector*, and which alters the primary spectrum. For trace analyses of granular residues, a simple quartz-glass block is sufficient in that

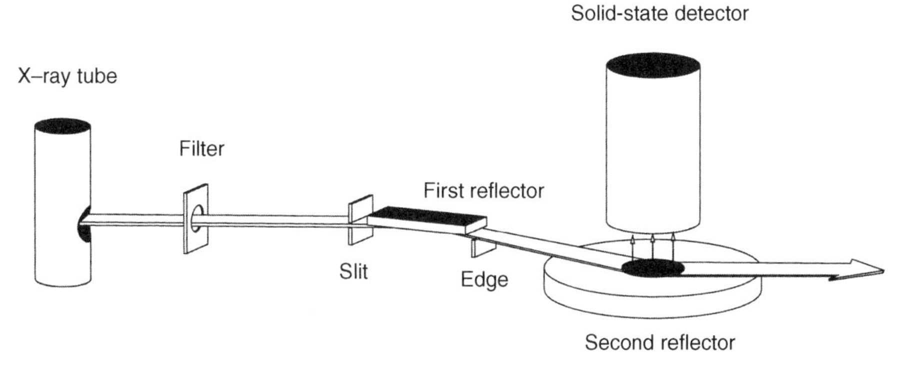

Figure 3.1. Basic design of a TXRF instrument. It differs from Figure 1.1b mainly by the addition of the first reflector. Figure from Ref. [2], reproduced with permission. Copyright © 1996, John Wiley and Sons.

role, acting as a totally reflecting mirror or low-pass filter. It only cuts off the high-energy part of the brems-continuum under grazing incidence. For surface- and thin-layer analyses, the first reflector has to be a real monochromator. Generally, natural crystals or synthetic multilayers are used, acting as Bragg reflectors (as described in Section 2.1.2).

After passing this first reflector, the primary beam hits the sample carrier as the second reflector under grazing incidence. The sample carrier may be loaded with some sample material or may be exchanged by a layered or unlayered substrate representing the actual object of analysis itself. For trace analyses, the carrier may simply be pressed against two parallel cutting-edges or three steel balls so that a fixed angle setting is guaranteed. For surface- or thin-layer analyses, however, a special device is needed for the positioning and tilting of the sample. This component has to control the angle of incidence, which is the governing free variable for angular intensity profiles.

The fluorescence intensity of the sample is generally recorded by an energy-dispersive solid-state detector, formerly a Si(Li) or a high-purity germanium (HPGe) detector and recently a Si-drift detector (SDD). It is mounted perpendicular to the carrier plane to obtain spectra with a minimum scattered background (as discussed in Section 1.4.2). The distance to the sample can be reduced to less than 3 mm in order to secure the detection of the fluorescence radiation within a large solid angle. This intensity is registered by a multichannel analyzer, leading to an energy-dispersive spectrum. Measurements are usually carried out in ambient air. In order to suppress the Ar peak from ambient air, the measuring chamber can be flushed with nitrogen. For the detection of low-energy peaks of lighter elements, a nitrogen or helium flush or even an entire vacuum chamber are recommended (see Section 7.2.1).

3.2 HIGH AND LOW-POWER X-RAY SOURCES

A strip-like primary beam is needed for excitation, showing a high intensity in a certain spectral region. Consequently, a high-power X-ray source must be chosen for TXRF, just as it is used for X-ray diffraction (XRD). But the emitted X-ray beam must still be adapted with respect to geometry and chromaticity.

The applied X-ray sources consist of a generator and a fine-focus X-ray tube, also known as a fine-structure tube. The generator supplies the X-ray tube with high-potential power for the anode and filament power for the cathode. Normally, the very stable generators of conventional XRF or XRD are applied. They deliver a rectified high-voltage between 5 and 60 kV (or even 100 kV) and a current of 5 to 80 mA (or even 100 mA), usually in increments of 1 kV and 1 mA, respectively. The output power can reach a maximum load of 3 or 4 kW. High-quality generators are stable to ±0.01% for both voltage and current provided that the line fluctuations are less than ±10%. The usual electrical safety precautions are generally taken.

The new generation of high-powered units can provide a maximum load of 18 kW (or even 30 kW), a maximum voltage of 60 kV, and a maximum current of 300 mA (or even 500 mA). The stability of these generators is ±0.1%, which is indeed remarkable, but they need to be cooled with water at a flow rate of about 6 l/min.

3.2.1 Fine-Focus X-Ray Tubes

X-ray tubes provide the primary X-ray beam by which the sample is excited to fluorescence. Fine-focus X-ray tubes with a fixed anode are generally used for TXRF [6]. As shown in Figure 3.2, they consist of a spiral filament acting as the cathode, and a water-cooled block of copper as the anode. Both electrodes are sealed off in an evacuated glass–metal cylinder. In contrast to Figure 1.4, the anode is not a beveled block but a right cylinder with a horizontal plane. The filament made of tungsten wire is embedded in a narrow steel grove, 1 mm wide and about 10 mm long. The copper block is plated with the actual anode material, such as chromium or tungsten. The cathode is operated at high negative potential, while the anode is grounded.

When the filament is fed by the heating current at white heat, it emits electrons. They are attracted and accelerated in the direction of the anode at a distance of about 20 mm. The bombarded area of the target is about 0.25 mm × 10 mm, representing a line focus of X-rays. These X-rays are emitted in all directions but only emerge through a thin side window. If the diameter is about 6 mm and the window-to-spot distance is nearly 30 mm, a beam will leave the tube as a cone with an aperture of about 12°. If the total tube is now tilted by 6°,

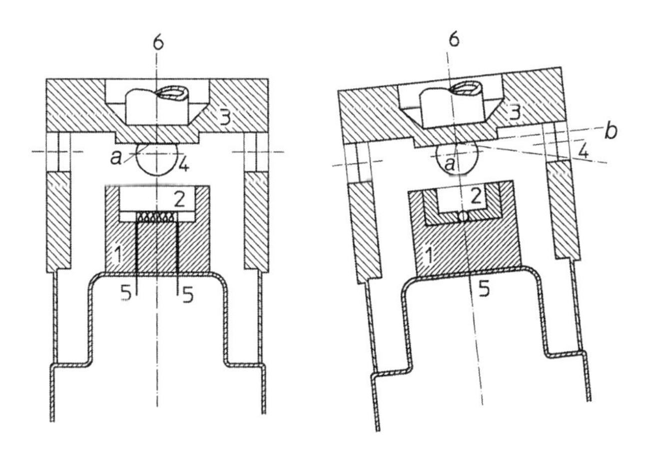

Figure 3.2. Schematic representation of tube SF 60 formerly produced by Rich. Seifert & Co., Ahrensburg, Germany [6]. A fine-focus X-ray tube is represented in two sectional views: (1) cathode block; (2) tungsten filament; (3) anode block; (4) thin window; (5) electrical connections; (6) cooling-water connection; (a) line focus; (b) radiation cone. Figure from Ref. [2], reproduced with permission. Copyright © 1996, John Wiley and Sons.

the beam axis runs horizontally. On the outside, the focus is *observed* under the small angle of 6° and *appears* as a line of 25 μm × 10 mm.

Fine-focus X-ray tubes are available with several different target materials: Au, W, Ag, Mo, Cu, Co, Fe, and Cr (listed with decreasing atomic number). The maximum permissible power is about 2 kW for high-*Z* anodes and about 1 kW for the low-*Z* ones. The exit window is made of beryllium, which is highly transparent for X-rays. Foils of 0.2–1 mm thickness are used, while the high-*Z* anode tubes require the thicker windows. The water consumption is 4 l/min at a pressure of 3–5 bar and a temperature of 20–30 °C. The tubes' life span is some 3000–6000 operational hours.

In order to achieve such a long life, some rules have to be observed. Loading of the tube as well as switching it off should never be performed suddenly but only as a careful step-by-step operation. The warm-up should take between 15 and 45 min, depending on the period of interruption, and the cooling-down phase should last for 10 min. The maximum power should never be exceeded. Voltage and current should be chosen so that their product is always less than a maximum rating. All this should be performed under automatic control. Tube changes, however, have to be carried out manually with utmost care.

Modern X-ray tubes are perfectly shielded and permit operation without radiation hazard. Nevertheless, participation in a monitoring program is recommended.

3.2.2 Rotating Anode Tubes

Conventional X-ray tubes have a fixed anode and are sealed off under vacuum. In contrast to these tubes, a new class of tubes with a rotating anode as demonstrated in Figure 3.3 is also available. The anode is bombarded by

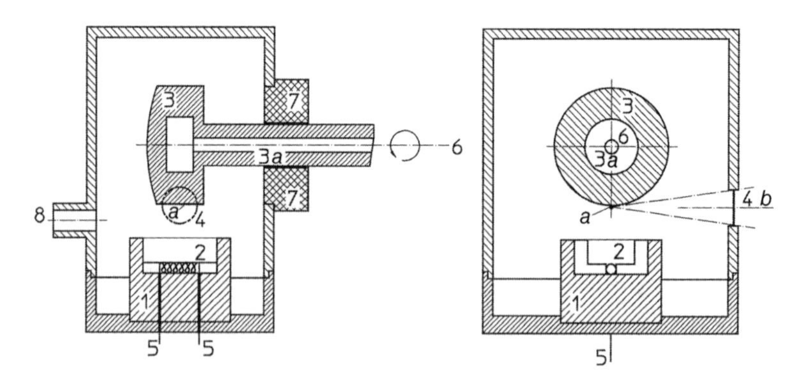

Figure 3.3. Rotating anode tube of Rigaku International Corporation [7] in two sectional views: (1) cathode unit; (2) tungsten filament; (3) cylindrical anode with (3a) rotary shaft; (4) thin window; (5) electrical connections; (6) cooling-water connection; (7) sealing gasket; (8) high-vacuum flange; (a) line focus; (b) radiation cone. Figure from Ref. [2], reproduced with permission. Copyright © 1996, John Wiley and Sons.

electrons on its rotating sidewall surface. The rotary shaft is sealed by a special gasket, and a high vacuum (<10^{-4} Pa) is maintained by a pump system consisting of a turbo-molecular pump with an oil-rotary backing pump. Rotating anode tubes can be operated with a maximum power of 18 up to 30 kW at a voltage of 60 kV, and a current of 300 to 500 mA [7]. Consequently, they provide an X-ray source of high brilliance. The emergent primary beam has a 9-fold or even 15-fold intensity compared to a conventional primary beam of a 2 kW tube.

The high thermal stress of the anode is reduced by rotation at 2500–6000 rpm and by intensive water-cooling with a flow rate of 8–15 l/min at a pressure of 2–3 bar. The size of the line focus typically amounts to $0.2 \times 10\,mm^2$, just as for conventional tubes. The stability is about 0.1% for voltage and current, not very much poorer than for conventional tubes. Of course, this high-performance X-ray source with a high-power generator, a rotating anode device, and a vacuum system is much more expensive than a conventional X-ray source. On the other hand, the new tubes are highly flexible because the anodes can be interchanged. Several anodes are available with target materials also applied for the former tubes. These exchangeable components are of course much cheaper than an entire conventional tube.

Rotating anode tubes are warranted for 2000 h of service-free operation. A safety circuit provides for an automatic shutdown in case of overloads and possible malfunctions. The rules given for running the conventional X-ray units should be observed nonetheless.

3.2.3 Air-Cooled X-Ray Tubes

The conventional X-ray tubes used for TXRF but originally developed for diffraction tasks have a power of 2 or 3 kW DC. These types of X-ray tubes have been known since the middle of the 1950s. They offer two exit windows, one with a line focus representing an excellent illumination for TXRF. As already mentioned earlier, the efficiency of conventional X-ray tubes is poor. Only about 0.1% of the electric power needed for operation is transformed into X-rays and more than 99% is converted into heat. Water cooling consumes about 5 l/min at a minimum pressure of the order of 7 bar. Many different so called "improvements" have been presented in the last decades (dual anodes, composite anodes) but substantial advantages were not observed to justify their use for TXRF analyses.

A more effective development started in the early 1960s. The concept was simple—a miniaturization of the complete X-ray tube offering an anode load up to only 50 W. The idea to reduce all distances, to adapt all dimensions, to avoid water cooling and to make the complete system portable was realized step by step. The stability and the lifetime of these new products were particularly improved. The first X-ray tubes of the new generation were presented at the end of 1980. Today, X-ray tubes using a small battery and a USB connection to a computer are produced giving the required X-ray

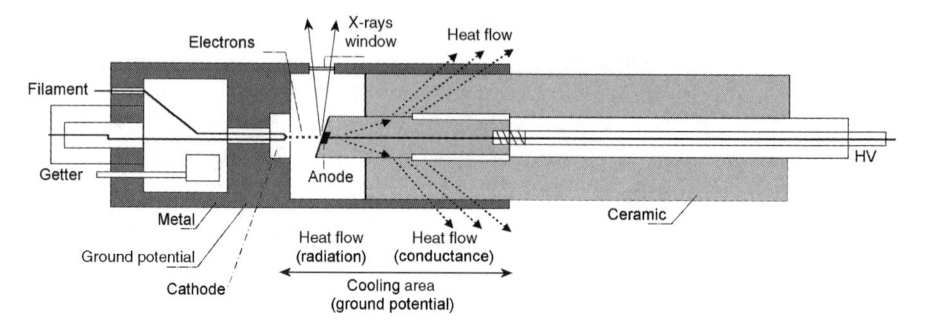

Figure 3.4. Sketch of a typical high-efficiency air-cooled X-ray tube made of metal–ceramic. The hot filament produces electrons that are accelerated toward the anode made of different metals. High voltage (HV) is applied up to 65 kV so that X-rays are emitted and pass a thin beryllium window (see Table 3.1). The heat is abducted by a special metal–ceramic casing via heat radiation and heat conductance. A getter pump provides the necessary vacuum inside the tube.

radiation. For other applications air-cooled X-ray tubes—even with rotating anodes—are available.

The production of small X-ray tubes was strictly connected to the developments in the domains of material sciences, electronics, and computer sciences. First, the newly developed "metal–ceramic" combination with a robust junction of both materials gave successful results. The cooling device was integrated in the ceramic part of the tube as shown in Figure 3.4 reaching a temperature of about 60 °C under operation [8,9]. The lack of intensity of the primary radiation required for X-ray fluorescence analysis was overcome by using modern capillary optics and/or multilayer mirrors, which were placed very close to the exit window of the X-ray tubes. For an adaptation, very precise and stable focal spots are absolutely essential. Today, the combination of a 30 W tube and X-ray optics is comparable to a 3 kW tube without optics.

Some characteristic numbers of modern metal–ceramic X-ray tubes are given in Table 3.1. The differences to conventional "diffraction" X-ray tubes are clear: watts instead of kilowatts, focal spot sizes down to some μm^2 instead of mm^2, and air cooling instead of water cooling. Other types of air-cooled X-ray tubes not specified in Table 3.1 can be produced on demand, for example, with specific geometric requirements, with other anode materials or even with specially shaped anodes and exit windows.

3.3 SYNCHROTRON FACILITIES

About 200 different institutions are producing synchrotron radiation (SR) worldwide. Many countries operate synchrotrons and most of them were built in the last 25 years. Major progress has been made by the development of superconducting magnets, by the addition of insertion devices like undulators,

TABLE 3.1. Typical Data for Metal-Ceramic Air-Cooled X-Ray Tubes with Anode Loads up to 50 W

Target Material	Focal Spot Size	Anode Angle (°)	Emission Angle (°)	Useful Beam Diameter (mm)	Be Window (mm)	Max DC Voltage (kV)	Anode Load (W)	Temperature at Cooling Area (°C)
W, Mo, Cu, Rh, others on request	4.5 mm × 0.5 mm	21	39	10	0.2	+20	50	60
W, Mo, Cu, Rh, others	1.5 mm × 0.1 mm	6	—	10	0.2	+50	40	60
W, Mo, Cu, Rh, others	Ø 0.2 mm	21	20	10	0.2	+65	40	60
W, Mo, Cu, Rh, others	0.2 mm × 0.2 mm	21	40	20	0.5	+65	40	60
W, Mo, Cu, Rh, others	50 μm × 50 μm	12	8	—	0.1	+50	30	60

wigglers, and FELs, and by the improvement of X-ray optics [10–13]. The scientific projects aim at material research with extremely small spatial structures (10^{-12} m or picometer) or with extremely fast time processes (10^{-15} s or femtosecond).

3.3.1 Basic Setup with Bending Magnets

The passageway for the electrons is a metallic pipe with an octagon cross-section. The pipe is shaped like a doughnut, some 100 m to some kilometers long and with a cross-section of about 50 cm^2. It is made of copper, stainless steel, or an aluminum alloy. The latter is nonmagnetic, corrosion resistant, and can be well processed by the extrusion method.

The ring is usually evacuated by several pumps, mostly ion pumps, down to 10^{-10} mbar (UHV). They are normally filled 1–2 times per day. Hundreds of electron bunches are injected at intervals of 1 s. In top-up operation, the lost electrons are replaced continuously and injected 1–3 times every 1 min or every 5 min. Normally, the beamline runs 10 months a year with a shut-down of twice a month. The storage-ring and a part of the beamline are enclosed in concrete walls to protect the staff and users from X-rays. Because of the relatively high magnetic field strength, people with a pacemaker have to stay away from the area around the storage ring. All users have to obey proper safeguards and fulfill respective safety requirements.

Usually, the bending magnets are dipole magnets with a horse-shoe shape and with a field strength of 0.1–2 T.[1] Electromagnetic dipoles, quadrupoles, and sextupoles are installed in the actual synchrotron ring because their field strength can be adjusted synchronously with the increasing energy of the electrons. In the case of superconducting coils, the magnets have to be cooled down below the boiling point of helium (<4.22 K or <269 °C). The costs for the electrical power supply are low but the supply with liquid helium is expensive. For the storage ring, permanent magnets can be used consisting of a sintered material made of neodymium, iron, and boron. Their bending radius is between 3 and 50 m.

The beamlines are tangentially attached to the storage ring and lead the light beam to the experimental stations covering a distance up to several 10 m. These light pipes made of metal, for example, stainless steel or copper, are shielded by lead and steel pipes. At the front end of each beamline, there is a water-cooled beam stopper that absorbs the "bremsstrahlung" and scattered electrons. A beam shutter is added for switching the beam on and off. Radiation-proof hutches with walls of sandwiches (aluminum, lead, aluminum) are built up along each line. Figure 3.5 gives a typical floor plan of an X-ray beamline. The first hutch usually contains specific optics like shutters, windows, filters, slits,

[1] A magnetic necklace or bracelet has a field strength of about 0.1–0.3 T. MRI (magnetic resonance imaging) devices of high-tech medicine have a strength of about 1.5 T. The magnetic field of our earth shows a much smaller strength of only some 0.00005 T.

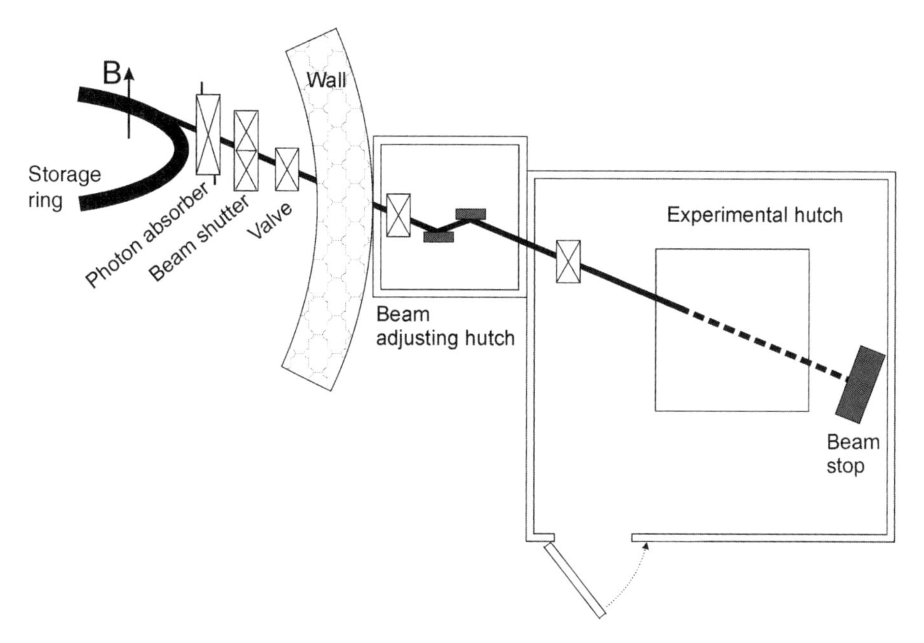

Figure 3.5. Schematic diagram of the set-up between synchrotron storage ring and experimental hutch. The beamline leaves the storage ring through a concrete wall and enters the beam adjusting hutch with, for example, double-crystal monochromator. Finally, it reaches the experimental hutch, which may contain a sample holder and X-ray detector. For hard X-rays, an interlock safety circuit is installed.

mirrors, and monochromators. It is especially used for the selection of a specific energy band used for excitation, particularly by Si (111), Si (311), and InSb (111). A second hutch with individual equipment follows. Such a station can be flexible or fixed, and can incorporate the experimental setup, for example, a sample holder, a goniometer with 6 degrees of freedom for the sample stage, a spectrometer, and a detector. A last block of some 10 cm lead is used as beam-stopper. A particular hutch may be used as a control center, where the intended experiments are controlled remotely by motors or robotic devices.

3.3.2 Undulators, Wigglers, and FELs

As mentioned before, bending dipole magnets were first used in the circular sections of a storage ring to produce synchrotron radiation. In order to generate an even stronger flux and brilliance, sophisticated insertion devices were employed in the straight sections of the storage ring—so-called wigglers and undulators. These periodic arrays are composed of 10–100 dipole magnets with spatially alternating north and south poles as demonstrated in Figure 3.6. Their field vectors are vertical to the straight-line path of the electrons and usually they are also vertical to the horizontal orbit plane. If the magnetic fields are within the orbit plane this leads to certain focusing problems. The magnetic

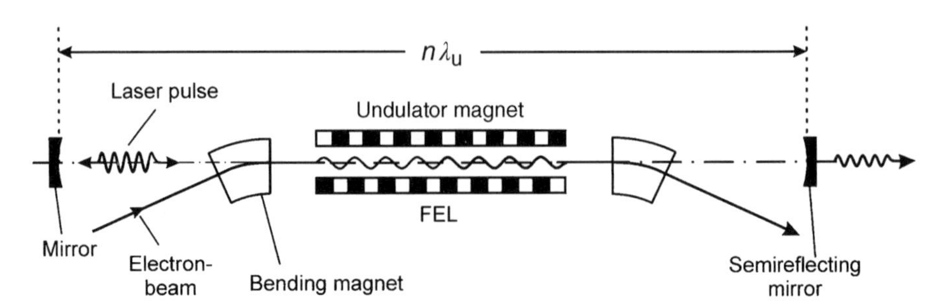

Figure 3.6. Instrumental arrangement of an undulator and free-electron laser. An essential part is the undulator with up to 100 alternating poles at a distance λ_u. It is placed between two outer mirrors in a distance of $n\lambda_u$, which can be up to several tens of meters. The bending magnets before and behind the undulator deflect the electron beam right into it and out of it, respectively, so that the second mirror will not be destroyed. This second mirror is semitransparent or has a small hole so that the synchrotron radiation can get out. Figure from Ref. [15], reproduced with permission from the author.

fields force the electrons on sinusoidal curves around their straight-line path, that is, the electrons weave on a slalom course within the horizontal orbit plane or vertical to this plane [14,15].

The distance of two neighboring like poles is called undulator wavelength, λ_u. Undulators cause up to 100 oscillations of the electrons around the straight line with small λ_u values and small amplitudes (vertical divergence $\leq 1/\gamma$ where γ is the Lorentz factor), while wigglers cause less deflections at larger λ_u values but higher amplitudes of the electron beam (vertical divergence $\geq 1/\gamma$). The spectrum of a wiggler is a broad continuum while the spectrum of an undulator shows narrow energy bands [15,16]. The so-called undulator parameter, $K = \lambda_u/2\pi\rho_m$, differs for multipole wigglers ($K \gg 1$) with a continuum and undulators with several discrete energy peaks, so-called harmonics ($K \approx 1$), or with only the first harmonic ($K \ll 1$). The radiation of an undulator is quasi-monochromatic. In comparison to bending magnets of the second generation, the flux, and moreover the brilliance of the radiation are increased by three to five orders of magnitude for wigglers and by four to eight orders of magnitude for undulators. N oscillations give a gain of N^2 for the brilliance; for example, 100 oscillations give 10 000-fold brilliance.

Undulators and wigglers are usually several meters long. If this length is increased to 10 or even 100 m, the photons of the radiation can interact with the relativistic electrons and can get "light amplification by stimulated emission of radiation." This effect is referred to as free-electron laser (FEL). It is the latest development in the field of relativistic particles [11,14,17,18].

Such a free-electron laser is also shown in Figure 3.6. It contains an undulator as the main component between two outer mirrors. The electron beam replaces the gas or solid medium of a conventional optical laser. Such a laser is consisting of a gain medium, a mechanism to supply energy to it, and a device to provide an optical feedback [19,20]. The highly accelerated electrons in the storage ring act as the amplifying medium when crossing the undulator. Conventionally, a laser consists of a medium with electrons bound in special quantum states of its atoms. A FEL contains only free electrons moving in and interacting with an electromagnetic field. The independence of quantum states offers the possibility to realize FELs in a large range of frequencies. Radiation obtained from FELs has several favorable characteristics, for example, the radiation is extremely brilliant, it is linearly polarized vertically to the orbit plane, it is coherent, and it has a special time structure.

Free-electron lasers will provide the most interesting X-ray source of the near future. Its brilliance is more than 10 orders of magnitude higher than SR of bending magnets. FELs are of particular interest as primary radiation for TXRF and grazing incidence XRF. A detailed description of the physical background can be found in the literature [14,17,18].

3.3.3 Facilities Worldwide

All synchrotron light sources are attached to advanced research facilities on a large scale and may be divided into three categories [21]. The first-generation machines were generally built for research in atomic physics regarding relativistic electrons or positrons. The production of synchrotron light was originally taken as parasitic. The second-generation facilities are connected with beamlines especially dedicated to produce synchrotron light. Bending magnets are usually inserted in the curved sections of the storage ring; undulators and wigglers may additionally be installed into the straight sections. The third generation includes SR sources that mainly use undulators, wigglers, and also FELs. Meanwhile, beamlines of the first and second generation have been reconfigured and developed. Furthermore, the fourth generation of SR facilities has been created, which is characterized by linear insertion devices like FELs and ERLs (energy recovery linacs). The European X-FEL may specially be mentioned. It is a linear electron accelerator with a length of 3.4 km, is highly brilliant (with an average of 10^{25} photons/s/mm^2/mrad^2/0.1% BW) in the range of 0.2–25 keV, and moreover is coherent like a laser beam. The spatial resolution is in the region of pm and time resolution of the pulses is in the fs range. Hereby, 3D images can be obtained of inner atomic structures, and extremely fast processes can be observed.

Table 3.2 gives an overview of 80 present facilities with name, institution, location, country, and website [12,22–25]. They are arranged in alphabetical order. Twenty-two facilities are located in the EU (European Union), 10 of them in Germany, 14 can be found in the USA, and 9 in the former SU. A further Table 3.3 lists commissioning dates (mostly within the last 20 years), circumference of the orbit (between 25 m and 3 km), number of beamlines (typically around 10), maximum electron energy (0.01 up to 18 GeV), and electron current (some 100 mA).

TABLE 3.2. Synchrotron Light Sources (incl. FELs) Around the Whole World in Alphabetical Order[2]

Facility	Institution	Location	Country	Website
ALBA	CELLS	near Barcelona	Spain	www.cells.es
ALS	LBNL	Berkeley	US-CA	www.als.lbl.gov
ANKA	KIT/ISS	Karlsruhe	Germany	http://www.anka.kit.edu/
APS	ANL	Argonne	US-IL	https://www1.aps.anl.gov/
AS		Melbourne	Australia	https://vbl.synchrotron.org.au/BeamLineAccess/eVBL/documents/eVBLAbout.pdf
ASTRID 2	ISA	Aarhus	Denmark	www.isa.au.dk
BESSY II	HZB	Berlin	Germany	www.bessy.de
BSRF	IHEP	Beijing	China	http://www.ihep.cas.cn/
CAMD	LSU	Baton Rouge	US-LA	www.camd.lsu.edu
CANDLE		Yerevan	Armenia	www.candle.am
CESLAB		Brno	Czech Rep.	www.synchrotron.cz
CESR	CHESS	Ithaca	US-NY	www.chess.cornell.edu
CLIO	LCP	Orsay	France	http://clio.lcp.u-psud.fr/clio_eng/clio_eng.htm
CLS	CSRF	Saskatoon	Canada	www.lightsource.ca
CTST	UCSB	Santa Barbara	US-CA	http://sbfel3.ucsb.edu/
DaΦfne	INFN - LNF	Frascati	Italy	www.lnf.infn.it
DELSY	JINR	Dubna/Moskow	Russia	http://wwwinfo.jinr.ru/delsy/
DELTA	ZfSy-TU	Dortmund	Germany	www.delta.uni-dortmund.de
DFELL	Duke Univers.	Durham	US-NC	www.fel.duke.edu
DLSD		Oxfordshire	England	www.diamond.ac.uk

[2] Unfortunately, some data of institutions seem not to be available.

DORIS II	HASYLAB	Hamburg	Germany	http://www.desy.de/
ELETTRA		Trieste	Italy	www.elettra.trieste.it
ELSA	University	Bonn	Germany	http://www-elsa.physik.uni-bonn.de/
ESRF		Grenoble	France	www.esrf.fr
European X-FEL	DESY	Hamburg	Germany	www.xfel.eu
FLASH	DESY	Hamburg	Germany	http://www.desy.de/
FELBE	HZDR	Dresden-Rossendorf	Germany	www.hzdr.de
FELIX	FOM	Nieuwegein	Netherlands	www.rijnh.nl/felix
FOUNDRY	LBNL?	Berkeley	US-CA	www.foundry.lbl.gov
HSRC	University	Hiroshima	Japan	www.hsrc.hiroshima-u.ac.jp
IFEL	University	Osaka	Japan	www.fel.eng.osaka-u.ac.jp
ILSF	IPM	Teheran	Iran	http://ilsf.ipm.ac.ir
INDUS II	Raimanna C.	Indore	India	www.cat.ernet.in
IR-FEL-SUT	Univ. of Science	Tokyo	Japan	www.rs.noda.sut.ac.jp/
ISI 800		Kiev	Ukraine	www.imp.kiev.ua
KEK	High En. Acc. Res. Org.	Tsukuba	Japan	http://legacy.kek.jp
Kharkow Institute	National Science Center	Kharkov	Ukraine	www.kipt.kharkov.ua
KSR - NSRF	University	Kyoto	Japan	www.kuicr.kyoto-u.ac.jp
KSRS	Russ.Res.Center	Kurchatov	Russia	www.kcsr.kiae.ru/en
LCLS	SLAC	Menlo Park/Stanford	US-CA	https://www6.slac.stanford.edu/

(*continued*)

TABLE 3.2. (*Continued*)

Facility	Institution	Location	Country	Website
LNLS		Campinas	Brasil	www.lnls.br
MAX III	MAX-lab	Lund	Sweden	www.maxlab.lu.se
MLS		Berlin	Germany	www.ptb.de
MSRF	NIRS	Inage-ku, Chiba	Japan	www.nirs.go.jp/ENG
Nano-Hana			Japan	
NSLS X-Ray	BNL	Brookhaven	US-NY	www.nsls.bnl.gov
NSRL		Hefei	China	www.ustc.edu.cn
NSRRC		Hsinchu, Taiwan	China	www.srrc.gov.tw
NewSUBARU	Spring-8 JASRI	Nishi-Harima	Japan	http://www-linac.kek.jp/
NUSRC	Nagoya University	Nagoya	Japan	www.nusrc.nagoya-u.ac.jp
PETRA III	DESY	Hamburg	Germany	www.hasylab.desy.de
Photon Factory	KEK	Tsukuba	Japan	www.kek.jp
PLS	PAL	Pohang	South-Korea	http://pal.postech.ac.kr/paleng/
SACLA	Spring-8	Kouto, Hyogo	Japan	http://xfel.riken.jp/information/index_en.html
SAGA-LS	KSLRC	Kyushu	Japan	www.saga-ls.jp
SESAME	Al-Balqa	Salt/Amman	Jordan	www.sesame.org.jo
SIBERIA II	Kurchatov Inst.	Moscow	Russia	www.kiae.ru
SLRI		Nakhon Ratch.	Thailand	http://www.slri.or.th/en/
SLS	Paul-Scherrer	Villigen	Switzerland	http://www.psi.ch/sls/swiss-light-source
SOLARIS	NSRC	Krakow	Poland	www.synchrotron.uj.edu.pl
SOLEIL		Saint Aubin	France	www.synchrotron-soleil.fr

SPEAR 3	SSRL/SLAC	Stanford	US-CA	http://www-ssrl.slac.stanford.edu/
SPring-8	RIKEN	Nishi-Har.,Kansai	Japan	www.spring8.or.jp
SRC		Stoughton/Madison	US-WI	http://www.src.wisc.edu/
SRS	SRS	Daresbury	England	http://www.diamond.ac.uk/Home.html
SSLS	Nation. Univ.	Singapore	Singapore	www.nus.edu.sg
SSRF		Shanghai	China	http://ssrf.sinap.ac.cn/english/
SSRC	Budker-Inst.	Novosibirsk	Russia	http://ssrc.inp.nsk.su/CKP/eng/
SuperB FEL	University	Rome	Italy	www.cabibbolab.it/index.php/en/instruments/the-light-source
SuperSOR	University of Tokyo	Kashiwa, Chiba	Japan	www.issp.u-tokyo.ac.jp/labs/sor/project/MENU.html
SURF	NIST	Gaithersburg	US-MD	www.physics.nist.gov
Tristan	KEK	Tsukuba	Japan	www.kek.jp/ja/Facility/IPNS/TRISTAN/
TNK	F.V. Lukin Inst.	Zelenograd/Moscow	Russia	www.niifp.ru/index_e.html
TSRF	Tohoku University	Tohoku	Japan	www.lns.tohoku.ac.jp/index.php
UCSB	CTST	Santa Barbara	US-CA	http://sbfel3.ucsb.edu/
UVSOR	NINS	Myodaji, Okazaki	Japan	www.uvsor.ims.ac.jp
VEPP-3	INP	Novosibirsk	Russia	www.ssrc.inp.nsk.ru
VEPP-4	INP	Novosibirsk	Russia	www.ssrc.inp.nsk.ru
VU FEL	Vanderbilt Univ.	Nashville	US-TN	www.vanderbilt.edu/fel/

Source: From Ref. [26], reproduced with permission from Center for X-Ray Optics and Advanced Light Source, Lawrence Berkeley National Laboratory (http://www.cxro.lbl.gov).

TABLE 3.3. Synchrotron Facilities with Characteristic Data[3]

Facility Abbreviation	Commissioning Date	Circumference of Orbit (m)	Number of Beam Lines	Energy E_{el} (GeV)	Current I (mA)	Lorentz factor γ
ALBA	2009	269	7 (31)	3.0	250 (400)	5871
ALS	1987; 1993	197	39	1.0 (1.9)		1957
ANKA	1996; 2003	110	10 (22)	2.5	200	4892
APS	1995; 2010	1104	35	7.0		13699
AS	2003; 2007	216	9 (30)	3.0	200	5871
ASTRID 2	2009; 2012	47	>9	0.58		1135
BESSY II	1993; 1998	240	46	1.7	270	3327
BSRF	—	—		2.2 (2.8)		4305
CAMD	—	—	15?	1.5		2935
CANDLE	2007; 2011	216	6 (50)	3.0	350	5871
CESLAB		270?	33	3.0		5871
CESR	1979; 1988; 2008	768	6	5.5 (12)		10763
CLIO	—	—		—		
CLS	2004; 2008	174	12	2.5 (2.9)	250	4892
CTST			3 FELs			
DaΦne	1957; 1969; 1993	105	3	1.5	100	2935
DELSY	—	136		1.2	300	2348
DELTA	1996; 1999	115.2	12	1.5	100	2935
DFELL	2005?	107.5		1.2	115	2348
DLSD	2006	561.6		3.0		5871

[3] Some data are missing. It has often been difficult to receive the data of orbit length, maximum electron energy, or electron current.

DORIS III	1974; 1980	289		4.5 (5.5)		8806
ELETTRA	1993	260	27	1.5 (2.4)	400 ?	2935
ELSA	1988?	164.4	10	3.5	200?	6849
ESRF	1992	844		6.0		11742
Europ. X-FEL	2009; 2015	[3400]	6	17.5		34247
FLASH	2004	[260]	5	1.0		1957
FELBE	2004			0.01–0.03	0.2	20–59
FELIX	—	—		—		
FOUNDRY			6?			
FEL-SUT						
HSRC			16	1?		
IFEL	—	—		—		
ILSF	being planned					
INDUS II	1999; 2005	36	16 (22)	2.5		4892
IR FEL						
ISI 800	1985	163		0.8		1566
KEK		3016		6.0 (12.0)		11742
Kharkow Institute						
KSR - NSRF	1998?	25.7		0.3	100	587
KSRS						

(continued)

TABLE 3.3. (*Continued*)

Facility Abbreviation	Commissioning Date	Circumference of Orbit (m)	Number of Beam Lines	Energy E_{el} (GeV)	Current I (mA)	Lorentz factor γ
LCLS	2009	X-ray laser at SLAC				
LNLS	1997	93.2	15	2.0 (1.37)		3914
MAX III	2008	36		0.7 (1.5)		1370
MLS	2004; 2008	48				
MSRF						
Nano-Hana	1999?	107.5	8?	2		3914
NSLS X-Ray	1982	170		2.5 (2.8)		4892
NSRL	—	—		—		
NSRRC	2008	518.4		3.3		6458
NewSUBARU				1.5		2935
NUSRC	—	—		—		
PETRA III	2008; 2009	2304	14/30	5.0 (6.5)		9785
Photon Factory	—	—	81?	2.5		4892
PLS	1994	280.6		2.5		4892
SACLA						
SAGA-LS	—			1.4		2740
SESAME	BESSI I	60/125		0.8 (2.5)		1566
SIBERIA II	1999	124.1	12;24	2.5	300	4892
SLRI	2004	81.4	9	1.2	150	2348
SLS	2001	288		2.1 (2.8)		4110

SOLARIS	2015	98	>4	1.5	500	2935
SOLEIL	2006	354		3.0		5871
SPEAR 3	1973	234	33	3.0		5871
SPring-8	1997	1436	56	8.0		15656
SRC	1987	121		0.8 (1.0)		1566
SRS				2.0		3914
SSLS	—	—		—		
SSRC	—	—		—		
SSRF	2007	432		3.5		6849
SuperB FEL		1200		3	400	5871
SuperSOR		93.5	19	1.8		3523
SURFIII		828.2	11	0.38	300	744
TNK	—	—		—		
Tristan				25 (30)		48924
TSRF						
UVSOR	—	—		—		
VEPP-3			12	2.2		4305
VEPP-4				5.0 (7.0)		9785
VU FEL						

Source: Data were collected from different websites on the Internet [21–24].

The Lorentz factor in the table was determined for the maximum electron energy:

$$\gamma \approx 1000 \frac{E_{\mathrm{el}}[\mathrm{GeV}]}{0.511} \tag{3.1}$$

Actual γ values lie between 20 and 34 000. The bending radius of a dipole magnet with a given maximum field strength B can be calculated after

$$\rho_m[\mathrm{m}] \approx 3.34 \frac{E_{\mathrm{el}}[\mathrm{GeV}]}{B[\mathrm{T}]} \tag{3.2}$$

Typical ρ_m values are between 3 and 10 m for B values of 0.5 T up to 1.5 T. The critical energy of photons emitted from the electrons passing this bending magnet can be determined according to the formula

$$E_{\mathrm{crit}}[\mathrm{keV}] \approx 2.96 \times 10^{-10} \frac{\gamma^3}{\rho_m[\mathrm{m}]} \tag{3.3}$$

These values are between a few eV and 30 keV. Some of the facilities are working in the spectral region of IR and UV, but most of them produce X-radiation. The largest and brightest beamlines are the APS in the USA, the ESRF in France, and Spring-8 in Japan. The APS at the Argonne National Laboratory with about 500 employees and 5000 scientists is located near Chicago, Illinois. It is called "the brightest X-ray source in the Western Hemisphere." The ESRF in Grenoble, France, is called "the most brilliant storage ring worldwide," distinguished by a brilliance of 10^{21} photons/s/mm^2/mrad^2/0.1% BW. Spring-8 in Hyogo, Japan, is called "the world's largest third-generation SR facility." Further sources have additionally been proposed around the world.

There is a distinct improvement in the spectral brilliance of synchrotron light in the last 25 years [21]. The term brilliance is used here as brightness per source area; brightness is said for flux per square angle of observation; and flux is defined as number of photons emitted per time and 0.1% BW. Figure 3.7 shows this characteristic quantity determined in-plane and depending on the photon energy [27]. The continuum and also characteristic lines of X-ray tubes yield a much smaller spectral brilliance of 10^7 up to 10^{10} photons/s/mm^2/mrad^2/0.1% BW. The continuum of our sun reaches a maximum of 2×10^{10} photons/s/mm^2/mrad^2/0.1% BW in the visible region. Beamlines with bending magnets have a brilliance that is five orders of magnitude higher than the continuum of usual X-ray tubes. Undulators and wigglers even give an improvement of 10–12 orders of magnitude while a FEL ultimately shows a brilliance of 10^{25} photons/s/mm^2/mrad^2/0.1% BW, which is one trillion times more brilliant (18 orders of magnitude or 10^{18} times) than conventional X-ray tubes.

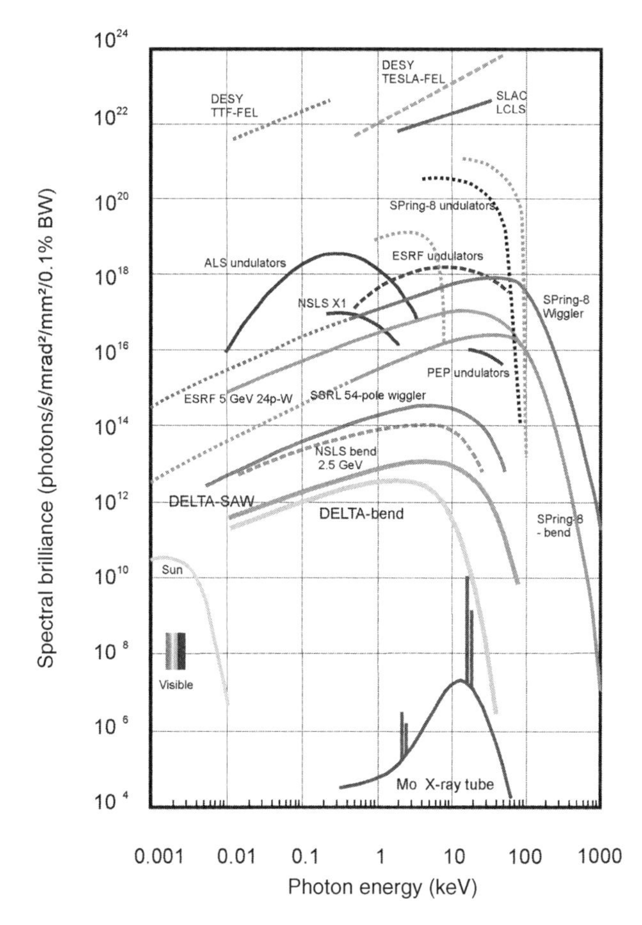

Figure 3.7. Spectral brilliance for several SR beamlines with bending magnets, with wigglers, with undulators, and with FELs depending on the energy of emitted photons. For the purpose of comparison, the brilliance of conventional X-ray tubes and of our sun is represented. The range of X-ray tubes is estimated to cover two orders of magnitude: for tubes with stationary anodes and with rotating anodes. In addition to the continuum, some characteristic X-ray lines are represented. Figure from Ref. [27], reproduced with permission. Copyright © 2008 by the Royal Society of Chemistry.

Nearly all synchrotron institutions are national facilities that are open to everyone. They provide access to beamlines supported by dedicated staff scientists, engineers, and technicians. The beamlines operate day and night during particular periods, that is, 24 h a day. Several beamlines are dedicated to EXAFS and XRD and some beamlines are permanently installed for TXRF, GI-XRF, XRR, and Micro-XRF. For most beamlines, interested scientists can file an application, which can be done online. Obtaining beam time and becoming a user, the applicant has to make a research proposal and to choose an experimental station that is suitable for realizing his/her plan. Such proposals are normally invited twice a year. The use of a beamline is free of charge if the results are published afterward.

3.4 THE BEAM ADAPTING UNIT

The primary X-ray beam must still be adapted with respect to its geometrical shape and spectral distribution. Shaping the beam is simply done by two collimator slits or at least by two metallic edges acting as diaphragms. Silver, steel, and platinum have been chosen with a thickness of about 1 mm and a width of about 20 mm. They are either fixed in entirely self-contained devices or have to be adjusted in modular components. In any case, their task is to mask out a strip-like beam of some 10 μm thickness.

Alteration of the spectral distribution is not as easy to do as is shaping. A metal foil is usually employed as a supplementary means of accomplishing this, but the real goal is achieved only by a first reflector. For *trace analysis*, a quartz-glass mirror is preferably used acting as a low-pass filter [28]. However, a multilayer can also be used, acting as a monochromator or rather as a broad bandpass filter. Even two components or combinations of them may be utilized. The combination quartz–multilayer and multilayer–multilayer was shown to be highly suitable [29], whereas a crystal monochromator seemed to be unsuitable [30]. The single or double multilayer arrangement allows selection of a specific excitation energy without the need to change the X-ray tube. This energy tuning can be performed by a vertical shifting and/or a horizontal displacing of the multilayers.

For *surface and thin-layer analyses*, a natural crystal acting as a true monochromator or small bandpass filter may be applied as a first reflector. However, multilayers have also been used successfully. They may possibly be chosen as a combination of two.

3.4.1 Low-Pass Filters

In order to achieve total reflection at the sample carrier, the primary beam must strike this carrier as a second reflector at a small angle of incidence. For X-ray photons of 60 keV, the glancing angle should be set at only 0.015°, while the divergence of the beam must be held to an even smaller angle. Such a narrow primary beam would provide only poor intensity. Besides, the high energy photons would scarcely be effective since excitation is preferably induced by lower energy photons, for example, by the characteristic radiation of the anode material. For this reason, the glancing angle and the divergence of the primary beam are increased threefold. In that case, the high-energy photons would not be totally reflected but would partly be scattered, causing a significant background. To avoid this effect, the high-energy photons must be eliminated first. This can be done by a first reflector acting as low-pass or cutoff filter. It only transmits the low-energy part of the primary spectrum but eliminates the high-energy part [28,31].

Such a filter can be realized by a simple quartz-glass block. It represents an ideal low-pass filter if applied as a totally reflecting mirror. At a given angle of incidence, low-energy photons are totally reflected whereas high-energy photons are absorbed or scattered. This angle is called the cutoff angle. If a cutoff is wanted for a given energy E_{cut}, a corresponding cutoff angle α_{cut} has to be set. In accordance with Equation 1.68, it is determined by

$$\alpha_{cut} = \frac{1.65}{E_{cut}} \sqrt{\frac{Z}{A}\rho} \tag{3.4}$$

where E_{cut} is given in keV and ρ in g/cm^3 in order to get α_{cut} in degrees. For quartz glass, this relation can be specified by

$$\alpha_{cut} = \frac{1.73}{E_{cut}[\text{keV}]} \tag{3.5}$$

Obviously, other materials can also be used as totally reflecting mirrors; however, quartz glass is highly effective and easy to machine. The efficiency of quartz glass mirrors is demonstrated by Figure 3.8. The reflectivity depending on the photon energy is shown for three different cutoffs. The step-like decrease from nearly 100% to about 0% with only weak tailings is evident. The

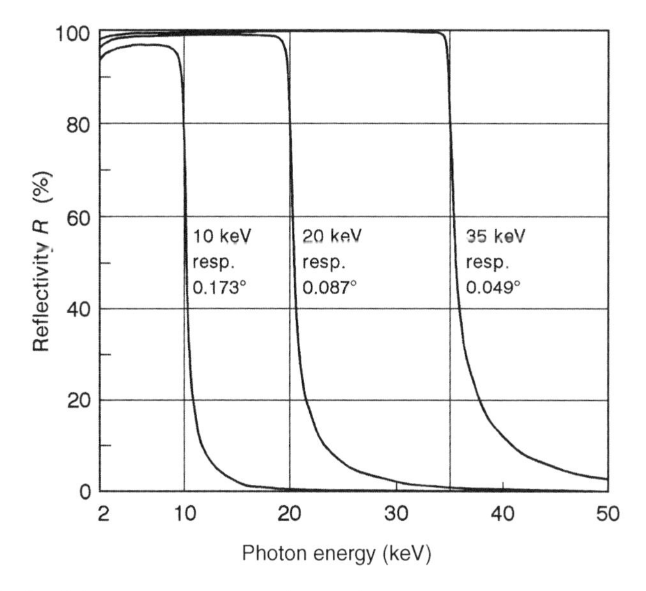

Figure 3.8. Reflectivity R of a quartz-glass mirror acting as a low-pass filter. Three different settings of the "cutoff" energy or the "cutoff" angle are represented. Figure from Ref. [2], reproduced with permission. Copyright © 1996, John Wiley and Sons.

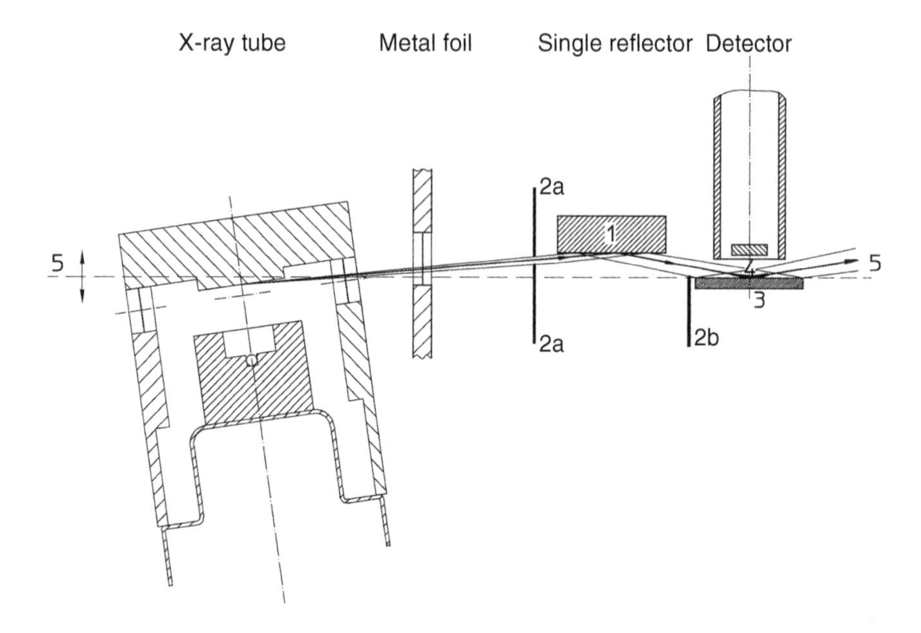

Figure 3.9. Schematic representation of EXTRA II formerly produced by Rich. Seifert & Co., Germany. A single reflector is shown acting as a totally reflecting mirror and a low-pass filter as well. Only the low-energy part of the primary tube spectrum passes through this filter and is used for excitation in TXRF: (1) single reflector; (2a) front slit; (2b) back diaphragm; (3) sample carrier; (4) sample; (5) reference plane. Figure from Ref. [2], reproduced with permission. Copyright © 1996, John Wiley and Sons.

low-energy part is preserved, whereas the high-energy part is annihilated to a large extent. Deviations from the ideal cutoff are caused by absorption.

In practice, an assembly of filters and diaphragms can be arrayed in a different manner. Figure 3.9 shows a simple arrangement with a single reflector or mirror, one front slit, and one back cutting edge. In contrast to this design, the devices of Figures 3.10 and 3.11 employ a double reflector, one front slit and one back edge or a second slit. They mainly differ from one another by the reflector length allowing either a twofold reflection (Figure 3.10) or a threefold reflection (Figure 3.11). The reflection blocks are made of quartz glass (mostly Suprasil®) and are about 10 mm thick, 15 mm wide, and 30–50 mm long in the version of Figure 3.10. They are mounted in a three-point bearing with three adjustable steel balls and are removable from their pedestal. The double reflector of Figure 3.11 is 100 mm long and is joined together with two spacers between the reflector blocks [32,33]. It can only be adjusted as a whole. Instead of a plane reflector, a slightly curved mirror can be applied for the simple arrangement of Figure 3.10. With a 50 m radius of curvature, the signal intensity is thus increased fivefold due to a larger divergence of the primary beam [34].

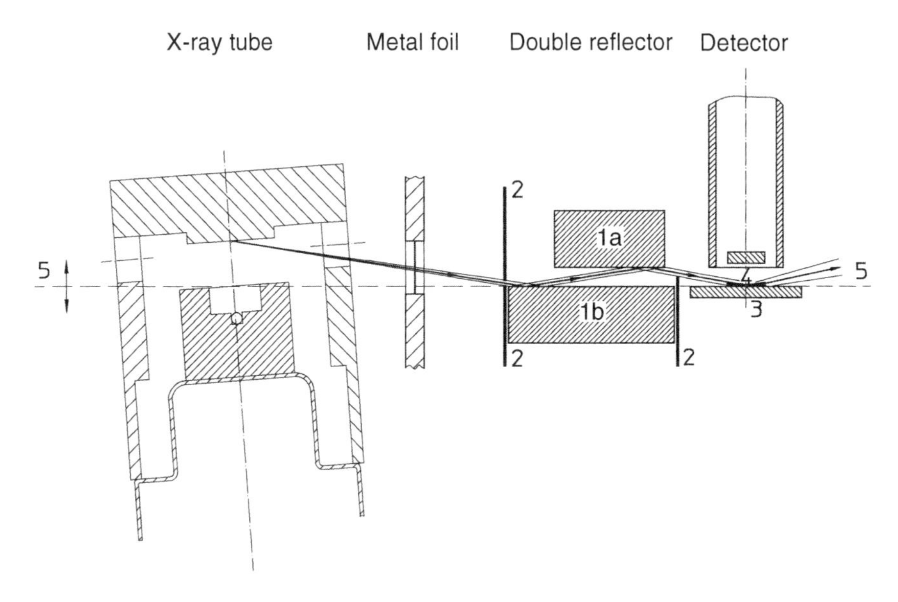

Figure 3.10. Sketch of an early EXTRA II of Rich. Seifert & Co. A double reflector acting as low-pass filter is used for TXRF. The beam-limiting device enables a twofold reflection at the upper and the lower blocks of the reflector before the primary beam reaches the sample: (1a) upper reflector block; (1b) lower reflector block; (2a) front slit and back diaphragm; (3) sample carrier; (4) sample; (5) reference plane. Figure from Ref. [2], reproduced with permission. Copyright © 1996, John Wiley and Sons.

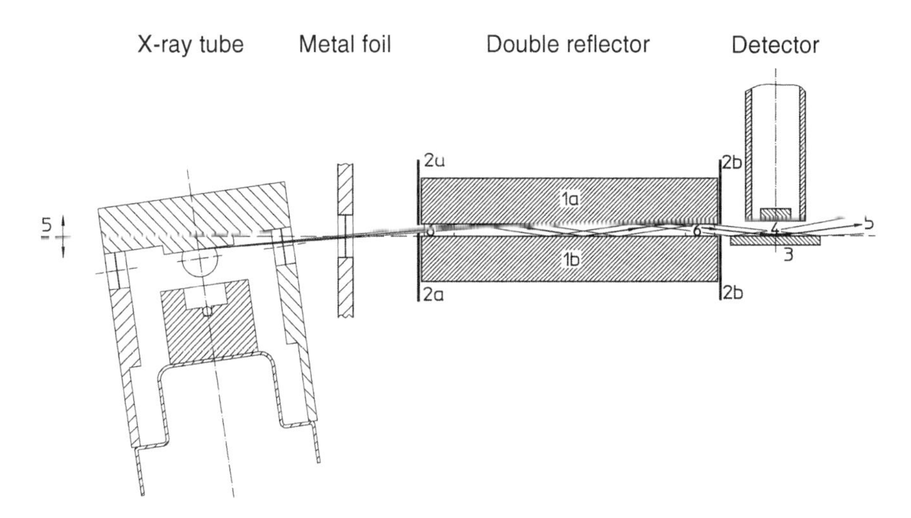

Figure 3.11. Double reflector acting as a low-pass filter used for TXRF. The beam-adapting device allows a threefold reflection of the primary beam before excitation of the sample: (1a) upper reflector block; (1b) lower reflector block; (2a) front slit; (2b) back slit; (3) sample carrier; (4) sample; (5) reference plane; (6) spacer. Figure from Ref. [31], reproduced with permission from Elsevier.

In all three cases, the primary spectrum can only be altered by a simple metal foil. Different metals with a thickness from 10 to 200 μm can each be mounted on a frame and positioned in front of the X-ray tube. They can easily be exchanged without any special adjustment. As noted in Section 1.4.3, certain spectral regions are attenuated thereby as a supplement to the low-pass effect of the first reflector.

All three arrangements may be prealigned as regards the reflector and collimator system, whereas the X-ray tube should be adjusted last. For this purpose, the tube is first shifted vertically and then tilted horizontally until a maximum intensity of the fluorescence signal is reached. This procedure is only necessary after installation of the total equipment or exchange of individual components.

An evaluation of the three systems has to consider the divergence, the intensity, and the spectral distribution of the adapted primary beam. Obviously a smaller divergence is realized with the double reflector, especially at threefold reflection. Consequently, the spectral cutoff is sharper, on the one hand, but the total intensity is reduced, on the other. Priority should be given to a higher intensity obtained with the single reflector. Besides, this simple version is easier to construct and to adjust as well.

The simple arrangement of Figure 3.9 will now be described in further detail. We define the horizontal distance of the X-ray tube to the first reflector as a_1, and the distance of the first to the second reflector or the sample carrier as a_2, in both cases with reference to the respective centers. The carrier will be hit at the defined angle α_2 if the first reflector is lifted above the reference plane by

$$h_1 = a_2 \tan \alpha_2 \tag{3.6}$$

Furthermore, the first reflector must be tilted by an angle $\alpha_1 - \alpha_2$ with respect to the reference plane so that the X-ray beam will be reflected under the chosen angle α_1. The line focus of the X-ray tube has to be lowered by

$$h_0 = a_1 \tan(2\alpha_1 - \alpha_2) - h_1 \tag{3.7}$$

In addition, the X-ray tube has to be tilted by $2\alpha_1 - \alpha_2 + 6°$.

The carrier will be irradiated with a width w_2 if the X-ray beam is limited to a thickness or height t at the sample position. This thickness is

$$t = w_2 \sin \alpha_2 \tag{3.8}$$

The width w_2 should be chosen smaller than the diameter of the sample carriers but wider than that of the detector crystal.

The thickness t of the X-ray beam determines the aperture or divergence of this beam. It can be calculated by

$$\Delta\alpha = \frac{w_2 \sin \alpha_2}{a_1/\cos(2\alpha_1 - \alpha_2) + a_2 \cos \alpha_2} \tag{3.9}$$

where a_1 and a_2 are horizontal distances. For small angles α_1 and α_2, it can be approximated by

$$\Delta\alpha \approx \frac{w_2}{a_1 + a_2}\alpha_2 \tag{3.10}$$

This quantity is independent of α_1 and only dependent on α_2 and the total distance in the denominator. Finally, the irradiated area of the first reflector can be determined by the width

$$w_1 = w_2\frac{\sin\alpha_2}{\sin\alpha_1} \tag{3.11}$$

For an actual example, a Mo-tube may be chosen as the X-ray tube and quartz glass as reflector material; α_2 shall be set to 0.07° in order to ensure total reflection at the sample carrier (recommended in Section 4.5.1), α_1 may be set to 0.09° in order to cut off the spectrum above 20 keV by means of the first reflector. The signal intensity is optimized by close distances between the X-ray tube, the first reflector, and the sample carrier. If the distances a_1 and a_2 are kept to a minimum of 100 mm and 40 mm, respectively, the first reflector has to be raised by $h_1 = 49$ μm and the X-ray focus lowered by about $h_0 = 143$ μm. The X-ray beam has to be limited to $t = 24$ μm in height, determining an aperture of $\Delta\alpha = 0.01°$. In this case, the first reflector is irradiated within a width of $w_1 = 16$ mm while the second reflector, that is, the sample carrier, is irradiated within a width $w_2 = 20$ mm.

3.4.2 Simple Monochromators

For surface- and thin-layer analyses, intensity profiles have to be recorded that are uniquely dependent on the glancing angle and not on the photon energy. Consequently, considerable demands on the spectral purity have to be made. The primary beam needs to be fairly monochromatic, which can be realized neither by a low-pass filter nor a foil filter but only by a monochromator. However, even trace analyses may profit from monochromatic excitation, as already mentioned earlier. For such cases, the first reflector of Figure 3.9 must be a natural crystal or a synthetic multilayer. Both types of crystals are used as Bragg reflectors with a definite energy band selected at a particular angle of reflection, as discussed in Section 2.1.2.

This angle should be set in accord with Equation 2.8. For a chosen photon energy E in keV, the angle can be calculated by

$$\alpha = \arcsin\left(\frac{0.620}{Ed}\right) \tag{3.12}$$

where d is the interplanar spacing (in nm) of the reflector in use. A lithium-fluoride, LiF(200), or a graphite crystal, C(002), can be used; the latter should preferably be a highly oriented pyrolytic graphite (HOPG). However, a set of multilayers is also available from different manufacturers (e.g., Ovonic Synthetic Materials Co. Inc., Troy MI, USA; Rigaku Innovative Technologies, Auburn Hills, MI, USA; Applied X-ray optics AXO, Dresden, Germany). Multilayers can be produced as flat or curved optics for laboratories or synchrotron beamlines. Such multilayers consist of about 150 up to 500 individual bilayers, that is, a combination of a heavy and a light element. Standard combinations are W/Si and Mo/B_4C, but Mo/Si, W/C, and several bilayers of a heavy element and B_4C as second layer are also commercially available. They are used to select a strong K or L peak of the anode material of an X-ray tube, for example, the Mo-Kα or W-Lβ peak, or photons of a particular energy of a beamline. Corresponding d values and glancing angles for a few crystals and multilayers are listed in Table 3.4.

The Bragg angle for common multilayers is about 1° and for natural crystals it is on the order of 10° due to the smaller spacing. As a consequence, the X-ray tube of Figure 3.9 has to be lowered by about 3.3 mm ($\alpha_1 = 1°$) or even 3.6 cm ($\alpha_1 = 10°$) in accord with Equation 3.7. The irradiated area of the first reflector thus shrinks to about 2.0 mm ($\alpha_1 = 1°$) or to 0.2 mm ($\alpha_1 = 10°$), in accord with Equation 3.11. Apart from the lowering of the tube and the larger tilt of both the tube and the first reflector, the geometric arrangement of Figure 3.9 can be maintained in principle. The variation of the glancing angle α_2 will be treated in Section 3.5.3.

Of course, the various crystals and multilayers differ in efficiency when acting as a monochromator [35]. The decisive characteristics are the peak reflectivity and the spectral bandwidth. The product of peak reflectivity and decisive divergence of the beam is an approximate value for the integrated reflectivity. Raw data for peak reflectivity, bandwidth of a few crystals and multilayers are also listed in Table 3.4. The peak reflectivity is some 10% for natural crystals, about 90% for multilayers, and nearly 100% for a totally reflecting mirror or low-pass filter. The bandwidth shows a similar trend. It can be estimated from

$$\delta E = E\frac{\delta\alpha}{\tan\alpha_1} \tag{3.13}$$

where δE is the bandwidth, E is the selected photon energy, α_1 is the adjusted angle of the first reflector, and $\delta\alpha$ is the decisive divergence of the X-ray beam. This divergence is determined by the aperture of the beam (between 0.01° and 0.05°) but also by a certain misorientation of the mosaic structure of natural crystals and by a certain variance of the spacing of multilayers.

The bandwidth of natural crystals has been estimated as ranging from about 5 to 200 eV and that of multilayers from about 200 to 1000 eV. The integrated reflectivity for crystals is quite low (about 10^{-4}) while it is higher for new multilayers (10^{-3} up to 10^{-2}). As demonstrated later in Figure 3.20 the Mo/B_4C multilayer can also be used for a WDS (wavelength-dispersive spectrometer) with a Soller collimator of $\delta\alpha = 0.15°$ (instead of 0.01° earlier).

TABLE 3.4. Natural Crystals and Synthetic Multilayers Frequently Used to Select the Mo-Kα or W-Lβ Line for Excitation in TXRF

Crystal	LiF(200)	C(002)	W/Si	Mo/B_4C
Spacing *d* (nm)	0.2014	0.3354	2.5	8.0
Bragg Angle α:				
For Mo-Kα	10.15°	6.08°	0.814°	0.254°
For W-Lβ	18.55°	11.01°	1.469°	0.459°
Peak:				
Reflectivity	40–50%	~20%	~90%	~80%
Bandwidth	10–25 eV	40–200 eV	400–1000 eV	250–800 eV

Source: From Ref. [2], reproduced with permission.

In that case, the bandwidth is as small as 0.3–5 eV for X-rays in the low-energy range of $0.050\,\text{keV} < E < 0.800\,\text{keV}$.

Consequently, multilayers can separate the Kα and Kβ or the Lα and Lβ peaks of a primary beam but not the doublets Kα_1 and Kα_2 or Lα_1 and Lα_2. These doublets, however, can be separated and selected by most natural crystals except HOPG. In consequence, natural crystals will be used if a high spectral selectivity is needed in preference to intensity. The lower integral intensity might be compensated by a powerful X-ray source, for example, a rotating anode tube. Multilayers provide a superior intensity at the expense of selectivity. Nevertheless, they can be applied for surface- and thin-layer analyses if their selectivity is just sufficient for a particular Kα or Lα excitation.

Moreover, multilayers can also be used as simple low-pass filters [33]. For that purpose, the glancing angle has to be reduced below the critical angle of total reflection, that is, from about 1° down to some 0.1°. The upmost layers will then act as a totally reflecting mirror with a low-pass effect. Multilayers thus have the advantage of being usable as both a monochromator and a low-pass filter.

3.4.3 Double-Crystal Monochromators

For the great demands of GI-XRF in surface- and thin-layer analyses, a highly monochromatic and coherent X-ray beam is needed. The relevant conditions cannot be fulfilled by a single monochromator but only by a double-crystal monochromator. A huge vacuum chamber containing the DCM is usually positioned in the experimental hutch of a synchrotron beamline as elaborated by Matsushita [36]. It consists of two adjustment stages with six degrees of freedom (two 6-axis goniometers) for a control of settings with a precision of sub-μm and sub-milli°. Each stage is equipped with a flat Bragg crystal, for example, Si (111) or Si (311), with 2*d* values of 0.6271 or 0.3275 nm, respectively. The planes of both crystals have to be parallel, that is, the two axes of both crystals have to be parallel on a scale of 0.1 mrad (≈0.0057° ≈ 0.34 arc minutes).

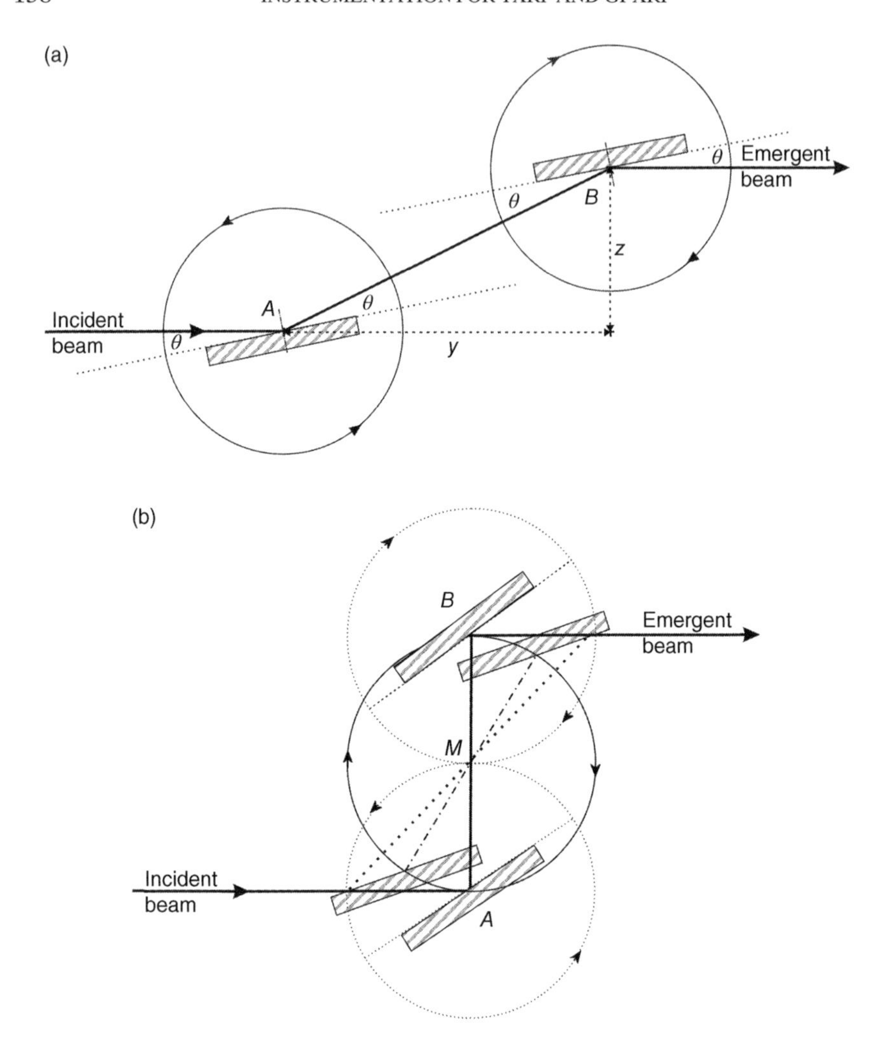

Figure 3.12. A double-crystal monochromator DCM with two parallel Bragg crystals affording a fixed exit beam. In order to select a specific energy band of the X-ray spectrum in accord with Bragg's law, the first crystal A is rotated anticlockwise and the second crystal B is rotated clockwise by an angle Θ. In both variants (a and b), the doubly reflected outgoing beam is shifted by a constant offset z in relation to the incoming beam independent of the glancing angle Θ. In the first arrangement (a) the axis of crystal A is stationary in its position, while the axis of crystal B is additionally translated by a distance y. In the second arrangement (b) both crystals A and B rotate around an axis through the center of their connecting line.

The arrangement of both crystals is demonstrated in Figure 3.12 (top and bottom). The X-ray beam is reflected from the first crystal A and the second crystal B, in both cases by an angle Θ (corresponds with glancing angle α) but in opposite directions. Consequently, the doubly reflected beam is parallel to the entering beam and the angular dispersion is zero. This parallel position of both crystals is called (1,–1) where 1 means the first order of reflection, which is the strongest one and which is usually chosen. The minus sign means the counter reflection of the second crystal. If a certain photon energy has to be selected from the total spectrum, the glancing angle Θ has to be set in accord with Bragg's law of Equation 1.62. This is usually done by simple motions of both crystals with the aim of a fixed parallel exit beam [36].

The simplest way is to rotate the first crystal anticlockwise while shifting the second crystal with a distance y and rotating it simultaneously with the first one but clockwise. The distance y is given by the relationship

$$y = \frac{z}{\tan 2\theta} \tag{3.14}$$

where z is the vertical distance of both crystals (see Figure 3.12a). When Θ is varied between 5° and 45° and z is chosen to be 3 cm the distance y has to be reduced from 17 to 0 cm.

Another simple way to get a fixed emergent beam is to rotate crystals A and B individually as mentioned beforehand by the Bragg angle Θ and to rotate them additionally and jointly around the center of its connecting line AB with constant distance z by an angle χ (see Figure 3.12b). The relation between Θ and χ is not quite easy:

$$\tan\theta = \frac{\cos\chi - 1}{\sin\chi - 1/\tan 2\theta} \tag{3.15}$$

However, it can be approximated by a linear function for Θ and χ in °:

$$\chi = 67.5° - 1.5 \cdot \theta \tag{3.16}$$

It may be mentioned that the adjustment of the arrangements is somewhat difficult; however, the setting of angles is no problem since all motions of rotation and shifting are computer-controlled. Furthermore, it may be underlined that the illuminated spot is fixed in the first case, but shifted by some millimeters along both crystals in the second case.

Thc spectral resolution of a DCM can be determined by rotating only the second crystal and recording the X-ray intensity versus the glancing angle. This so-called rocking curve (which looks like a Lorentzian profile) shows a width δΘ, which is determined by the structure of a nonideal Bragg crystal. It is usually built of small crystallites that are not all ideally parallel but somewhat tilted with respect to each other (scc Section 1.5.2). The spectral resolution or monochromaticity expressed by the figure of merit $E/\Delta E$ is determined by the ratio

$$\frac{E}{\Delta E} = \frac{\tan\theta}{2\delta\theta} \tag{3.17}$$

The width $\delta\Theta$ of the rocking curve is usually inversely proportional to E. For Si crystals and photons of 10 keV it is of the order 35 μrad [36]. Consequently, the spectral resolution is approximately

$$\frac{E}{\Delta E} \approx 1500 \cdot \frac{hc_0}{2d} \tag{3.18}$$

It is dependent on d but independent on E and amounts to the order of 3000–6000. The relevant energy band ΔE is as small as only 1 to 10 eV and can be used for excitation in the range of 5 to 25 keV. For photon energies below 5 keV artificial multilayers instead of natural crystals can be applied (see Table 3.4). The corresponding device is called DMM (double multilayer monochromator) and the energy band can be 0.01 to 0.1 eV. For the selection of even smaller bandwidths, a four-crystal monochromator is installed at the Physikalisch-Technische Bundesanstalt (PTB) in Berlin, Germany.

Because of the high brilliance of synchrotron radiation, the heat load on the crystals is rather high. It amounts to the order of 1 W/mm^2 up to 500 W/mm^2. In order to prevent thermal strain of the crystals and drift of their position the heat has to be dissipated. For that purpose the crystals have to be cooled—at least the first crystal because of its higher heat load produced by the white synchrotron radiation. This can be done either by electrical Peltier-cooling, by water cooling, or even by cryogenic cooling with liquid nitrogen. The crystal may have a metallic fixture with fins for heat dissipation.

3.5 SAMPLE POSITIONING

The sample to be analyzed may be presented in a small amount or volume on a solid and compact carrier. Various materials are in use as sample carriers. They have to be optically flat and even, in order to ensure the total reflection of X-rays. However, they also should meet some further requirements, for example, a high resistance to acids and solvents and a low price. There is no ideal carrier for *all* purposes, but quartz glass and Plexiglas can be recommended for a lot of applications.

The loaded carriers are inserted into a sample changer, and from there they are sequentially brought into the fixed measuring position. When a sample changer is lacking, the samples are placed in this position manually and directly. Analyses are carried out at a fixed angle smaller than the critical angle of total reflection.

The sample to be investigated can also be presented as a flat disk that totally reflects the X-ray beam itself. Wafer material is mostly analyzed in this way. Generally, the wafers are loaded from a sample changer into the start position. Subsequently, they are slightly tilted for stepwise angle variation and recording of the fluorescence intensity.

3.5.1 Sample Carriers

For trace analyses of granular residues, a carrier is required that serves as a sample support and as a totally reflecting mirror. Of course, only solid-state media are suitable. They should be highly reflective because the spectral background is reduced in proportion to $1 - R$, in accord with Equation 2.20 where R is the reflectivity. Besides, the carrier has to be optically flat to ensure its high reflectivity, as predicted by theory. The roughness should be less than 5 nm within an area of about 1 mm^2, corresponding to about $\lambda/100$ where λ is the mean wavelength of the visible light. The waviness should be less than 0.001° within an area of about 1 cm^2, corresponding to a radius of curvature of about 600 m. Commercially available carriers guarantee these characteristics. They can be checked by means of a profilometer and a contour meter.

Furthermore, carriers should be free of impurities so that no blank values appear for elements to be detected. The carrier material itself must not have a fluorescence peak in the spectral region to be considered. In addition, the carriers have to be chemically inert when strong inorganic acids and organic solvents are to be analyzed. The carriers should be easy to clean so that they can always be reused. Finally, they should be commercially available and inexpensive.

Several materials have been applied as sample carriers, especially quartz glass (mostly Suprasil®), as well as polymethyl methacrylate (generally Perspex® or Plexiglas) [37], and glassy carbon [38], and boron nitride [39]. One-element materials like silicon or germanium have also been used [40]. Table 3.5 gives an overview of different carrier materials [41].

Most carriers are on sale as circular discs with a diameter of 30 mm and a thickness of 2 or 3 mm, but rectangular carriers are also available. Their critical angle of total reflection amounts to about 0.1° for the Mo-Kα radiation as

TABLE 3.5. Important Properties of Various Materials Used as Sample Carriers

Carrier Material	Plexiglas	Glassy Carbon	Boron Nitride	Quartz Glass
Critical angle for Mo-Kα	0.08°	0.08°	0.10°	0.10°
Reflectivity at 0.07°	99.8%	99.8%	99.9%	99.4%
Flatness	Good	Fair	Good	Excellent
Purity	Good	Fair	Good	Excellent
Fluorescence	None	None	None	Silicon
Resistance	Insufficient	Good	Excellent	Good
Cleaning	Not Necessary	Difficult	Easy	Easy
Price for one[a]	$0.10	$30	$60	$28

[a] US currency.

Source: From Ref. [2], reproduced with permission. Copyright © 1996, John Wiley and Sons.

shown in Table 3.5. The reflectivity is between 99.4% and 99.9% at a fixed glancing angle of 0.07°. According to Equation 2.20, the spectral background could be *higher* by a factor of 3–6 for the *less* reflecting quartz glass in comparison to the other materials. However, this effect is compensated by a better flatness and purity of commercial quartz-glass carriers. In addition, quartz glass is chemically resistant against most acids besides hydrofluoric acid.

The other materials still benefit from the absence of a Si peak so that silicon and related light elements can be determined. The latter is not possible if quartz-glass carriers are applied. The most resistant material is boron nitride, which is even suitable for the analysis of strong acids and solvents. Unlike these carriers, Plexiglas carriers are only applicable to aqueous solutions or suspensions. Plexiglas, however, is an extremely low-priced material so that it is not necessary to clean and reuse these carriers. Glassy carbon is preferentially used for electrochemical applications because of its electrical conductivity [38]. In general, quartz-glass and Plexiglas carriers are the carriers most used for micro- and trace analyses.

3.5.2 Fixed Angle Adjustment for TXRF ("Angle Cut")

The sample carriers are usually inserted in plastic holders and are either put manually in their final measuring position or first loaded into a sample changer. From here they can be brought into the measuring position automatically one after another. The carriers are slightly pressed—preferably upward—either against two parallel cutting edges or four ball-points with a clearance of about 20 mm. These reference lines or points are fixed and define the plane of reflection. They also determine the angle of incidence for the primary beam. This beam should pass between the cutting edges or ball points and should be reflected under an angle of about 70% of the critical angle of total reflection. For quartz glass and Plexiglas carriers this angle amounts to about 0.07°. The fixed position of the carriers is the *only* position necessary for measurements, that is, for recording the spectra.

A sample changer is recommended for the investigation of a large number or series of samples. Changers with a capacity of up to 35 carriers are commercially available. Sample changing can be carried out under computerized control. The total device is usually incorporated in a plastic chamber for dust protection. This chamber can be flushed with a gas, such as helium or nitrogen, and can be evacuated if necessary.

3.5.3 Stepwise-Angle Variation for GI-XRF ("Angle Scan")

For surface- and thin-layer analyses, the sample has to be present as a flat disk capable of total reflection of the primary beam. Wafers are especially suited for this kind of investigation, for example, Si wafers or GaAs wafers. For that purpose, sample holders are constructed that grip wafers with a diameter between 100 and 200 mm. The wafers may be stacked in a magazine and may be loaded into the sample holder one at a time and then set back after the measurement. This process can be carried out automatically, even by a robot.

The sample positioning device is the core of those instruments that are suitable for surface- and thin-layer analyses. The sample should be capable of being adjusted in the reference plane and of being tilted around a horizontal axis. The angle of incidence should be variable, preferably stepwise. This quantity is the key parameter for recording the angular-dependent intensity profiles and is the basis for surface- and thin-layer analyses. Apart from being adjustable, the sample should be displaceable in order to set each spot of a larger sample in the measuring position and to check the total surface.

Figure 3.13 depicts schematically a convenient device for an angle variation and sample positioning with 6 degrees of freedom. However, an even simpler device with ball-points and stepper motors may be sufficient. The wafer or sample S is placed on a flat carrier C (made of float glass). As an attachment, either a mechanical chuck is used, applying a slight pressure, or an electrostatic chuck is designed to apply a specific voltage. The sample can be adjusted to the tilt center, T, by a vertical shift, z, and this center can be driven into the reference plane (see Figure 3.9) determined by the two horizontal axes, a and b, at right angles. For that purpose, the base B is shifted by z_0. The sample surface can further be adjusted with respect to the reference plane by a tilt correction of Θ and χ. Each spot of the sample can be placed into the measuring position or tilt center T by a lateral movement x and a rotation φ on the vertical axis. Finally, the angle of incidence can be varied by a tilt Θ around the axis a lying on the wafer surface.

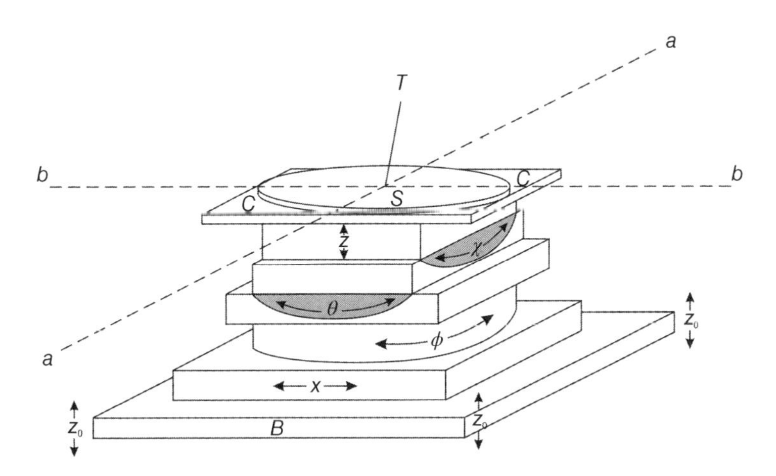

Figure 3.13. Schematic diagram of an angle variation and sample positioning device suitable for thin-layer analysis by GI-XRF: B = solid base; C = flat carrier; S = flat sample or wafer; a and b = horizontal axes spanning the reference plane, T = tilt center; x = lateral movement; Φ = rotation on the vertical axis; Θ = tilt around axis a; χ = tilt correction for axis b; z = vertical shift of the sample, z_0 = vertical shift of the base. Figure from Ref. [2], reproduced with permission. Copyright © 1996, John Wiley and Sons.

All six movements can be driven by stepping motors. The z-shift should have a range of 10 mm with a step-size of 1 μm. The tilts Θ and χ should be variable between 0° and 3° at a step width of 0.001°. Coarse steps are satisfactory for the movement χ and rotation φ, but a large range of adjustment is necessary in this case (>100 mm for χ and >180° for φ). The height z_0 of the base has to be corrected very rarely, and this can be done manually.

The adjustment of a sample for thin-layer analysis is obviously not as easy as fixed angle positioning for micro- or trace analysis. The angle of incidence has to be set with an absolute *accuracy* of <0.005° and has to be varied in a range of about 2–3°. The stepwise variation is needed to record an angle-dependent intensity profile, which is described in detail in Section 4.6.1.

3.6 ENERGY-DISPERSIVE DETECTION OF X-RAYS

The fluorescence radiation of the sample has to be detected and recorded as an X-ray spectrum. This problem can generally be solved either by a wavelength-dispersive or an energy-dispersive spectrometer (WDS or EDS). For TXRF instruments, the EDS had usually been applied so far. While a WDS requires a goniometer, an EDS needs no mechanically moving parts. It simply uses an electronic system based on a special solid-state or semiconductor detector. There are no problems of a mechanical alignment or replacement apart from cooling the detector. Si(Li) and HPGe detectors have to be provided with liquid nitrogen (about 10 l/week) and that is troublesome. However, they need not be cooled permanently; they can be warmed up and returned to the temperature of liquid nitrogen (77 K or −196 °C) repeatedly. Silicon drift detectors, which have made a breakthrough in the last 10 years, can easily be cooled by a thermoelectric Peltier element (down to −20 °C or −40 °C) or can even work at room temperature.

The complete spectra of an EDS are registered simultaneously and not sequentially. Thus, the time-consuming mechanical scan of a WDS is avoided. Furthermore, the semiconductor detector of an EDS can be placed very close to the sample, thereby receiving a wide cone of the fluorescence radiation. The considerably increased intensity is used to substantially reduce the counting time. An EDS can record a total X-ray spectrum on the order of seconds instead of minutes, as is necessary for a WDS.

Because of the simplicity, speed, and convenience of operation, the EDS is adapted to nearly all TXRF instruments. Some essential limitation regarding its spectral efficiency must therefore be accepted. The main disadvantage of an EDS is poorer spectral resolution for photon energies below 15 keV. This causes troublesome peak overlaps (line interferences) especially below 3 keV. Furthermore, the efficiency of an EDS strongly diminishes for photon energies below 2 keV, leading to a weaker detection of lighter elements. Nevertheless, an EDS is a convenient and economical system for complex multielement analyses.

In such an EDS, the X-ray photons emitted from the sample are directly collected by a semiconductor detector. This special detector does not merely count the individual photons but can also determine their different energies. For any collected photon, it produces an individual voltage pulse, the amplitude of which is proportional to the energy of this photon. All the detector pulses are processed in an electronic measuring chain and finally sorted by a multichannel analyzer. The content of this counter storage can be represented as the particular X-ray spectrum and already observed during measurement. Afterward it can be processed directly by a dedicated computer [42,43].

Reflectivity measurements may be taken by an additional detector, which can simply be incorporated in the arrangement just described [44,45]. The primary beam monochromatized by a first reflector is then reflected at the layered or unlayered substrate. The intensity of this reflected beam can be measured by a second simpler detector. It may be a photodiode or a scintillation detector—much simpler and cheaper than a semiconductor detector used for TXRF. However, it must be ensured that this second detector is tilted at the double angle 2α when the layered sample is tilted at the single angle α—both around the same axis *a* of Figure 3.13.

3.6.1 The Semiconductor Detector

The heart of any energy-dispersive spectrometer is a special solid-state detector—or rather a semiconductor detector [46–49]. It basically consists of a pure silicon or germanium crystal. This crystal ought to be several millimeters wide and thick and should be extremely resistive. Germanium can be purified by zone refining to achieve the necessary ohmic resistance. In the last decade, silicon wafers have also been produced with such a high degree of purity that impurities of only several parts per billion (ppb) are left. The most common impurity is boron, which modifies silicon to a *p*-type semiconductor with decreased resistivity and increased conductivity. In order to suppress this effect, another impurity can artificially be added to the crystal. Usually, the boron “acceptors” are compensated or neutralized by lithium “donors.” The lithium diffuses into the crystal at elevated temperature and “drifts” under the influence of an electric field. In this way, a crystal with a high intrinsic resistivity is produced with a thin *p*-type layer and *n*-type layer at the end planes and a large intrinsic region between them. Such a crystal is termed a lithium-drifted silicon crystal, or a Si(Li). There are also lithium-drifted germanium crystals, or Ge(Li)s. However, these are being substituted by HPGe. Today, both are largely replaced by high-purity silicon flats with a sideward voltage drift—by so-called silicon drift detectors (SDDs).

As demonstrated in Figure 3.14 for a Si(Li), the frontal areas of the Si crystal are coated with thin layers of gold serving as electrodes [47]. An inverse DC voltage is applied, called a reverse bias (*p*-type layer negative, *n*-type layer grounded). It defines the direction of low conductivity, that is, of a small leakage current in spite of the high voltage (−500 to −1000 V). The total

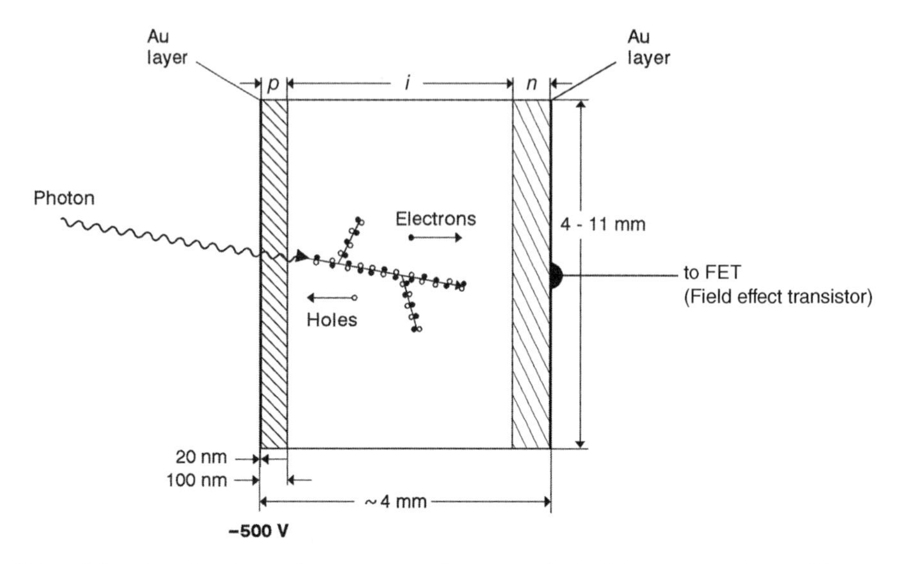

Figure 3.14. A semiconductor detector operated as a *p-i-n* diode with a reverse voltage or bias. An incident X-ray photon ultimately produces a series of electron-hole pairs. They are "swept out" by the bias field of −500 V: electrons in the direction of the *n*-layer; holes in the direction of the *p*-layer. Thus, a small charge pulse is produced. Figure from Ref. [2], reproduced with permission. Copyright © 1996, John Wiley and Sons.

configuration is termed a *p-i-n* diode with a reverse bias. It is cooled to the temperature of liquid nitrogen (−196 °C or 77 K) for two reasons: (i) to reduce the thermal leakage current even further, and (ii) to prevent a reverse diffusion of the lithium ions, that is, to freeze-in the compensation state. The second reason does not apply to HPGe when it is used as the intrinsic material. But the first reason requires that the germanium crystal be cooled as well.

The mode of operation is the same for both detectors—the Si(Li) detector and the HPGe detector. As illustrated in Figure 3.14, an incident X-ray photon interacts with the crystal and ionizes the crystal atoms creating photoelectrons and Auger electrons. These electrons then pass on their energy in several steps and raise outer electrons from the valence band into the conduction band of the crystal lattice. Simultaneously, electron holes are created in the valence band. A total track of electron-hole pairs is produced until the energy of the incident photon is used up. Because of the applied high voltage, the electron-hole pairs separate and the electrons and electron holes rapidly drift to the positive and negative electrodes, respectively. A charge pulse is produced usable for single-photon counting. Because the number of electron-hole pairs is directly proportional to the energy of the particular photon detected, the magnitude of the charge pulse is proportional to the photon energy as well. Consequently, the charge pulse gives a measure of the energy of the detected photon. Hence, the solid-state detector is capable of counting single X-ray photons and of reading their different energies as well. This feature is the prerequisite for an EDS.

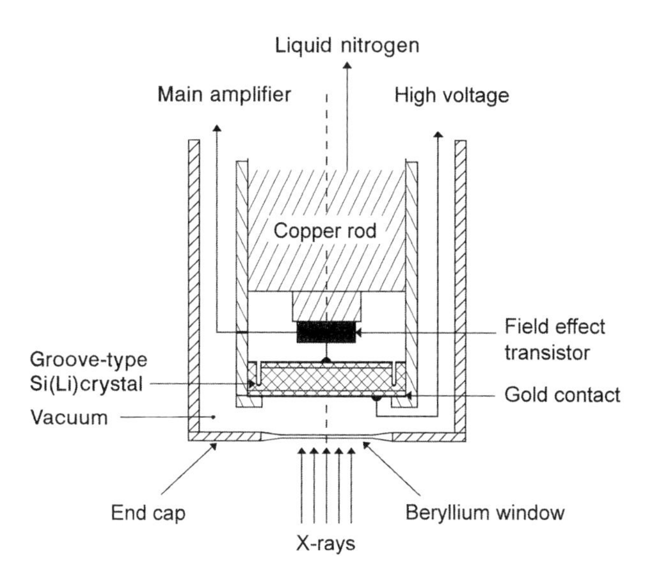

Figure 3.15. Cross-section of the front end of a solid-state or semiconductor detector, here with a grooved Si(Li) crystal. Crystal and preamplifier are connected with a cooled copper rod and shielded by a case with end cap and Be window.

Figure 3.15 shows a sectional view of the front end of a Si(Li) detector. A grooved silicon crystal is coated with gold layers serving as electrodes for the high voltage [49]. It is connected to a copper rod immersed in liquid nitrogen. A field-effect transistor (FET) is used as preamplifier and installed between the detector and the cooled copper rod. The inner device is encapsulated in an outer metallic case locked by a thin beryllium window. This case is evacuated for good thermal insulation of the cooled inner device. The consumption of liquid nitrogen is kept low by this means, and the window does not become covered with moisture. The beryllium window itself can be traversed by X-rays with a fairly low attenuation and it protects the crystal against air, dust, moisture, and light.

3.6.2 The Silicon Drift Detector

This kind of detector was introduced by Gatti and Rehak between 1984 and 1987. Kemmer from the Technical University of Munich was the first to build a commercial instrument. Today, SDDs are fully developed and well established. As already mentioned, the essential part of an SDD is a silicon flat or wafer of high purity and extremely high resistivity of 3 kΩ·cm [50,51]. Figure 3.16 shows such a wafer disc, several millimeters in diameter but only 0.4 mm thick. The front area exposed to the X-ray beam usually has an area of 5 mm^2 up to 100 mm^2. The sensor chip is mostly cut as a hexagon and consists of an n-type silicon covered with a homogeneous shallow p^+-type layer on the front side.

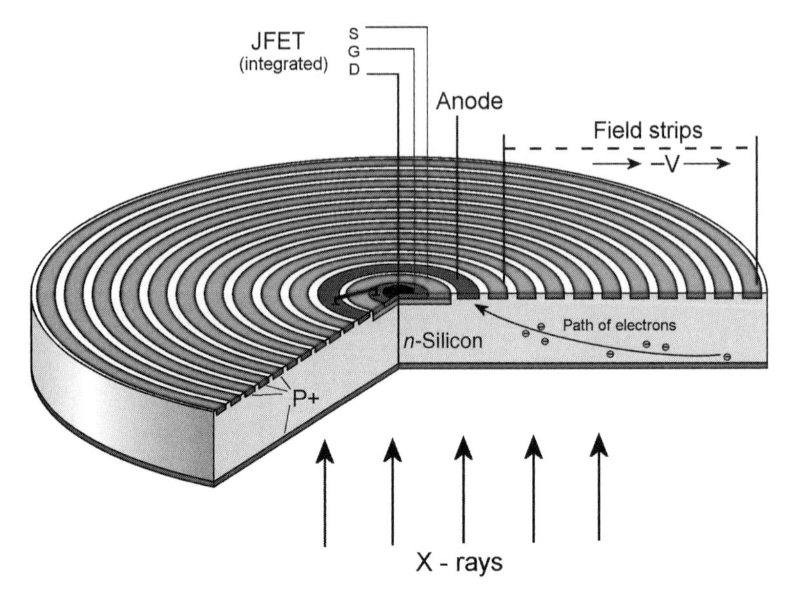

Figure 3.16. Cross-section of a silicon drift detector or SDD. It consists of a small disc of a high-purity Si wafer with a thin P^+-doped cathode at the front area and several concentric drift electrodes on the opposite side (on top). X-rays coming from the front area (from below) produce several electron-hole pairs in the Si crystal. The electrons are drifted down the field gradient kept by the drift rings and finally are collected at the small central anode ring. The charge pulse is converted into a voltage pulse either by a closely attached FET or even by an integrated "on-chip" JFET. From Ref. [53], reproduced with permission from PNDetector GmbH, Munich, Germany.

The rear shows a series of several concentric p^+-type rings with a small anode ring in the center (area <0.5 mm^2). The homogeneous front side may be covered by an ultrathin layer of Al (30 nm) instead of a gold layer, which is normally used for Si(Li) detectors. Consequently, the absorption of the front layer is strongly reduced for soft X-rays. The front layer is switched as cathode with negative voltage while the rings on the rear are provided with an increasingly negative voltage from the inside to the outside [52–55].

Each X-ray photon hitting the crystal produces electrons within the penetration depth of silicon, which is <0.5 mm for $E < 17$ keV. These electrons drift sideward to the central anode via the horizontal voltage field created by the concentric "drift" rings and produce a chain of electron-hole pairs on its sideward path to the anode, which is lengthened up to some 5 mm. The small anode in the center guarantees for an input capacitance that is extremely low in comparison to that of conventional Si(Li)s leading to a small electronic noise and good spectral resolution. A capacitance of only 35 fF instead of 100 pF for gas-filled detectors means an improvement of 3000-fold.

The charge pulses at the anode have to be converted into voltage pulses and linearly amplified. For that purpose, a junction field effect transistor (JFET) is monolithically integrated into the center of the sensor chip by implantation [53].

Source, gate, and drain are developed as concentric rings inside the small anode ring while the anode and gate of the JFET are connected by a micrometer-short metal strip. By this means, parasitic electric capacitances of conventional bonds are avoided. The minimized capacitance between anode and JFET allows shorter shaping times and higher count rates and even a better spectral resolution.

The sophisticated fabrication of the sensor chips leads to a very low leakage current [56,57], which is proportional to the small detector volume so that traditional cooling by liquid nitrogen quite normal for Si(Li)s is not necessary. Consequently, the vibrational noise generated by the boiling bubbles of the liquid nitrogen is completely avoided. A simple Peltier cooler ensures a moderate thermo-electrical cooling down to −20 °C or −40 °C. Consequently, SDDs are maintenance-free and guarantee an absolutely vibration-free operation. The moderate cooling is sufficient for good spectral resolution (<130 eV for Mn-Kα) and a high count rate efficiency (up to 1×10^6 cps).

The sensor chip is usually mounted on a socket with a ring-shaped collimator, electron trap, cooling studs, and electrical pins and bonds. This module is enclosed in a cylindrical case with a thin window in order to protect the sensor against visible light, air, dust, and moisture. A 7.5 μm thin beryllium window is normally used. However, for the detection of light elements with $5 \le Z \le 15$ or X-ray photons with low energies between 0.18 and 2 keV, respectively, an ultrathin window of a 0.5 μm polymer foil is used, which hardly absorbs these X-rays (polyimide or Kapton® of Du Pont). It may be supported by a fine Si grid so that it becomes stable, leakage-free, and pressure resistant. Furthermore, a medium vacuum with about 100 Pa can be applied. For a typical air-path of 0.5 cm the absorption is reduced below 3% for photons with energies ≥0.1 keV (Kα peak of beryllium).

The head of both detectors is shown in Figure 3.17a and b, demonstrating the technical development within the last 45 years. Figure 3.18a shows a spectrum of a type letter recorded by a Si(Li) detector. Figure 3.18b represents a spectrum of a pigment mixture taken by an SDD. Elements like C, O, Cu, As, S, and Ca can be detected by the SDD at photon energies between 0.2 and 4 keV. The peak width is in the range of only 40–110 eV so that the Kα peaks of these elements are clearly separated. Even boron at 0.184 keV has been identified, namely, as boride inclusions of stainless steel [58].

3.6.3 Position Sensitive Detectors

A direct imaging of the information provided by X-rays is of general interest and has been developed in the last decades. The most prominent 2D (two-dimensional) X-ray detectors are applied for imaging in medicine and for monitoring of X-ray diffraction patterns [59]. The simplest way to obtain information about spatially resolved X-ray data is to use a Si photodiode and to scan the area of interest. More comfortable arrangements are CCDs (charge-coupled devices) normally used as image sensors for visible light.

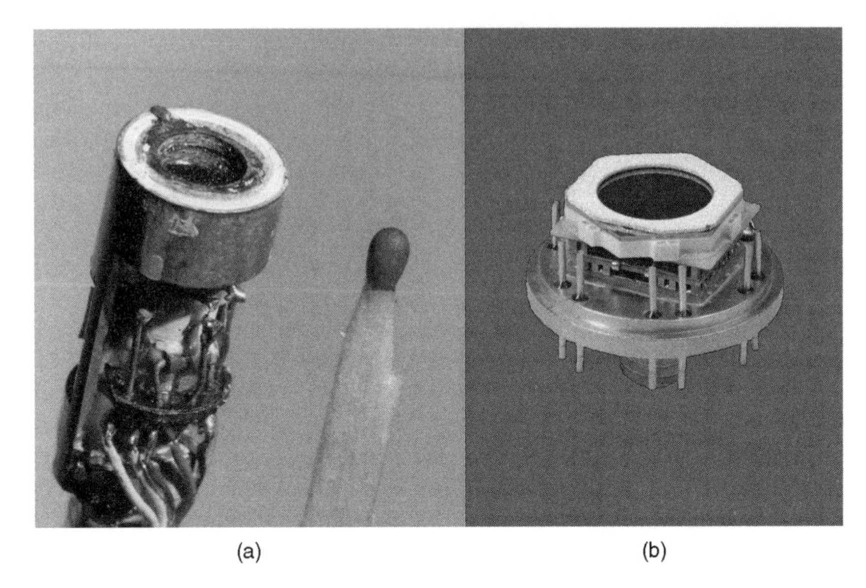

(a) (b)

Figure 3.17. Head of a classical Si(Li) detector made in 1967 (a) and of a modern SDD made in 2012 (b). The diameter of both collimators is about 5 mm, the front area of both sensors is 20 mm^2. The electronic wiring of the old detector head is handmade and coarse-textured while it is monolithically integrated in the new detector head with pins and pads. Figure 3.17a; private property of the authors; Figure 3.17b from Ref. [53], reproduced with permission from PNDetector GmbH, Munich, Germany (See colour plate section).

These sensors are also sensitive to X-rays. Superior results can be obtained by 2D sensors like CMOS (complementary metal oxides) and FPS structures (flat panel sensors). The energy range suitable for these sensors is 1 to 10 eV if the sensor is directly illuminated without an additional image converter like a scintillator plate. Higher photon energies between 10 eV and 10 keV are recorded by special windowless front- and/or back-illuminated CCDs. The highest energy range above 10 keV is covered by CCDs with FOS (fiber optic plates) and FPS [60]. CCDs need to be protected by scintillation plates to prevent damage by direct illumination with X-rays of such high photon energies. An additional advantage of the scintillation devices is the enhanced spatial resolution for imaging tasks.

Si photodiodes are available as front- and back-illuminated types. For higher sensitivity and lower noise, devices with a larger active area are preferred. Higher sensitivity can also be achieved when combining these products with ceramic or crystal scintillator plates. Instead of a scan performed by a single diode, an array of single diodes can be used. The number of diodes and their distribution over the active area can be chosen as required.

Charge-coupled devices used for imaging in the X-ray region are offered with diverse sizes. Chips with 512 × 512 pixels up to 1024 × 252 pixels and pixel sizes of 24 μm × 24 μm are offered for direct illumination. Chips of 1536 × 128 pixels with 48 μm × 48 μm pixel size using CsI scintillators are available for

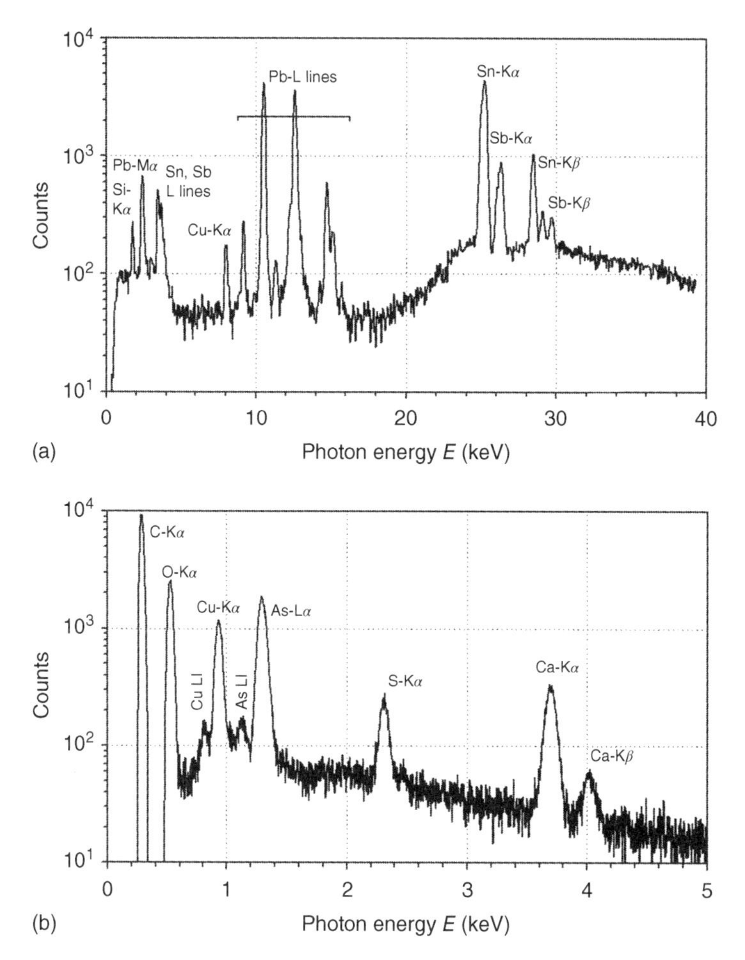

Figure 3.18. Energy-dispersive spectra of two samples as a semilogarithmic plot. (a) Spectrum of a type metal with Pb, Sb, and Sn recorded by a classical Si(Li) detector. Several broad peaks in the range <40 keV have an FWHM of some 150 eV (silicon) to 490 eV (tin), which are close together. (b) Spectrum of a mixture of two pigments (realgar: As_4S_4, and Poss blue: (CuCa)$(CH_3COO)_2 \cdot 2H_2O$) taken by a modern SDD. This spectrum shows a few peaks in the range <5 keV, which are clearly separated. Even the Kα peaks of oxygen at 0.525 keV and of carbon at 0.277 keV are recorded. The FWHM is between 40 (carbon) and 100 eV (calcium).

material sciences and medical tasks. For nondestructive inspections, CCDs with 1700 × 1200 pixels and pixels of 20 μm × 20 μm are on offer. They are combined with different types of scintillators. A typical arrangement for CMOS chips with CsI(Tl) scintillators is presented in Figure 3.19. Chips with 1.5 or 2 megapixels are available, for example, from Hamamatsu [60,61]. The spectral resolution of CCDs is mostly dependent on the operating temperature. If the device is sufficiently cooled a charge transfer efficiency of less than 1×10^{-5} is possible

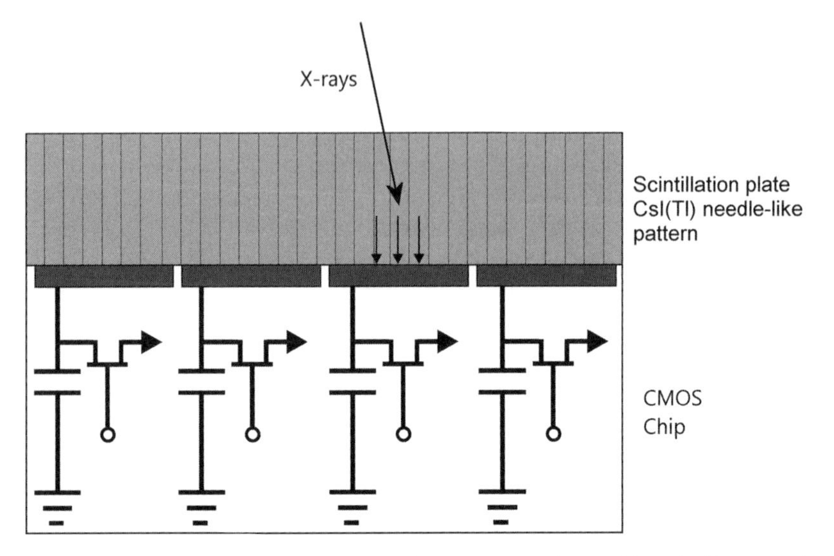

Figure 3.19. Position-sensitive detector. X-rays hit upon a plate with sensitive scintillator needles of CsI (Tl), which are some 2 μm thin and 20 μm long and have a divergence angle of about 5°. About 10–30 needles are bundled up to the CMOS chips. Light pulses induced by the incoming X-rays are registered by these CMOS chips and indicated depending on the chip position in two dimensions.

and the energy resolution is limited only by the readout noise. For an optimized CCD the energy resolution is about 140 eV for Mn-Kα (5.9 keV).

Flat panel sensors are modules of two-dimensional photodiode arrays for X-ray imaging based on large-area CMOS image sensors combined with scintillators. The complete device is manufactured as an integrated circuit on a single-crystal silicon chip. Flat panels for medical applications with 2400×2400 pixels and a size of $50\,\mu m \times 50\,\mu m$ each are state of the art [60].

Currently, these techniques are not used in routine applications of TXRF. There is no necessity for spatially resolved information when performing chemical analysis. However, several applications are possible in terms of 2D imaging. For example, the homogeneity of a sample can simply be studied by TXRF with a position-sensitive detector placed in a distance of some millimeters directly above the sample. This way the distribution of elements in a dry residue of liquids or contaminations on a wafer surface can be recorded directly. Moreover, such detectors may also be advantageous for surface- or thin-layer analysis by GI-XRF. In that case, the rotation of a conventional SDD can be avoided and the position-sensitive detector can directly record the necessary angle scan.

A last point to be mentioned is the quantity of information and data to be handled in the case of a 2D full spectrum. Every single pixel of an array with about 1700×1200 pixels may give an X-ray spectrum with more than 1024 channels (energy bands) and intensities up to 5×10^4 counts. That means that about 2×10^6 single spectra are obtained for an area of 34 mm × 24 mm. With 20 bits per spectrum, the total information amounts to the order of 40 GB.

3.7 WAVELENGTH-DISPERSIVE DETECTION OF X-RAYS

In contrast to an EDS, a WDS needs parts of a goniometer, which have to be mechanically moved with a high precision. A total spectrum is not recorded simultaneously but sequentially, that is, the spectral lines of the individual elements of a sample are registered one after the other [48,49,62–64]. Consequently, the WDS method is time-consuming and needs some 10 min or even more instead of only 1 min or less. However, a WDS has particular advantages:

- For photon energies up to 15 keV, the spectral resolution is much better in comparison to an EDS. Consequently, line identification is much easier because troublesome line interferences (peak overlaps) appear more seldom by far.
- In the low-energy region <1 keV, the spectral resolution is sufficient for the detection of the light elements like Be, B, C, N, O, F, and Ne, which is impossible by means of an EDS with Si(Li) or HPGe.
- A better spectral resolution generally leads to better detection limits since the spectral background is reduced in comparison to the spectral lines or peaks.
- Detectors usually applied for WDS have a higher capacity of count rates (photon flux), compared to an EDS detector. Consequently, irritating sum peaks do not appear. Besides, detection limits can be lowered by increasing the primary intensity of an X-ray tube or using the high brilliance of synchrotrons.

So far, only a few scientists have implemented these advantages: Sakurai *et al.* [65], Schwenke *et al.* [66], and Awaji [67]. That may change in the future and for that reason, the wavelength-dispersive detection of X-rays will be described and its advantages will be emphasized here.

A very simple variant of a WDS is shown in Figure 3.20. The main part is a goniometer, that is, a mechanical device for angle measurement. It consists of two collimators, a Bragg crystal with a known interplanar spacing and an X-ray detector, mostly a gas-filled proportional detector. X-rays coming from the sample can be reflected from the crystal at a certain glancing angle and detected at twice this angle if their wavelength meets Bragg's law of Equation 1.62. That means that a monochromatic part of the sample's spectrum can be detected at a fixed angle arrangement when sample, collimators, crystal, and detector are set up permanently. This mode of operation is called "fixed." The second mode is the "scanning" or "sequential" operation, when a broad X-ray spectrum of the sample is recorded. In that case the crystal is turned around an axis on top of its

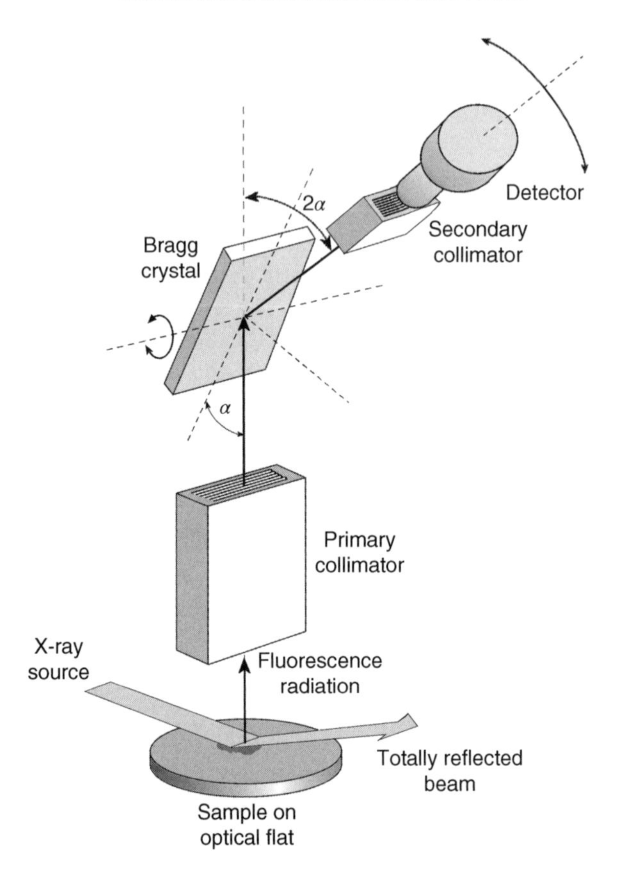

Figure 3.20. Wavelength-dispersive spectrometer with flat crystal and gas-filled detector. The centimeter-wide sample is excited to X-ray fluorescence at total reflection or under grazing incidence by the primary X-ray source. The secondary radiation of the sample is directed to the Bragg crystal by a first collimator and reflected under an angle α according to Bragg's law. The reflected monochromatic radiation is guided by a second collimator and reaches the detector under an angle 2α. The complete spectrum or a part of it is recorded if the Bragg crystal rotates around its center on a vertical axis and the second collimator plus detector rotates with double angular speed.

surface and vertical to the fluorescence beam, while the detector is turned simultaneously around this axis at a double angular speed. This speed can be chosen between slow (0.1°/min) and fast (180°/min). Crystal and detector have to be adjusted with a precision of 0.01°.

In order to present the spectrum as dependent on the photon energy, Bragg's law must be rewritten:

$$E = m\frac{hc_0}{2d\sin\alpha} \tag{3.19}$$

where E is the photon energy, m is a natural number called "order of reflection," h is Planck's constant, c_0 is the light velocity, and α is the corresponding glancing angle. Usually, only the first and most intensive order ($m = 1$) is recorded while the higher orders ($m \geq 2$) are eliminated by an electronic device, called pulse-height discriminator, which is usually connected to the detector. For this first order we have the simple relationship

$$E \approx \frac{1.240/2d}{\sin \alpha} \tag{3.20}$$

where $2d$ is to be given in nanometers so that E results in keV. This equation represents the photon energy as a nonlinear function of the glancing angle. For a certain crystal with spacing $2d$, only photon energies $E \geq 1.240/2d$ can be detected ($\sin \alpha \leq 1$). In practice, α is only varied between $\alpha_{min} \approx 10°$ and $\alpha_{max} \approx 70°$. The lower limit is determined by the low performance of Bragg crystals due to a poor reflectivity and energy resolution. The upper limit is given by a spatial restriction of sample and detector. Consequently, photons with energies between E_{min} and E_{max} can be detected by means of only one single crystal.

The device of Figure 3.20 represents a nonfocusing flat-crystal spectrometer as used by Awaji [67]. It is the simplest type of a WDS, but suitable for the investigation of flat samples with centimeter-extent by TXRF or GI-XRF. Spectrometers of the focusing curved-crystal type with Rowland arrangement are more complicated. They additionally require the axis of the Bragg crystal to be moved on a straight line or on a Rowland circle, respectively. At the same time the detector has to be moved on an epicycloid or also on the Rowland circle, respectively. In addition, such focusing spectrometers need bent crystals of the Johann type (only bent) or of the Johansson type (bent and ground) instead of flat ones. Focusing spectrometers are rather useful for electron microprobes where microspots are investigated, but for extended flat samples analyzed by TXRF their operation is too complicated.

Different natural and artificial crystals are suitable for Bragg reflections. They differ in their interplanar spacing, in their reflectivity, and energy resolution. As a consequence of the mentioned restriction $E_{min} < E < E_{max}$, different crystals may be necessary for different photon energies and elements to be detected. If only one single element has to be detected, only one single crystal is sufficient, and the goniometer has to be adjusted only for one respective glancing angle α. This is the case for surface- and thin-layer analyses of a single element performed under grazing incidence of the primary X-ray beam (GI-XRF). If, however, several elements in a sample must be determined, more than one crystal may be necessary and several parts of X-ray spectra have to be recorded. This is the case for micro- and trace analyses of several elements carried out under total reflection of the primary X-ray beam (TXRF). The goniometer has to be turned around its axis and several spectral bands have to be recorded dependent on α or E, respectively.

In general, four different crystals are necessary and sufficient to cover the total X-ray spectrum between 0.1 up to 40 keV. For this purpose, a crystal changer is designed to take up to four crystals. Furthermore, a changeable collimator is provided with, for example, two different openings or divergences. And finally, provision is made to interchange a gas-filled detector and a scintillation detector. To make use of the high performance of a WDS in the low energy range of an X-ray spectrum, a vacuum chamber is necessary containing collimators, crystals, and detectors. A fine vacuum with a pressure of about 1 Pa is sufficiently low for an air path of 50 cm. In that case the transmission is greater than 97% for photons with energies above 0.1 eV (Kα peak of beryllium).

3.7.1 Dispersing Crystals with Soller Collimators

Flat-crystal spectrometers first need one or two collimators in order to align the divergent radiation of the sample and to separate a straight beam, which can be used for Bragg reflection at a definite glancing angle α [67]. The most suitable collimator is the Soller-type collimator consisting of an array of several metal sheets, each with a length l and a thickness Δ. They are arranged parallel to each other with a closed spacing δ. Such a collimator can be passed by X-ray beams in the direction of the collimator axis with an angular divergence $\leq 2\delta/l$ whereas X-ray beams with a larger divergence are absorbed by these plates [48,49,62–64].

Usually, two collimators that are interchangeable are provided: a coarse collimator with a divergence of about 0.5° and a fine collimator with a divergence of about 0.15°. The fine collimator leads to a better spectral resolution at the expense of a higher absorption and lower intensity. The sheets are usually made of copper or steel (covered with Bi microcrystals). They are 3–10 cm long, about 0.1 mm thick, and spaced 0.2–0.8 mm apart. The cross-section of the collimator should be a square of about 2 cm × 2 cm. Consequently, a stack with 20–80 metal sheets and spacers is necessary. The divergence of the primary collimator before the crystal is <0.3° and its absorption is up to 50%. The secondary collimator before the detector is coarse-spaced and mainly used for the elimination of stray X-rays scattered from the crystal or its holder and from the primary collimator.

Several flat crystals can be used as dispersing elements or Bragg crystals, respectively, and are for sale [48,49,62–64]. A list of several different crystals with double interplanar spacing is given in Table 3.6. In the case of inorganic compounds, naturally occurring crystals can often be used. Other dispersing elements are "pseudocrystals" or synthetic multilayers. Such multilayered thin soap films (esters of carboxylic acids where heavy metal atoms are separated by long fatty acid chains) are *organic*. Stacks of thin layers (heavy metals and light spacers arranged in pairs) are *inorganic*. The spacing values reach from about 0.16 nm up to 20 nm; their precision is <0.1% for natural crystals (about 3×10^{-5}% for silicon) and <0.5% for synthetic crystals.

TABLE 3.6. Several Bragg Crystals Commonly Used for WDS in Ascending Order of 2*d* Values

Bragg Crystal	Acronym	Chemical Formula	(*hkl*)	2*d* [nm]
Lithium fluoride	LIF4	LiF	422	0.1652
Lithium fluoride	LIF2	LiF	220	0.2848
Silicon	SI2	Si	220	0.3840312
Germanium	GE2	Ge	220	0.4000
Lithium fluoride	LIF1	LiF	200	0.4028
Rock salt	ROSA	NaCl	200	0.5641
Calcite	CAL	$CaCO_3$	200	0.6071
Silicon	SI1	Si	111	0.62712
Germanium	GE1	Ge	111	0.6532
α-Quartz	QUA	SiO_2	1011	0.6686
Pyrolytic graphite	PYRG	C	002	0.6715
Indium antimonide	INSB	InSb	111	0.7480
Pentaerythritol	PET	$C{\cdot}(CH_2{\cdot}OH)_4$	002	0.8742
Ammonium dihydrogen phosphate	ADP	$NH_4{\cdot}H_2PO_4$	101	1.0648
Muscovite or mica	MIC	$K_2O{\cdot}(Al_2O_3)_3{\cdot}(SiO_2)_6{\cdot}(H_2O)_2$	002	1.9840
Thallium hydrogen phthalate	TLHP	$TlOOC{\cdot}C_6H_4{\cdot}COOH$	1010	2.575
Rubidium hydrogen phthalate	RBHP	$RbOOC{\cdot}C_6H_4{\cdot}COOH$	1010	2.61
Potassium hydrogen phthalate	KHP	$KOOC{\cdot}C_6H_4{\cdot}COOH$	1010	2.66
Inorganic multilayer	WSI	W/Si		5.0
Lead myristate	PBMY	$[CH_3{\cdot}(CH_2)_{12}{\cdot}COO]_2Pb$		8.06
Lead stearate	PBST	$[CH_3{\cdot}(CH_2)_{16}{\cdot}COO]_2Pb$		10.0
Lead melissate	PBME	$[CH_3{\cdot}(CH_2)_{28}{\cdot}COO]_2Pb$		16.5
Inorganic multilayer	MOC	Mo/B_4C		20.4

Source: Data from Ref. [48], compiled with permission from Plenum Press.

The natural inorganic crystals can be manufactured by an easy basal cleavage, their shape can be cut and polished, and their surface can possibly be finished. The orientation of their lattice planes is highly accurate. Normally, a dispersing crystal is made up of a mosaic of small blocks or domains, which are misaligned by less than 0.05°. The whole crystal is usually shaped as a thin plate with a length of about 5 cm, a height of about 2 cm, and a thickness below 3 mm. Artificial multilayers are prepared on millimeter-thick substrates of quartz glass. All these dispersing elements can be mounted into holders.

Table 3.7 lists the range of detectable photon energies between $E_{min} \approx 1.32/2d$ and $E_{max} \approx 7.14/2d$ according to Equation 3.20. The respective peaks or lines appear with a certain peak width ΔE_{min} and ΔE_{max} on condition that a collimator with an opening of 0.15° is used. This width is quite low (0.1–8 eV) for peaks at E_{min} but increases to the 80-fold (5–640 eV) for peaks at E_{max}. Details are discussed in Section 3.8.2.2.

TABLE 3.7. Characteristics of Bragg Crystals Used for the Detection of Peak Energies with Different Peak Width for Corresponding Elements

Crystal Abbrev.	Detection E_{min} [keV]	Peak Width ΔE_{min} [eV]	Detection E_{max} [keV]	Peak Width ΔE_{max} [eV]	Elements by K lines	Elements by L lines
LIF4	8.0	7.6	43.2	642	$_{29}$Cu–$_{64}$Gd	$_{69}$Tm–$_{92}$U
LIF2	4.6	4.4	25.1	372	$_{23}$V–$_{49}$In	$_{57}$La–$_{92}$U
SI2	3.4	3.3	18.6	276	$_{20}$Ca–$_{43}$Tc	$_{50}$Sn–$_{92}$U
GE2	3.3	3.1	17.9	265	$_{19}$K–$_{42}$Mo	$_{50}$Sn–$_{92}$U
LIF1	3.3	3.1	17.7	263	$_{19}$K–$_{35}$Br	$_{50}$Sn–$_{92}$U
ROSA	2.3	2.2	12.7	188	$_{17}$Cl–$_{36}$Kr	$_{43}$Tc–$_{89}$Ac
CAL	2.2	2.1	11.8	175	$_{16}$S–$_{34}$Se	$_{42}$Mo–$_{86}$Rn
SI1	2.1	2.0	11.4	169	$_{16}$S–$_{34}$Se	$_{41}$Nb–$_{84}$Po
GE1	2.0	1.9	10.9	162	$_{16}$S–$_{33}$As	$_{40}$Zr–$_{83}$Bi
QUA	2.0	1.9	10.7	159	$_{15}$P–$_{34}$Se	$_{40}$Zr–$_{82}$Pb
PYRG	2.0	1.9	10.6	158	$_{15}$P–$_{33}$As	$_{40}$Zr–$_{82}$Pb
INSB	1.8	1.7	9.6	142	$_{15}$P–$_{31}$Ga	$_{38}$Sr–$_{78}$Pt
PET	1.5	1.4	8.2	121	$_{14}$Si–$_{29}$Cu	$_{36}$Kr–$_{73}$Ta
ADP	1.2	1.2	6.7	100	$_{12}$Mg–$_{26}$Fe	$_{33}$As–$_{66}$Dy
MIC	0.67	0.63	3.6	53	$_{9}$F–$_{19}$K	$_{26}$Fe–$_{51}$Sb
TLHP	0.51	0.49	2.8	41	$_{8}$O–$_{17}$Cl	$_{24}$Cr–$_{45}$Rh
RBHP	0.51	0.48	2.7	41	$_{8}$O–$_{17}$Cl	$_{23}$V–$_{45}$Rh
KHP	0.50	0.47	2.7	40	$_{8}$O–$_{17}$Cl	$_{23}$V–$_{44}$Ru
WSI	0.26	0.25	1.4	21	$_{7}$N–$_{12}$Mg	$_{20}$Ca–$_{34}$Se
PBMY	0.16	0.16	0.89	13	$_{5}$B–$_{10}$Ne	$_{20}$Ca–$_{28}$Ni
PBST	0.13	0.13	0.71	11	$_{5}$B–$_{9}$F	$_{20}$Ca–$_{26}$Fe
PBME	0.080	0.08	0.43	6	$_{4}$Be–$_{7}$N	$_{20}$Ca–$_{21}$Sc
MOC	0.065	0.06	0.35	5	$_{4}$Be–$_{6}$C	$_{20}$Ca

Table 3.7 also gives the range of elements with atomic numbers that can be detected by their most intensive K or L lines. For the light elements with $3 \leq Z \leq 12$ (from lithium to magnesium), this line is the Kα line. Elements with $13 \leq Z \leq 35$ (from aluminum to bromine) are detected via their K$\alpha_{1,2}$ doublet if it is nonresolved. For medium heavy elements with $36 \leq Z \leq 49$ (from krypton to indium), the Kα_1 line is the most intensive line, which can be separated from their Kα_2 partner by a WDS and therefore is mostly used for detection. If only a low excitation energy <10 keV is available, the Lα_1 line may be preferred. The remaining heavy elements with $50 \leq Z$ (from tin upward) will generally be detected by their Lα_1 lines. At a low excitation energy <10 keV, heavy elements may also be detected by their Mα lines.

3.7.2 Gas-Filled Detectors

The detector applied to X-ray fluorescence analyses is responsible for the registration of X-ray spectra, that is, of characteristic spectral lines or peaks of the elements and the continuous spectral background. That means that individual X-ray photons emitted by the sample within a finite period of time have to be registered by the detector. As far as possible, the energy of these photons should also be determined. Generally, the detector gives an electronic pulse as a response to the incoming photon, which is counted. If the pulse height is proportional to the photon energy, it can be used as a measure for this energy after calibration.

Two main types of detectors can be used for WDS, and both of them are based on the principle that atoms of a particular medium will be ionized by photoelectric processes [47,49,62]. The gas-filled detectors use an inert gas; the scintillation detectors use a solid phosphor material. A gas-filled detector is represented in Figure 3.21. X-ray photons are directed into the detector through the secondary collimator and enter it through a thin window. They can hit now individual atoms of the gas and produce several photoelectrons and gas ions. The photoelectrons are accelerated to the central thin wire held at a high positive potential serving as the anode. The heavy ions are collected by the

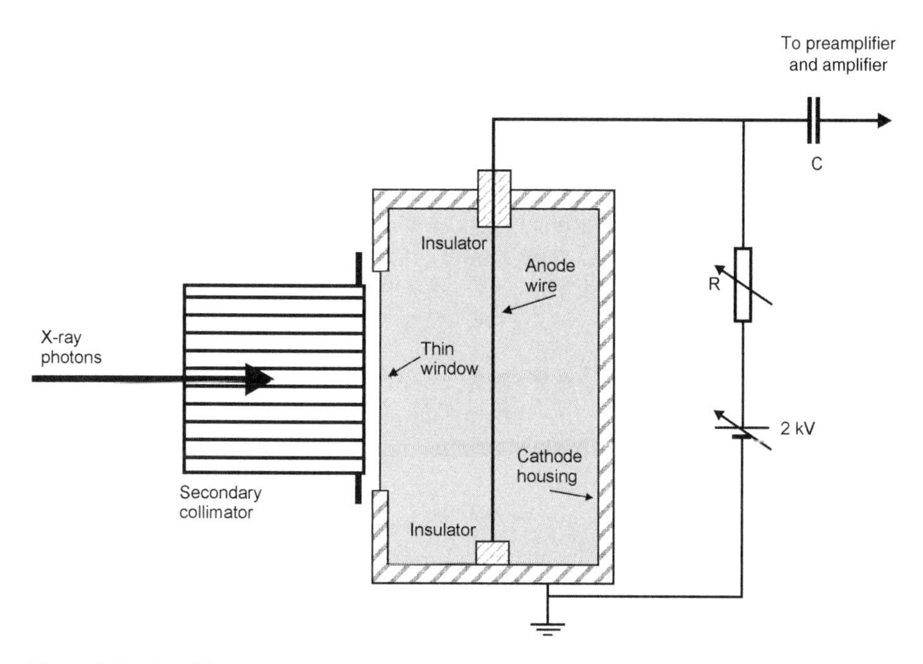

Figure 3.21. Gas-filled detector with an insulated thin wire tightened along the axis of a metal cylinder. The wire with about 80 μm diameter is held at a high positive voltage of about 2 kV and used as anode. The metal cylinder filled with argon or xenon gas is grounded and is used as cathode. X-ray photons enter the detector through a collimator and a thin window. They ionize gas atoms and produce photoelectrons. The electrons are accelerated toward the wire producing further electron–ion pairs, the ions are discharged at the casing. The gas amplification can be controlled by the variable ohmic resistor. The electron–ion pairs produce a short voltage drop at the capacitor, which is proportionally amplified and passed to the counting circuitry.

casing of the detector used as the grounded cathode. The accelerated electrons produce an electrical pulse in the form of a rapid voltage drop at a capacitor. This voltage pulse is amplified and registered by the attached counting circuitry. The potential of the anode is set by an ohmic resistor so that the amplitude of the voltage pulses is proportional to the energy of the detected photons.

Two types of gas-filled detectors are in use:

1. *A sealed gas detector* with argon or xenon as filling gas. The pressure inside the casing is not much below normal pressure (400–900 hPa). In order to prevent leakage of the gas, the metal casing is locked by a relatively thick and airtight window, for example, a 50–100 μm thick beryllium window. Such a sealed gas detector is applicable for the detection of photons with energies ≥3 keV. The transmission for such photons is ≥75%, the transmission of the air path inside the WDS is ≥97% if the pressure is reduced to 1 Pa (fine vacuum). This can be reached by a simple mechanical pump, for example, by a rotary vane pump.
2. *A gas-flow detector* with a relatively thin window. For example, a 0.6–6 μm thin Mylar^{TM} foil (polyethylene terephthalate) or a 0.5–3 μm thin polypropylene foil may be chosen, which are not totally airtight. However, these windows are transmissible for photon energies down to 0.1 keV, that is, for the detection of the light elements. The foils can be supported by a strengthening grid, for example, a loosely woven nylon mesh. They can easily be replaced if they are broken, which frequently happens. Argon is used as filling gas and an organic quenching agent is normally added, mostly 10% of methane. This mixture flows through the detector at a slow constant rate of about 0.5 l/h and keeps the pressure inside the detector at atmospheric pressure even if the window foil is somewhat permeable to air. Such gas-flow detectors can detect photons with energies down to 0.1 keV (Kα of boron at 0.185 keV, Kα of beryllium at 0.110 keV). For the detection of such photons, the pressure inside the whole WDS system is additionally reduced to fine vacuum.

The basic principle of gas-filled detectors is the photoelectric effect represented by the chemical relation

$$\mathrm{Ar} + h\nu \rightarrow \mathrm{Ar}^{+} + \mathrm{e}^{-} \tag{3.21}$$

where Ar is the argon atom, $h\nu$ is the energy of the photon to be detected, Ar^{+} is the positive ion of the argon gas, and e^{-} is the electron removed from an outer shell of the gas atom. Argon gas can also be replaced by xenon gas. The photoelectric effect is not the only process induced by incoming photons because there are also nonionizing collisions. Consequently, the average energy required for the photoelectric effect is above the ionization potential of 16 eV for argon and 12 eV for xenon, respectively. Values of 26.4 eV for Ar atoms, and about 21 eV for Xe atoms have been reported [47–49,62]. If this average energy for the production of an ion–electron pair is called e_{pair} and the initial photon energy is E, the number of electrons and ions on the average produced by one single photon is given by

$$N = E/e_{pair} \tag{3.22}$$

Usually, N is about 20 up to 1500 electron–ion pairs per photon. The effect is controlled by Poisson statistics and can be approximated by a Gaussian distribution.

The voltage pulse of a single photon is of the order of some millivolts and microseconds. It has to be linearly amplified to the V-region strictly proportional to the energy of the incoming photon. All these pulses of photons with the same energy are registered by the counting circuitry and are summed up to a peak in the X-ray spectrum. Photons of different energies give different peaks or lines and yield the total spectrum. The peak width, ΔE, usually is determined by the FWHM or "full width at half maximum." It can be approximated following Poisson statistics with a so-called "Fano"-factor, F:

$$\Delta E \approx 2.355\sqrt{FN} \cdot e_{pair} \tag{3.23}$$

For argon gas, F is about 0.22, and for xenon it is about 0.12. Actual ΔE values for elements with $Z \geq 22$ (Ti) are between 700 and 2000 eV for the Ar-gas detector. For lighter elements down to $Z = 4$ (beryllium), ΔE lessens to about 100 eV. For a Xe gas detector, respective values are half of the Ar values. Such a spectral resolution is much poorer than the resolution of dispersing crystals in a WDS; however, it is sufficient to electronically discriminate higher spectral orders of the Bragg spectrometer, to eliminate electronic noise as well, and to include escape peaks of the detector.

For that purpose, an electronic window is set up around the pulse corresponding with the first order peak of a certain element, which should solely be registered because of its high intensity. Higher spectral orders ($m > 1$) from other elements with lower intensity, which are reflected by the same Bragg crystal at the same or nearly the same glancing angle, have a shorter wavelength λ_1/m, or a higher photon energy mE_1, respectively. These disturbing lines or peaks, however, give electronic pulses that are m-times higher than the actual pulse of the evident peak reflected in the first order. Consequently, these pulses can be filtered out electronically so that only the actual pulse passes to the counting circuitry. In this way, peaks with higher orders are eliminated and discriminated from the peak E_1 in the first order. This method is called "pulse-height discrimination." For an automatic control, the electronic filter with lower and upper limit is adapted to the crystal spectrometer by a sin-potentiometer.

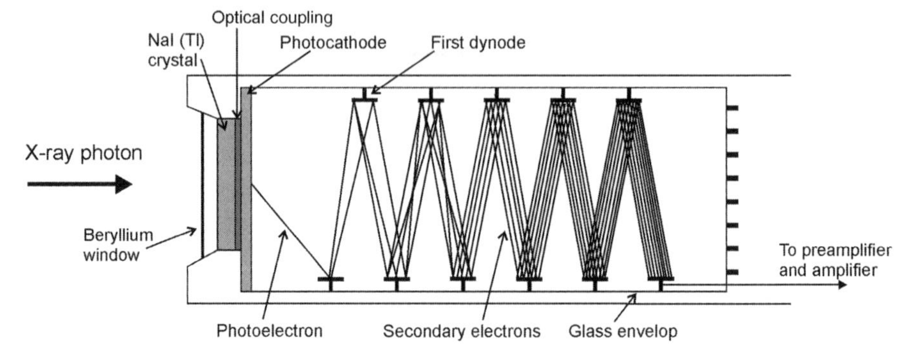

Figure 3.22. Scintillation detector with a scintillation crystal made of a doped phosphor material and a photomultiplier. Each X-ray photon entering the scintillation crystal through a thin beryllium window produces a series of light pulses in the blue part of the visible spectrum. These light pulses produce a voltage pulse at the photocathode of the attached photomultiplier, which is further amplified by the multiplier and recorded by a counter. The photon energy, the number of light pulses, and the amplitude of the voltage pulse have to be strictly proportional.

3.7.3 Scintillation Detectors

A scintillation detector always contains two essential components as demonstrated in Figure 3.22:

1. A transparent crystal suitable as phosphor material. It consists of atoms the electrons of which are promoted from valence band states to a higher conduction band by an incident photon. When these electrons fall back into the stable ground state, energy is emitted within the visible light range. That means that the energy of the incident photon is converted into a series of light pulses.
2. A photomultiplier directly attached to the scintillation crystal that converts the light pulses into voltage pulses with a height of some millivolts and a length of some microseconds. These pulses are furthermore enhanced by a preamplifier and amplifier.

Usually sodium iodide (NaI) is chosen as phosphor or scintillator crystal [48,49]. It is doped with thallium as an impurity element (0.1%). When an X-ray photon hits the crystal, a valence band electron of the halide can be lifted into the conduction band and simultaneously produces a positive hole in the valence band. If the lifted electron falls back from its excited state to the valence band, a light photon is emitted. This photon can again lift an electron from the valence band so it is not emitted from the crystal. In close proximity of an impurity atom, however, the photon can transfer only a part of its energy to this atom. The respective electron can fall down to the valence band and emit a photon (or a phonon) of a smaller energy that can escape now from the crystal.

The average energy of a single photon emitted from NaI, doped or activated with thallium, is 3 eV (blue light). The number of such light pulses is strictly proportional to the energy of the incident photons according to Equation 3.22.

The scintillation material is usually presented as a transparent disc, about 3 cm wide and 2 mm thick. Because this phosphor material is hygroscopic, it must be enclosed in an airtight covering. Furthermore, it must not be hit by environmental light. Hence, it is coated with an aluminum layer, about 1 μm thin, and covered by a 0.2 mm thick beryllium window. As a consequence, scintillation detectors are only applicable in the region above 6 keV. The rear of the scintillation crystal is optically coupled to a thin glass window followed by the photocathode and a series of individual dynodes within the photomultiplier. The impact of the first photoelectron releases further electrons at the first dynode. At each following dynode, an amplifying effect leads to a higher current pulse avalanche-like. The gain factor of the multiplier is of the order of 10^6.

Scintillation detectors are used in the region from 3 keV upward. The efficiency of light pulse generation and electron production is quite low; only 20% of X-ray photons create a light pulse and only 5% of the light photons create a first photoelectron on the photocathode of the multiplier. Consequently, the total efficiency is only about 1% and the average energy for an ionization process is increased to $e_{ion} = 200$ eV. As a result, the pulses and the peak widths are $\sqrt{200/26.4/0.8}$ or about three times wider in comparison to gas filled detectors. However, scintillation detectors have a somewhat higher count rate capability, up to 2×10^6 counts per second. It is common practice, to use both detectors, either separately with interchange or in tandem. In the latter case, a gas-flow detector is fitted with an additional window at its rear, and the scintillation detector is mounted to this window of the gas-filled detector.

3.8 SPECTRA REGISTRATION AND EVALUATION

The different charge pulses produced by the detector have to be processed by an elaborate electronic system. They are amplified, shaped, and sorted according to their amplitudes. Finally, all pulses with certain amplitudes are counted. A strong proportionality between the amplitude of the initially produced and subsequently processed pulses should accurately be maintained. As long as this condition is met, the pulse amplitude remains an accurate measure for the energy of the detected photons. Ultimately, the result of the photon counting can be demonstrated as an energy-dispersive spectrum representing the number of photons depending on their energy.

3.8.1 The Registration Unit

The electronic systems employed usually have the components shown in the block diagram of Figure 3.23: a FET-based preamplifier, a main linear amplifier with baseline restorer and pulse-pileup rejector, an analog-to-digital converter

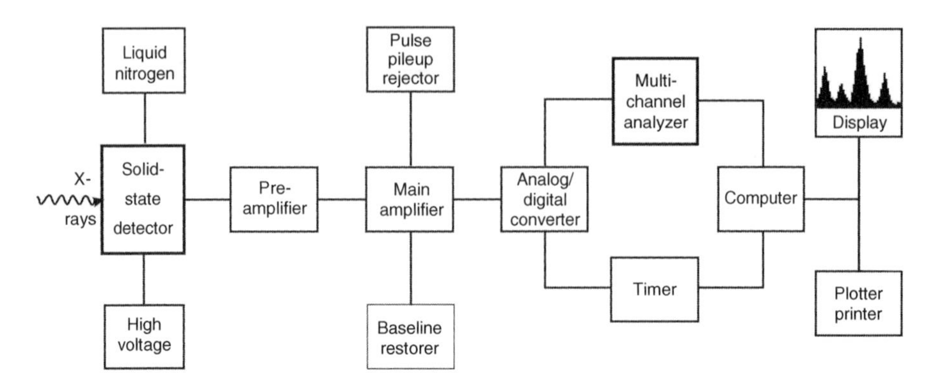

Figure 3.23. Major components of an energy-dispersive spectrometer with solid-state detector and multichannel analyzer. Figure from Ref. [2], reproduced with permission. Copyright © 1996, John Wiley and Sons.

(ADC), a multichannel analyzer (MCA), a timer, and a dedicated computer with a color video display and a plotter/printer output device.

An FET is the most used preamplifier operated in the mode of pulsed optical feedback for discharge. Since around 1988, a Pentafet[4] has been used for direct charge neutralization [68,69]. The preamplifier converts the charge pulses produced by the detector to low-voltage pulses. The amplitudes or pulse heights are strictly proportional to the number of electron-hole pairs and hence to the energies of detected X-ray photons. The preamplifier is installed quite close to the detector and also cooled down (at least to −20 °C or even to the temperature of liquid nitrogen) in order to reduce the electronic noise.

The millivolt pulses are further increased to voltage levels by a high-gain linear amplifier. In addition, the pulses are shaped to a particular form with different amplitudes but with a constant shaping time. Effective noise suppression is achieved by selecting a sufficiently high shaping time (6 or 8 μs rather than 1 or 2 μs). On the other hand, the probability of a pulse overlap increases with higher shaping time. Such an overlap of two coinciding pulses leads to the registration of one pulse the amplitude of which is the sum of the individual pulse amplitudes. This incorrect registration is called *pulse-pileup effect.* It can be avoided by a device called *pulse-pileup rejector* at least in those cases when the two pulses are not exactly coincident in time but arrive at least 1 μs apart. These nearly coincident pulses are "rejected" and get lost in the counting. However, the loss can be compensated by an appropriate extension of the measuring time.

The pileup effect becomes serious at high count rates. In such cases, not only coincident but also successive pulses will overlap at least with their "tails." Consequently, the baseline of the amplifier is shifted to a higher voltage and the output pulses appear reduced or "depressed" below their true amplitude. This incorrect reading can be avoided electronically by another device called a *baseline restorer.*

[4] Registered Trademark of Oxford Instruments, High Wycombe, England.

The output pulses of the main amplifier are transferred to an ADC. The analog information of the pulse amplitudes is converted here into digital form. The resultant number serves as an address for the connected MCA. It sorts the different pulses according to their address and counts their numbers, which are stored in the MCA. The different "channels" correspond to small energy ranges. Usually, an MCA contains 2048 up to 8192 channels that are assigned to consecutive increments of 5 eV up to 40 eV. A total range of 10, 20, 40, or 80 keV is indicated for the energy of X-ray photons. The storage capacity of today's MCA is nearly unlimited. Usually, it is chosen to be 2^{32}, that is, 4×10^9 counts per channel.

The digital form of an MCA's content is well adapted to computer operation. The raw data can easily be processed by a dedicated minicomputer or a personal computer (PC). The relevant contents can be shown as an energy-dispersive spectrum and can be displayed on a monitor and plotted by a printer. The spectrum is usually represented as a histogram indicating counts versus energy and can already be observed during the measurement. The computer can control the output devices and also a device called a timer, which is used to start and to stop the measurement or data collection in a preset time. In most systems, the timer is based on a "live time" clock. It stops any further pulse counting during the dead time of the system when an input pulse is still being processed.

3.8.2 Performance Characteristics

The performance of an EDS with semiconductor detector and electronic pulse processor and of a WDS with a gas-filled or scintillation detector is characterized by four main features: the spectral efficiency of the detector, the spectral resolution of the system, the input–output yield of the processor, and the troublesome interference phenomena caused by neighboring lines, by sum peaks, and by escape peaks resulting from the detector.

3.8.2.1 Detector Efficiency

The efficiency of a detector is defined as the percentage of detected photons with respect to all incident photons. For a semiconductor detector, the efficiency is nearly 100% for photons with energies between 5 and 10 keV but is reduced for lower and higher energies by several absorption effects, which are described by the formula [70]

$$\varepsilon = \exp\left[-(\mu/\rho)_{\mathrm{Be}}\rho_{\mathrm{Be}}t_{\mathrm{Be}} - (\mu/\rho)_{Au}\rho_{Au}t_{\mathrm{Au}} - (\mu/\rho)_{\mathrm{Si}}\rho_{\mathrm{Si}}t_{\mathrm{Si}}\right] \times \left\{1 - \exp\left[-(\tau/\rho)_{\mathrm{Si}}\rho_{\mathrm{Si}}d_{\mathrm{Si}}\right]\right\} \quad (3.24)$$

where (μ/ρ) is the total mass-absorption coefficient; ρ is the density of the respective layer; t is its thickness; (τ/ρ) is the photoelectric mass-absorption

TABLE 3.8. Data for the Absorbing Components of Typical Si(Li), SDD, or HPGe Detectors

Absorbing Layer or Medium	Density ρ (g/cm^3)	Typical Thickness (nm or μm or mm)
Beryllium window	1.85	7.5 μm
Polymer foil	1.20	0.5 μm
Gold contact	19.3	20 nm
Al contact	2.7	30 nm
Silicon dead layer	2.33	0.1 μm
Si(Li) intrinsic region	2.33	3 mm
Highly pure silicon	2.33	0.5 mm
Germanium dead layer	5.32	1 μm
Highly pure germanium	5.32	5 mm

Source: From Ref. [2], reproduced with permission. Copyright © 1996, John Wiley and Sons.

coefficient; and d is the thickness of the detector crystal. The lower efficiency is caused by a reduced transmission of X-rays in the beryllium window, in the gold-layer, and in an inactive or dead layer of the Si crystal expressed by the first bracket of Equation 3.24. Furthermore, it is caused by the photoelectric absorption of X-rays in the intrinsic region of the Si(Li) taken into account by the curly bracket of formula 3.24. This formula is written for a Si(Li) detector with Au layer and Be window. For a HPGe-detector, the silicon values have to be replaced by germanium values.

The individual data of a typical Si(Li) and HPGe detector and also for an SDD are listed in Table 3.8 and used as the basis for Figure 3.24. Values of (μ/ρ) for the calculation were taken from Ref. [26]. The calculated curves only serve for a general assessment. If true values for the detector efficiency are required, an experimental determination has to be carried out.

At photon energies below 3 keV, the absorption of the Be window is the decisive factor. At photon energies above 10 keV, the transmission through the intrinsic or active crystal becomes dominant. Different steps of the efficiency curves can be identified: five small steps at the M edges of gold between 2.2 and 3.5 keV, one small step at the K-absorption edge of silicon at 1.839 keV, one larger step at the K edge of germanium at 11.103 keV, and a smaller one at the LIII edge at 1.217 keV.

As shown in Figure 3.24, Si(Li) and HPGe detectors differ strongly in the region above 11 keV. After a stepwise decrease, the Ge detector attains 100% efficiency for energies above 20 keV, whereas the Si(Li) detector steadily loses efficiency. In the region below 11 keV, the detectors have similar characteristic curves.

For today's SDDs with a thinner intrinsic region and with another contact layer and window material, the formula 3.24 has to be modified. Standard media with respective t and d values are listed in Table 3.6. As can be read from Figure 3.25, the SDD is restricted to photon energies below 25 keV instead of

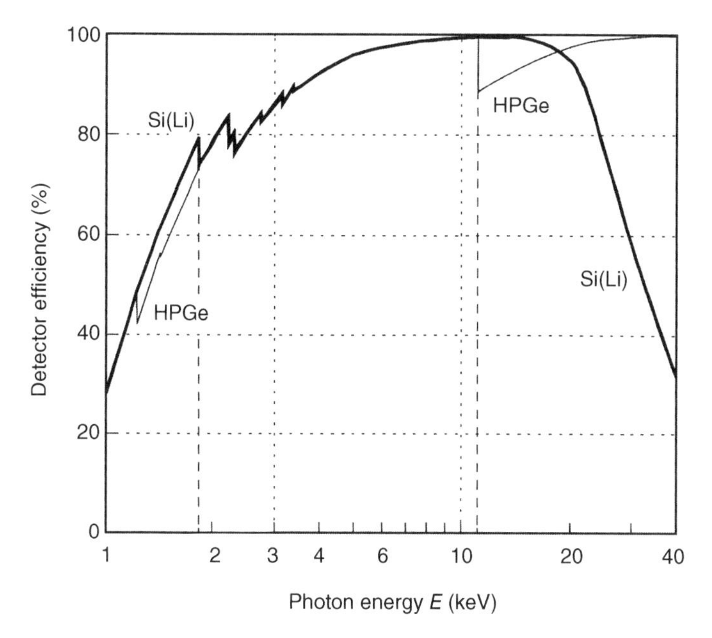

Figure 3.24. Efficiency of typical Si(Li) (———) and HPGe detectors (———) calculated as dependent on the energy of the indicated photons. The Si(Li) detector was assumed to have a 3 mm thick Si intrinsic region and the HPGe detector to have a 5 mm high-purity Ge crystal. Both detectors are assumed to be provided with a 20 nm thin gold contact and a 7.5 μm thin Be window. Figure from Ref. [2], reproduced with permission. Copyright © 1996, John Wiley and Sons.

50 keV.[5] On the other hand, the replacement of the gold layer by an aluminum layer leads to a somewhat better efficiency for low photon energies <8 keV. The low efficiency at photon energies below 1 keV is caused by the beryllium window even if it is only 7.5 μm thick. This window, however, can be replaced by an ultrathin polymer foil, for example, a 0.5 μm thin polyimide foil [53,54]. Figure 3.25 indicates a shifting of the 10% efficiency from 0.7 keV photons down to 0.2 keV photons (see also Ref. [58]). The steps in the curves can be assigned to the M edges of gold, to the K edges of aluminum and of silicon, and to the K edges of C, N, and O of the window foil.

If the thin polymer foil has to be strengthened by a silicon grid the efficiency may be reduced to about 80% in total. At photon energies below 5 keV, the low efficiency can be diminished still further by an air path between the sample and the detector. For a typical distance of about 5 mm, the overall efficiency is reduced to 72% for 2 keV photons and to 13% for 1 keV photons. However, this reduction can be avoided by applying a medium vacuum (100 Pa) by a simple water-jet pump or by a helium-flush.

[5] New detectors have been developed with an intrinsic region of 1 mm thickness so that the efficiency for high-energy photons is extended to about 35 keV.

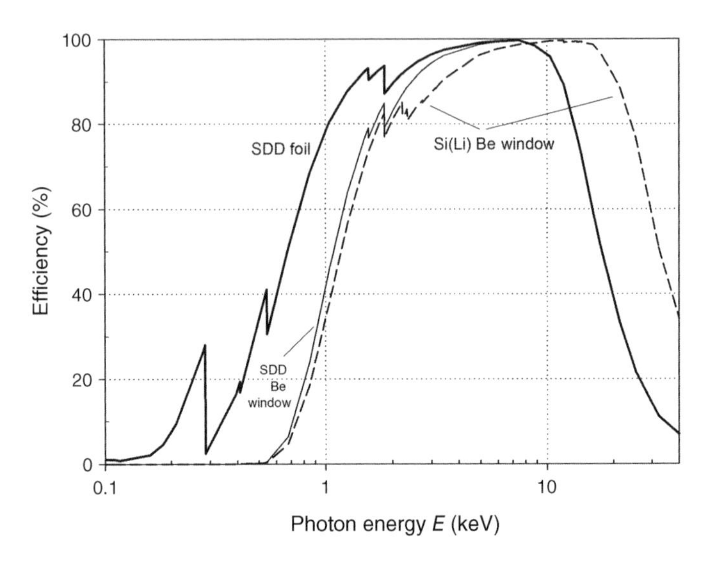

Figure 3.25. Efficiency of a new SDD in comparison to that of a Si(Li) detector (see Figure 3.24) dependent on the energy of the indicated photons. The SDD is made of a 0.5 mm thick high-purity Si crystal. Its front side may be metallized by a 30 nm thin Al layer. The whole sensor is usually protected by a 7.5 μm thin Be window (------) or by a 0.5 μm thin polyimide foil (———). The Al-K edge at 1.559 keV of the front layer and the Si-K edge at 1.839 keV of the dead layer can clearly be recognized. The edges at 0.284 keV, at 0.410 keV, and at 0.544 keV come from the elements C, N, and O of the polyimide foil.

The efficiency of gas-filled and scintillation detectors is essentially unity or 100% in the energy range between 20 and 100 keV. For gas-filled detectors it remains >10% for lower energies down to 0.1 keV. For a typical WDS the path length may be about 20 cm and the transmission of X-rays in such an air path is <4% for energies below 3 keV but can distinctly be increased by applying a high vacuum of 1 Pa to >90% for energies >0.1 keV.

3.8.2.2 Spectral Resolution

The different characteristic X-ray peaks of an EDS are not infinitely small but cover an energy range of some 100 eV. The histograms usually recorded show peaks that span about 10–20 channels, with nearly a Poisson or Gaussian distribution. Each peak can be characterized by a width defined as FWHM.

The peak width is mainly caused by the production of electron-hole pairs because of the incident photons. This process is not the only one possible; rather, it competes with the generation of lattice vibrations or the emission of phonons (heat radiation), for example. Without such a competing process, the necessary energy for the generation of an electron-hole pair would amount to the band gap energy of 1.1 eV for silicon and about 0.7 eV for germanium. Because of the competing processes, however, the average energy consumed per electron-hole pair is greatly increased: to about 3.8 eV for silicon and

2.96 eV for germanium at the operating temperature of 77 K and significantly exceeds the band gap [42,47]. Moreover, the number for electron-hole pairs is no longer constant but fluctuates around a statistical mean. A corresponding frequency distribution may be found for the pulse amplitudes, which leads to the observed peak width.

Such a distribution would be a Poisson distribution if independence and a small probability for the observed events could be taken for granted. However, the processes responsible occur fairly frequently rather than rarely, and not independently. Consequently, the distribution is not a true Poisson distribution [48], but it can be approximated by what is called a quasi-Poisson distribution. The deviation can be characterized by the Fano factor, which is usually smaller than 1. It determines a significantly smaller peak-width ΔE in the spectrum, usually measured by the FWHM of the peak and covering about 76% of the total peak integral. It can be found by combining Equation 3.23 with 3.22:

$$\Delta E \approx 2.355\sqrt{F e_{\mathrm{pair}} E} \tag{3.25}$$

where e_{pair} is the average energy per electron-hole pair, and E is the energy of the detected photons. For a NaI scintillation detector, e_{pair} has to be replaced by e_{ion} (see Section 3.7.3).

Figure 3.26 shows the dependence of ΔE on the photon energy for different detectors with different e_{pair} and F values suitable for EDS. It is an unavoidable random statistical fluctuation characteristic for photon counting by electronic devices. It is called "quantum noise" or "shot noise" and is proportional to the square root of E. Unfortunately, there is a further independent source of random fluctuations, the electronic noise $\Delta E_{\mathrm{electronic}}$, which is largely independent of the photon energy and therefore is called "white noise." It arises from the detector, the FET, and amplifier even at low-temperature cooling and presents a lower limit relevant for small photon energies. This inherent electronic noise includes unforeseeable temporal fluctuations of current and voltage [50,54,55,71–74] and has three primary components:

1. The first component is called "current noise" and increases with shaping or peaking time $\Delta E_1 = a_1\sqrt{\tau}$. It can be represented by a noise source parallel to the detector and is proportional to its leakage current $a_1 \sim \sqrt{i_{\mathrm{leak}}}$, which is <75 pA [50]. The gain for modern SDDs is fivefold.
2. The second component is called "1/f noise" formerly called "Flicker noise," and is independent of the peaking time. Today, it is mostly called "low-frequency noise." Its origin is not completely understood but it is mainly caused by impurities of the sensor crystal. This kind of electronic noise has two subcomponents, which can be recognized when the time-dependent signals of current or voltage are Fourier-transformed into oscillations of different frequencies f. The first

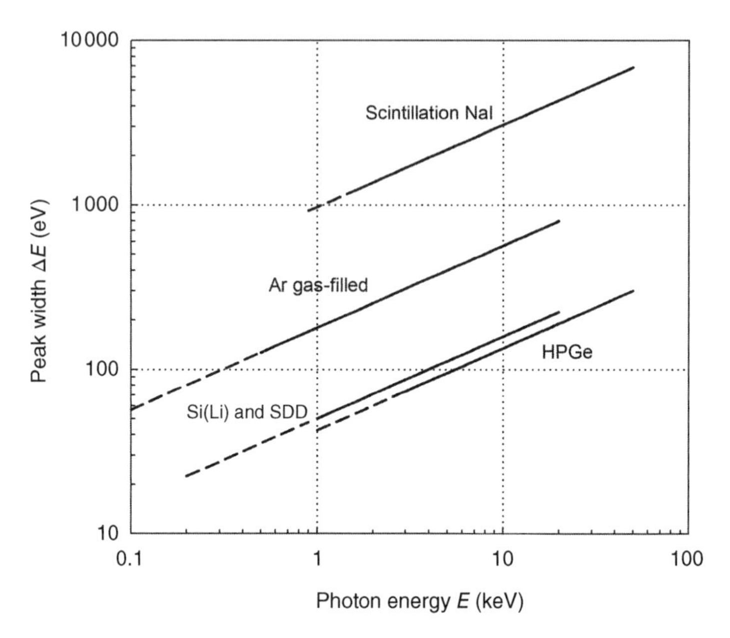

Figure 3.26. Peak width in eV caused by statistical fluctuations of counting photons with different energy. In the double logarithmic plot we get straight lines for different detectors suitable for EDS. These straight lines are asymptotes or lower limits for high energies and are called the Fano limit. They bend upward at low energies for the reason of electronic noise (see Figure 3.28).

subcomponent is independent of the frequency f but is proportional to the input capacitance $\Delta E_{21} = a_2 \sim C$. For solid-state detectors it is rather low though it is generally relevant [50]. For SDDs with integrated JFET and short connections, the capacity could be reduced by a factor 10 down to less than 50 fF and this subcomponent was reduced at the same rate. The second *subcomponent* is proportional to $1/f$ and becomes relevant below a corner frequency f_C. It can be described by the formula $\Delta E_{22} = a_2\sqrt{f_C/f}$. Both subcomponents are combined by the sum of their squares, which is demonstrated in Figure 3.27. In the literature of X-ray detection, this subcomponent is ignored or neglected because the corner frequency is quite low (<2 kHz for JFETs instead of 1 GHz for MOSFETs).

3. The third component is called "voltage noise," is in series with the detector, and decreases with peaking time $\Delta_3 = a_3/\sqrt{\tau}$. It is caused by a random thermal motion of charge carriers and increases with the temperature of the device. For that reason it is also referred to as "thermal noise" or "Johnson noise." Moreover, it is proportional to the input capacitance $a_3 \sim C \cdot \sqrt{T}$. This means that the higher temperature of SDDs can be compensated by a lower capacitance. The overall gain is threefold.

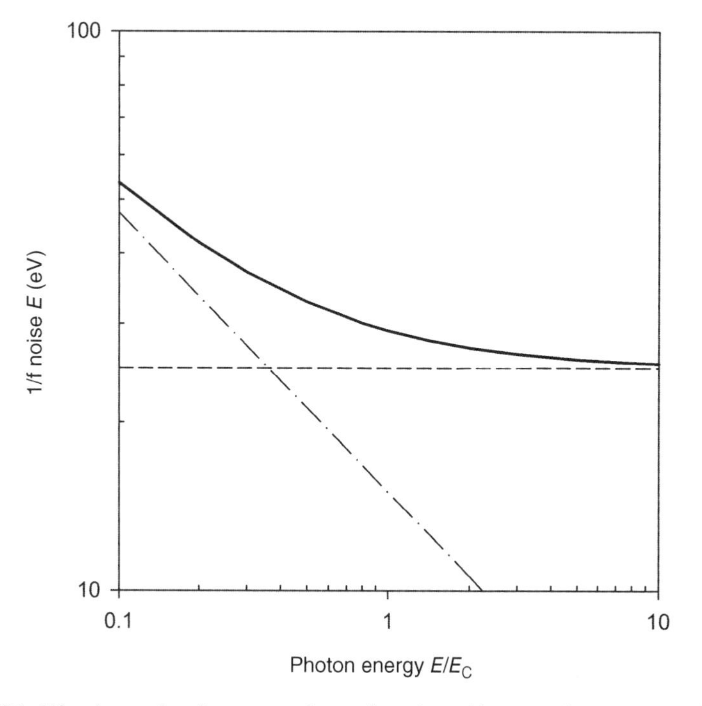

Figure 3.27. $1/f$ noise or low-frequency electronic noise with two subcomponents. The first subcomponent is decisive above the corner frequency, f_C, and is independent on f. The second subcomponent is dominant for $f < f_C$ and decreases with $\sqrt{1/f}$. The first subcomponent is called "white," the second is called "pink." In the double logarithmic plot, the first subcomponent is constant and amounts to only about 25 eV for SDDs, the second component shows a straight line with a slope of −0.5.

These three components—statistically uncorrelated—lead to a total electronic noise determined by the square root of the sum of variances [46–48,50]:

$$\Delta E_{\text{electronic}} = \sqrt{\left(a_1^2 \tau + a_2^2 + a_3^2/\tau\right)} \tag{3.26}$$

It is demonstrated in Figure 3.28 for a Si(Li) with liquid nitrogen cooling (−196 °C) and for an SDD with Peltier cooling (−20 °C) under the condition that photon energies are >1 keV, that is, the mentioned second subcomponent of the flicker noise was first neglected. Respective data a_1, a_2, and a_3 can be taken from different sources [50,56,57,71–74] and give similar pictures. The final curve consists of three components demonstrated in this figure. As can simply be checked by differentiation of Equation 3.26, the curve has a local minimum at $\tau_{\min} = a_3/a_1$ when "current noise" and "voltage noise" are equal. The minimum itself is $\Delta E_{\min} = \sqrt{a_2^2 + 2a_1 a_3}$. For Si(Li) the peaking time should be about 100 μs for an electronic noise of about 50 eV. The maximum count rate is between 10 and 40 kcps. For an SDD the peaking time

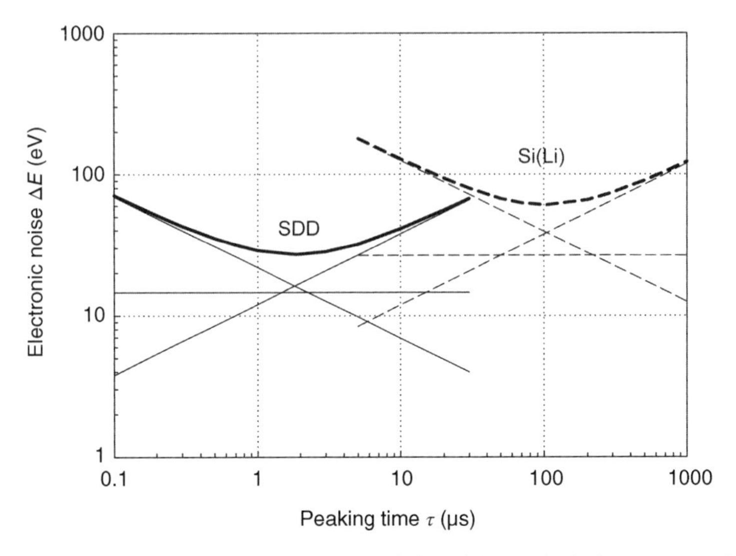

Figure 3.28. Electronic noise $\Delta E_{\text{electronic}}$ in eV consisting of three principal components: "parallel current" noise, "1/*f*" noise and "serial voltage" noise. These components differ mainly by their dependence on the peaking time τ in μs. They are proportional to $\tau^{1/2}$, τ^0, and $\tau^{-1/2}$ and lead to the typical profile of a hammock. For a Si(Li) detector (right) the electronic noise is strongly reduced by nitrogen cooling. Nevertheless, the noise of a new SDD (left) is two times smaller even if it is only Peltier-cooled. Because of the potentially smaller peaking times the count rate of SDDs can be distinctly higher.

can be about 2 μs and the electronic noise is reduced to about 30 eV at a count rate of about 200 kcps.

The electronic noise leads to an additional spread of the peaks. The total width becomes

$$\Delta E_{\text{total}} \approx \sqrt{\left(\Delta E^2_{\text{electronic}} + \left[2.355\sqrt{F e_{\text{pair}} E}\right]^2\right)} \tag{3.27}$$

Even though it is quite small, $\Delta E_{\text{electronic}}$ represents the lower limit of the spectral resolution of an EDS at low photon energies. Its amount can be read from a peak at $E = 0$ keV in the X-ray spectra. This "zero peak," also called "noise peak," has a width of about 35–50 eV. Table 3.9 gives data of e_{pair} or e_{ion} and F for five different detectors: Si(Li), SDD, HPGe, Ar gas-filled, and NaI scintillation detector as known from literature [48,49,62–64]. Data for ΔE_{total} based on $\Delta E_{\text{electronic}}$ were compared with measured values typical for good quality detectors and found to be in good agreement.

The total peak width is frequently used as a measure of the spectral resolution and is represented in Figure 3.29 as a function of the photon energy. The influence of the electronic noise causes a weak rise for SDDs below 1 keV. The figure is representative for detectors currently available but the results can

TABLE 3.9. Characteristic Data of Different Energy Dispersive X-Ray Detectors

Detector	e_{pair} or e_{ion} (eV)	Fano Factor F	$\Delta E_{electronic}$ (eV)
Ar gas-filled	26.4	0.8	100
NaI scintillation	200	1.0	200
Si(Li)	3.86	0.115	55
SDD	3.86	0.115	25
HPGe	2.96	0.10	60

differ for each individual detector on a 10% level. The solid-state detectors Si (Li), SDD, and HPGe show very similar characteristics. Note that the HPGe detector exhibits a better resolution than the Si(Li) or the SDD detector but the differences are on the 10% level. While the HPGe is better suited for the detection of higher photon energies, the SDD can even detect photons with energies down to 0.2 keV. For comparison, a gas-filled or a scintillation detector has a much poorer spectral resolution. For that reason, they normally are not used alone but only in combination with a Bragg crystal of a WDS.

The characteristics of five Bragg crystals are also illustrated in Figure 3.29. Their peak or line width is only between 0.3 and 100 eV for photons with energies between 0.1 and 10 keV. In this spectral region, their spectral resolution is very good. The energy-dispersive detectors are much poorer in resolution than a WDS with Bragg reflectors; however, their resolution is amply sufficient to separate the Kα peaks of neighboring elements in the periodic system. A good spectral resolution is important for the separation of neighboring peaks (see Section 4.3.2). It is also recommendable for a reduction of the spectral background with respect to low detection limits (see Section 6.1.2).

Conventionally, the spectral resolution[6] is only specified by the Mn-Kα peak with a photon energy of 5.9 keV. This quantity can simply be measured by means of the radioactive isotope iron-55 or ^{55}Fe, which decays to ^{55}Mn (half-life 2.7 years) emitting the Mn-Kα peak. Of course this is a simplification of the complex relationship represented by Figure 3.29. However, the single value of the peak width of Mn-Kα is well suited to characterize the influence of the size of the detector crystal. The size of the frontal area determines its capacitance and influences the noise component $\Delta E_{electronic}$ [57,68]. Table 3.10 presents a survey of crystals of different sizes and their influence on the resolution for Si (Li) and HPGe detectors [68].

The spectral resolution is usually determined at a low photon flux or count rate below 1000 cps. However, the resolution of an EDS is also dependent on this flux of the incident X-ray photons. Section 3.8.2.3 considers this dependence.

[6] The smaller the peak width, the better is the spectral resolution. For that reason, the peak width may not be suited as an ideal measure of the spectral resolution. The ratio of photon energy and peak width, defined as a relative measure, $R(E) = E/\Delta E$, may be better suitable.

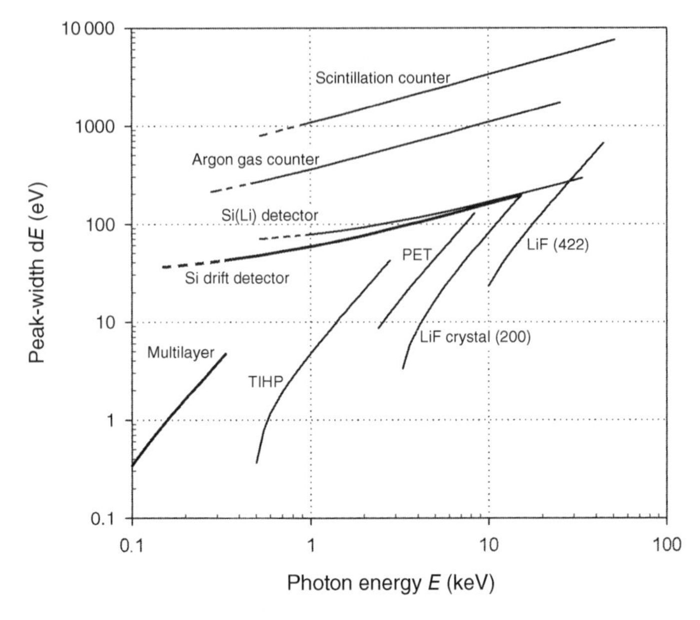

Figure 3.29. Total peak width or line width as a measure of the spectral resolution dependent on the photon energy or peak position in a double logarithmic plot. The peak width was calculated for four different proportional detectors: a NaI(Tl) scintillation detector, an Ar gas-filled detector, a Si (Li), and an SDD used for EDS. They are compared to the peak width of a WDS with Bragg crystals: LiF(422); LiF(200); PET; TlHP; and Mo/B_4C, each with a collimator of 0.15° divergence.

TABLE 3.10. Influence of the Crystal Size of Si(Li) and HPGe Detectors on the Spectral Resolution ΔE of the Mn-Kα Peak

Size of Detector Crystal		Spectral Resolution		
Diameter (mm)	Frontal Area (mm^2)	Noise ΔE_{noise} (eV)	Si(Li) ΔE (eV)	HPGe ΔE (eV)
3.6	10	50	128	115
6.2	30	62	133	120
8.0	50	75	140	127
10.1	80	95	151	140

Source: From Ref. [2], reproduced with permission. Copyright © 1996, John Wiley and Sons.

3.8.2.3 Input–Output Yield

The number of pulses produced by the detector crystal within a certain time interval is called the *input rate*. The number of pulses being processed by the electronic chain and indicated by the MCA is defined as the *output rate*. The input and output rates differ from one another due to the dead time effect of the electronic system.

The pulses coming from the detector have a certain peaking or shaping time that can be chosen between 0.1 and 100 μs. The pulses are transmitted to the MCA if they do not overlap within a period of 5–40 μs. Otherwise, their registration is refused by the pulse-pileup rejector. This period of rejection is the governing dead time of the system. Because of this dead time, the indicated output rate is always smaller than the given input rate. It falls more and more behind an increasing input rate due to increasing dead time losses.

All input–output curves follow an exponential equation defined by

$$n_{out} = n_{in} \cdot \exp(-\tau_{dead} \cdot n_{in}) \tag{3.28}$$

where n_{out} is the output rate, n_{in} is the input rate, and τ_{dead} is called dead time [51,72]. This dead time is a further characteristic quantity of the detector system and is composed by the sum of the peaking time or rise time of the electronic pulses, of their fall time, and of the conversion time of the ADC. It is about the three- to fivefold of the peaking time of the pulses.

The formula represents a set of curves with the dead time as parameter. All these curves are upside-down with a wide base for small τ_{dead} values. They show a vertex at $n_{in} = 1/\tau_{dead}$. The *maximum* output rate is $n_{out} = 1/e \cdot 1/\tau_{dead} \approx 0.37/\tau_{dead}$, which is inversely proportional to the dead time. The deviation from a linearity between input and output rate is caused by dead-time losses, which can be given as percentage values:

$$D = (n_{in} - n_{out})/n_{in} \cdot 100\% \tag{3.29}$$

The detector system will be "dead" during a percentage D of the total time and it is "live" for the remaining $(100 - D)$ percent. Inserting Equation 3.28 leads to

$$D = 1 - \exp(-\tau_{dead} \cdot n_{in}) \tag{3.30}$$

Characteristic input–output curves are presented in Figure 3.30 in a double logarithmic plot for a Si(Li) detector [46,47,51,69]. For small input rates, the output rate is at first linearly increasing. But then the increase slows down. After reaching a maximum, the output rate even decreases. Output rates with a constant D value are represented by dashed straight lines. They are parallel to the *ideal* straight line for which the output and input rates are equal, that is, $D = 0\%$. Four different sets are chosen here for the dead time and the dead time losses. They produce differences in the spectral resolution serving as parameter of the curves, that is, the width ΔE of the Mn-Kα peak can be assigned to the four different curves. Figure 3.31 illustrates the higher output yield for an SDD according to Refs [51–57,72,74].

With a shorter dead time or peaking time, the dead time loss can be reduced and a higher count rate can be recorded. On the other hand, the spectral resolution of the detection system deteriorates by increased electronic noise that leads to widening of the peaks. Naturally, a reasonable compromise has to

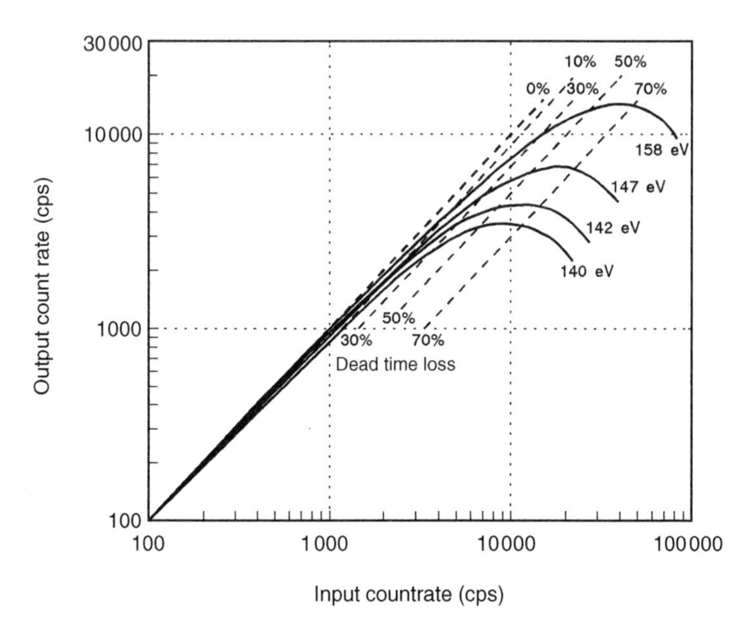

Figure 3.30. Indicated output rate dependent on the true input rate of an EDS system with a Si(Li) detector of 50 mm^2 front area. Due to the dead time losses, the curves deviate from the ideal straight line and finally show a choking of the detector. Each curve corresponds to a certain spectral energy resolution ΔE (in eV) and a dead time τ_{dead} (in μs), respectively. Constant dead time losses are indicated by dashed straight lines with D values in percent. Figure from Ref. [2], reproduced with permission. Copyright © 1996, John Wiley and Sons.

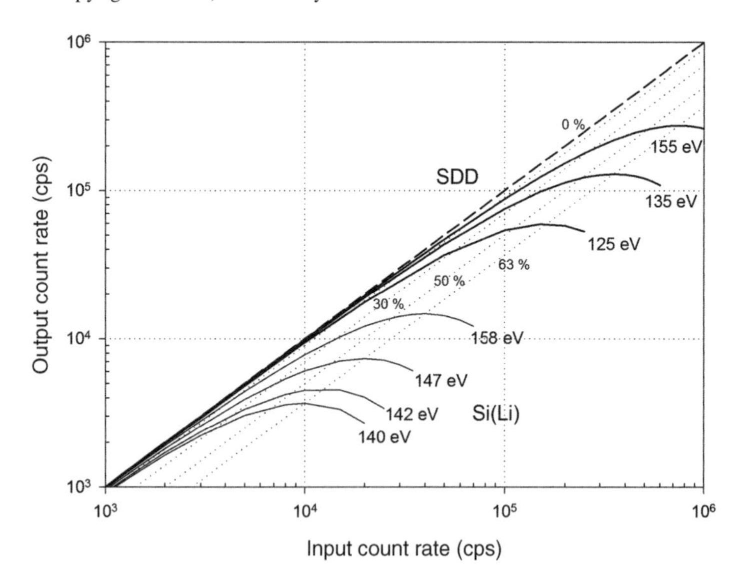

Figure 3.31. Indicated output rate dependent on the true input rate of an EDS system with a recent SDD of 25 mm^2 front area. These curves are similar to those of a Si(Li) detector (see Figure 3.30) but with a choking at higher input rates. They follow the set of functions given by Equation 3.28. The maxima are reached for a dead time loss of 63.2%. The input count rate for a Si(Li) can be up to 40 kcps; for an SDD it can reach about 1 Mcps.

be made between a high count rate and a small peak width, that is, "good" spectral resolution.

To determine the actual input rate from the recorded output rate, the dead time losses have to be corrected for. Up to the respective maximum, the actual input rate can be read unambiguously from the curves of Figures 3.30 and 3.31. Beyond this maximum, however, a measured output rate would lead to a second value for the unknown input rate. In order to avoid this ambiguity, output registration and input reading have to be restricted to values below the maximum.

In practice, the dead-time losses are compensated by a prolongation of the measurement. This can be achieved by stopping the timer for the period of dead time. If a "live" time t is preset for the measurement, the actual time for the acquisition of pulses will be extended to an "acquire" time T. For a given dead time loss D, it must be equal to

$$T = \frac{100}{100 - D} t \tag{3.31}$$

If the D value amounts to 50%, for instance, the acquire time is extended to the double of the preset live time. For this case, the input rate is just double the output rate.

The dead-time losses can be indicated by a special loss meter. As can be seen from Equation 3.29 and Figures 3.30 and 3.31, the 63% level determines the maximum output rate. This limit should not be exceeded as mentioned earlier, but for a high count rate the dead time loss should come close to this value. This rule can be a guideline for setting up excitation parameters and for achieving the optimum performance of an EDS.

3.8.2.4 The Escape-Peak Phenomenon

A peculiar phenomenon arises by the appearance of spurious peaks, called *escape peaks*, in X-ray spectra. They can be a real nuisance when they coincide with the small peaks of trace elements.

Escape peaks arise when a strong element peak is recorded by the detector. Accordingly, they can be regarded as "daughter" peaks produced by a strong "mother" peak. Their formation occurs within the sensor material of the detector, that is, the detector crystal, the gas, or the phosphor. When an incident X-ray photon is passing through these materials and its energy is sufficiently high, it can produce a photoelectron from an inner shell of a sensor atom. As a result, the excited atom can emit an X-ray photon by fluorescence, mostly a Kα or Kβ photon. It is normally reabsorbed in the sensor, creating a chain of electron-hole pairs and thus producing a charge pulse of the detector. However, this Kα or Kβ photon can also escape from the sensor with a certain probability. In that case, it carries off the quite definite energy $E_{K\alpha}$ or $E_{K\beta}$ of the element the sensor is composed of and does not produce its own chain of electron-hole pairs. On the one hand, this photon is lost for detection and the detector efficiency is subject to the discontinuities already shown in Figures 3.24 and 3.25. On the other hand, the residual energy shows up as an individual photon of the actual energy $E - E_{K\alpha}$ or $E - E_{K\beta}$. Such "packages" of energy or "photons" appear as a separate peak in the spectrum. Figure 3.32 shows a few such escape peaks due to some strong mother peaks. As is shown in the figure, their appearance is quite different for a Si(Li) or SDD detector on the one hand and for an HPGe detector on the other hand.

An effect similar to the escape-peak phenomenon will occur if secondary electrons instead of X-ray photons escape from the sensor volume, mainly from a near-surface layer. This effect is called *incomplete charge collection*. It leads to a reduction and tailing of the mother peak. The spectral background on the low-energy side of the mother peak is thereby lifted but only slightly (<0.1% of the peak height in a distance >500 eV).

The position of escape peaks is dependent on the position of the mother peak. Their peak height is mainly dependent on the fluorescence yield and the mass-absorption coefficient of the sensor material. This is demonstrated in Figure 3.33 for a Si(Li) detector and an HPGe detector. Escape peaks will only arise if the mother peak lies "above" the energy of the respective absorption

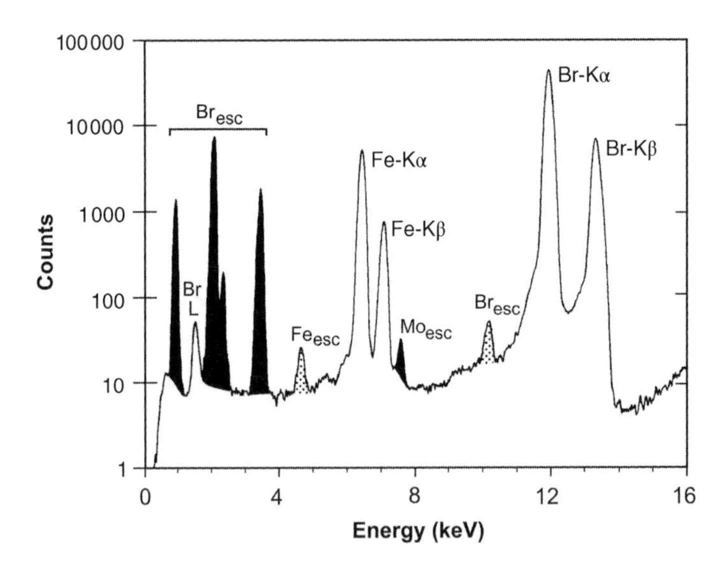

Figure 3.32. Energy-dispersive spectrum of $FeBr_2$ excited by a Mo X-ray tube and recorded by a Si(Li) detector (semilogarithmic). The Si escape peaks of iron and bromine are filled with dots. If an HPGe detector is used instead of the Si(Li) detector, four escape peaks of bromine and even one of molybdenum will distinctly appear whereas the two Si escape peaks will vanish. The Ge escape peaks are marked in black. Figure from Ref. [2], reproduced with permission. Copyright © 1996, John Wiley and Sons.

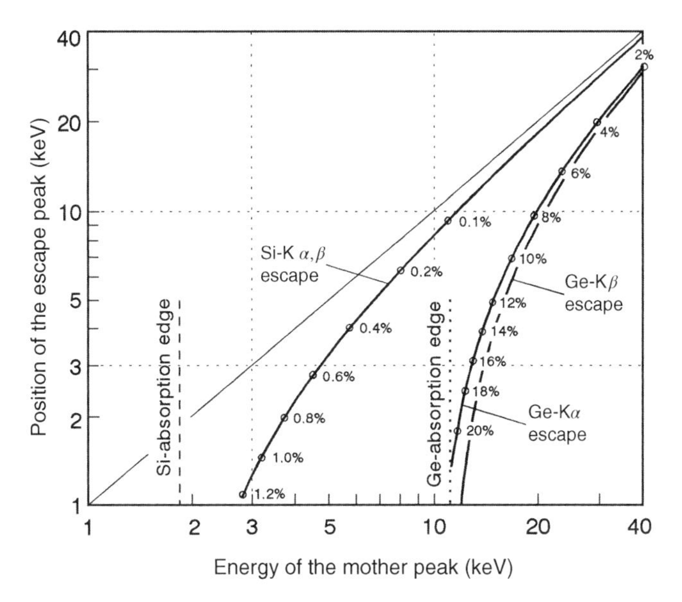

Figure 3.33. Position of *escape* peaks of a Si(Li) or SDD (left), and a HPGe detector (right) dependent on the energy of their *mother* peak. The percentage values noted on the curves give the peak height of the *escape* peaks in relation to that of the *mother* peaks. Figure from Ref. [46], reproduced with permission from Thermo Fisher Scientific.

edge of the crystal (1.839 keV for silicon and 11.103 keV for germanium). The position of the escape peak is given by

$$E_{\mathrm{Esc}} = E_{\mathrm{mother}} - E_{\mathrm{K}\alpha} \quad \text{or} \quad E_{\mathrm{Esc}} = E_{\mathrm{mother}} - E_{\mathrm{K}\beta} \tag{3.32}$$

$E_{\mathrm{K}\alpha}$ is 1.739 keV for silicon and 9.883 keV for germanium; $E_{\mathrm{K}\beta}$ is 1.831 keV for silicon and 10.981 keV for germanium.[7] The Si escape peaks are much smaller than the Ge escape peaks. For Si(Li) detectors, the escape peaks attain a height of only about 1% of the mother peak and each Kβ escape peak is strongly overlapped by the respective Kα escape peak. For Ge detectors, not only the Kα escape but also the relevant Kβ escape peak can be observed separately. Their peak heights reach up to 20% of the mother peak, leading to several annoying interferences.

For a gas-filled detector and a scintillation detector, we have similar relationships as illustrated in Figure 3.34. The K-absorption edge of argon is at 3.204 keV, the Kα peak lies at 2.957 keV and the Kβ peak at 3.190 keV. Respective values for sodium are 1.076, 1.041, and 1.067 keV. For iodine, the relevant K edge is at 33.164 keV, the Kα peak is at 28.607 keV, and the Kβ peak

[7] Escape peaks do not appear at $E_{\mathrm{mother}} - E_{\mathrm{edge}}$. The difference between E_{edge} and $E_{\mathrm{K}\alpha}$ or $E_{\mathrm{K}\beta}$ is not lost for detection but is used for the production of further electron-hole pairs.

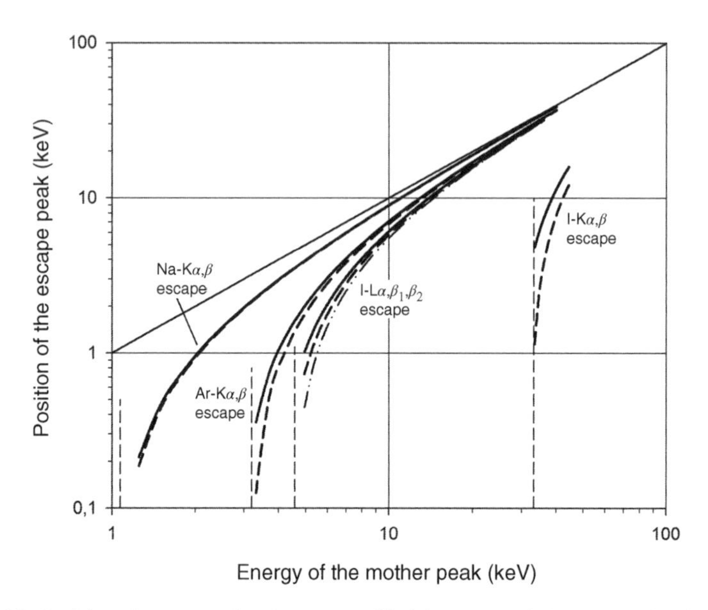

Figure 3.34. Position of *escape* peaks of an Ar gas-filled detector and of a NaI scintillation detector dependent on the energy of their *mother* peak (Na-K escape, Ar-K escape, I-L escape, and I-K escape). These annoying peaks can electronically be processed by "pulse-height discrimination."

lies at 32.289 keV. In normal practice, the excitation energy hardly lies above the K-absorption edge of iodine. However, the LIII edge of iodine is exceeded in usual excitation. The energy of 4.559 keV is required to excite the $L\alpha_1$ peak at 3.937 keV and the $L\beta_2$ peak at 4.507 keV. The respective escape peaks may arise in the spectra (if taken without Bragg crystals); however, they are usually suppressed by "pulse-height discrimination."

REFERENCES

1. Marten, R., Rosomm, H., and Schwenke, H. (1977). German Patent pend., Nr. P 26 32 001.4.
2. R. Klockenkämper (1997). *Total-Reflection X-Ray Fluorescence Analysis*, 1st ed., John Wiley & Sons, Inc., New York.
3. Iida, A., Yoshinaga, A., Sakurai, K., and Gohshi, Y. (1986). Synchrotron radiation excited X-ray fluorescence analysis using total reflection of X-rays. *Anal. Chem.*, **58**, 394–397.
4. Rieder, R., Wobrauschek, P., Ladisich, W., Streli, C., Aiginger, H., Garbe, S., Gaul, G., Knöchel, A., and Lechtenberg, F. (1995). Total reflection X-ray fluorescence analysis with synchrotron radiation monochromatized by multilayer structures. *Nucl. Instrum. Methods*, **A355**, 648–653.

5. Iida, A., Gohshi, Y., and Matsushita, T. (1985). Energy-dispersive X-ray fluorescence analysis using synchrotron radiation *Adv. X-Ray Anal.*, **28**, 61–68.
6. Rich. Seifert & Co. (1985). X-Ray Tubes. Technical booklet.
7. Rigaku International Corporation (1992). Rotating Anode X-Ray Generator Systems. Technical Brochure.
8. Windeck, C. and Warrikhoff, H. (1998). *A New Series of Metal-Ceramic X-Ray Tubes for Low Power Applications*. 7th ECNDT European Conference for Non-Destructive Testing, Vol. 3, Broendby Copenhagen/Denmark, 2790–2795.
9. Toshiba X-Ray Tube History 1915–2015. 1385 Shimoishigami, Otawara-shi, Tochigi, 324-8550 Japan.
10. Kwang-Je, K. (2009). Characteristics of synchrotron radiation. In: Thomson, A.C. (editor) *X-Ray Data Booklet*, 3rd ed., Center for X-Ray Optics and Advanced Light Source, Lawrence Berkeley National Laboratory, University of California, Berkeley, CA, pp. 2-1–2-16.
11. Yanwei, L. (2009). Free electron lasers (FELs) at extreme ultraviolet and X-ray wavelengths. In: Thomson, A.C. (editor) *X-Ray Data Booklet*, 3rd ed., pp. 2-17–2-20.
12. Winick, H. and Attwood, D. (2009). Synchrotron radiation facilities. In: Thompson, A.C. (editor) *X-Ray Data Booklet*, 3rd ed., pp. 2-29–2-32.
13. Robinson, A.L. (2009). History of synchrotron radiation. In: Thomson, A.C. (editor) *X-Ray Data Booklet*, 3rd ed., pp. 2-21–2-28.
14. Luchini, P. and Motz, H. (1990). *Undulators and Free-Electron Lasers*, Oxford University Press: Oxford.
15. Wille, K. (1992). *Physik der Teilchenbeschleuniger und Synchrotronstrahlungsquellen*, Teubner Studienbücher: Physik, Stuttgart.
16. http://en.wikipedia.org/wiki/Synchrotron/Synchrotron, Wikimedia Foundation, INC, USA, 2014, Synchrotron, 2014.
17. Edwards, G.S., Austin, R.H., Carroll, F.E., Copeland, M.L., Couprie, M.E., Gabella, W.E., Haglund, R.F., Hooper, B.A., Hutson, M.S., Jansen, E.D., Joos, K.M., Kiehart, D.P., Lindau, I., Miao, J., Pratisto, H.S., Shen, J.H., Tokutake, Y., van der Meer, A.F.G., and Xie, A. (2003). Free-electron-laser based biophysical and biomedical instrumentation. *Rev. Sci. Instrum.*, **74**, 3207–3246.
18. Feldhaus, J., Arthur, J., and Hastings, J.B. (2005). X-ray free-electron lasers. *J. Phys. B: At. Mol. Opt. Phys.*, **38**, 799–819.
19. Maiman, T.H. (1960). Stimulated optical radiation in ruby. *Nature*, **187** (4736), 493–494.
20. Siegman, A.E. (1986). *Lasers*, University Science Books: Mill Valley, CA, USA, pp. 2.
21. Jones, K.W. (2002). Synchrotron radiation-induced X-ray emission. In: Van Grieken, R.E. and Markowicz, A.A. (editors) *Handbook of X-Ray Spectrometry*, 2nd ed., Marcel Decker: New York, pp. 501–558.
22. www.lightsources.org, 2014, Light Sources Worldwide, 2014.
23. http://en.wikipedia.org/wiki/List_of_synchrotron_radiation_facilities, Wikimedia Foundation, INC, USA, 2014, List of Synchrotron Facilities, 2014.

24. www-elsa.physik.uni-bonn.de/accelerator_list.html, Physikalisches Institut der Universität Bonn, Germany, 2014, Particle Accelerators Around the World, 2014.
25. http://www.ira.inaf.it/~ddallaca/P-Rad_3.pdf, INAF Istituto Nazionale di AstroFisica, Italy.
26. http://www.cxro.lbl.gov, A U.S. Department of Energy National Laboratory Operated by the University of California, 2014, X-ray Data Base, 2014.
27. von Bohlen, A. and Tolan, M. (2008). Editorial for synchrotron radiation. *J. Anal. At. Spectrom.*, **23**, 790–791.
28. Schwenke, H., Knoth, J., Marten, R., and Rosomm, H. (1978). German Patent pend., Nr. P 27 36 960.4.
29. Knoth, J., Schneider, H., and Schwenke, H. (1994). Tunable exciting energies for total reflection X-ray fluorescence spectrometry using a tungsten anode and band-pass filtering. *X-Ray Spectrom.*, **23**, 261–266.
30. Ladisich, W., Rieder, R., Wobrauschek, P., and Aiginger, H. (1993). Total reflection X-ray fluorescence analysis with monoenergetic excitation and full spectrum excitation using rotating anode X-ray tubes. *Nucl. Instrum. Methods*, **A330**, 501–506.
31. Schwenke, H. and Knoth, J. (1982). A highly sensitive energy-dispersive X-ray spectrometer with multiple total reflection of the exciting beam. *Nucl. Instrum. Methods*, **193**, 239–243.
32. Kregsamer, P. and Wobrauschek, P. (1991). Total reflection X-ray fluorescence analysis of the rare earth elements by K-shell excitation. *Spectrochim. Acta*, **46B**, 1361–1367.
33. Wobrauschek, P., Kregsamer, P., Ladisich, W., Rieder, R., and Streli, C. (1993). Total-reflection X-ray fluorescence analysis using special X-ray sources. *Spectrochim. Acta*, **48B**, 143–151.
34. Schwenke, H. and Knoth, J. (1993). Total reflection XRF. In: van Grieken, R. and Markowicz, A. (editors) *Handbook on X-Ray Spectrometry*, Vol. **14**, Practical Spectroscopy Series, Dekker: New York, pp. 453–490.
35. Schuster, M. (1991). A total reflection X-ray fluorescence spectrometer with monochromatic excitation. *Spectrochim. Acta*, **46B**, 1341–1349.
36. cheiron2010.spring8.or.jp/text/lec/4_X-ray_Monochromators_T.Matsushita.pdf, Photon Factory High Energy Accelerator Research Organization, 2010, X-ray Monochromators, 2014.
37. Schmitt, M., Hoffmann, P., and Lieser, K.-H. (1987). Perspex as sample carrier in TXR. *Fresenius Z. Anal. Chem.*, **328**, 594–595.
38. Kollotzek, D. (1980). Beitrag zur Elementspurenbestimmung durch Röntgenfluoreszenzspektrometrie mit Anregung unter Totalreflexion der Primärstrahlung. Diploma thesis, University of Stuttgart.
39. Prange, A., Kramer, K., and Reus, U. (1993). Boron nitride sample carriers for total-reflection X-ray fluorescence. *Spectrochim. Acta*, **48B**, 153–161.
40. Wobrauschek, P. and Aiginger, H. (1979). Totalreflexions-Röntgenfluoreszenzanalyse. *X-Ray Spectrom.*, **8**, 57–62.
41. Prange, A. and Schwenke, H. (1992). Trace element analysis using total-reflection X-ray fluorescence spectrometry. *Adv. X-Ray Anal.*, **35B**, 899–923.

42. Ellis, A.T. (2002). Energy-dispersive X-ray fluorescence analysis using X-ray tube excitation. In: Van Grieken, R.E. and Markowicz, A.A. (editors) *Handbook of X-Ray Spectrometry*, 2nd ed., Marcel Decker: New York, pp. 199–238.
43. Piorek, St. (2002). Radioisotope-excited X-ray analysis. In: Van Grieken, R.E. and Markowicz, A.A. (editors) *Handbook of X-Ray Spectrometry*, 2nd ed., Marcel Decker: New York, pp. 433–500.
44. Philips Analytical (1991). Characterization of Thin-Layered Samples by Glancing Incidence X-ray Analysis. Technical Brochure.
45. Hüppauf, M. (1993). Charakterisierung von dünnen Schichten und von Gläsern mit Röntgenreflexion und Röntgenfluoreszenzanalyse bei streifendem Einfall. Ph.D. Thesis, RWTH Aachen; Berichte des Forschungszentrums Jülich, Jül-2730.
46. Woldseth, R. (1973). *All You Ever Wanted to Know About X-Ray Energy Spectrometry*, Kevex Corp.: Burlingame, California.
47. Gedcke, D.A. (1972). The Si(Li) X-ray energy analysis system: Operating principle and performance. *X-Ray Spectrom.*, **1**, 129–141.
48. Bertin, E.P. (1975). *Principles and Practice of Quantitative X-Ray Fluorescence Analysis*, 2nd ed., Plenum Press: New York, pp. 163–213; 220–246; 981–1002.
49. Williams, K.L. (1987). *An Introduction to X-Ray Spectrometry*, Allen & Unwin: London, 370 pages. pp. 49–55; 88–91; 101–110.
50. Iwanczyk, J.S., Patt, B.E., Segal, J., Plummer, J., Vilkelis, G., Hedman, B., Hodgson, K.O., Cox, A.D., Rehn, L., and Metz, J. (1996). Simulation and modelling of a new silicon X-ray drift detector design for synchrotron radiation applications. *Nucl. Instr. Meth. in Phys. Res.*, **A380**, 288–294.
51. Lechner, P., Fiorini, C., Hartmann, R., Kemmer, J., Krause, N., Leutenegger, P., Longoni, A., Soltau, H., Stötter, D., Stötter, R., Strüder, L., and Weber, U. (2001). Silicon drift detectors for high count rate X-ray spectroscopy at room temperature. *Nucl. Instrum. Methods Phys. Res.*, **A458**, 281–287.
52. Oxford Instruments (2010). Silicon Drift Detectors Explained. Technical Brochure, 27 pages.
53. PNDetector GmbH (2013). Silicon Drift Detector. Product Brochure, 4 pages; http://www.pndetector.de/brox and PND_Products_Update_2013_s_pdf.
54. Bruker-AXS (2012). Working Principle – Silicon drift chamber principle (SDD). Technical Brochure. http://www.bruker-axs.com/xflashsddx-raydet0.html, 2 pages.
55. Amptec (2012). Super SDD. Application Note. http://www.amptek.com/, 11 pages.
56. Spieler, H. (2002). Pulse processing and analysis. IEEE: Short course; radiation, detection and measurement. http://www-physics.lbl.gov/~spieler/NSS_short-course/NSS02_Pulse_Processing.pdf
57. Pahlke, A. (2003). Einfluss der Oxidqualität auf die Stabilität von Halbleiterdetektoren bei Röntgenbestrahlung, Dissertation an der Technischen Universität München, 184 pages.
58. Berlin, J. (2011). Analysis of boron with energy dispersive X-ray spectrometry. *Imaging & Microscopy*, **13**, 19–21.
59. Eikenberry, E.F., Brönnimann, Ch., Hülsen, G., Toyokawa, H., Horisberger, R., Schmitt, B., Schulze-Briese, C., and Tomizaki, T. (2003). PILATUS: a two-dimensional X-ray detector for macromolecular crystallography. *Nucl. Instrum. Methods Phys. Res.*, **A501**, 260–266.

60. http://www.hamamatsu.com, Hamamatsu Photonics K.K., 2014, Image Sensors, 2014.
61. http://www.studiodentisticovenuti.it/wp-content/uploads/2012/06/x-ray-detectors.pdf, Studio Dentistico Venuti, Italy, 2012, X-ray Detectors, 2014.
62. Jenkins, R. (1977). *Einführung in die Röntgenspektrometrie*, Heyden: London, pp. 71–91.
63. Klockenkämper, R. (1981). Röntgenspektralanalyse am Rasterelektronenmikroskop II Wellenlängendispersive Spektrometrie. In: Bock, R. Fresenius, W., Günzler, H., Huber, W., and Tölg, G. (editors) *Analytiker Taschenbuch, Band 2*, Springer Verlag: Berlin, Heidelberg, New York, pp. 182–186.
64. Helsen, J.A. and Kuczumow, A. (2002). Wavelength-dispersive X-ray fluorescence. In: Van Grieken, R.E. and Markowicz, A.A. (editors) *Handbook of X-Ray Spectrometry*, 2nd ed., Marcel Decker: New York, pp. 95–198.
65. Sakurai, K., Eba, H., Inoue, K., and Yagi, N. (2002). Wavelength-dispersive total-reflection X-ray fluorescence with an efficient Johansson spectrometer and an undulator X-ray source: detection of 10^{-16} g-level trace metals. *Anal. Chem.*, **74**, 4532–4535.
66. Schwenke, H., Beaven, P.A., Knoth, J., and Jantzen, E. (2003). A wavelength-dispersive arrangement for wafer analysis with total reflection X-ray fluorescence spectrometry using synchrotron radiation. *Spectrochim. Acta*, **B58**, 2039–2048.
67. Awaji, N. (2003). Wavelength dispersive grazing incidence X-ray fluorescence of multilayer thin films. *Spectrochim. Acta*, **B59**, 1133–1139.
68. Ellis, A.T. (1991). PICXAM-Workshop, Hawaii. Instruction material.
69. Oxford Instruments (1993). Link EDX Detectors. Technical booklet.
70. Klockenkämper, R. and von Bohlen, A. (1989). Determination of the critical thickness and the sensitivity of thin-film analysis by total reflection X-ray fluorescence spectrometry. *Spectrochim. Acta*, **44B**, 461–470.
71. Goulding, F.S. and Landis, D.A. (1982). Signal processing for semiconductor detectors. *IEEE Trans. Nucl. Sci.*, **NS-29**, 1125–1141.
72. Eggert, T. (2004). Die spektrale Antwort von Silizium-Röntgendetektoren. Dissertation an der Technischen Universität München, 165 pages.
73. Amptec (2010). Amptec Silicon Drift Detectors (SDD). Application Note AN-SDD-003. http://www.amptek.com/ansdd003.html
74. Haschke, M., Waldschläger, U., and Scheller, S. (2012). *Energieauflösung vs. Zählraten – Was ist der beste Kompromiss?* 19. Anwendertreffen Röntgenfluoreszenz- und Funkenemissionsspektrometerie, Dortmund.

CHAPTER

4

PERFORMANCE OF TXRF and GI-XRF ANALYSES

An analytical strategy has to be based on the prerequisites of the method to be applied. Abovc all, total reflection X-ray fluorescence (TXRF) is restricted to small sample amounts. Only micrograms of a solid material and less than 100 μl of a liquid can be analyzed at one time. Consequently, TXRF is a method of microanalysis, as is defined by the International Union of Pure and Applied Chemistry (IUPAC) [1], and samples can seldom be analyzed as received.

Total-Reflection X-ray Fluorescence Analysis and Related Methods, Second Edition.
Reinhold Klockenkämper and Alex von Bohlen.

A certain pretreatment is generally required, in contrast to conventional XRF. Samples have to be prepared as solutions, suspensions, fine powders, or thin sections. Solids must be ground or dissolved. For a determination of ultratrace components, the matrix of the sample should first be separated and removed. For that purpose, all techniques that have already been tested and combined with other methods of atomic spectroscopy, for example, with atomic absorption spectrometry (AAS) or ICP-OES, can be used. Certain precautions have to be taken in dealing with small samples, and working with a clean-bench (a laminar flow cabinet) is mandatory for critical steps of sample preparation.

On the other hand, TXRF is a variant of energy-dispersive X-ray spectrometry and shares all the convenient features. The complete spectrum is recorded simultaneously within seconds; it is displayed on a screen, and the registration can be observed continually during the measurement. A dedicated computer is usually incorporated for advanced processing of the spectra. Automatic peak identification is made possible, enhancing the speed and ease of a qualitative analysis. A visual comparison of two complete spectra enables a fingerprint analysis.

Quantitative analysis by TXRF is essentially facilitated by the use of only small amounts of sample. Troublesome matrix effects do not arise—neither absorption nor enhancement effects. Quantification can therefore be carried out after the addition of an element serving as the internal standard. For a single-element analysis, the analyte can itself be used as the standard. For multielement analyses, any element not present in the sample can be chosen as the standard against which all the other elements are to be determined. In this case, the different sensitivity values for these elements are needed. They have to be determined prior to analysis, but only once for each new instrument.

Surface and thin-layer analyses can be carried out by TXRF only in combination with a stratified etching of flat and even samples. However, another simple variant for a *direct* analysis is given by GI-XRF (grazing incidence XRF). The glancing angle of the primary beam has to be varied in the region of total reflection and the peak intensity of concerned elements has to be recorded simultaneously. The angle-dependent intensity profiles give a first qualitative picture of contaminants, layers, and/or a substrate. However, quantification by an internal standard is not possible because the primary beam not only passes through a thin upmost layer but also penetrates to a greater depth when the critical angle of total reflection is exceeded. To get a quantitative description of the layered system, an algorithm has to be applied that is already known from conventional XRF and is called fundamental parameter method. It is based on a simple model in order to calculate fluorescence intensities of the individual elements while allowing for matrix effects. Only one external standard is needed. The fundamental data can be obtained from tables or partly be calculated by use of equations.

4.1 PREPARATIONS FOR MEASUREMENT

As stated earlier, TXRF is a method of microanalysis and is directly applicable when only a small sample amount is available or when only a small part of a larger amount can be taken. So when a large sample amount is received for analysis, an appropriate sampling of a smaller part has to be carried out first. This part may be called the *specimen* in contrast to the total amount, which is called the *sample* [2]. In order to get a representative result by means of a specimen, the sample has to be homogeneous from the beginning or thoroughly homogenized prior to sampling. At the same time, the main constituent or the matrix of the sample should be removed if possible so that the analyte is enriched and essentially present in the actual specimen. Finally, this specimen is placed on a flat carrier for analysis. All these steps of preparation and presentation have to be carried out very carefully. Only clean or specially cleaned vessels, instruments, and of course carriers should be employed. In general, only analytically pure or suprapure reagents may be used. Acids should be finally purified by subboiling; water should be prepared by a double-stage deionization or bidistillation. Beginners in the field of micro- and trace analyses have to be warned against negligence. The utmost care and frequent regular checks are indispensable even though it is time-consuming.

Highly pure acids needed for sample preparation are commercially available in the p.a. grade ("p.a." = pro analysi) or suprapure. They may have, however, some residual impurities of the order of ng/ml. Especially, elements like Mg, Al, Fe, Cu, Zn, and Pb may show high blank values. The same problem arises for water even if it is deionized or bidistilled; Cl, K, Ca, Br, and Sr are the troublesome elements here. However, purification of these liquids is possible by *subboiling* distillation. The liquid is vaporized by heating below the boiling point. The vapor is condensed at a cooling finger, and the purified condensate is collected in a small clean flask. Impurities are significantly below 1 ng/ml.

4.1.1 Cleaning Procedures

For the preparation of samples, only vessels made of quartz glass (Suprasil or Synsil), PTFE (polytetrafluorethylenes, e.g., Teflon), or PP (polypropylenes) should be used. Due to their high purity, contamination by these materials is greatly reduced. Also, vessels made of PFA (perfluoroalkoxy polymers) can be recommended because of their especially smooth walls. Before the vessels are used, they must be cleaned by boiling and additional steaming (except for PP). Figure 4.1 represents a *steaming* device as it is usually employed [3]. Pure nitric acid and pure water are recommended as cleansing agents and should be used successively. The liquids are vaporized, and the respective extremely clean vapor is condensed at the cool vessels in the steaming chamber. The condensate takes up the impurities on the

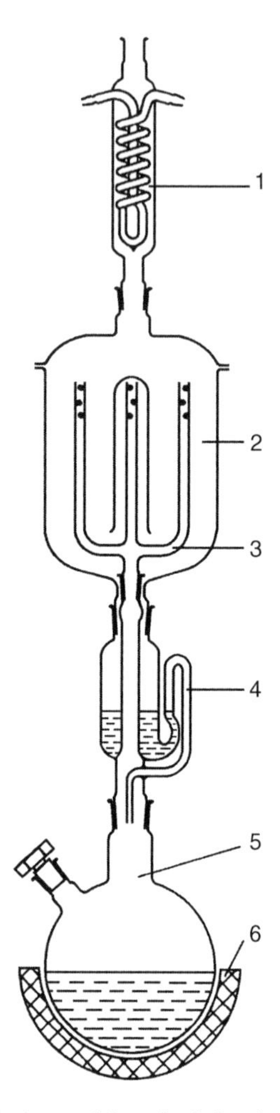

Figure 4.1. Set-up of a steaming device used for a final cleaning of glass vessels and instruments: (1) condenser; (2) steam chamber; (3) rack of quartz glass as support for vessels; (4) overflow; (5) glass flask; (6) heating jacket. Figure from Ref. [3], reproduced with permission. Copyright © 1980 by Springer.

surface of the vessels and moves them into the liquid phase. This process is continued for a period of up to 12 h.

The specimen should be placed on highly clean carriers. Plexiglas carriers can be applied without cleaning. These cheap carriers are used only once. All the other more expensive carriers are used frequently for cost savings. They must to be cleaned even before their first use because they are not delivered in sufficiently clean condition.

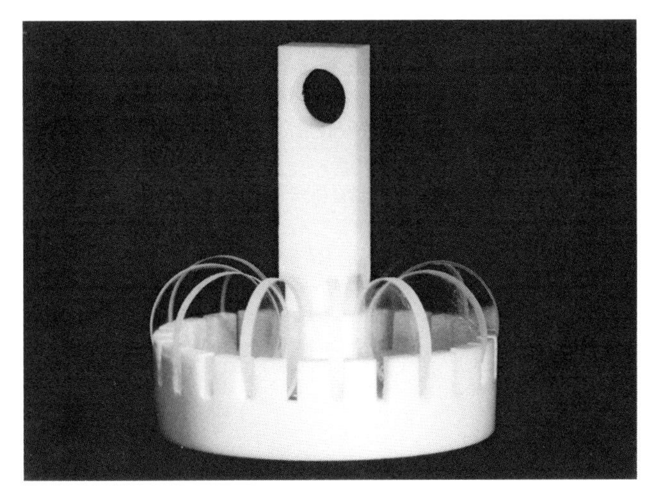

Figure 4.2. A special support of carriers used in the cleaning process. The support is made of Teflon® and is designed for 24 carriers. Figure from Ref. [4], reproduced with permission. Copyright © 1996, John Wiley and Sons.

A simple *cleaning* process can be recommended that works even without a time-consuming steaming step. This process is preferably carried out for a set of several carriers. Figure 4.2 shows a special support made of PTFE and designed to take 24 different carriers with a diameter of 30 mm. This support is immersed in a large beaker of 800 ml, half filled with a cleaning agent (e.g., RBS 50 in a 10% dilution; Carl Roth GmbH & Co., Karlsruhe, Germany), which is brought to a boil and cooled down. The support with the carriers is rinsed with distilled water and put into a second beaker with Milli-Q water (Millipore Corp., Bedford, Massachusetts). This water is boiled and then cooled down to about 40 °C, after which all carriers and the support are dried with fluff-free precision wipers (Kimwipes, Kimberly-Clark Corp., Northop, UK). Thereafter, the support with the carriers is placed in a third beaker with concentrated nitric acid (p.a. grade, E. Merck, Darmstadt, Germany), boiled for 1 h, cooled down, and again placed in the second beaker with fresh ultrapure water. It is warmed up to about 60 °C for 1 h. Each bath has to be prepared freshly, and the beakers must be covered appropriately. The boiling must take place in a clean fume cupboard only used for such a purpose in order to preclude any source of contamination. The support and carriers must not be touched by hand.

After a last cooling to 40 °C in a clean bench, the support with the carriers is lifted out and remaining droplets are wiped down. The dried carriers are put into clean Petri dishes. Covered by the top, they are kept in a drawer until needed for analysis. The total procedure takes about 4 h. The clean carriers can be used for deposition of solid samples without restrictions. For deposition of liquid samples they must be hydrophobic, otherwise the deposited droplets run out. Silicon, glassy carbon, boron nitride, and Plexiglas are hydrophobic by nature but quartz glass is not. It is initially to be coated with a hydrophobic film. For that purpose, a 2 μl droplet of a silicone solution (Serva GmbH & Co., Heidelberg, Germany) is usually pipetted onto quartz glass carriers. It is spread over a circular area of 1 cm^2 and dried in a small laboratory oven at 100 °C within 1 h. This oven should be used exclusively for that purpose.

The result of the carrier cleaning can be checked by TXRF itself. A spectrum of each carrier is recorded within 100 s and inspected for impurities. For example, Figure 4.3 shows element peaks arising from residual contaminations of about 10 pg. Such carriers should be sorted out and cleaned again in order to

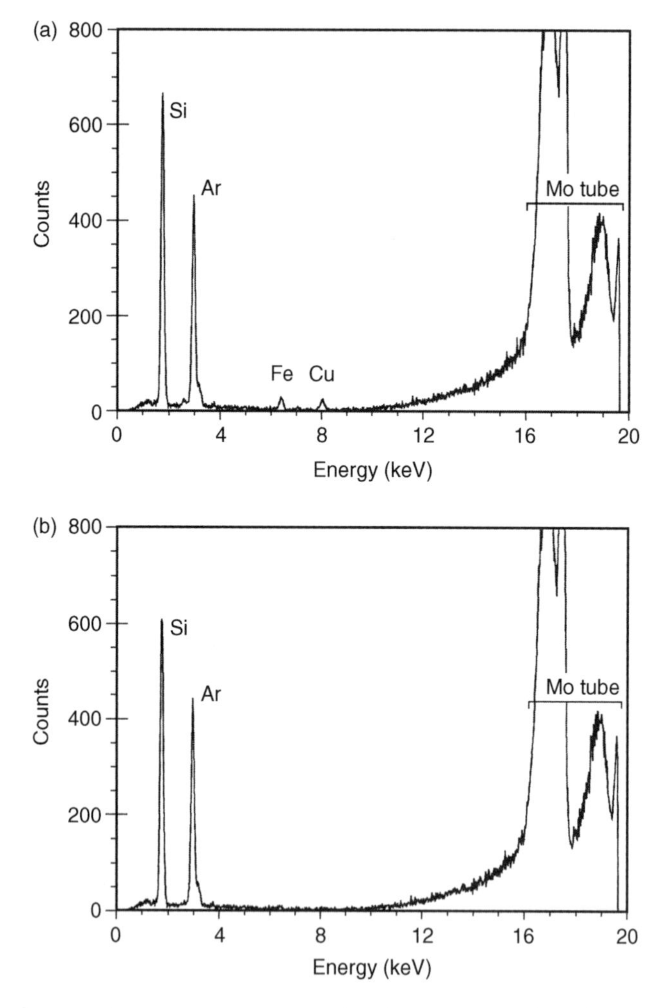

Figure 4.3. Spectra of a quartz-glass carrier after cleaning. (a) A first attempt shows small impurities of Fe and Cu. The peak intensities of about 20 counts per 100 s come from an amount of only 10 pg. (b) A second attempt turns out to be successful. Figure from Ref. [4], reproduced with permission. Copyright © 1996, John Wiley and Sons.

be suitable for ultratrace analysis. The success rate will be above 95% if a skillful operation is carried out.

4.1.2 Preparation of Samples

When the analysis is to be performed of a small specimen representative of a larger amount of sample material, this material must first be homogenized. Furthermore, when a trace analysis is to be performed, the sample matrix should preferably be separated. In general, these operations are more difficult for solids than for liquids and more difficult for inorganic than for organic or biomaterials [5,6]. A diagram of preparatory steps taken prior to TXRF analysis is outlined in Figure 4.4.

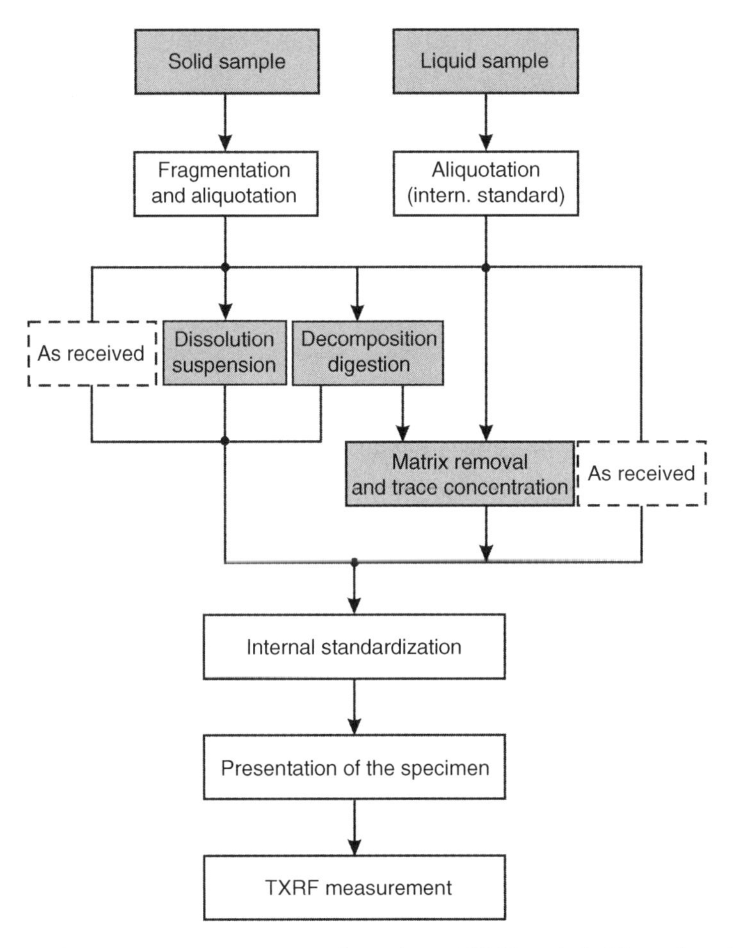

Figure 4.4. Diagram of preparatory steps taken prior to TXRF analysis. Figure from Ref. [4], reproduced with permission. Copyright © 1996, John Wiley and Sons.

- *Solid bulk materials* can first be fragmented by cutting, sawing, crumbling, shredding, and so on. The small pieces can then be ground down in a mill or mortar. The *fine powder* prepared in this way or received already pulverized is poured into a solution of water and ethanol. A fairly homogeneous suspension is achieved by shaking or thoroughly mixing with a magnetic stirrer or by ultrasonication. Aliquots can be taken by a pipette and used as specimens for TXRF.

  The solid material preferably in the *pulverized* form can be dissolved either in total or in parts. Different inorganic and organic solvents are suited for that purpose but have to be ascertained for the individual material. After dissolution, an aliquot can be taken, pipetted on a carrier, and dried. In this way, the previous dilution of the sample is reversed and the detection of traces is not affected.
- *Biomaterials* in particular can be subjected to various methods of sample preparation. Such methods are chosen to decompose the sample matrix and to transform it into a clear colorless solution. Samples can be ashed in an oxygen plasma driven by a radiofrequency discharge at a moderate temperature (possibly under high pressure). The ashed sample can be dissolved in nitric acid, for instance, and aliquots can be applied as specimens. A corresponding process can even be carried out on quartz glass carriers [7].
- Furthermore, *mineral samples* and *biomaterials* can be digested in either an open or a closed vessel. The sample material is usually inserted into a Teflon or quartz glass vessel, and a strong acid or mixture of acids is added. Usually, nitric acid is chosen as are hydrogen peroxide, hydrofluoric acid, and hydrochloric acid. Perchloric and sulfuric acid are not suited for TXRF because they do not evaporate afterward. Because of its shorter decomposition time, microwave heating is preferred to conventional heating. If the vessel is closed, the pressure will be increased and the decomposition accelerated. The temperature may come up to 300 °C and the pressure to 10^7 Pa. The digestion in a closed "bomb" is often called *pressure digestion*. It is particularly advantageous because element losses caused by volatilization are avoided. The decomposition is normally complete in 1 h or less, and the final solution can be used for analysis.

  The methods just considered are intended for homogenization of solid bulks. They usually transform the solid materials into a liquid phase. The final solution contains the original sample in a certain dilution. For trace analyses, this solution should be concentrated, that is, main compounds of the sample should be separated and the traces should be collected for analysis. Various methods already tested for other methods of elemental analysis, for example, for AAS and ICP-OES, have been utilized for this purpose. They are suitable for digested samples (e.g., metal digests) and for samples received as liquids.

- *Liquids* can be concentrated by evaporation at elevated temperatures. Such drying is easily feasible with aqueous solutions. Other liquids are more or less suitable according to their boiling point and vapor pressure. A very gentle evaporation is made possible by freeze-drying. The sample is frozen just below the freezing point and simultaneously exposed to a fine vacuum of 0.1 Pa. When such an evaporation is performed for hours, it can raise the concentration of trace components up to 10^6-fold.
- Another simple method is based on volatilization of the matrix. By reduction of the volume and/or by addition of a liquid reagent, a chemical reaction is initiated whereby the matrix evaporates as a gaseous compound. The process usually runs at an elevated temperature.
- The extraction of traces via different separating phases has also been performed. Such a procedure is based on the addition of an inorganic acidic solution to the solution of the sample in an organic solvent. After a thorough mixing, the two phases separate locally. The traces should thus have been transferred from the organic to the inorganic phase quantitatively while the sample matrix should have been left in the organic phase. The enriched inorganic phase is used for TXRF. In the case described here the traces are transferred from an organic to an inorganic phase, but this course can also be vice-versa.

A more generally applicable technique is the separation of traces by a chelating agent. After its addition, the metal traces are chemically bound as chelates. The solution is filtered by a special material whereby the metal complexes are adsorbed. Afterward, these complexes are eluted by an organic solvent and analyzed.

A summary of various methods of homogenization and preconcentration is given in Table 4.1 [8–26]. These have all been employed for TXRF analyses of different kinds of samples. All these methods of sample preparation must be carried out very carefully, and certain rules should be followed scrupulously. It is especially necessary to guard against *losses of* and *contaminations by* elements, which lead to serious systematic errors. Losses due to evaporation of volatile elements can be avoided by using closed vessels. Losses caused by adsorption on the vessel surface are reduced by use of quartz-glass containers. Contamination by reagents can be diminished by use of highly pure materials and only small quantities. Contamination by vessels or instruments can be avoided by the exclusive use of quartz glass or Teflon and by careful cleaning. Contamination from the ambient air can be prevented by working in a clean-bench. As a matter of principle, only one and the same vessel should be used for preparation, and pouring from one vessel into another should be avoided. For the decomposition of microquantities (<10 mg), only small vessels with a volume of 1–3 ml should be used.

TABLE 4.1. Different Methods of Sample Preparation Already Tested for TXRF Analyses of Various Kinds of Samples

Objective	Method	Kind of Sample	Physicochemical Process
Formation	Suspension	Aerosols, dust, ashes, sediments, pulverized biomaterials	Fine distribution of insoluble particles in an aqueous solution (water + ethanol) by stirring or ultrasonication [8,9]
	Dissolution	Essential oils, air dust, filters	Solution in a solvent (water, tetrahydrofurane, chloroform) at moderate temperatures [10–12]
Decomposition	Ashing	Air dust on filters, blood serum, edible oils, mineral oils, digested fine roots	Ashing of the organic matrix (cellulose, protein) in an oxygen stream by high-frequency heating [13–16]
	Open digestion	Suspended matter, aerosols on filters, algae, blood, serum, tissue, ashes, mud, sediment	Solution of the sample in concentrated acids or mixtures (HNO_3, HF, HCl) by a chemical reaction at moderate temperature with infrared or microwave heating [5,16–19]
	Pressure digestion	Dust on filters, lichen, minerals, soils, mussels, and fish	Like open digestion but in a closed vessel (PTFE-bomb) at high temperatures (150–300 °C) and high pressures of 10^6–10^7 Pa [13,17,20]
Matrix removal	Drying	Rainwater, ultrapure acids	Evaporation of the matrix by warming up, possibly in a nitrogen stream [16,21,22]
	Freeze-drying	Drinking water, ultrapure water, acids, organic solvents	Freezing of the matrix and evaporation under high vacuum [5,21]
	Volatilization	Digests of Si and SiO_2, sulfuric acid	Volatilization of silicon as SiF_4 after addition of HF [23]; volatilization of sulfur as SO_2 after addition of HI [23]
	Extraction	Dissolved high-purity iron, digested blood	Extraction of the iron matrix by MIBK [24] Extraction of the Fe-component by MIBK [18]
	Complexation	Rain- and seawater Digests of river water, blood and aluminum	Separation of the alkaline, alkaline earth or aluminum components by complexation as carbamates, adsorption at chromosorb/cellulose, elution of the traces [17,20,21,25,26]

Source: From Ref. [4], reproduced with permission. Copyright © 1996, John Wiley and Sons.

The addition of an internal standard is a further aid to eliminating certain systematic errors. With reference to a standard element, various nonspecific errors of preparation are compensated for, for example, nonspecific losses. In order to compensate for a maximum of errors, the standard should be added to the sample as early as possible. This step, called *spiking*, should be done after dissolution or suspension of the sample but before a matrix separation or trace concentration if possible.

Different types of complex samples can be prepared for TXRF analysis by ashing, pasting, and plotting [27]. Consequences can be summarized by the following statements:

- Edible oils (fatty acids) can be prepared by fast microwave ashing and analyzed down to sub-ppb.
- Solar cells can be pasted onto an aluminum disc with hot glue and contaminants of the cells that get stuck on the disc can be determined by TXRF. Iron and copper can be detected with amounts of some ng down to 35 pg.
- Seawater can be diluted with PVA (polyvinyl solution) in order to avoid so-called coffee rings after drying. By means of the automatic dispenser TWO described in Section 4.1.3.4, a grid of 12×12 droplets of 16 nl each can be plotted on a quartz carrier. Plotting can be repeated several times and the dried residue can be analyzed by TXRF.
- A tutorial review on sample pretreatment strategies was given by De La Calle *et al.* [28]. Acid digestion, extraction, slurry preparation, *in situ* microdigestion, microflow online preconcentration, and lab-on-a-chip are represented. Their advantages as well as drawbacks are considered. Statistical revision of the different strategies published in papers between 2008 and 2013 is also given.

4.1.3 Presentation of a Specimen

Small amounts of a sample should be placed on a cleaned carrier and presented as a specimen. As already mentioned in Section 3.5.1, quartz glass is the most frequently used carrier for aqueous solutions and the most acidic and basic solutions. Although Plexiglas is a cheaper carrier, it is only suitable for aqueous solutions. After an electrostatically protective foil is stripped off, the clean carriers can be used directly.

As an upper limit, micrograms of solids (1–200 μg) and microliters of liquids (1–200 μl) yielding micrograms of dry residue can be used as specimens. The appropriate amount can easily be determined by means of the detector. The specimens should be restricted to an amount keeping the dead time loss of the detector below 63% (see Section 4.2.3). As a lower limit, even picograms of a solid and picoliters of a liquid can be sufficient for analysis. A more detailed consideration needed for quantitative analyses of three typical sample matrices is given in Section 4.4.3.

4.1.3.1 Microliter Sampling by Pipettes

In most cases, the samples are received or prepared as liquids. These *liquids* can be pipetted on the center of a circular carrier with a high degree of reproducibility. Micropipettes with a volume between 5 and 50 μl are normally used (Eppendorf-AG, Hamburg, Germany). The pipette tips, made of polypropylene, must not be touched and are used only once. The single *droplets* are dried by a simple evaporation in air, which needs some 2 h for a 10 μl droplet. To speed up the evaporation, the carriers can be placed on a hot plate or under an infrared lamp. Droplets can also evaporate in a desiccator coupled with a diaphragm pump but this technique is rather inconvenient. The hot plate seems to be best suitable, reducing the drying time to about 20 min. By this means, several droplets can be evaporated in a rapid succession on top of each other. Organic solvents with a volume of 100 to 300 μl can be dried with the help of a small PTFE cylinder pressed against the carrier [25].

The residue may be seen as a bright spot of about 1–5 mm in diameter or may be hardly visible. Figure 4.5 demonstrates the simple technique of pipetting and additionally shows a small residue. This residue should be fairly dry, stable, and homogeneous and should stick to the carrier. For that reason, appropriate reagents are added either to the total sample or to the presented specimen. Multivalent alcohols and chelating agents are recommended [29,30]. Nevertheless, the disruptive effect of evaporation usually leads to a nonuniform morphology of the dried residue. It represents a small spot with a certain ring-shaped or crater-shaped wall. Figure 4.6 shows microscope images of some residues resulting from 10 and 50 μl droplets on hydrophobic silicon substrates. The outer diameter of the respective walls is about 0.5–2.5 mm and the wall itself has a width on the order of 5–100 μm and a height of 0.1–10 μm. Such

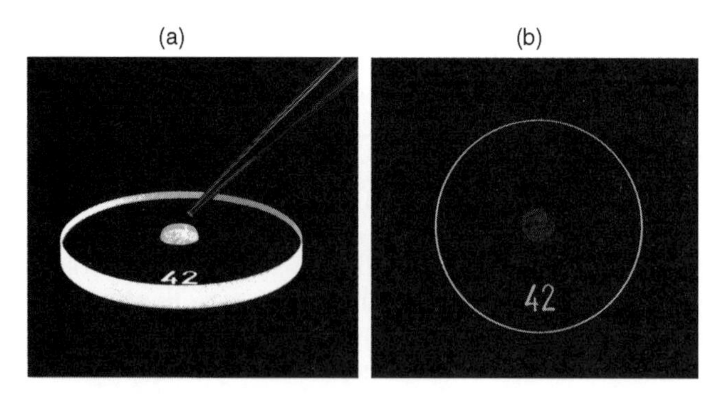

Figure 4.5. (a) A droplet of 10 μl is pipetted on a carrier with a diameter of 3 cm. (b) The droplet leaves a dry residue after evaporation. Figure from Ref. [4], reproduced with permission. Copyright © 1996, John Wiley and Sons.

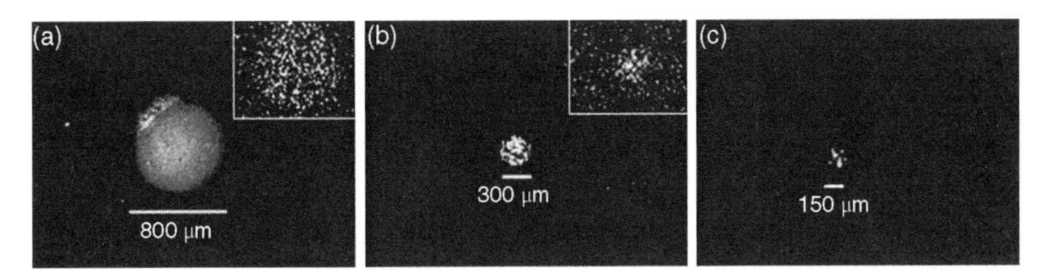

Figure 4.6. Some dried residues of microliter droplets represented by a light microscope: (a) conventional VPD residue of a 100 μl droplet; (b) 50 nl dried residue; (c) 10 nl dried spot. The small insets of elemental area maps present grain conglomerates of iron-salt. Figure from Ref. [31], reproduced with permission. Copyright © 2004 by Elsevier.

rather inhomogeneous specimens will reduce the accuracy of a quantitative analysis.

Besides real liquids, *fine powders* or pulverized solid materials can be presented via a droplet. They must first be prepared as a suspension that should be stirred thoroughly. After that, droplets of about 10 μl can be presented as specimens.

4.1.3.2 Nanoliter Droplets by Capillaries

The production of nanoliter-sized droplets (10^{-9} l) was developed by Miller *et al.* [31,32] in order to reduce the lengthy drying times and inhomogeneous evaporation associated with microliter-sized droplets. Smaller droplets of 10–50 nl can be deposited on flat substrates, such as silicon wafers (TIP10XV119 of World Precision Instruments, Sarasota, FL), using glass capillary pipettes. The pipettes have an internal diameter of only 10 μm and a length of some 10–50 mm. The droplets are deposited by an injector, which is manually positioned by a micromanipulator. The pipette is nearly brought into contact with the substrate with the aid of a magnifying glass. The nanoliter-sized droplets dry in ambient air within some 20 s to 2 min. The dried residues have a diameter of some 10–300 μm depending on the volume of the droplet and the substrate (hydrophobic or hydrophilic).

Nanoliter droplets of liquid samples as well as of standard solutions can be prepared containing several elements with a mass between 5 and 100 ng. The whole procedure should be carried out in a cleanroom environment. Figure 4.7 illustrates the effect by different spots of dried residues produced either by 1 μl or by 10 nl droplets. The residues of nanoliter-sized droplets represent more homogeneous samples that are film-like in character and better suitable for quantification. In order to cover an area of some mm^2 as necessary for TXRF, uniform arrays with about a hundred of small spots can be generated. The repeatability, however, is insufficient because of the varying contact between the tip of the pipette and the substrate.

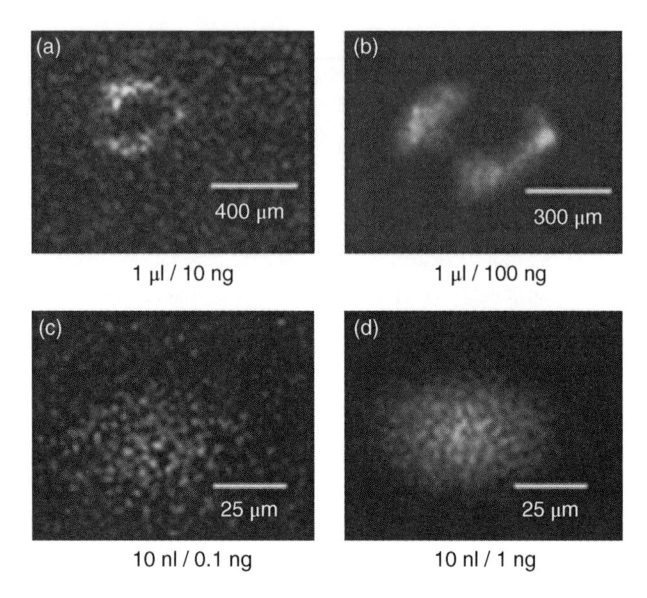

Figure 4.7. Dried residues of 1 μl sized droplet (top) in comparison to a 10 nl sized droplet (bottom). The pictures were taken by an X-ray microprobe and show the distribution of copper by means of the Cu-Kα peak. (a) 1 μl with 10 ng Cu; (b) 1 μl with 100 ng Cu; (c) 10 nl with 0.1 ng Cu; (d) 10 nl with 1 ng Cu. Figure from Ref. [32], reproduced with permission. Copyright © 1996 by John Wiley and Sons.

4.1.3.3 Picoliter-Sized Droplets by Inkjet Printing

For even minor deposits, even smaller volumes of liquid samples should be used; that is, picoliter-sized droplets (10^{-12} l). Furthermore, they should be deposited in a well-defined programmable pattern or specific array that matches the detector's view field. The multifold repeated depositions should be carried out automatically, not manually, in a short time of a few minutes with a high repeatability.

For that purpose the technique of inkjet printing was applied by Fittschen and coworkers [33–36]. Inkjet printers have been developed for computers since 1970 and represent a fairly mature technique. They either use piezoelectric pulses to generate droplets of ink and to force them from different nozzles or they use a tiny heater to rapidly vaporize such droplets propelling them onto a substrate, usually paper. Because of the high velocity (10–40 m/s) the distance between head and substrate can be up to 1 cm. The ink filled into a cartridge is usually an aqueous solution but may have a volatile component.

Commercially available inkjet printers (e.g., the thermal inkjet picofluid system or TIPS of Hewlett-Packard) can be modified slightly in order to use pico-sized droplets for TXRF analysis. The droplets of an aqueous standard solution have a volume of 1–200 pl containing several elements, mostly trace metals with a mass between 1 pg and 2 ng. They are deposited on different carriers, such as silicon wafers or quartz glass (coated or uncoated). Residual

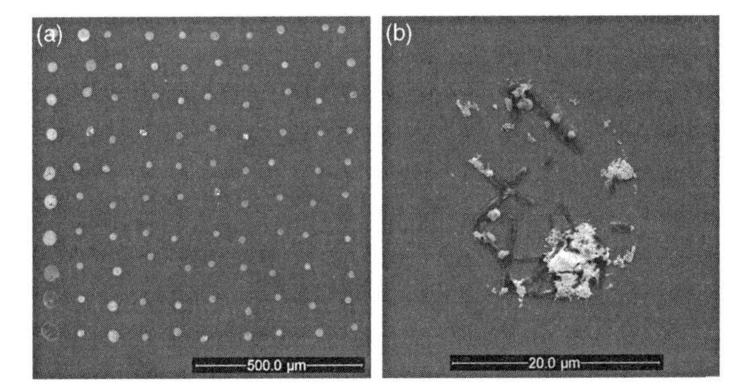

Figure 4.8. (a) A square array of 100 spots of picoliter droplets with an area of 1 mm^2 and (b) a single spot of the dried residue of a 1 pl droplet with 20 μm in diameter represented by an SEM (scanning electron microscope). Photo by A. von Bohlen, private property.

spots with a diameter of 5–100 μm are created, their thickness ranging between 1 nm and 2 μm. These spots are round, have nearly the same size, and do not show satellite droplets as demonstrated in Figure 4.8.

A lot of similar spots can be arranged in an array of 10 mm length or diameter each consisting of 10 × 10 up to 90 × 90 residues in a periodical pattern. The individual droplet with 10 pl volume may consist of a single-element or multielement standard solution. In order to prevent evaporation from the nozzles, the frequency of shooting should be > 10/s (the time between two shots should be about 100 ms, the spotting velocity about 1 mm/s). For a consistently homogeneous deposition it is recommendable to use a lower concentration of about 0.002 up to 0.02 g/l. Higher concentrations usually lead to nonuniform conglomerates and clusters of salts. After drying within tenths of a second, the droplets give spots of 5–10 μm in diameter that are spaced 100 μm apart. The residues are thin-film-like in character with an average height of about 2 nm (see Section 4.5.1).

A droplet cannot be deposited on top of a previous residue because it would redissolve this residue leading to a larger, less homogeneous residue. However, several uniform arrays can be printed on a single wafer next to each other, for example, arrays of a standard solution with increasing concentration. These arrays can serve as custom-made external standards, for example, for the contamination control of silicon wafers. A precision of better than 5% can be reached for TXRF analysis on the level of 10^{11} up to 10^{13} atoms/cm^2. Furthermore, inkjet printing of picodroplets can be applied for standard addition of aerosol samples collected on the different stages of a Berner impactor.

The dried residues and their casting by shadows were observed by the color X-ray camera CXC SL cam of the IFG (English: Institute for Scientific Instruments) in Berlin [37]. Different uniform microscopic deposits were designed with ring-shaped and array-shaped pattern of individual picoliter droplets. A pattern of 64 picoliter droplets, each with 19 ng of nickel in a distance of about 180 μm, was deposited by means of a Sonoplot printing device and dried. The camera was positioned directly above the sample in order to monitor the shading of the dried residues at usual small angles of incidence. By a rotation of the sample, the shading could be minimized and the fluorescence intensity be maximized.

4.1.3.4 Microdispensing of Liquids by Triple-Jet Technology

A new technique for automatically dispensing a liquid sample on a flat carrier was developed by the company M2 Automation in Berlin, Germany, and is commercially available. The instrument, TWO, is a 2D imaging system for droplet detection and imaging and combines three different microdispensers: (1) a piezo-driven dispenser for picoliter droplets; (2) an M2 dispenser with shockwave generator for low nanoliter drops; and (3) a solenoid-valve-driven dispenser for midnanoliter drops. The first produces single droplets between 20 and 300 pl, the second produces droplets between 5 and 200 nl, and the third produces drops between 30 nl to 1 µl. Dispensing tips are made of glass, stainless steel, and plastic material. The latter tips are cheap and can simply be disposed after use instead of washing them. The maximum driven range is 20 cm in the *xy* directions and 5 cm in *z* direction. The speed is about 0.1 m/s, and the accuracy is ±20 µm in *xy* directions. Dispensing is carried out with a maximum velocity of five depositions per second. A substrate designer assists in arraying patterns of some 10 to 400 dots via a mouse click.

This instrument was tested by Mages *et al.* for TXRF analysis [38]. Biological samples (benthic invertebrates) were analyzed after cold-plasma ashing and digestion by HNO_3 and H_2O_2. A manual pipetting of 10 µl droplets lead to the notorious ring-shaped residues after drying and to a reduced precision and accuracy of analytical determinations. By means of the microdispensing technique, 80 single droplets with only 10 nl volume were pipetted in six concentric circles on a carrier automatically. After drying at 50 °C, all the single dots were mutually analyzed by TXRF. The relative standard deviation of all 13 investigated elements determined by three replicates could be reduced to about 1 to 2.5%. Five elements K, Mn, Fe, Cu, and Zn were determined with a concentration between 11 µg/ml (ppm) for potassium and 45 ng/ml (ppb) for copper. For these elements, the relative standard deviation, resp. the precision, was improved by a factor between 3 and 12 (geometric mean 6.3). The deviation of the mean values determined after manual pipetting of one drop and automatic dispensing of 80 droplets was below 3% for seven elements.

4.1.3.5 Solid Matter of Different Kinds

Solid samples in the form of fine powders with a grain size of <1 µm can be applied directly in a most simple way. A spatula made of PTFE can be used to take a small amount of the powder and dust it on a carrier. The small grains usually stick to the carrier but can be removed by knocking the carrier when too large an amount has been taken.

Furthermore, *pigments* of oil paints can be sampled by means of a cotton bud, often called Q-tip. Some material is rubbed off the painted surface, and the loaded Q-tip is dabbed onto a carrier [39]. Microgram amounts can be transferred in this way. Of course, this technique is also applicable for other mineral or oxide powders.

Air dust, or more generally *aerosols*, can be collected on filters, which have to be ashed or digested prior to analysis. However, they can also be collected directly on the carriers recommended for TXRF. These carriers can be used as impaction plates in a device called an *impactor*. The dust particles taken by an airstream are deposited according to their inertia. The loaded carriers can be directly used for analysis [11,40].

Individual particles like crumbs, grains, fibers, or splinters of a few micrograms can simply be put right onto the carrier. A wooden toothpick should be used for the manipulation. Such samples are especially suitable for a qualitative analysis by TXRF. Small amounts of *solid bulk* samples can be removed by laser ablation [41,42]. The laser beam is focused by an objective with a long focal length. Material is molten, even evaporated, and is thrown out from a small crater of about 10 μm in diameter. The emitted material can be deposited on a quartz glass carrier that should be positioned between the objective and the sample, just above the latter, and that should be penetrated centrally by the laser beam. The material deposited on the carrier can be used as a specimen for TXRF. This sampling technique is capable of providing a microdistribution analysis, for example, by a line scan. It can also be used for depth-profiling analysis. For that purpose, the sample should first be abraded with an angle of $<1°$.

When only a survey but not a local analysis is required, another simple technique can be applied. *Solid samples* can be rubbed on a hard quartz glass carrier in a single stroke; or, the other way around, a quartz glass carrier can be rubbed on a fixed object that has a lower degree of hardness. In either case, a microgram amount of sample material will be smeared on the carrier and can be used as specimen [8]. This technique is especially useful for large finished products that are hardly accessible by other means.

Organic materials and *biomaterials* are normally decomposed by ashing, combustion, and/or wet chemical digestion. However, such materials can also be freeze-cut by a microtome. The frozen sections are placed on a carrier, dried, and directly analyzed. For quantification, an internal standard can be added to the section afterward [43].

The liquid and solid samples enumerated so far are presented for micro- or trace analysis by TXRF. In addition, there are samples suitable for surface or thin-layer analysis by TXRF. Moreover, liquid or solid samples can be investigated by GI-XRF, for example, when nanoparticles deposited on a flat sample support like a silicon wafer have to be analyzed. *Wafers* not yet patterned can directly be applied as flat disks provided that an appropriate sample holder is available. Uncoated wafers can be subjected to a surface analysis; coated wafers, to a thin-layer analysis. One must be aware that the actual specimen is defined by the detector's range of vision, that is, by the projection of its front area onto the wafer disk. Only this region is analyzed. In order to examine the total wafer, a displacement device is necessary. Such a thin-layer analysis is also possible when *thin films or layers* are deposited on a flat substrate, for example, quartz glass. The samples can directly be used as specimens for TXRF and GI-XRF. The analysis can also be of use for monitoring a sputter or a vapor decomposition process, for example.

4.2 ACQUISITION OF SPECTRA

The actual measurement consists in recording the spectra. The carriers prepared for that purpose are put into sample holders. These plastic slides are pushed either directly into the fixed measuring position or first loaded in a sample changer and afterward pushed into that position automatically one after another. The X-ray tube, filter, first reflector (monochromator), and carrier must be set in a certain combination and geometry as recommended for analysis. The tube voltage, tube current, and preset live time have to be selected appropriately, in accord with definite rules. After these preparations, the spectra are recorded and stored. A subsequent interpretation aiming at a qualitative analysis is based on the detection of individual peaks, the identification of these peaks, and finally determination of elements inferable from these peaks.

4.2.1 The Setup for Excitation with X-Ray Tubes

X-ray tubes, foil-filters, and reflectors (monochromators) are components that can usually be chosen or exchanged for the particular operation. Of the several combinations that have already been tested, some are commonly used whereas others are appropriate for specific applications. The aim is an effective excitation of the sample. However, the different elements require different conditions for an optimum excitation. Consequently, a compromise solution is generally necessary for multielement analyses.

Excitation is usually performed by line-focus X-ray tubes. The continuous radiation as well as the characteristic radiation of the anode material can contribute to excitation, but with differences in efficiency. This is illustrated in Figure 4.9. Excitation is only possible by photons with an energy that exceeds the respective absorption edge of the element. The continuum is effective for excitation within a wide range of photon energies. It can generate either K or L peaks of nearly all elements if the operating voltage is above 25 kV. Maximum efficiency will be achieved if the energy that corresponds to the applied voltage is between three and five times the edge energy.

The characteristic radiation, on the other hand, will be highly effective if the peak energy of the tube anode just equals the absorption edge of the analyte.

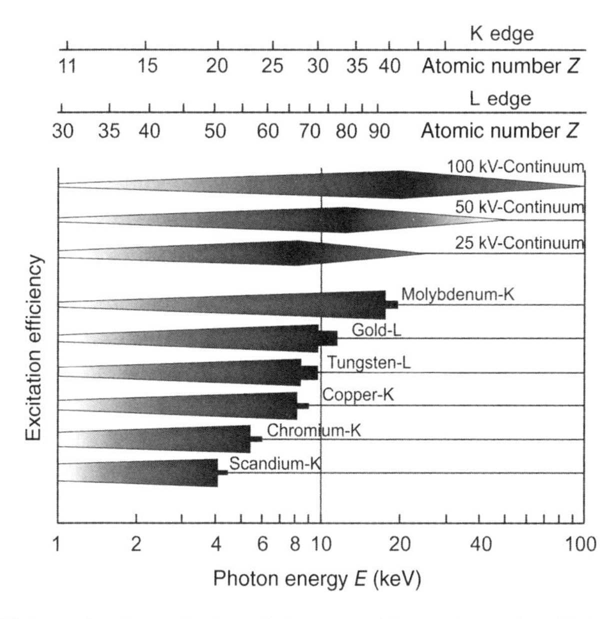

Figure 4.9. Efficiency for the excitation of elements with atomic number Z, dependent on the photon energy E. The excitation can be performed by the continuum (100, 50, 25 kV) and by the K and L peaks of different X-ray tubes (Mo, Au, W, Cu, Cr, Sc). The upper two scales show the energetic position of the K and the L absorption edges of the analytes. Figure from Ref. [4], reproduced with permission. Copyright © 1996 by John Wiley and Sons.

Excitation above this value is not possible; excitation of all elements with absorption edges below this value decreases steadily. As can be seen in Figure 4.9, a Mo tube operated above 50 kV is suitable for the excitation of all elements. However, a second tube is desirable to excite the lighter elements ($Z < 20$), with their K peaks, and the medium heavy elements ($40 < Z < 60$), with their L peaks, more effectively. The tubes may be exchangeable, but it is more convenient when both are permanently installed.

The tube spectrum is usually processed by a metal foil acting as a specific absorption filter, by a totally reflecting mirror acting as a low-pass filter, or even by a natural crystal or multilayer acting as a Bragg monochromator (see Section 3.4.2). These components are used to further reduce the background of the spectra or to avoid a blurring of angle-dependent intensity profiles. For micro- and trace analyses, the spectral background has to be lowered, which can be achieved by a combination of filters. For surface and thin-layer analyses, blurring must be suppressed, which is preferably done by a natural crystal or multilayer. Metal foils clamped in holders can easily be positioned and replaced, but mirrors or Bragg reflectors have to be carefully adjusted and should be left in this position continuously.

For micro- and trace analyses by TXRF, three different excitation modes are recommended:

(a) A W tube operated at 50 kV, a Ni foil of 100 μm, and a quartz glass mirror set at a glancing angle of 0.05° (pass energy 35 keV)
(b) A Mo tube operated at 50 kV, a Mo foil of 50 μm, and a quartz-glass mirror set at 0.09° (pass energy 20 keV)
(c) A W tube operated at 25 kV and a Cu foil of 10 μm; with respect to mode (a), only the foil has to be exchanged and the voltage reduced (the tube and the mirror can be kept unchanged).

The three modes can easily be interchanged when the two required tubes are installed. Three further but similar modes of excitation were tested by Knoth *et al.* [44]. They only need a single W tube and a combination of two multilayers. However, the arrangement has to be adjusted back and forth in order to change the mode.

The primary spectrum processed in the three modes (a)–(c) is sketched in Figure 4.10. It illustrates that excitation is mainly realized by a broad spectral band (a), by the Mo-Kα and -Kβ peak (b), and by the W-Lα peak (c). Mode (a) can be used for excitation preferably in the energy range between 9 and 30 keV; mode (b) is preferable for excitation between 4 and 16 keV; and mode (c) is suitable for excitation energies between 2 and 8 keV. Figure 4.11 demonstrates the efficiency of the three modes for the detection of elements with atomic number, Z; they are detected either by their K peaks or their L peaks.

For surface and thin-layer analyses by GI-XRF, only a single peak or a small band of the continuum is cut out of the spectrum by means of a real monochromator. Natural crystals or multilayers tried and tested for this purpose are listed in Table 3.1. If two X-ray tubes, for example, a Mo and a W tube, are at your disposal, the two peaks Mo-Kα and W-Lα may be used

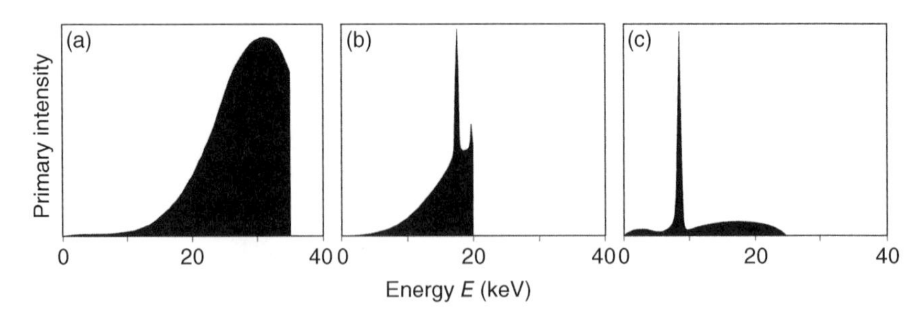

Figure 4.10. Primary intensity of three recommendable modes of excitation: (a) W tube, Ni foil, low-pass filter; (b) Mo tube, Mo foil, low-pass filter; (c) W tube, Cu foil. Figure from Ref. [4], reproduced with permission. Copyright © 1996, John Wiley and Sons.

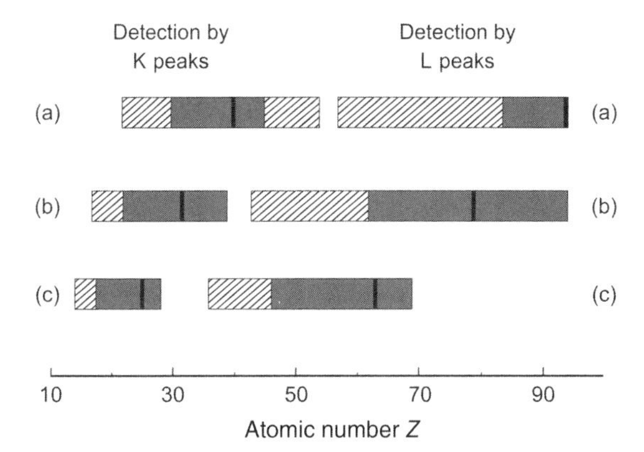

Figure 4.11. Efficiency of three special modes of excitation applied for the detection of elements with atomic number Z. Lower-Z elements are detected by their K peaks; higher-Z elements, by their L peaks. The three modes (a), (b), (c) are described: *hatching*, moderate efficiency; *dotting*, high efficiency; *solid black*, optimum efficiency. Figure from Ref. [4], reproduced with permission. Copyright © 1996, John Wiley and Sons.

again. However, the W-Lβ peak is normally preferred to the W-Lα peak because it has the advantage of additionally exciting the elements Cu and Zn. If only one X-ray tube is available, two separate monochromators may be useful, installed in the same instrument [45,46]. In this way, lighter and heavier elements can be detected with a comparable efficiency.

4.2.2 Excitation by Synchrotron Radiation

Necessary devices belong to the usual setup of the experimental hutch of a synchrotron beam line [47]. Several slits reduce the beam profile to about 10–100 μm in height and 1 mm in width. The vertical divergence is mostly <1 mrad or <0.06° (=1/γ), the horizontal divergence is some 15 mrad or about 1°. A tiltable absorber may reduce the intensity of the synchrotron beam. For sample holding and positioning a six-axis goniometer is installed for x, y, and z translations as well as Θ, Φ, and χ rotations. Several detectors are provided for recording of X-rays, at least a NaI scintillation counter (see Section 3.7.3) and a Si(Li) or silicon drift detector (see Section 3.6.2). Finally, a fast shutter may cut off the synchrotron beam.

For micro- and trace analyses by simple TXRF, a low-pass filter or even a simple monochromator (see Sections 3.4.1 and 3.4.2) is sufficient. For surface and thin layer analyses carried out by GI-XRF at a synchrotron beamline a double monochromator with Si Bragg crystals or multilayer crystals is needed (see Section 3.4.3). A spectral band of only some 1–10 eV is selected out of the continuum of the synchrotron radiation. For that purpose the double monochromator has to be adjusted but usually it is ready for a computer-controlled selection of an energy band. A precise alignment and adjustment, however, is always necessary for the sample stage with the attached sample. This can be done by means of a scintillation counter, which is placed in the direction of the synchrotron beam at a distance of about 10 cm behind the sample. The Si(Li) detector or SDD, however, is placed about 1 cm above and vertical to the sample.

Grazing incidence XRF measurements are carried out by a rotation of the sample around an axis a that is perpendicular to the synchrotron beam and to the horizontal axis b of the sample (see Figure 3.13). At first the zero position of the sample has to be adjusted with a glancing angle of 0° and an accuracy of 0.001°. The mentioned scintillation counter is used for that task. For an intensity-angle scan the sample is tilted between 0.001° and 1° (or even 2°), frequently in more than 50 steps of <0.02°, and generally computer-controlled. At each individual step an X-ray spectrum is recorded by the Si(Li) detector or SDD, normally with a measuring time between 10 and 60 s. Usually, the spectrum may be observed in total but only some spectral peaks are recorded. This can be done by setting only one, or two, or even three regions of interest (ROIs). If only 50 individual angle settings are chosen with a measuring or live-time of 10 s, about 10 min are needed for the analysis. For 100 angles and a 60 s live time, nearly 2 h are required.

For a quantitative evaluation two corrections of the spectral intensity have to be made: (1) a correction for the reduced current of the synchrotron beam during the measurements and (2) a correction for the increased area, which is illuminated by the exciting synchrotron beam at smaller glancing angles. The first correction is of the order 2%–10% per hour, the second is only necessary for glancing angles <0.6°. The so-called footprint effect is treated in Section 4.6.2. The correction factor given by Equation 4.38 is inversely proportional to the relevant glancing angle.

One has to be careful with some specific peaks to avoid misinterpretations of the spectra. The light elements N, O, and Ar from ambient air may lead to peaks at 0.4, 0.5, and 3 keV. Peaks of silicon at 1.7 keV and germanium at 10 keV may come from the substrate or the sensor material of the detector. Peaks of metals like iron or copper may result from the sample stage, sample holder, or metallic parts of the goniometer.

It is common practice to observe the specular reflection of X-rays (XRR) and the scattered radiation—either elastic Rayleigh or inelastic Compton scattering—simultaneously to GI-XRF. The scattered radiation is directly recorded by the Si(Li) or SDD. For XRR measurements, a simple scintillation counter behind the sample can be used. The angle scan may be extended to 5° instead of 1°; with a step size >0.002° a period of 1 to 4 h is needed for a single sample. A detailed evaluation is given in Section 7.3.1.

4.2.3 Recording the Spectrograms

Once chosen, the instrumental set-up is usually kept unchanged, but some devices have to be put in operation over and over again, and the operational parameters may be varied according to the sample and the analytical problem.

The operating voltage of the X-ray tube will usually be set to a fixed value if a quantitative analysis is also planned. The sensitivity values needed for quantification are not applicable to any but a fixed operating voltage. As already mentioned, values of 50 and 25 kV are commonly used. The operating current of the tube, on the other hand, can be selected in a wide range, normally between 5 and 60 mA. Its particular setting is irrelevant even for quantification. However, an upper limit is determined by the dead time of the detection unit. The current has to be limited to such a value that the dead time losses are kept below 63%. For trace analyses, this current should be chosen if at all possible. If the applied current is first set too high, it should be lowered; otherwise, the sample mass has to be reduced. Finally, the count rate meter should indicate a value between 300 up to 1 000 000 cps.

4.2.3.1 Energy-Dispersive Variant

Normally, the intensity is measured in counts or number of photons depending on the photon energy, $N(E)$. For energy-dispersive spectrometry of X-rays (EDS), this measurement is carried out simultaneously for all indicated photons and shown on a display. Only a few parameters have to be set for EDS. The energy scale can normally be restricted to 10, 20, and 40 keV. The decision depends on the analytical problem and its approach. Excitation mode (a) requires a 40 keV recording; mode (b), the usual 20 keV setting; and mode (c), the lower 10 keV setting. If the multichannel analyzer (MCA) has the usual number of 2000 channels, the channel width will consequently be fixed to 20, 10, and 5 eV, respectively. A shaping time of 50 μs is usually selected for a Si(Li) detector; a shaping time of only 2 μs is preferable for an SDD. It should be increased when a better spectral resolution is wanted and should be decreased when a higher count rate is desirable. Finally, the energy axis has to be calibrated or recalibrated. This procedure should be be repeated once or twice a day. For that purpose, some special carriers are prepared and always reused. They are loaded with the residue of an aqueous solution of certain standard elements. Molybdenum is preferred for excitation mode (a), iron for mode (b), and titanium for mode (c).

After a warm-up phase of about 1 h, the individual spectra of the different samples can be recorded. A live-time is preset between 10 and 1000 s; 100 s setting is commonly selected, and 1000 s setting is only chosen for the detection of ultratraces. The increasing spectra can be observed on a color display during measurement. At the end, the spectra are coded and stored for the subsequent evaluation and analysis. Of course, samples are first examined in one certain excitation mode. After such examination, the mode may be changed and the samples may be examined in the next mode.

4.2.3.2 Wavelength-Dispersive Mode

For wavelength-dispersive spectrometry of X-rays (WDS), a spectral range has to be chosen first, which should be investigated for a sample or a set of samples.

A Bragg crystal, Soller collimators, and a detector have to be selected. When the crystal is rotating along with the exit collimator and detector, the intensity or number of photons is determined then as dependent on the Bragg angle, $N(\alpha)$. A conversion of α into the respective wavelength λ is carried out by the Bragg relation $m\lambda = 2d \sin\alpha$, where d is the spacing of the Bragg crystal in use (see Section 3.7.1) and m is the spectral order. The most intensive order with $m = 1$ is mostly preferred. A conversion of α into the respective photon energy is not usual in classical XRF with WDS but recommendable here because all the calculations for TXRF and GI-XRF have been made for photon energies rather than wavelengths. This can easily be done in accord with $E = hc_0/\lambda$, where hc_0 is 1239.84 eV × nm.

A spectrum between 1 and 20 keV can be recorded in about 10 min up to 1 h. For such a wide energy range, usually four crystals, two collimators, and two detectors have to be selected and exchanged against another (see Section 3.7.1). The pulse-height discriminator also has to be adjusted in order to discriminate the higher orders of a spectrum. This is automatically done with a $\sin\alpha$ potentiometer. Smaller parts of the spectrum can be covered with less crystals and only one detector in correspondingly shorter periods. For TXRF analysis, a wider range of the spectrum is usually chosen. For GI-XRF investigations, in general, only a few elements are relevant so that only a few spectral peaks have to be investigated.

4.3 QUALITATIVE ANALYSIS

The capability of performing a simultaneous multielement analysis is the most obvious virtue of an EDS and is highly useful when one needs to get an idea of the composition of a sample that is initially completely unknown. It enables a further analytical strategy without overlooking any essential element. The necessary qualitative analysis is based on the interpretation of the spectra. Each spectrum is treated and evaluated as a whole. A data processing system is generally used to provide a rapid and extremely convenient evaluation. Hardware and software already used for conventional EDS are similarly used here. The spectra can be spread or compressed on both the energy axis and on the intensity axis, and can be rolled back and forth along the energy axis. They can also be smoothed and corrected for escape peaks. A second spectrum can be adjusted to the intensity of a first spectrum and stripped in order to cancel out certain overlappings or interferences. Markers can be recalled in order to indicate the position of the peaks of any selected—or expected—element.

4.3.1 Shortcomings of Spectra

In comparison to other methods of atomic spectrometry, X-ray spectra are generally simple. Nevertheless, some failings or artifacts may appear caused by the EDS—as well as by the WDS—system. WDS spectra seldom suffer from troublesome interferences of neighboring peaks because of a higher spectral resolution. If these interferences result from Bragg reflections of higher orders, they can be eliminated by pulse-height discrimination. EDS spectra show many interferences in the lower energy region, which cannot easily be avoided. Furthermore, pulse pileup effects of the detector at high count rates can lead to failures by sum peaks. Also, both methods show annoying *escape peaks*, which have to be eliminated or taken into account.

4.3.1.1 Strong Spectral Interferences

In general, X-ray spectra show only a few peaks, much less than visible or ultraviolet (UV) spectra do. There are only about 500 distinct peaks for 90 elements that can be detected (atomic number ≥ 3). All these peaks are in accord with Moseley's law with regard to their position, and peaks of the K, L, or M series keep to certain intensity ratios (see Section 1.3.2.1). Table 4.2 lists 494 principal peaks with a high intensity. More extensive tables that also include further 600 weak peaks are available in the literature [48–50]. Main absorption edges needed for choosing suitable excitation energies of principal peaks are listed in Table 7.1.

Each element has at least one or two characteristic peaks like the light elements with $3 \leq Z \leq 19$. The heavier elements with $Z \geq 20$ show up to 20 different characteristic peaks in a spectral range between 3.6 and 50 keV. Consequently, the statistical probability of a line interference for two elements is 20%. The peaks have a natural width of only some 0.1 eV, which is enlarged by the detector to some eV or even to some 10 or 100 eV. The major peaks are the Kα_1 and Kβ_1 peaks, the Lα_1 and Lβ_1 peaks, and the Mα and Mβ peaks. These peaks are used for the detection of elements, preferably the higher energetic peaks Kα_1, Lα_1, and Mα because of their higher intensity. The probability that they may interfere with each other depends on the composition of the sample. For a sample with only one or two elements, only a few interferences can be expected. But for multielement samples composed of 10–20 elements, interferences become a serious problem.

Interferences of two spectral lines or peaks can be figured out from Table 4.2. A peak of some element with atomic number Z_1 may interfere with a peak of another element with atomic number Z_2 at a small energy difference, ΔE in eV. Figure 4.12 shows a two-dimensional matrix of those elements. This distance of neighboring peaks is characterized by different colors for the most intensive interferences, that is, if their distance is below 240 eV. If we consider the six major peaks—Kα, Kβ, Lα, Lβ_1, Mα, and Mβ—we have 21 possible interferences for each pair of two elements. However, we have only seven interferences for the most intensive peaks of two elements: Kα and Kα, Kα and Kβ; Kα and Lα; Kα and Mα; Lα and Lα; Lα and Mα; and finally Mα and Mα.

TABLE 4.2. Principal Peaks of X-ray Spectra[a] Obtained from Elements with Atomic Number Z

Z	Elem.	$K\alpha_1$	$K\beta_1$	$L\alpha_1$	$L\beta_1$	$L\beta_2$	$L\gamma_1$	$M\alpha$	$M\beta$
3	Li	0.052							
4	Be	0.110							
5	B	0.185							
6	C	0.282							
7	N	0.392							
8	O	0.523							
9	F	0.677							
10	Ne	0.851							
11	Na	1.041	1.067						
12	Mg	1.254	1.297						
13	Al	1.487	1.553						
14	Si	1.740	1.832						
15	P	2.015	2.136						
16	S	2.308	2.464						
17	Cl	2.622	2.815						
18	Ar	2.957	3.192						
19	K	3.313	3.589						
20	Ca	3.691	4.012	0.341	0.344				
21	Sc	4.090	4.460	0.395	0.399				
22	Ti	4.510	4.931	0.452	0.458				
23	V	4.952	5.427	0.511	0.519				
24	Cr	5.414	5.946	0.571	0.581				
25	Mn	5.898	6.490	0.636	0.647				
26	Fe	6.403	7.057	0.704	0.717				
27	Co	6.930	7.649	0.775	0.790				

28	Ni	7.477	8.264	0.849	0.866		
29	Cu	8.047	8.904	0.928	0.948		
30	Zn	8.638	9.571	1.009	1.032		
31	Ga	9.251	10.263	1.096	1.122		
32	Ge	9.885	10.981	1.186	1.216		
33	As	10.543	11.725	1.282	1.317		
34	Se	11.221	12.495	1.379	1.419		
35	Br	11.923	13.290	1.480	1.526		
36	Kr	12.648	14.110	1.587	1.638		
37	Rb	13.394	14.960	1.693	1.752		
38	Sr	14.164	15.834	1.806	1.872		
39	Y	14.957	16.736	1.922	1.996		
40	Zr	15.774	17.666	2.042	2.124	2.219	2.302
41	Nb	16.614	18.621	2.166	2.257	2.367	2.462
42	Mo	17.478	19.607	2.293	2.395	2.518	2.623
43	Tc	18.367	20.619	2.424	2.538	2.674	2.792
44	Ru	19.278	21.555	2.558	2.683	2.836	2.964
45	Rh	20.214	22.721	2.696	2.834	3.001	3.144
46	Pd	21.175	23.816	2.838	2.990	3.172	3.328
47	Ag	22.162	24.942	2.984	3.151	3.348	3.519
48	Cd	23.172	26.093	3.133	3.316	3.528	3.716
49	In	24.207	27.274	3.286	3.487	3.713	3.920
50	Sn	25.270	28.483	3.443	3.662	3.904	4.131
51	Sb	26.357	29.723	3.605	3.843	4.100	4.347
52	Te	27.471	30 993	3.769	4.029	4.301	4.570
53	I	28.610	32.292	3.937	4.220	4.507	4.800
54	Xe	29.779	33.624	4.111	4.422	4.720	5.036

(continued)

TABLE 4.2. *(Continued)*

Z	Elem.	$K\alpha_1$	$K\beta_1$	$L\alpha_1$	$L\beta_1$	$L\beta_2$	$L\gamma_1$	$M\alpha$	$M\beta$
55	Cs	30.970	34.984	4.286	4.620	4.936	5.280		
56	Ba	32.191	36.376	4.467	4.828	5.156	5.531		
57	La	33.440	37.799	4.651	5.043	5.384	5.789	0.834	0.854
58	Ce	34.717	39.255	4.840	5.262	5.612	6.051	0.883	0.902
59	Pr	36.023	40.746	5.034	5.489	5.849	6.321	0.929	0.949
60	Nd	37.359	42.269	5.230	5.722	6.088	6.601	0.978	0.996
61	Pm	38.784	43.826	5.431	5.956	6.336	6.891		
62	Sm	40.124	45.400	5.636	6.206	6.587	7.180	1.081	1.100
63	Eu	41.529	47.027	5.846	6.456	6.842	7.478	1.113	1.153
64	Gd	42.983	48.688	6.059	6.714	7.102	7.788	1.185	1.209
65	Tb	44.470	50.391	6.275	6.979	7.368	8.104	1.240	1.266
66	Dy	45.985	52.178	6.495	7.249	7.638	8.418	1.293	1.325
67	Ho	47.528	53.934	6.720	7.528	7.912	8.748	1.348	1.383
68	Er	49.099	55.690	6.948	7.810	8.188	9.089	1.405	1.443
69	Tm	50.730	57.513	7.181	8.103	8.472	9.424	1.462	1.503
70	Yb	52.360	59.352	7.414	8.401	8.758	9.779	1.521	1.567
71	Lu	54.063	61.282	7.654	8.708	9.048	10.142	1.581	1.631
72	Hf	55.757	63.209	7.898	9.021	9.346	10.514	1.644	1.698
73	Ta	57.524	65.210	8.145	9.341	9.649	10.892	1.709	1.765
74	W	59.310	67.233	8.396	9.670	9.959	11.284	1.775	1.835
75	Re	61.131	69.298	8.651	10.008	10.273	11.683	1.842	1.906
76	Os	62.991	71.404	8.910	10.354	10.596	12.094	1.910	1.978
77	Ir	64.886	73.549	9.173	10.706	10.918	12.509	1.980	2.053
78	Pt	66.820	75.736	9.441	11.069	11.249	12.939	2.050	2.127
79	Au	68.794	77.968	9.711	11.439	11.582	13.379	2.123	2.204

80	Hg	70.821	80.258	9.987	11.823	11.923	13.828	2.195	2.282
81	Tl	72.860	82.558	10.266	12.210	12.268	14.288	2.271	2.362
82	Pb	74.957	84.922	10.549	12.611	12.620	14.762	2.345	2.442
83	Bi	77.097	87.335	10.836	13.021	12.977	15.244	2.422	2.525
84	Po	79.296	89.809	11.128	13.441	13.338	15.740	2.502	2.618
85	At	81.525	92.319	11.424	13.873	13.705	16.248	2.582	2.706
86	Rn	83.800	94.877	11.724	14.316	14.077	16.768	2.663	2.795
87	Fr	86.119	97.483	12.029	14.770	14.459	17.301	2.746	2.882
88	Ra	88.485	100.14	12.338	15.233	14.839	17.845	2.828	2.968
89	Ac	90.894	102.85	12.650	15.712	15.227	18.405	2.913	3.054
90	Th	93.334	105.59	12.966	16.200	15.620	18.977	2.996	3.146
91	Pa	95.851	108.41	13.291	16.700	16.022	19.559	3.082	3.239
92	U	98.428	111.29	13.613	17.218	16.425	20.163	3.171	3.336
Intensity ratios for the two K-lines 100: 15, for the four L-lines 100: 50: 20: 10 and for the two M-lines 100: 50									

[a] The photon energy is given in keV. The peaks appear with an approximate intensity ratio given in the last line of the table.

Source: Data from Ref. [50], reproduced with permission from Plenum Press.

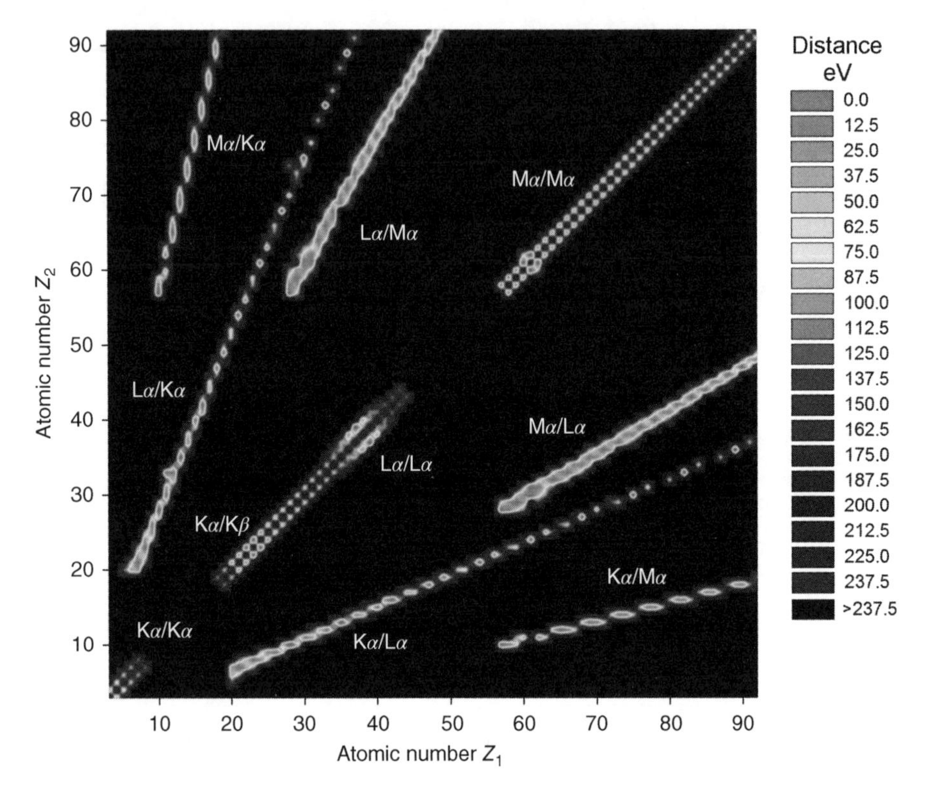

Figure 4.12. Matrix of peak or line interferences. The detection of an element with atomic number Z_1 characterized by its characteristic spectral peaks Kα, Lα, and Mα may be disturbed by another element with atomic number Z_2 via spectral peaks Kα, Kβ, Lα, or Mα. Strong interferences with $\Delta E < 12.5$ eV are marked in light red, those with $225\,\text{eV} < \Delta E < 237\,\text{eV}$ are marked in dark blue. Further interferences are indicated with colors of the visible spectrum from red to blue. (See colour plate section)

Altogether $90^2 = 8100$ different interferences are generally possible. The most troublesome coincidences between 0 and 9 eV fortunately occur with a frequency of only 0.3% and are marked in red. Interferences between 10 and 30 eV can be found at 0.6%, marked in yellow. Less close interferences between 30 and 90 eV and between 90 and 250 eV, marked in green and blue, respectively, occur with about 3% each. In total we find about 7% of the mentioned peak interferences. Most of them can be resolved by a WDS; however, this is not possible with help of an EDS.

It can be realized that a first Kα peak may be disturbed by a second Kα peak if two neighboring elements are present, that is, if $Z_2 = Z_1 \pm 1$. This interference can only happen for light elements with $Z \leq 8$. It is not very close and can be resolved by WDS. A first Lα peak may overlap with a second Lα peak if $Z_2 = Z_1 \pm 1$ and if $21 \leq Z_1 \leq 44$. An interference of Mα with Mα happens for neighboring elements if $Z_1 \geq 58$. All the interferences of Lα and Lα and Mα and

Mα, respectively, can be separated by a WDS, but not by means of an EDS. The combination of Kβ and Kβ does not occur at all.

The interference of Kα and Kβ and vice versa can happen for neighboring elements with $Z_2 = Z_1 \pm 1$ if $19 \leq Z_1 \leq 28$. The coincidence of Cr-Kα and V-Kβ ($\Delta E = 12$ eV) and of V-Kα and Ti-Kβ ($\Delta E = 20$ eV) are well-known examples. A further overlap is possible for elements after next, that is, if $Z_2 = Z_1 \pm 2$ and $35 \leq Z_1 \leq 42$. As an example, Y-Kα and Rb-Kβ are separated by only 3 eV. More complicated relations exist for the coincidence of Kα peaks and Lα peaks: $Z_2 \approx 25/11 \cdot Z_1 + 5$ if $5 \leq Z_1 \leq 30$. Close interferences are produced by Ne-Kα and Ni-Lα ($\Delta E = 3$ eV), by Al-Kα and Br-Lα ($\Delta E = 6$ eV), by S-Kα and Mo-Lα ($\Delta E = 15$ eV), and by As-Kα and Pb-Lα ($\Delta E = 8$ eV). The interference of Lα by Mα occurs if $Z_2 \approx 26/15 \cdot Z_1 + 8.5$ for $31 \leq Z_1 \leq 49$. Notorious examples are Y-Lα and Os-Mα ($\Delta E = 8$ eV) as well as Zr-Lα and Pt-Mα ($\Delta E = 8$ eV). The Kα lines can even be disturbed by Mα lines if $Z_2 \approx 25/6 \cdot Z_1 + 15$ for $10 \leq Z_1 \leq 19$. Particular examples are the line or peak coincidences of S-Kα and Pb-Mα ($\Delta E = 38$ eV) and Mg-Kα and Tb-Mα ($\Delta E = 13$ eV). The coincidences mentioned cannot even be separated by a powerful WDS with its good spectral resolution.

4.3.1.2 *Regard of Sum Peaks*

At high count rates different pulses may be added and yield so-called sum peaks. Such pulse pileup effects can be reduced by the *pulse-pileup rejector* as already mentioned. If, however, the count rate exceeds the level of 10^5 cps, two pulses may arrive less than 1 μs apart. These nearly coincident pulses appear as one single sum peak. In that case, the spectrum shows an additional peak that corresponds to the sum of two photon energies E_1 and E_2.

Of course, such peaks generally arise for main constituents of a sample, either for two principal peaks of two main constituents or for one or two principal peaks of only one main constituent. In general, they can be identified as such and the net intensity of the individual peaks can be determined by a simple mathematical approach.

4.3.1.3 *Dealing with Escape Peaks*

The total effect is more of a problem during the identification of small trace-element peaks overlapped by escape peaks caused by a strong mother peak of a main component. It is less troublesome for silicon detectors than for germanium detectors, where it poses difficulties throughout the important region of X-ray spectra between 1 and 30 keV. However, these problems disappear if the sample consists only of elements with mother peaks below 11 keV, for example, if only lighter elements ($Z < 31$) are present.

Escape peaks cause weaker problems as regards counting losses of the mother peak. A correction is possible by the addition of the daughter peak to the mother peak. As can be read from Figure 3.33, this correction is only about

1% for EDS with Si(Li) detectors or SDDs but is up to 10% or even 20% for EDS with HPGe detectors. For a gas-filled or scintillation counter attached to a Bragg crystal spectrometer, that is, for WDS, the problem becomes much easier. Escape peaks arise if the Bragg crystal is adjusted for a strong mother peak; however, they are only produced within the detector. In the following counter they can generally be separated by pulse-height discrimination and can be added to the mother peak. Fortunately, escape peaks do not occur at their respective angle position of the Bragg spectrometer so that overlaps with small peaks of trace elements do not crop up.

4.3.2 Unambiguous Element Detection

A qualitative analysis is intended to detect elements by means of their characteristic peaks in the spectrum of a specimen. This problem is generally solved in three steps. First, the peaks are detected by their appearance, and their centroid positions are localized (e.g., peaks at 8.05 and 8.91 keV). Second, these peaks are identified as X-ray peaks by comparison with tables or data files (for the given example: Cu-Kα or Ir-Ll and Cu-Kβ or Os-Lα, respectively). Third, from these data the presence of particular elements is deduced (for the given example: copper). These three steps—peak localization, peak identification, and element deduction—can also be carried out automatically by a search program. However, the results should always be checked by an experienced operator.

The first step—*peak localization*—is made complicated by the poor resolution of an energy-dispersive detector and by the statistical fluctuations of X-ray spectra. The position of the peaks cannot therefore be determined exactly. Strong peaks are localized with an uncertainty of ±10 eV in energy; weak peaks, with as much as ±50 eV. Weak peaks may be overlapped by strong peaks and may even be overlooked. Overlappings or interferences, frequently occur in the energy range below 10 keV. However, weak peaks of trace elements may also disappear in the fluctuations of the spectral background. This is a problem having to do with detection limits and is further treated in Section 6.1.2.

The second step—*peak identification*—is hampered as a consequence of the aforementioned uncertainty. Frequently, multiple tabulated peaks fit an energy position, and may be assigned in the permissible bandwidth. In such cases, too many peaks may be classified—except for those cases where two peaks really exist with a strong interference. Such interferences result in the identification of more than one peak during this second step, *peak identification*. A detailed description of such coincidences has been given in Section 4.3.1.1.

More seldom, not even one peak may be found with help of any table in use. These generally weak peaks may be artifacts of the detector, for example, sum peaks or escape peaks (see Section 3.8). Moreover, they can represent "satellite" and "forbidden" peaks of main constituents, which may be missing in some tables (see Section 1.3.2.1). Satellite peaks arise from doubly ionized instead of simply ionized atoms and can possibly be observed as two small peaks near a strong Kα_1, Kα_2 doublet (e.g., Kα_3, Kα_4, or Kα_5, Kα_6). Forbidden peaks arise from outer energy levels if these are close together and the selection rules for quantum transitions are no longer stringent. Forbidden peaks can be observed as small peaks near strong Kβ or Lβ peaks (e.g., Kβ_4 or Lβ_8).

The last step—*element deduction*—is actually rather easy. A first necessary condition for the detection of an element is the appearance of its strongest principal peak (Kα or Lα). However, an element will only be considered to be *unambiguously* detected if two of its strong principal peaks are identified (Kα and Kβ or Lα and Lβ or Kα and Lα) and are in a reasonable intensity proportion. This rule can be followed for main and additional constituents of the specimen but not for traces. In trace analysis, only the strongest peak (Kα or Lα) should be called for. Attention has to be paid to satellite and forbidden peaks of main constituents. They must not be confused with a weak peak of a trace element.

It should further be taken into account that some strong peaks can also come from other sources and not only from the specimen itself. The Si peaks can arise from a quartz glass or silicon carrier, the Ar peaks are normally caused by ambient air, and the peaks of Mo and W can be due to the X-ray tube. Under such circumstances, these peaks must not be used for the detection of elements of the actual specimen. One must be aware of all these phenomena, but the difficulties should not be overestimated. A careful operator with some experience will find no undue problem and will make no grave mistakes.

4.3.3 Fingerprint Analysis

A particular useful mode of qualitative analysis is called *fingerprint* or *signature* analysis, which discerns the similarity of two complete spectra by superposition. It is a first approximation to a *quantitative* analysis. Two complete spectra are compared for that purpose and are checked for equality *in toto* or at least in part. Energy-dispersive spectra are highly suitable for this type of analysis since the complete spectra are always recorded. Of course, both spectra to be compared must be recorded in the same excitation mode with the same energy range and must be stored by a computer memory (hard disc, compact disc, USB stick).

Fingerprint analysis is carried out in a "compare" mode when the two spectra are jointly represented on a display screen. The first spectrum may be displayed in bars; the second, in dots. With regard to the intensity, both spectra can be adapted to each other by proper scaling. After that, the two spectra are compared by means pattern recognition, and equality or similarity is accepted or rejected. This simple yes/no decision can be relevant for certain parts of the spectrum or even for the total spectrum.

In a most simple case of fingerprint analysis, the presence of a certain element may be ascertained by overlapping the spectra of the unknown sample and of the pure element. For that purpose, a spectral library of individual pure elements is useful. It can be set up by aqueous solutions, single grains, and metal filings of the pure elements or even by pure gases pouring into the sample chamber. If a complete library of (nearly) all the elements is available, this technique can provide a qualitative analysis. To this end, the presence of all these elements has to be checked one after the other.

In most cases, however, the presence of only a few elements is confirmed and even their approximate concentrations are estimated. Usually, the spectra of the specimen in question and of a specimen defined as the reference are compared. Equality of the spectra indicates the identity of both specimens; dissimilarity means just the opposite. Such fingerprint analyses are useful for some typical applications. They are used to prevent a mix-up of alloys before processing, to distinguish between genuine art objects and fakes, and to scrutinize suspect materials in forensics.

4.4 QUANTITATIVE MICRO- AND TRACE ANALYSES

Intensity calculations for a thin homogeneous layer on a substrate have been carried out in Section 2.4.2 using the complex Fresnel formalism. However, a simpler approximation for the excitation of a specific element x with monochromatic radiation of energy E_0 can be derived [50]. It leads to the background corrected fluorescence intensity or net intensity N_x, which is dependent on the mass fraction or concentration c_x of the relevant element in the layer:

$$N_x = c_x S_x \frac{1 - \exp[-(\mu/\rho)_{\text{matrix}} \rho\, d]}{(\mu/\rho)_{\text{matrix}}} N_0 \tag{4.1}$$

S_x is the so-called sensitivity of the element according to Equation 2.37 for the mentioned excitation. $(\mu/\rho)_{\text{matrix}}$ is the mass-absorption coefficient of the layer matrix with respect to the excitation energy E_0. The energy of a principal element peak is E_x, the density of the layer matrix is ρ, its thickness is d, and N_0 is the intensity of the exciting monochromatic beam. The fraction of Equation 4.1 represents the mass attenuation effect of the layer matrix. For photon energies below 20 keV, it is mainly determined by mass-absorption and not by scattering.

The quantity $(\mu/\rho)_{\text{matrix}}$ includes the mass-attenuation of the primary incoming beam and of the secondary emerging beam, corrected for geometry. It can be calculated from

$$(\mu/\rho)_{\text{matrix}} = \sum c_i \left[(\mu/\rho)_{i,E_0}/\sin\alpha + (\mu/\rho)_{i,E_i}/\sin\beta\right] \tag{4.2}$$

where c_i is the mass fraction of the different elements i in the layer matrix, (μ/ρ) is its tabulated mass-absorption coefficient for photon energy E_0 and E_i, respectively, α is the glancing angle of the exciting beam, and β is the respective take-off angle of the fluorescence radiation.

The geometry of TXRF with α about 0° and β about 90° leads to the simplified formula

$$(\mu/\rho)_{\text{matrix}} = \sum c_i \left[(\mu/\rho)_{i,E_0}/\alpha + (\mu/\rho)_{i,E_i}\right] \tag{4.3}$$

For thin layers, that is, for small values of the thickness, the nominator of Equation 4.1 can be expanded by a Taylor series. The first two terms lead to the approximation

$$N_x \approx c_x \rho\, d\, S_x\, N_0 \tag{4.4}$$

The dependency of intensity and mass fraction is represented by a straight line as an approximation for small values of $(\mu/\rho)_{\text{matrix}}\, \rho\, d$. It is normally valid for dried residues of aqueous micro- or nanoliter droplets. For thin layers, this relationship can be rewritten as

$$N_x \approx c_{\text{A}}\, S_x\, N_0 \tag{4.5}$$

where c_{A} is the area related mass of the element x within the layer given by the product $c_x\, \rho\, d$. Instead of the fluorescence intensity I of Chapter 2, emitted from a sample, this net intensity is indicated by N as it means total counts recorded in a preset live time of the detector.

A plot of Equation 4.4 or 4.5 gives the so-called *calibration straight line*. Its slope is determined by the sensitivity S_x. Individual elements differ by their particular sensitivity. The conditions of excitation and detection have to be clearly defined and kept constant, and the chemical composition and the physical state of the sample matrix should not affect the calibration.

In XRF analysis, this ideal case is fairly well approximated for thin layers or thin films, while notorious deviations occur for thick layers or bulk samples. Such deviations are called *matrix effects*. In TXRF, only small residues of liquids, small amounts of powders, thin sections or layers, and individual particles are subject to analysis. Such samples with a tiny mass and thickness also meet the conditions for the ideal case. Quantification can then be done by simply adding a standard to the sample and by subsequently using it as an *internal standard*. When only one element has to be determined, this element itself can be used as the standard element. This method is well known as the standard addition method from AAS. When, however, several elements must be determined, an additional element is added that was previously not present in the sample. This method is appropriate for a multielement determination and is normally used for TXRF. It is based on the sensitivity values of the different elements and already known from other methods of atomic spectroscopy.

Both variants of quantification first require the determination of the net intensity of principal peaks in the X-ray spectrum. They may arise from one or more analyte elements and from the internal standard element. Besides, the sensitivity values of these elements should be known from preparatory measurements. However, this work has to be carried out only once as a kind of calibration. Each subsequent quantification is generally easy and reliable, and consequently TXRF has great advantages over conventional XRF.

Of course, there are some conditions for the applicability of internal standardization, but the limitations can easily be observed. They apply to the sample amount used for micro- and trace analyses. On the other hand, a real surface or thin-layer analysis should be carried out in accord with its own regulations, which are treated in Section 4.5.

4.4.1 Prerequisites for Quantification

The net intensities of the principal peaks have to be determined in each spectrum. This can be done following the common practice in EDS. The sensitivity values for these peaks of the respective elements, however, need to be determined quite seldom. This is done for any particular excitation mode and is only repeated after a repair or replacement of instrumental components.

4.4.1.1 Determination of Net Intensities

An X-ray spectrum generally shows several peaks on a structured background, some of which may partly overlap. These peaks have already been identified and assigned to certain elements by means of a previous qualitative analysis. Each element is consequently represented by two or three principal peaks, or at least by one. The net intensity N_x of the strongest peak or possibly of some additional peaks, must be determined, for each analyte x. This intensity is defined by the area under the peaks but should be corrected for the spectral background and for the overlapping of neighboring peaks. It may be determined by summation of all the counts being accumulated in the individual channels representing the peaks and by a subsequent subtraction of the background and overlapping peaks. This can be done semiautomatically after setting of "windows" that limit each peak to a lower and upper border of the photon energy. The correction is made by subtraction of a trapezoidal area under the peak and between the borders. This method is a rough approximation that is only suitable for strong peaks on a low background and without strong overlapping. In this case, the window width can be chosen to be two or three times the peak width (FWHM = full width at half maximum) so that the correction is appropriate. In most cases, however, a severe error will result from use of this method, which is unacceptable.

For that reason, other methods have been developed that can also be applied automatically. The first corrects for the background and is based on a Fourier transformation of the spectrum. This mathematical procedure results in three transformed but significantly different parts of the spectrum. The low-frequency part represents the background, the medium-frequency portion corresponds to the peaks, and the high-frequency part is equivalent to the noise of the background. The low- and high-frequency parts are removed by a mathematical filter and only the medium-frequency part is retransformed. This method, also called a filter technique, gives a background-corrected and smoothed spectrum only containing the spectral peaks. The net intensity can easily be determined now, but overlaps still exist and are troublesome.

For the latter reason, a second method was developed that can be applied to background-free spectra in order to correct for the overlaps. It is based on the principal peaks of pure elements and fits them to each spectrum in question. Consequently, it needs a spectral library like that required for fingerprint analyses in Section 4.3.3. Each spectrum of those pure elements detected by a previous qualitative analysis is reduced in peak height until its principal peaks approximately fit with the corresponding peaks of the spectrum in question. After that, the single spectra are added and a last adaptation to the relevant spectrum is made by a least-squares fit. Finally, the intensity values of the adapted spectra of the pure elements can be read. Each intensity value will preferably include all principal peaks of the respective element, for example, the K peaks and/or the L peaks. The final result is free from the spectral background and from any overlapping peaks, and also escape peaks are taken into account. Both methods of correction successively applied give highly reliable values of net intensities.

4.4.1.2 Determination of Relative Sensitivities

The linear relationship 4.5 between the XRF intensity of an analyte element and its concentration will be valid if the analyte is part of a small specimen deposited on a glass carrier. A plot of the measured net intensity versus the concentration of this element gives a calibration straight line. Its slope is called *absolute sensitivity*. As noted earlier, different elements generally have different slopes or sensitivities. The ratios of these absolute sensitivities with reference to a specific element are called *relative sensitivities* and can be determined by calibration. The mode of excitation chosen for that purpose has to be precisely defined. The X-ray tube, applied voltage, filter, and/or monochromator as well as the geometry of the equipment in use have to be fixed exactly. Any altered mode of excitation, for example, by synchrotron radiation, requires a new calibration and leads to a new set of relative sensitivities.

For the determination of relative sensitivities, different standard solutions are used that are commonly employed for calibration in AAS or ICP spectral analysis. They are obtainable at various suppliers (e.g., Merck KGaA, Darmstadt, Germany; Sigma Aldrich-Chemie GmbH, Steinheim, Germany; Alfa Aesar A Johnson Matthey Company, Karlsruhe, Germany). For TXRF, either complete multielement standards are chosen or single-element standards are mixed together. The stock solutions should be diluted by an aqueous solution (5% nitric acid, suprapure) to a concentration level of 1–10 μg/ml for the individual elements. The finally applied standard solutions should have a volume of about 1 ml and should contain two to six different elements. One element, for example, cobalt or copper, which is chosen as reference element, must be present in all standard solutions.

A droplet of each solution with a volume of about 1–10 μl (possibly 100 nl) is pipetted onto a cleaned carrier and dried by evaporation. From the residues, spectra are recorded in one or several definite excitation modes, as demonstrated in Figure 4.13a–c. The net intensity for the individual elements is determined and the relative sensitivities are calculated by the formula

$$S_j = \frac{N_j/c_j}{N_{\mathrm{ref}}/c_{\mathrm{ref}}} S_{\mathrm{ref}} \tag{4.6}$$

where S is the relative sensitivity, N is the net intensity, and c is the concentration of either the different elements j or the reference element *ref* as indicated. The quantity S_{ref} can generally be set to 1 because only *relative* sensitivities have to be determined. Actually, a single determination per element is sufficient. However, it is preferable to apply several solutions with different concentrations and to repeat the measurements. The averaged values of each element should have a coefficient of variation <5%.

Figure 4.14 provides an overview of relative sensitivities of detectable elements, determined for the three excitation modes defined in Section 4.2.1. The semilogarithmic plot describes their dependence on the atomic number of the elements, detected by either their K or their L peaks. The six curves span three orders of magnitude with a steady course. Each curve shows a maximum sensitivity reached for a particular element in accordance with Figure 4.11.

Above all, it should be emphasized distinctly that the relative sensitivities are independent of the matrix of the applied samples. For the case just described, the matrix water was separated by evaporation and only oxides, hydroxides, and further light compounds of the analyte elements remained as residues. If the original solution, however, is a saline or a gelatinous solution, a mineral or an organic matrix, respectively, will form the residue. Nevertheless, the relative sensitivities of elements determined with these matrices correspond to each other with a deviation of <8% [51].

Relative sensitivities can even be calculated from theory, which is exceptional. According to Equation 2.37, the sensitivity can be expressed by

$$S_j = K\, g_j\, \omega_j\, f_j (\tau/\rho)_{j,E_0} \tag{4.7}$$

where S_j is the sensitivity of the analyte, g is the relative emission rate of the respective element peak in its series, ω is the fluorescent yield of the selected K or L peaks, f is the jump factor of the respective absorption edge, and $(\tau/\rho)_{E_0}$ is the photoelectric absorption coefficient for the primary beam with a photon

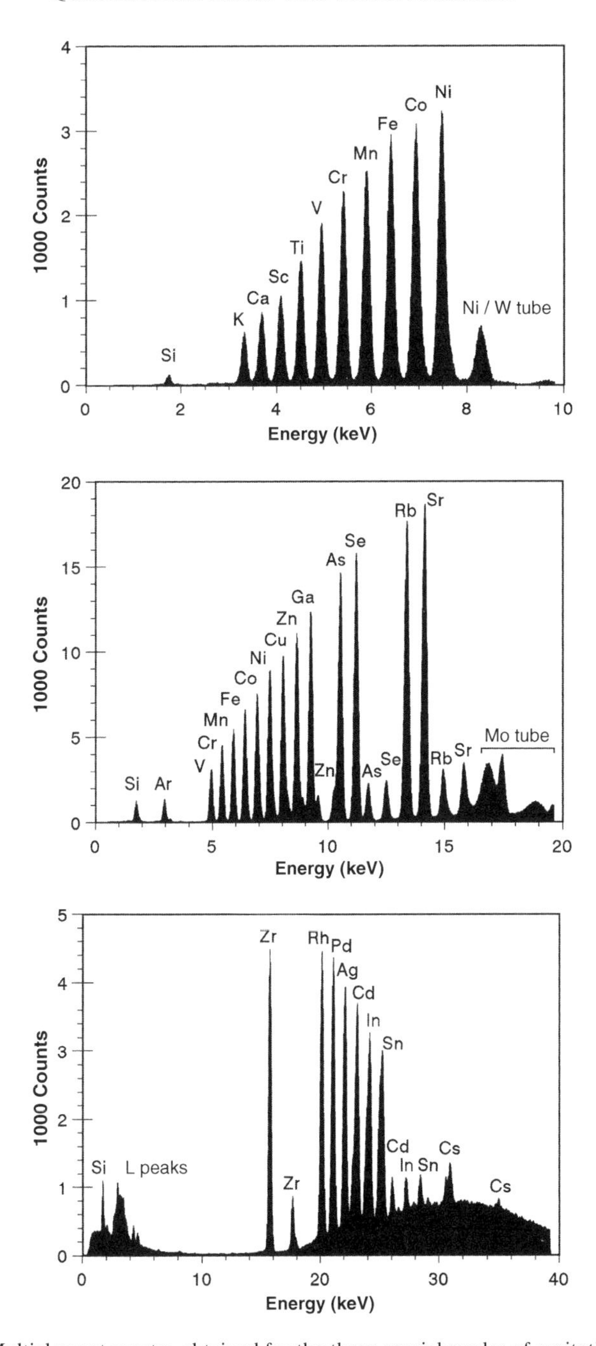

Figure 4.13. Multielement spectra obtained for the three special modes of excitation described in Section 4.2.1. The peaks of all standard elements arise from an applied amount of 1 ng per element. The different peak heights reflect the different sensitivities. Excitation is mainly performed by (a) the W continuum at 30 keV; (b) by the Mo-K peaks; and (c) by the W-Lα peak. Figure from Ref. [4], reproduced with permission. Copyright © 1996, John Wiley and Sons.

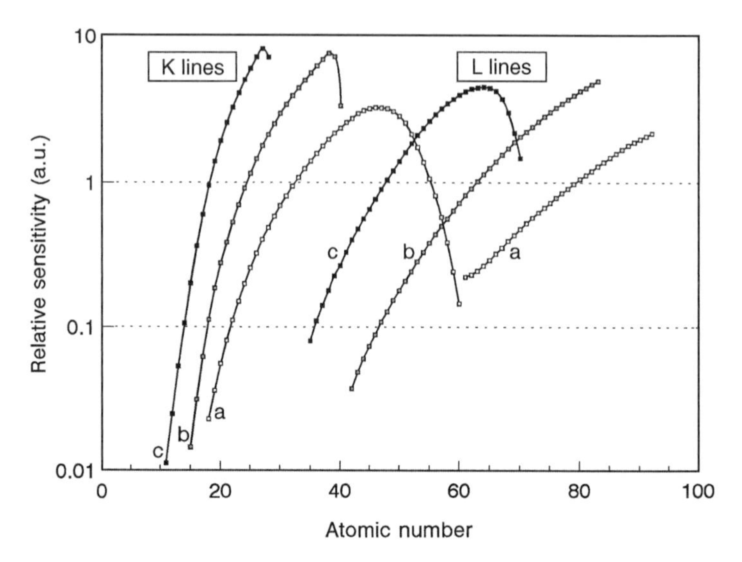

Figure 4.14. Relative sensitivities of different elements in dependence on their atomic number Z. Three excitation modes—(a), (b), and (c)—were applied as described in Section 4.2.1. The elements were detected either by their K peaks (medium Z values) or by their L peaks (high Z values). Figure from Ref. [4], reproduced with permission. Copyright © 1996, John Wiley and Sons.

energy E_0. The constant, K, can be determined with the help of a reference element for which S_{ref} is set to be 1. The fundamental data—g, ω, f, and (τ/ρ)—can be taken from tables. The efficiency of the detector and the transmission of the air path between the sample and the detector are assumed here to be equal for the individual elements, but they can easily be determined and taken into consideration. The results correspond to the values experimentally determined within a margin of 9%, in evidence of the validity of the theory and the absence of matrix effects [51].

It can be concluded that the sensitivity values of those elements for which standards are not available can be obtained by calculation or at least by interpolation between known values. Relative sensitivities can even be transferred from one instrument to another if the equipment is the same.

4.4.2 Quantification by Internal Standardization

In TXRF, quantification is generally carried out by internal standardization. This easy and reliable method can be employed because small or minute specimens are used for analysis. Two different methods can be chosen to this end: the first is suitable for a quantitative determination of a single element and the second is generally recommended for a multielement determination, which is the actual task of TXRF.

4.4.2.1 Standard Addition for a Single Element

The first method is known as *standard addition* and is generally used for single-element determinations in AAS [52]. A few aliquots (three up to five) are taken from the sample solution with a defined volume V (about 1 ml). They are spiked with a fixed small volume υ (50 or 100 µl) of a standard solution (mostly aqueous solutions) containing the single analyte element in different concentrations c_i ($i = 0, 1, \ldots, 4$). Usually, the first concentration c_0 is chosen to be zero (ultrapure water), the second one c_1 should be of the order of the unknown concentration. This value can roughly be estimated according to the respective peak height in the spectrum of the original solution. A certain amount of experience is of course valuable for such an estimation. The further concentrations c_2, c_3, and c_4 should be multiples (respectively about twofold, threefold, and fourfold) of c_1. The final solutions should be thoroughly mixed to ensure homogeneity.

Finally, a certain volume υ_{sp} (about 10 µl) of the individual spiked solutions is used for recording the spectra. The net intensities N_i of the respective peaks of the analyte are plotted against the concentration c_i as is demonstrated in Figure 4.15. Actually, the mass m_i of the analyte element taken by the aliquot υ_{sp} of the final solution should have been plotted on the abscissa. However, both quantities c_i and m_i can be exchanged according to the definite relationship

$$c_i = m_i \frac{\upsilon + V}{\upsilon \cdot \upsilon_{sp}} \tag{4.8}$$

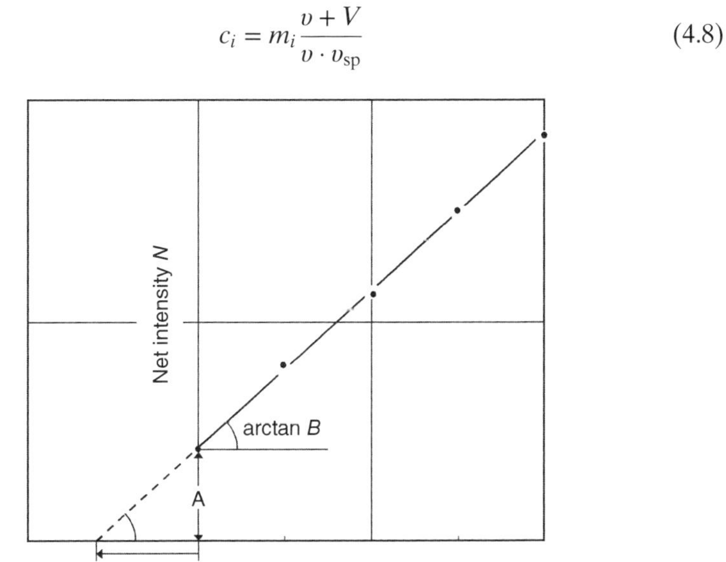

Figure 4.15. Demonstration of the standard addition technique. Aliquots of the original solution (volume V) are spiked with a few standards (volume υ) of increasing concentrations $c_0, \ldots, c_4$. The net intensities lead to a fitted straight line, which can be extrapolated. Its intersection with the abscissa yields the unknown concentration c_x of the original sample multiplied with the factor V/υ. Figure from Ref. [4], reproduced with permission. Copyright © 1996, John Wiley and Sons.

While the applied mass m_x is on the order of nanograms, the concentration c_x of the standard solutions is on the order of μg/ml.[1]

A linear relationship between net intensity and mass or concentration can be expected because of the small amounts of the specimens. It is represented by a calibration straight line in Figure 4.15 and can be determined by a linear regression according to

$$N = A + B\,c \tag{4.9}$$

where A is the ordinate offset, and B is the slope or absolute sensitivity. The unknown concentration c_x of the analyte element can now be determined by an extrapolation of the straight line. It is calculated by the formula

$$c_x = \frac{A}{B}\frac{\upsilon}{V} \tag{4.10}$$

where the second fraction υ/V is the dilution factor (1/10 to 1/20).

The uncertainty that is inevitably due to this determination can be estimated from a confidence interval. If three or more additions are carried out ($n \geq 3$) this interval will be of the order of

$$\Delta\,c = t(P, n-2)\frac{s_R}{B}\frac{\upsilon}{V} \tag{4.11}$$

where t is the student factor for a chosen significance level P, n is the number of standard additions, and s_R is the residual scatter of the measuring points around the straight line, and B its slope.

This method of quantification is very easy, and relative sensitivity values are not even required. Nevertheless, it is time-consuming and therefore seldom used [53,54]. It may be worthwhile when several more samples with the same matrix or composition have to be analyzed. In that case, the original straight line can be parallel shifted until it runs through the origin and can henceforth be used for external standardization. Furthermore, the standard addition technique is suitable for a later check of one result or the other of a multielement determination.

4.4.2.2 Multielement Determinations

The second method of internal standardization is based on the addition of an element initially not present in the sample. Generally, rare elements are chosen, for example, gallium or yttrium in acidic solutions and germanium

[1] TXRF primarily determines the minute absolute amounts m_x of different elements present in a tiny sample. Conventional XRF, however, determines the concentration or mass fraction c_x in a larger sample exclusively and directly.

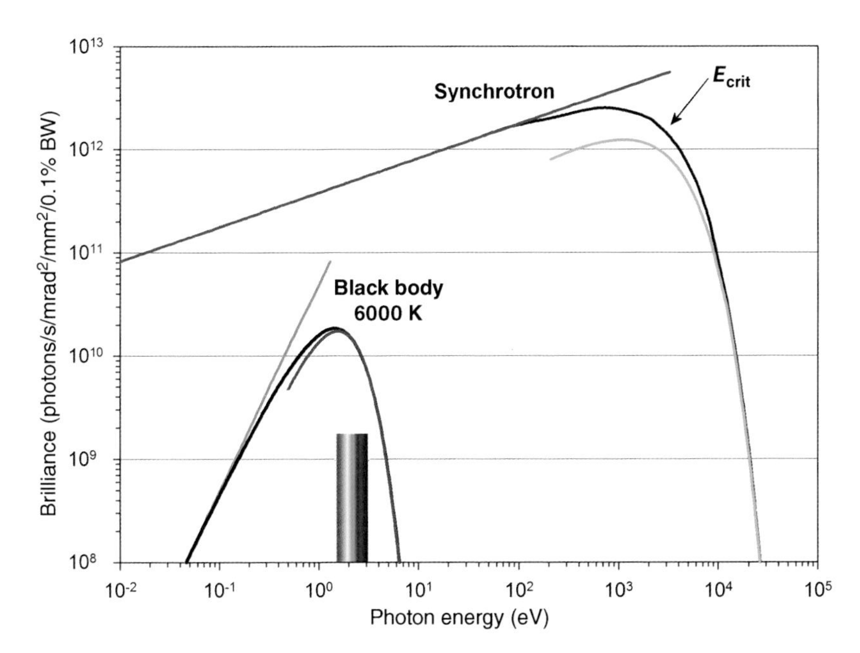

Figure 1.20. The continuous spectrum of Figure 1.19 as a double-logarithmic plot. The ordinate was converted into the brilliance or number of photons/s/mrad^2/mm^2/0.1% bandwidth, respectively. Again, the two asymptotes are presented for low and for high photon energies (red straight line and blue curved line). The range of transition is marked in black, the maximum is at 0.29 E_{crit}. For comparison, the photon flux of a black-body is plotted according to Planck's law. At a temperature $T = 6000$ K, the black curve on the left shows a maximum at $E_{max} \approx 1.5$ eV in the visible color-coded region.

Total-Reflection X-ray Fluorescence Analysis and Related Methods, Second Edition.
Reinhold Klockenkämper and Alex Von Bohlen.

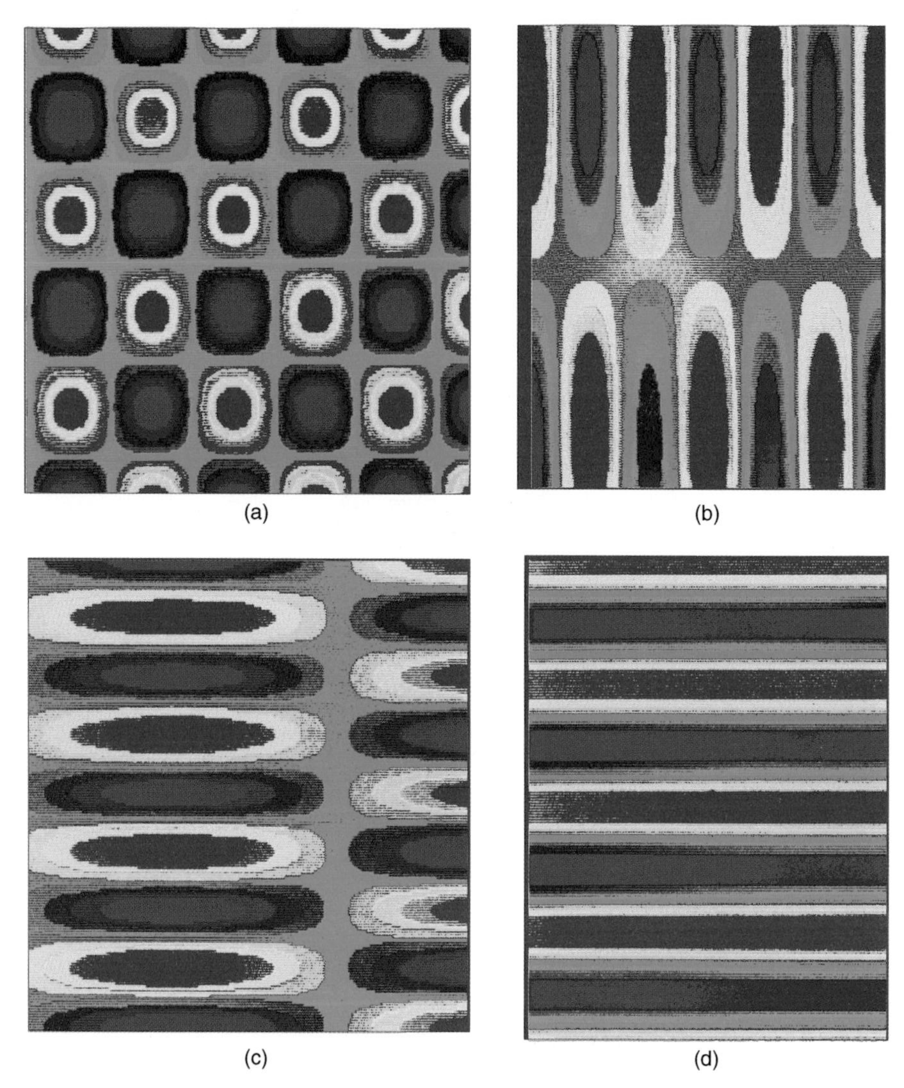

Figure 2.8. Interference patterns for different glancing angles α of the incident plane wave: (a) $\alpha=45°$, (b) $\alpha=10°$, (c) $\alpha=80°$, (d) $\alpha=90°$. Figure from Ref. [1], reproduced with permission. Copyright © 1996, John Wiley and Sons.

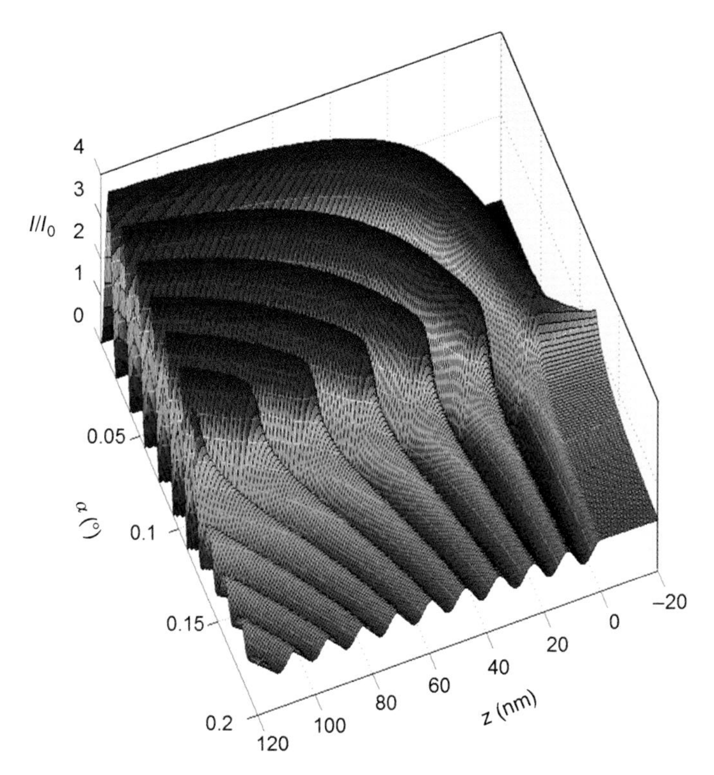

Figure 2.9. The normalized intensity I_{int}/I_0 is presented here as a three-dimensional graph depending on the glancing angle α and on the height z. (*See text for full caption.*)

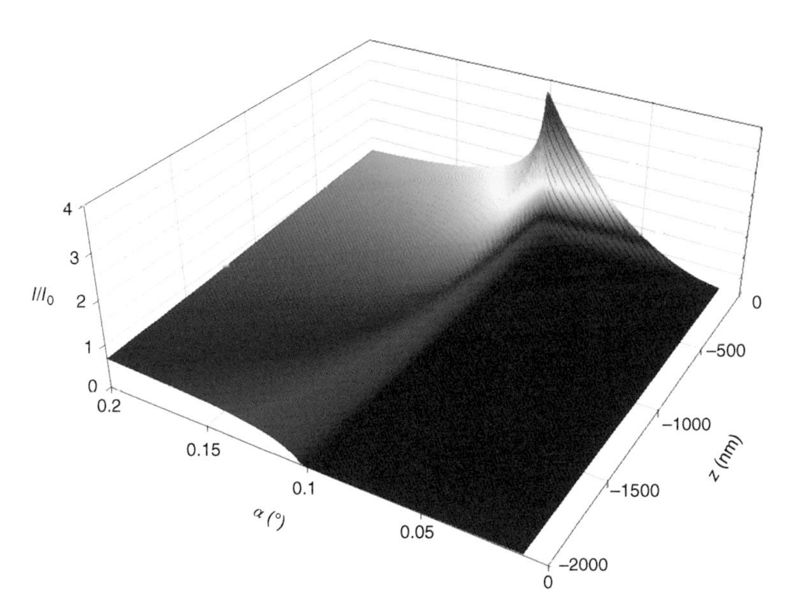

Figure 2.10. The normalized intensity I_{int}/I_0 depending on glancing angle α and depth z. (*See text for full caption.*)

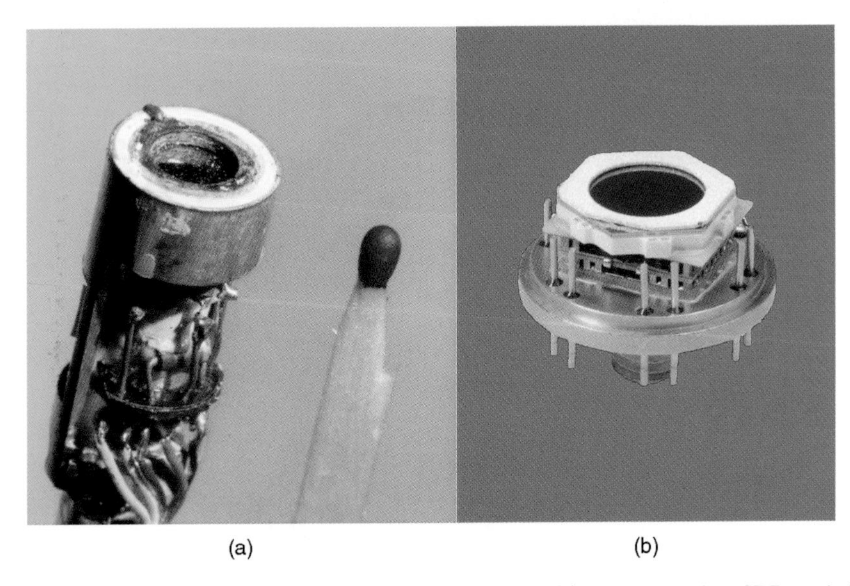

Figure 3.17. Head of a classical Si(Li) detector made in 1967 (a) and of a modern SDD made in 2012 (b). (*See text for full caption.*)

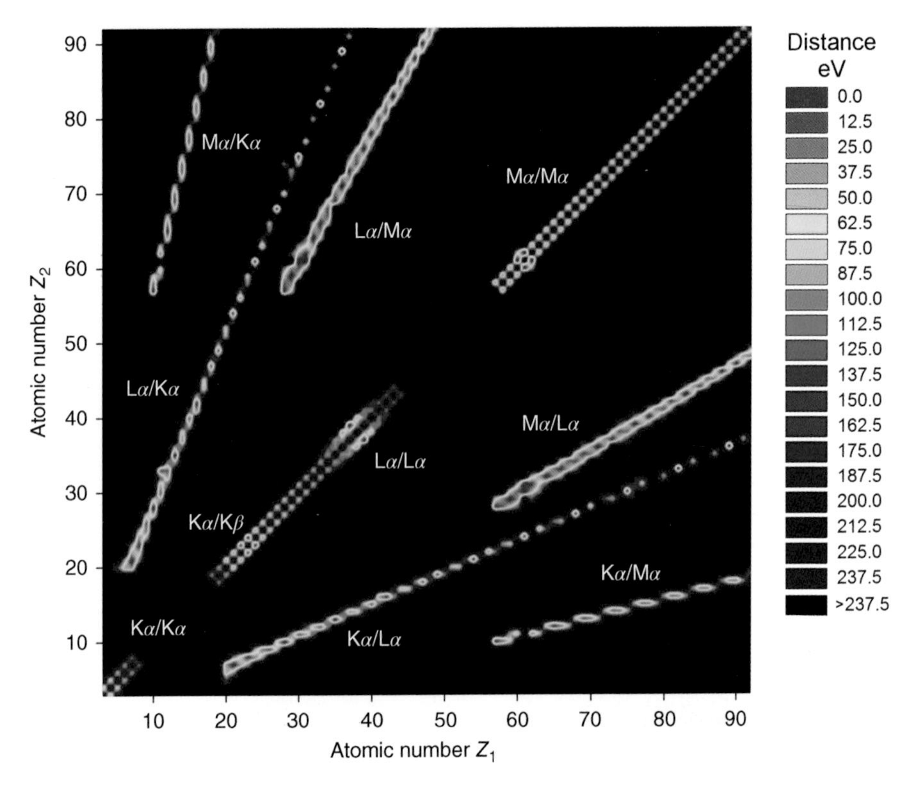

Figure 4.12. Matrix of peak or line interferences. (*See text for full caption.*)

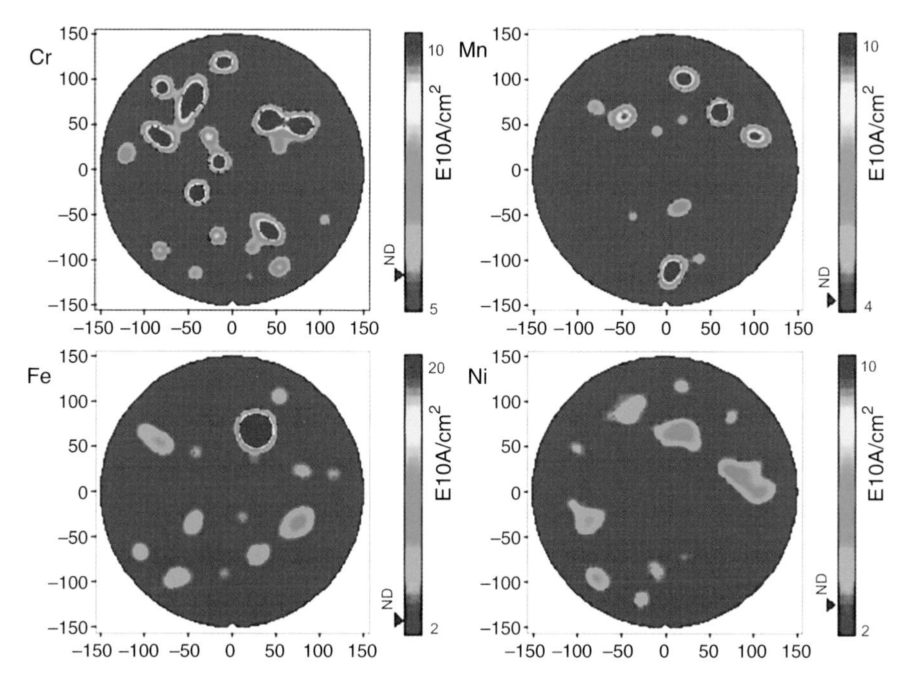

Figure 5.14. Area map of four elements Cr, Mn, Fe, and Ni on a 300 mm Si wafer. (*See text for full caption.*)

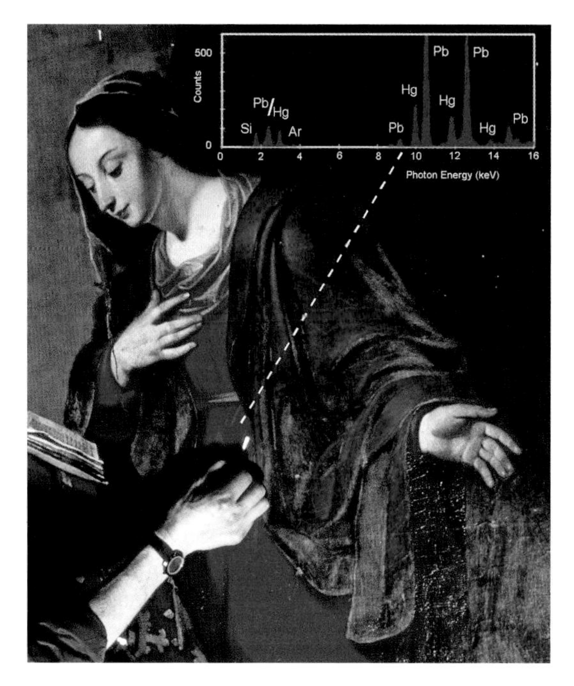

Figure 5.23. Soft sampling of an oil painting by a small cotton-wool bud (Q-tip). (*See text for full caption.*)

Figure 5.24. Two illuminations in the Flemish breviarium "Meyer van den Bergh": (a) f° 536v, Birth of Blessed Virgin Mary; (b) f° 552v, St. Michael. (*See text for full caption.*)

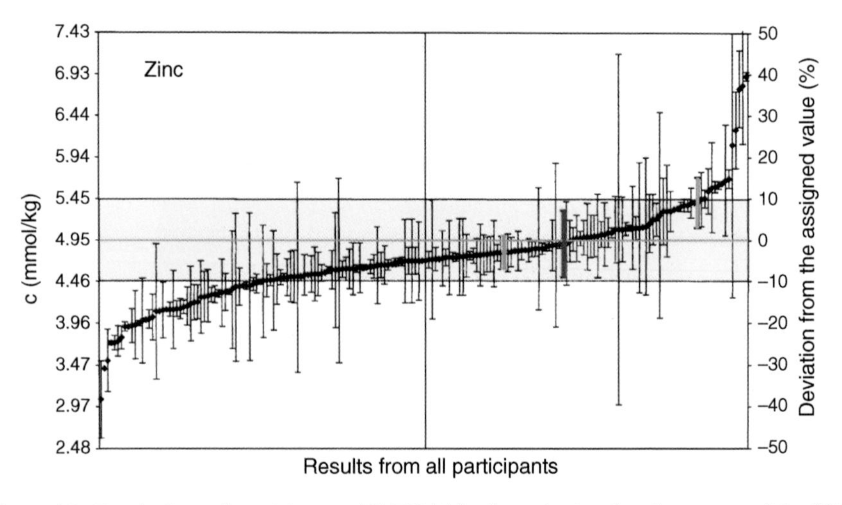

Figure 6.6. Results from all participants of IMEP-14 (sediment) related to the content of zinc [82]. (*See text for full caption.*)

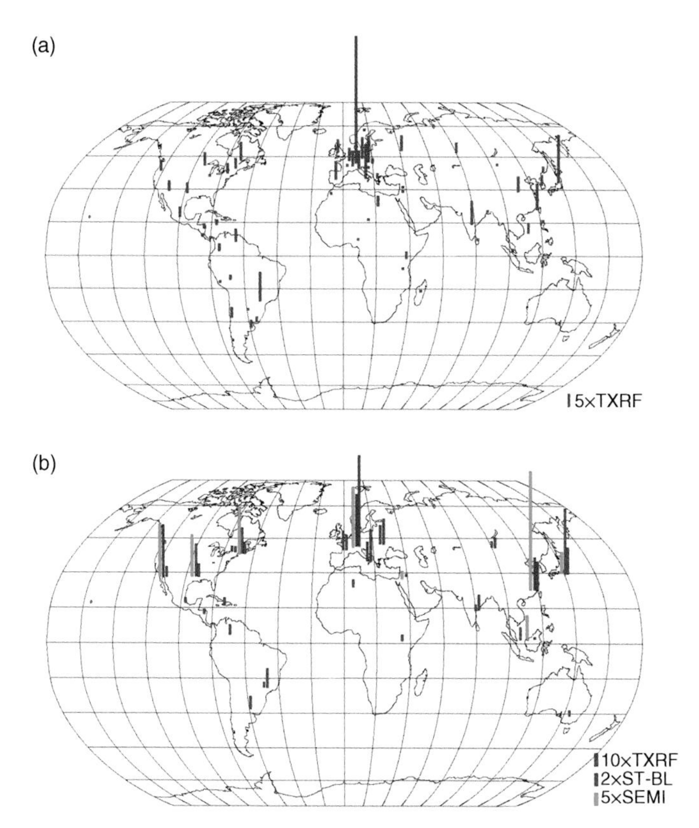

Figure 6.11. Worldwide distribution of TXRF instrumentation after a survey of conference attendees in 2013. The maps are shown in Robinson projection. (a) The red bars represent TXRF instrumentation in 54 countries. (b) TXRF equipment as shown in Figure 6.11a with red bars, but accumulated in 24 subregions according to the United Nations. The blue bars show synchrotron beamlines with TXRF facility, and the green bars stand for semiconductor fabs that probably apply TXRF for contamination control. The height of the bars is a measure for the number of devices. Red, blue, and green bars are scaled so that the total sum is equalized. Maximum for TXRF instruments is 34 in Europe, for the beamlines it is 10 in the USA, and for the semiconductor fabs it is 28 in Taiwan.

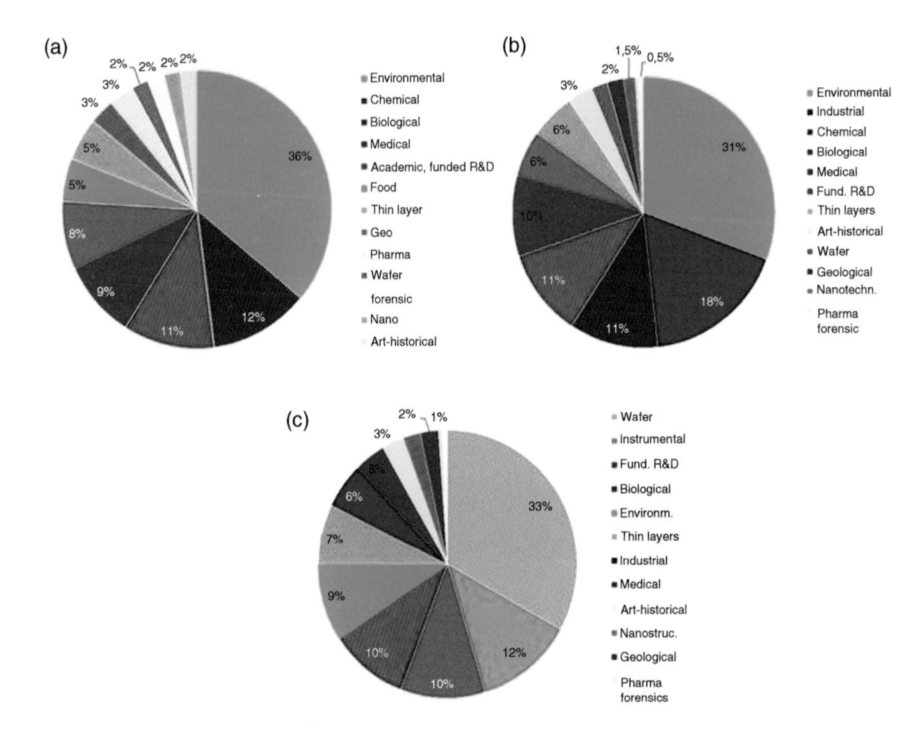

Figure 6.12. Pie charts with percentage of different fields of investigations or applications carried out by TXRF and related methods. (*See text for full caption.*)

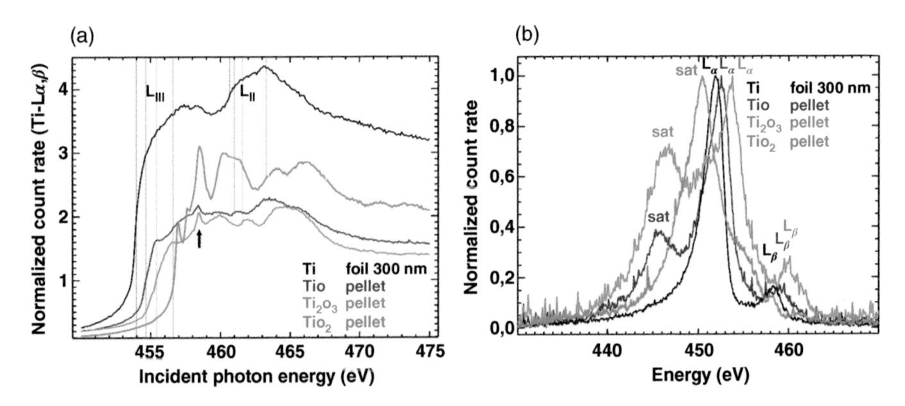

Figure 7.18 (a) L_2 and L_3 edge of a pure titanium foil and pellets of three different titanium oxides. (*See text for full caption.*)

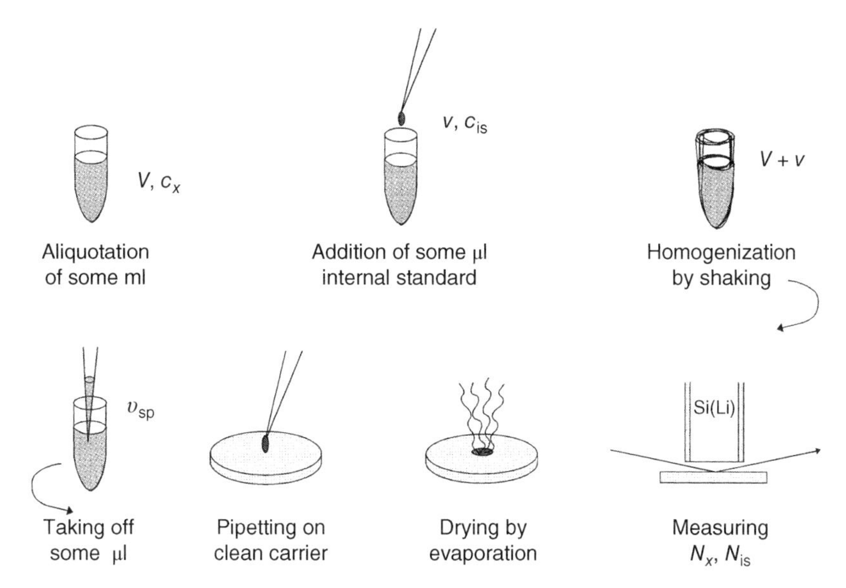

Figure 4.16. The different steps of quantification by internal standardization, applied for multi-element analyses by TXRF. Figure from Ref. [4], reproduced with permission. Copyright © 1996, John Wiley and Sons.

in basic solutions, possibly diluted with ultrapure water and ethanol. Medium-heavy elements with K detection are favored over heavy elements with L detection because of the minor number of peaks. Lighter elements with $Z \leq 21$ are not suited as standards because of particle-size effects that become troublesome in the low-energy range $E < 4\,\text{keV}$.

As illustrated in Figure 4.16, a larger aliquot V of the sample (some ml) is taken and a smaller aliquot v of the standard (some µl) is added to it and thoroughly mixed. The added element henceforth serves as an internal standard for all other elements of the sample already detected and now to be quantitatively determined. For that purpose, a small volume v_{sp} (0.1–10 µl) of the final spiked solution is pipetted on a clean carrier and dried by evaporation. The X-ray spectrum is recorded, providing the net intensities of detected elements with a preset live time. The concentration is then calculated by the simple formula

$$c_x = \frac{N_x/S_x}{N_{\text{is}}/S_{\text{is}}} c_{\text{is}} \tag{4.12}$$

where N is the net intensity, S is the relative sensitivity, and c is the concentration either of analyte, x, or internal standard, is, as indicated. This formula is independent of the volume of the spiked final solution v_{sp}, but the concentration values are both related to the small volume v of the standard. Because c_x is to be determined with respect to the original sample volume V, Equation 4.12 must still be multiplied with the dilution factor v/V.

This method of internal standardization can be applied for solutions or suspensions and for samples that are prepared as solutions or suspensions. The standard is preferably added to the sample at an early stage of sample preparation and is homogeneously mixed. However, it can even be added to a specimen already deposited on the sample carrier, although with the risk of an inhomogeneous distribution. For solid samples that are deposited as thin sections, the standard can only be added afterward [55]. The concentration c_{is} of the internal standard is given by the quotient of the pipetted mass m_{is} and the total mass m_0 of the section. This total mass can be determined by a difference weighing with a microbalance. The concentration c_x of an analyte element again follows from Equation 4.12.

Instead of the concentration c_x, the mass m_x has to be determined for airborne particulate matter deposited on a filter or on a carrier in a certain time t or from a certain volume V, respectively. The pipetted mass m_{is} of the internal standard must then be used instead of the concentration c_{is}, and Equation 4.12 can be applied appropriately. A concentration value can be derived by the ratio of analyte mass and sample volume. For the analysis of contaminations, for example, on a wafer surface, the mass m_x has to be related to the area limited by the detector's field of vision.

For microanalysis of powders, single grains, splinters, fibers, crumbs, metallic smears, or metalline covers deposited on the carriers, a weighing of the minute amount of micrograms is not possible. In that case, the concentration of elements with respect to this minute mass cannot be determined. However, the *detected* elements can be quantified with respect to their sum c_0, which may be set to 100%. A consequent modification of Equation 4.12 results in the formula

$$c_x = \frac{N_x/S_x}{\sum_j N_j/S_j} c_0 \tag{4.13}$$

The summation must include all detected elements.

The reliability of multielement determinations can be estimated from the results obtained for multielement standards or from intercomparison tests (see Section 6.1.5.). The precision of repeated determinations is about 1–5%, and the accuracy of the mean is only slightly poorer. Further details are given in Section 6.1.3.

4.4.3 Conditions and Limitations

Some conditions have to be fulfilled to enable the methods of quantification to be applied. The specimens on a carrier must especially be limited in thickness and consequently in area-related mass. Both quantities should be restricted to a range with an upper and a lower limit each (d_{max} and d_{min} as well as m_{max} and m_{min}).

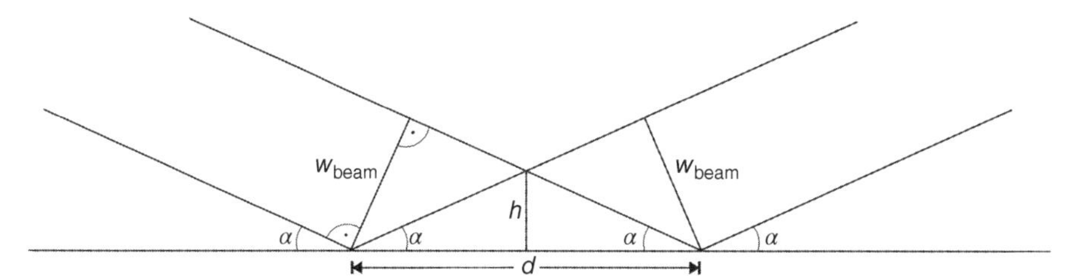

Figure 4.17. Cross-section of an X-ray beam reflected by a flat sample under an angle α. Incident and reflected beam interfere in a triangle with height h. It can simply be shown that h is $w_{beam}/(2\cos\alpha)$. For small angles, h is approximately $w_{beam}/2$ (relative deviation <0.06% for angles < 2°).

A first limitation is given by the geometry of excitation as illustrated in Figure 4.17. The width of the primary beam w_{beam} is usually limited by a slit that is less than 30 µm for X-ray tubes and less than 100 µm for synchrotron beam lines. The height of the triangular region $h_{triangle}$ where the standing wave field appears is approximately $w_{beam}/2$ for grazing incidence. For that reason, the thickness of a specimen should be restricted to several tens of micrometers.

The second reason for a restriction arises from the limited counting capability of the detector. A count rate of about 12 000 cps for a Si(Li) detector and of 1 000 000 cps for an SDD is a maximum value where the dead time percentage is 63%. This high count rate is achieved for a specimen amount mainly depending on the matrix. Upper values have been determined experimentally for three typical matrices characterized by their density ρ: organic tissues (dried, ~0.2 g/cm^3), mineral powders (~1–2 g/cm^3), and covers of metal salts (~2.5 g/cm^3) or metallic smears (~8 g/cm^3). The upper limits, m_{max}, are 250 µg/cm^2 for organic tissues, 140 µg/cm^2 for mineral powders or salts, and 8 µg/cm^2 for metallic smears. This is a rough estimate from unpublished measurements and can differ by a factor of two for the different excitation modes mentioned in Section 4.2.1.

The corresponding thickness d_{max} can be determined according to

$$d_{max} = m_{max}/\rho \tag{4.14}$$

provided that the carrier is uniformly covered. This quantity d_{max} is 12 µm for organic tissues, 0.7 µm for mineral powders, and 10 nm for metallic smears.

4.4.3.1 Mass and Thickness of Thin Layers

As already mentioned, a further limitation can be caused by a strong selective X-ray absorption of the matrix for the analyte element. For a thin layer the absorption can be expressed by Equation 4.1. It causes an intensity deficit, which will remain below a certain permissible level a_{rel} in percent, if the thickness of the specimen is restricted to a maximum value d_{max}. This further limit can be determined from Equation 4.2:

$$d_{\max} = \frac{2\,a_{\mathrm{rel}}/\rho}{(\mu/\rho)_{\mathrm{matrix}}} \tag{4.15}$$

$(\mu/\rho)_{\mathrm{matrix}}$ is the total mass-attenuation coefficient defined by Equation 4.3. A deficit a_{rel} of 5% is considered here to be acceptable. Respective values for $d_{\max}$ were calculated after Klockenkämper and von Bohlen [51]. Excitation mode (b) was provided, and K-Kα or Cd-Lα (E $\geq$ 3.1 keV) was chosen as fluorescence radiation. For the three aforementioned matrices, the upper limit was found to be 8 μm, 0.1 μm, and 4 nm, respectively.

This upper limit was deduced from the intensity of the analyte element alone, regardless of an internal standard. For that reason, the corresponding values are very restrictive. In practice, however, quantification is mostly performed by internal standardization. The net intensity of the analyte is related to that of an internal standard element in accord with Equation 4.12. Variations of the two intensity values are largely compensated by the intensity ratio. The upper limit $d_{\max}$ can be deduced from the condition that the intensity deficit caused by matrix absorption is less than 5% or 0.05, respectively, for the analyte and internal standard. Condition 4.15 can then be replaced by the formula

$$d_{\max} = \frac{0.10/\rho}{\left|(\mu/\rho)_x - (\mu/\rho)_{\mathrm{is}}\right|} \tag{4.16}$$

It can be shown that the respective values are significantly above the limits determined by Equation 4.15, generally by a factor of 10 to 400. The photon energy of the analyte and standard may even differ by threefold. However, it has to be provided that the analyte and standard are homogeneously distributed in the specimen and do not segregate during drying.

A lower limit for the area-related mass or covering is determined by the detection power, which sets a limit $m_{\min}$ of about 10 pg/cm^2. In addition, there is a lower limit for the thickness, determined by the modulated wave field above the carrier [56]. This troublesome modulation can be leveled out if the specimen includes *several* nodes and antinodes. For that purpose, the thickness must exceed a certain minimum value $d_{\min}$. It was determined by de Boer [57] for the three typical matrices to be 0.2 μm, 0.05 μm, and 15 nm, respectively.

Because of the internal standardization, the lower limit can also be reduced. Instead of several nodes and antinodes, only the first antinode above the carrier should contribute to excitation. The specimen should be as thick as half the node distance given by Equation 2.12. If the glancing angle is chosen to be 70% of the critical angle α_{crit}, the lower limit is approximately

$$d_{\min} \approx 22/\sqrt{\rho} \tag{4.17}$$

TABLE 4.3. Upper and Lower Limits of the Covering and Thickness for Three Typical Matrices

Matrices	Organic Tissues	Mineral Powders	Metalline Covers
Constituents	$_1$H . . . $_8$O	$_8$O . . . $_{20}$Ca	$_{24}$Cr . . . $_{30}$Zn
Covering:			
m_{max} [μg/cm^2]	250	140	8
m_{min} [μg/cm^2]	10^{-5}	10^{-5}	10^{-5}
Thickness:			
d_{max} [μm]	12	0.7	0.01
d_{min} [μm]	0.015	0.015	0.015

Source: From Ref. [4], reproduced with permission. Copyright © 1996 by John Wiley and Sons.

The density of the carrier is to be given in g/cm^3 so that the thickness d_{min} is calculated in nanometers. This quantity is independent of the excitation mode, the analyte, and the matrix. For quartz-glass carriers, d_{min} is about 15 nm.

The decisive values for lower and upper limits of the covering and thickness of a specimen are summarized in Table 4.3 for the three typical matrices mentioned. The lower limits are generally determined by the detection power and the height of the first antinode. The upper limits are set by the capability of the detector, that is, by its maximum count rate. Matrix absorption leads to the same order of magnitude. Its influence, however, may be reduced if internal standardization is carried out. The thickness of a specimen can even exceed 10- or 100-fold of the tabulated value of d_{max} if the covering m_{max} is kept constant. This can be achieved by an incomplete and nonuniform covering.

Even slighter restrictions may be necessary if the quantification is performed by standard additions. In this case, analyte and internal standard are the same and the limitation by matrix absorption, expressed in Equation 4.16, becomes irrelevant. But also here, a homogeneous distribution of the analyte and standard is desirable in order to get accurate results.

4.4.3.2 Residues of Microliter Droplets

The deposition of microliter droplets from diluted standard solutions on flat substrates like quartz glass or silicon leads to ring-shaped residues rather than thin homogeneous layers as shown in Section 4.1.3.1. After Hellin *et al.* the dried residues can be represented by wide hollow cylinders [57,58]. The outer diameter of a 50 μl droplet is typically 1.5 mm, the width of the wall can be 50 μm and its height may be 1 μm while the central bottom is only 0.15 μm thin. The density of the metal-salt residues is about 2.5 g/cm^3 with a large variance. The linearity of calibration curves was experimentally proved for samples of Ni and Ge salts—excited by W-Lβ and Mo-Kα radiation at 70% of the critical angle of total reflection. The linearity was also checked for hollow cylinders consisting of a matrix of metal-salts in accord with a formula similar to Equation 4.1.

A deviation from linearity of 20% was generally reached for a deposition less than 1×10^{14} metal atoms and a deviation of 10% for an amount less than 2×10^{13} atoms. With an average area of 1.5×10^{-3} cm^2 for the circular wall the area-related mass is about 6 μg/cm^2 for a systematic deficit $a_{rel} = 10\%$, and about 1 μg/cm^2 for $a_{rel} = 5\%$. Both specifications roughly correspond with the restrictions cited for the maximum mass of thin layers in Section 4.4.3.1. In general, the deposition of microliter droplets is not well suited for quantitative TXRF; the dynamic range of a linear calibration is restricted to less than 3 μg/cm^2 for thin layers of metallic coverings and less than 2 μg/cm^2 for ring-shaped metal-salt residues of microliter droplets. However, the dynamic range can be extended for film-like residues of nano- and moreover of picoliter droplets already dealt with in Sections 4.1.3.2, 4.1.3.3 and 4.1.3.4.

4.4.3.3 Coherence Length of Radiation

Constructive and destructive interference lead to the standing wave fields treated in Chapter 2. This two-beam interference, however, is only possible under the condition of coherence, that is, both beams, or strictly speaking their waves, should have a phase difference that is constant with regard to time and space. The first condition, "temporal constancy," is always met for TXRF experiments. The second condition, "spatial coherence," however, is limited for two reasons:

1. The two beams do not come from a punctual source but from a source with a certain extension w_{beam}.
2. The two beams are not completely monochromatic but have a certain bandwidth ΔE or $\Delta\lambda$, respectively. The monochromaticity can be defined by $E/\Delta E$.

As a result, the interference pattern of Figure 2.8 with maxima and minima is not completely sharp within the total triangle above the surface but is restricted in height.

The two kinds of spatial coherence are called transversal and longitudinal coherence and are represented in Figure 4.18 [47,59]. For Figure 4.18a we assume two waves with the same wavelength but originating from different points A and B of the source. Their wave fronts with distance λ cross over in direction of the observation point P, with a small deviation $\Delta\alpha$. In a lateral distance $\pm\xi_t$ from point P, both wave fronts coincide and are completely out of phase. $\Delta\alpha$ is given by

$$\tan \Delta\alpha = \frac{\lambda}{2\xi_t} \tag{4.18}$$

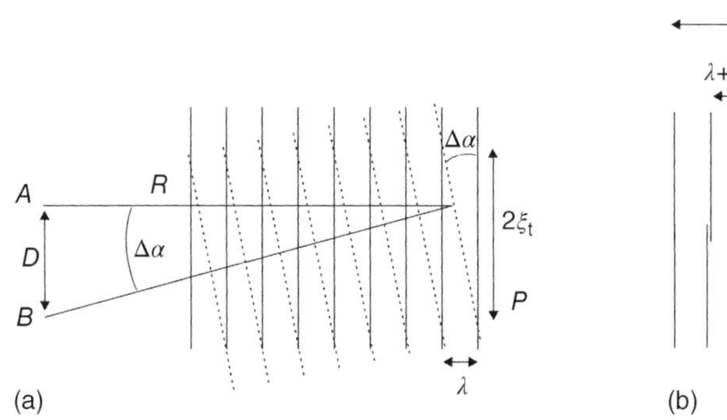

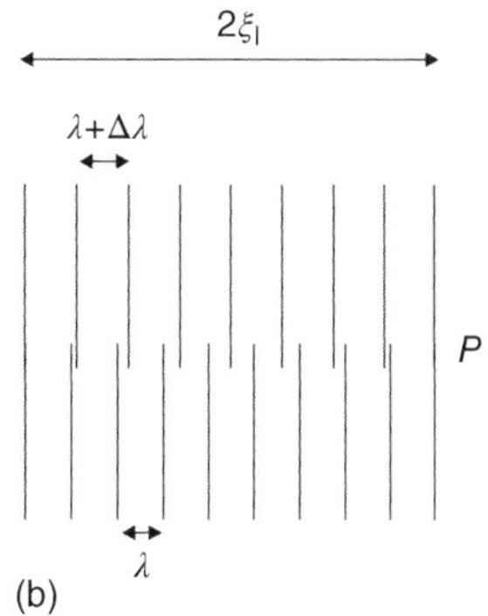

Figure 4.18. Spatial coherence for X-rays coming from a source that is not infinitely small and representing a beam that is not infinitely monochromatic. (a) An extended X-ray source between A and B emits two beams with the same wavelength λ, but their wave fronts include a small angle $\Delta\alpha$. In a distance $\pm\xi_t$ of point P, both wave fronts are out of phase with a difference of $\pm\lambda/2$. It leads to a limited "transversal" coherence length ξ_t. (b) Two waves are emitted from the same point A but with slightly different wavelengths λ and $\lambda + \Delta\lambda$. The two waves are out of phase in a distance $2\xi_l$ of point A, which is N times the wavelength λ and $(N-1)$ times $\lambda + \Delta\lambda$. Their phase difference λ is the reason for a limited "longitudinal" coherence. Figure from Ref. [47], reproduced with permission from the author.

Besides, this angle meets the relationship

$$\tan \Delta\alpha = \frac{D}{R} \tag{4.19}$$

where D or w_{beam} is the width of the beam and R is the distance between source and observation plane. Combining both equations leads to the transversal coherence length

$$\xi_t \approx \frac{\lambda}{2} \cdot \frac{1}{\Delta\alpha} = \frac{hc_0}{2E} \cdot \left(\frac{R}{D}\right) \tag{4.20}$$

with $hc_0 = 1.240\ \mu\mathrm{m}\cdot\mathrm{eV}$.

For Figure 4.18b, we assume two waves originate from the same point A, but with slightly different wavelengths λ and $\lambda + \Delta\lambda$. They coincide in the direction of observation and are completely out of phase in a distance $\pm\xi_l$ of point P. The value $2\xi_l$ is the Nth multiple of λ and besides the $(N-1)$th multiple of $(\lambda + \Delta\lambda)$. From here, the longitudinal coherence length can be approximated by

$$\xi_l \approx \frac{\lambda}{2} \cdot \frac{\lambda}{\Delta\lambda} = \frac{hc_0}{2E} \cdot \left(\frac{E}{\Delta E}\right) \tag{4.21}$$

For *transversal* coherence, the factor R/D is decisive. It can vary between 5×10^3 and 5×10^5. For a medium value, $R/D=5\times10^4$ and $E=5$ keV, this length amounts to 6.2 μm but decreases inversely to E (1.5 μm for $E=20$ keV). For *longitudinal* coherence, the relative spectral resolution $E/\Delta E$ is the key factor. It may differ between 5×10^1 and 5×10^4. For a medium value $E/\Delta E=5\times10^3$ and $E=5$ keV, this quantity is limited to 0.6 μm, which is a tenth of the mentioned transversal coherence length.

The structures to be analyzed—particulate contaminations, islands, or thin layers—have to be within the triangle of Figure 4.17 with a base d determined by the detector's field of vision and by the height h determined by the width of the incoming beam w_{beam}; for small angles h is about $w_{\text{beam}}/2$. Additionally, the structures should have a height h_{coh}, which corresponds to the smaller value of ξ_l and ξ_t. In accord with Bragg's law, we find for the longitudinal coherence

$$h_{\text{coh}} = \xi_l/2\sin\alpha \approx \frac{hc_0}{2E}\cdot\left(\frac{E}{\Delta E}\right)\cdot\frac{1}{2\alpha} \tag{4.22}$$

Of course it is doubtful whether the coherence of the two interfering beams is not destroyed or at least diminished by a specimen positioned in the relevant triangle. This important question may be answered by GI-XRF experiments.

Remembering the vertical period of a standing wave above a substrate defined by Equation 2.12a we find

$$h_{\text{coh}} \approx a_{\text{vertical}}\left(\frac{E}{\Delta E}\right) \tag{4.23}$$

This relationship is illustrated in Figure 4.19 with varying angle α and parameters E and $E/\Delta E$. A small glancing angle, a high spectral resolution, and a low excitation energy is beneficial for coherence. In the case of TXRF where α is chosen to be somewhat smaller than the critical angle (0.07° for a Si substrate and hard X-rays of 20 keV) the height of coherence in Figure 4.19 is about 65 μm for a medium spectral resolution of 5×10^3. In the case of GI-XRF with angles up to 1°, the layer thickness should be as small as 5 μm for hard X-rays.

These conditions can be moderated significantly for soft X-rays and smaller glancing angles as is shown in Figure 4.20. It is a supplement to Figure 4.19 and shows the photon energy versus the glancing angle in a double logarithmic plot. The straight lines represent the product in accord with the formula

$$E\cdot\alpha \approx 6\times10^{-3}\left(\frac{E}{\Delta E}\right) \tag{4.24}$$

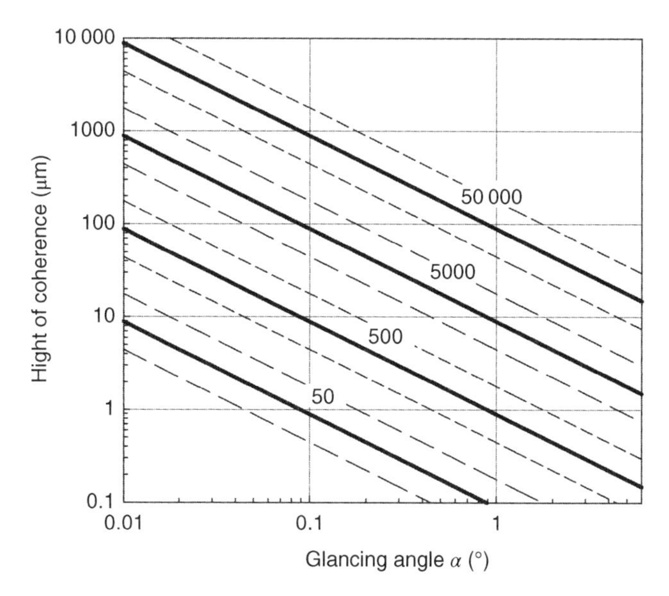

Figure 4.19. Height above a flat substrate where coherence of incident and reflected beam is guaranteed and the standing wave field is not disturbed by incoherence. Different straight lines are represented in the double logarithmic plot according to the excitation energy $E = 10$ keV and the monochromaticity $E/\Delta E$ of the exciting beam (4 solid lines). E can be selected between 5 keV for excitation of light elements and 20 keV for heavy elements (dashed lines); $E/\Delta E$ may vary between 50 for X-ray tubes with a simple monochromator and 50 000 for synchrotron beamlines with a double-crystal monochromator. The height of the triangle in Figure 4.17 is usually about 50 µm, the glancing angle for TXRF at silicon substrates is about 0.1°.

for different values of $(E/\Delta E)$. Coherent radiation and an undisturbed standing wave field are guaranteed in the mentioned triangle with a height up to of 50 µm if the product of both quantities E and α is below the given limit.

The effect of coherence was ignored in TXRF and even in GI-XRF for a long time. In 2009, von Bohlen *et al.* warned against a blurring of the interference pattern by incoherent beams or waves [47,60]. The effect is demonstrated in Figure 4.21 for the case that the area of the lower trapezoid with coherence is 50% of the area of the whole triangle. In that case, the height of the lower trapezoid is only 30% of the height of the given triangle above the reflecting surface. The interference pattern is distinct in the trapezoid but is pale in the upper triangle. The antinodes decrease with height above the surface from the fourfold to the double intensity and the nodes disappear.

Today, the prerequisite of coherence for the appearance of standing waves is usually noted. For qualitative analyses by simple TXRF the effect has no consequence; the fluorescence intensity is doubled in any case. The background, however, is increased if incoherence spoils the effect of standing waves

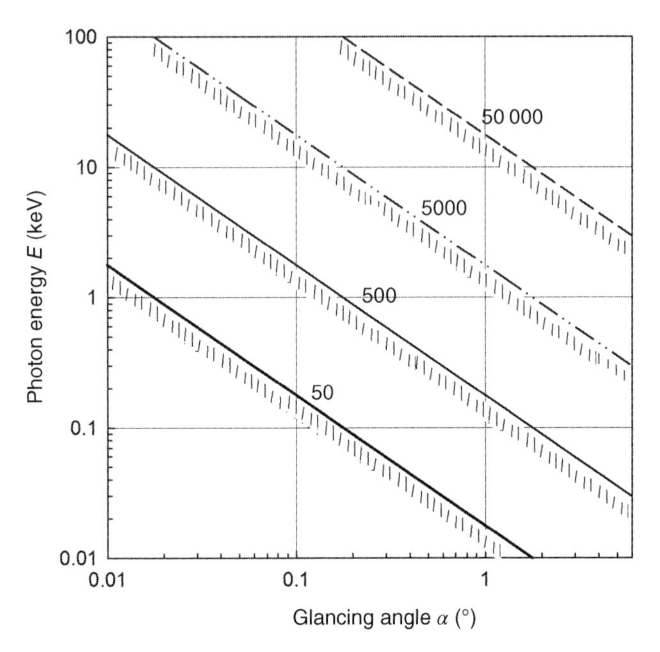

Figure 4.20. Photon energy of the exciting beam E versus the glancing angle α for an undisturbed XSW-field. In the double logarithmic plot, four different straight lines are shown with constant products of E and α. This product is characterized by the monochromaticity of the exciting beam $E/\Delta E$. Coherence of incident and reflected beam and distinct XSW is ideally guaranteed up to a height of 50 μm above the surface if values are chosen for E and α *below* the four respective straight lines with values $E/\Delta E$ between 50 and 50 000.

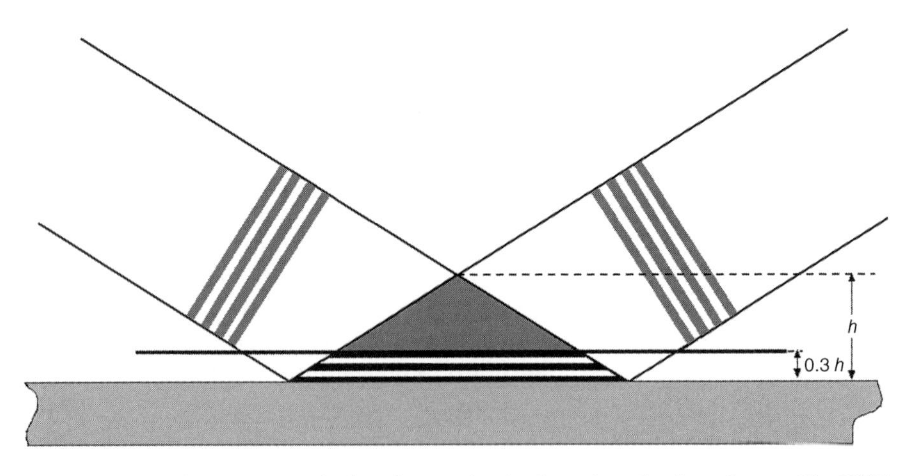

Figure 4.21. Interference pattern in the relevant triangle above the reflecting substrate. The XSW with nodes and antinodes is distinctly marked in the lower part of the triangle and fades in the upper part. Lower and upper region (trapezoid and upper triangle) have equal area if the height of the trapezoid is 30% of the total height. Figure from Ref. [60], reproduced with permission. Copyright © 2009 by the Royal Society of Chemistry.

so that detection limits get worse. For quantitative TXRF analyses, coherence and standing waves can be annoying if the internal standard is not distributed homogeneously. Consequently, for lowest detection limits, coherence is a must; for a reliable quantification, a homogeneous distribution of sample and internal standard is necessary.

For GI-XRF, an undisturbed standing wave field is generally aimed at. It will be ensured by excitation with a beam as small and as monochromatic as is appropriate. If coherence cannot be completely guaranteed, its limitation has to be taken into consideration.

4.5 QUANTITATIVE SURFACE AND THIN-LAYER ANALYSES BY TXRF

In addition to micro- and trace analyses, TXRF can be applied to surface and near-surface layer analyses provided that the samples to be examined are even and optically flat in order to ensure total reflection. This condition is largely fulfilled for glasses and wafers—flat and polished disks of silicon, germanium, or gallium arsenide. Such wafers designated for electronic components such as ICs are ideally suited as samples, as long as they are not patterned. TXRF is highly suitable (i) for the detection of surface contamination of such wafers, and (ii) for the characterization of different types of contamination. This control is highly important and has made TXRF an indispensable analytical tool for the semiconductor industry.

4.5.1 Distinguishing Between Types of Contamination

In the region of total reflection, the fluorescence intensity is strongly dependent on the glancing angle of incidence. First it is assumed or presupposed that an X-ray standing wave field (XSW) is created and is not disturbed in the relevant triangle above the substrate. From Section 2.3 we know that some special kinds of samples show a certain typical dependence, which may be characteristic. Three of them will now be examined in detail: infinitely thick and flat substrates, granular residues *on* a substrate, and buried layers *in* a substrate. This inspection aims at the determination of trace impurities either homogeneously distributed in thick samples, located in granular particles, or evenly deposited in thin near-surface layers. TXRF is capable of such a determination and especially of distinguishing among these three contaminations [61,62]. They may be called *bulk type* (>1 μm), *particulate type* (>50 nm), and *thin-layer type* (≤5 nm).

4.5.1.1 Bulk-Type Impurities

The three different angle-dependent intensity profiles are demonstrated in Figure 4.22. The fluorescence intensity of some impurity element is plotted

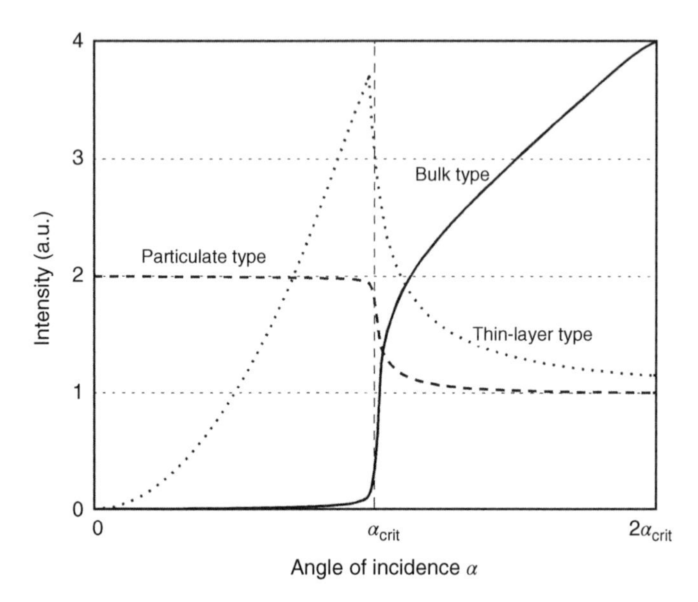

Figure 4.22. Characteristic intensity profiles for three different kinds of contaminations: (a) bulk type; (b) particulate type; and (c) thin-layer type. The critical angle α_{crit} is determined by total reflection at the flat substrate. Figure from Ref. [62], reproduced with permission. Copyright © 1991 by Springer.

against the glancing angle α. The curve valid for a bulk-type contamination is similar to the response of a blank carrier. The intensity is very small *below* the critical angle of the substrate but increases rapidly *above* this angle in accord with Equation 2.27. It can be written in a modified notation as

$$N_B(\alpha) = c_{VB}[1 - R(\alpha)]\alpha\, N_{0B} \tag{4.25}$$

where N_B is the net intensity of the analyte, c_{VB} is the volume-related concentration in atoms/cm^3, R is the reflectivity of the substrate given by Equation 1.76, α is the glancing angle of incidence, and N_{0B} is a reference value used for normalization at a larger angle. Such a profile will be recorded if impurities are homogeneously distributed within the substrate. With increasing angle, the penetration and information depth increases from several nanometers up to several micrometers.

4.5.1.2 Particulate Contamination

In most cases, contamination of Si wafers by metallic particulates was determined, while Hellin *et al.* [63] also investigated Ge wafers as substrates. For a particulate-type contamination, the intensity *above* the critical angle is nearly constant, because the particles are completely excited independent of the angle of incidence. Because of the total reflection, the intensity doubles *at* the critical angle in nearly steplike fashion and remains at the twofold value down to very small angles. This dependence already expressed by Equation 2.21 can be written as

$$N_P(\alpha) = c_{AP}[1 + R(\alpha)]N_{0P} \tag{4.26}$$

where c_{AP} is the area-related concentration in atoms/cm^2 and N_{0P} is a second reference value. This profile can be recorded if one or more grains of a certain size are located on a substrate. Their diameter has to be some 20 nm at a minimum and several micrometers at a maximum. Also, a total set of grains is permissible that may have a Gaussian or a logarithmic Gaussian distribution of the grain size with an appropriate width [50].

4.5.1.3 Thin-Layer Covering

For a thin-layer type contamination, the third curve of Figure 4.22 is valid. The intensity far *above* the critical angle is constant, just as for particulate contamination. The asymptotic intensities of both curves will even be equal if both concentration values c_{AL} and c_{AP} are the same. Obviously, the particulate and thin-layer types do not differ in fluorescence at larger angles. However, the intensity of the thin-layer type steadily increases to a three- to fourfold value just *at or below* the critical angle and thereafter decreases to zero in a parabolic curve. This profile can be described by a modified version of Equation 2.23:

$$N_L(\alpha) = c_{AL}[1 - R(\alpha)] \cdot \frac{\alpha}{z_n(\alpha)} \cdot N_{0L} \tag{4.27}$$

where c_{AL} is the area-related concentration in atoms/cm^2, z_n is the penetration depth of the primary beam in the substrate, normal to its surface, which is given by Equation 1.78, and N_{0L} is a reference value that must be equal to N_{0P} of Equation 4.26. Such a profile will only be recorded if an ultrathin layer of approximately 1 nm is deposited *above* or embedded *below* the surface of a thick substrate. No further differentiation of these two layers is possible [64].

Only the latter curve presupposes an XSW in the relevant triangle above the substrate. If the coherence height is less than 1 μm instead of about 50 μm the XSW disappears, the curve "thin-layer type" will alter and approximate the curve "particulate type."

4.5.1.4 Mixture of Contaminations

It can be concluded from the preceding discussion that each of the three contaminations can be identified by its characteristic profile. Even a mixture of

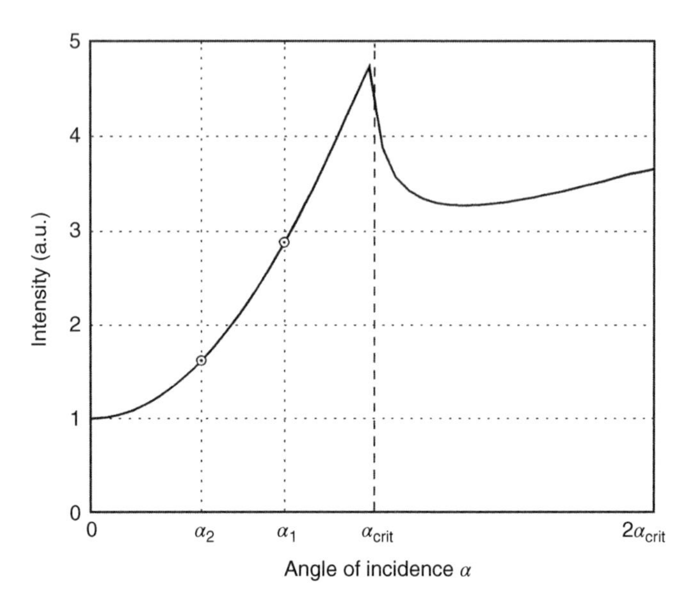

Figure 4.23. Intensity profile for a flat substrate contaminated by an element that is as well embedded throughout the substrate, as located in small grains above the substrate, as deposited like an ultrathin layer on top of or just below the substrate. The critical angle α_{crit} is determined only by the substrate itself. α_1 and α_2 are significant angles of operation. Figure from Ref. [4], reproduced with permission. Copyright © 1996, John Wiley and Sons.

two or three types of contamination can be analyzed and the different fractions can be determined [60]. For that purpose, actually an angle scan should be performed by GI-XRF and intensity profiles should be plotted for all elements under consideration. Such a profile is demonstrated in Figure 4.23 as an example. It can be concluded directly from this profile that all three types of contaminations are present.

In order to obtain quantitative results, the intensity sum is derived from Equations 4.25, 4.26, and 4.27, and fitted to the measured curve. The concentration values c_{VB} for the bulk-type, c_{AP} for the particulate-type, and c_{AL} for the layer-type contaminations are varied until a best fit is reached via a χ-square test. The ultimate values are treated as the results.

The method, however, must first be calibrated. This task consists of determining the calibration factors N_{0B}, N_{0P}, or N_{0L}. The three quantities essentially are a product of the relative sensitivity of the analyte and the intensity of the primary beam. They are further determined by the total mass-attenuation coefficient of the matrices in accord with Equation 4.2. Calibration by internal standardization is not permissible because any additional contamination must be avoided here. Consequently, an external calibration becomes necessary, to be carried out by means of an external standard. An overview of different methods is given by Hockett [65]. Some users applied a particulate-type standard presented by a microdroplet of a standard solution after drying [66]. Other users have recommended a layer-type standard produced by an immersion and/or spin-drying of the wafer with a spiked solution [67,68]. Others have used bulk-type standards in the form of polished pure metals or plated wafers [69,70]. Single- or multielement standards produced by inkjet printing of picodroplets lead to about 100 residues per square millimeter and only some nanometer thickness. Such standards are thin-layer-like with an increased dynamic range of linear calibration.

All types of standards are commercially available, but it is strongly recommended that their intensity profiles be inspected prior to use [70]. Standards that do not show the appropriate profile should be rejected. For practical work, only one external-standard element is necessary. Measurement at one appropriate angular position gives the calibration factors N_{0B}, N_{0P}, or N_{0L} for the given element after application of Equations 4.25, 4.26, or 4.27, respectively. Calibration factors for other elements can be deduced from this value by relative sensitivities.

The determination of contaminants and their differentiation by recording a total intensity profile via GI-XRF is rather time-consuming and not suitable for a mapping of an entire wafer. There is, however, a fast method that limits the measurements to one or two angular positions below the critical angle of total reflection, that is, by a simple application of TXRF [61,71]. If only the sum of a particulate- and layer-type contamination has to be determined, a single measurement will suffice. The appropriate angle of operation can be found at the intersection of the curves for particulate-type and thin-layer type in Figure 4.22. It can be derived by comparing Equations 2.21 and 2.24. Assuming a reflectivity of nearly 100%, we find that this angle of operation is approximately

$$\alpha_1 = \alpha_{crit}/\sqrt{2} \tag{4.28}$$

This is the only angle below α_{crit} where the intensity sum is proportional to the concentration sum of both contaminations. It may therefore be called iso intensive angle or shortly, iso-angle. The sum $c_{AL} + c_{AP}$ can be determined directly from the intensity reading N_1 at the angle α_1:

$$c_{AL} + c_{AP} = 0.5\, N_1/N_{0L} \tag{4.29}$$

If the percentage of the particulate or the layer type has to be determined as well, a second measurement will be necessary. It may be carried out at a smaller angle of operation, for example,

$$\alpha_2 = \alpha_{crit}/\sqrt{6} \tag{4.30}$$

Both angles of operation are indicated in Figure 4.23. The respective intensity readings N_1 and N_2 lead to the percentage c_L of the layer type defined by the ratio $c_{AL}/(c_{AL} + c_{AP})$. From Equation 2.24 it can be shown that this quantity is given by

$$c_L = \frac{3}{2}\left(1 - \frac{N_2}{N_1}\right) \cdot 100\% \tag{4.31}$$

The corresponding percentage of the particulate type c_P can be defined by the ratio $c_{AP}/(c_{AL} + c_{AP})$, given by

$$c_P = \frac{3}{2}\left(\frac{N_2}{N_1} - \frac{1}{3}\right) \cdot 100\% \tag{4.32}$$

Both values sum up to 100%. The calculations presuppose that samples or wafers are free from bulk-type contamination and ideally flat and even. Otherwise, the results only represent approximate values.

4.5.2 Characterization of Thin Layers by TXRF

Thin layers embedded in or deposited on a flat and even substrate can be characterized by recording angle-dependent intensity profiles, that is, by GI-XRF, which is nondestructive, but requires a mechanical tilting and above all it is indirect. However, there are two variants combined with *classical* TXRF that can compete without such drawbacks. They are based either on wet-chemical etching or dry-physical etching. Both methods are destructive but only a small piece of a rectangular sample is needed (about 1 cm^2).

4.5.2.1 Multifold Repeated Chemical Etching

A flowchart in Figure 4.24 demonstrates the method of wet-chemical etching and TXRF applied to Si wafers [72,73]. It starts with the determination of the wafer area, which is measured by a graduated sample stage of an optical microscope (e.g., Ortholux, Ernst Leitz GmbH, Wetzlar, Germany) at a magnification of 100-fold. The values of 2–3 cm^2 can be determined with high precision (RSD < 0.01%). Chemical etching is performed by oxidizing this piece with a solution of hydrogen peroxide for some minutes (30% H_2O_2 solution of high purity, some milliliter). Afterward, the thin oxidized sublayer is removed while dipping the sample in hydrofluoric acid (4% HF dilution of high purity, some milliliter). These steps are similar to vapor-phase decomposition (VPD, see Section 5.4.7.2). However, in contrast to VPD, oxidation is induced by H_2O_2 and not by ambient air. Moreover, etching is carried out by an aqueous solution and not by the vapor of HF.

Before oxidation and after etching, the wafer piece is dried and weighed thoroughly on a calibrated microbalance (see Section 4.5.2.2). The difference gives the mass of the removed sublayer (0.1–1 μg). With help of the density of

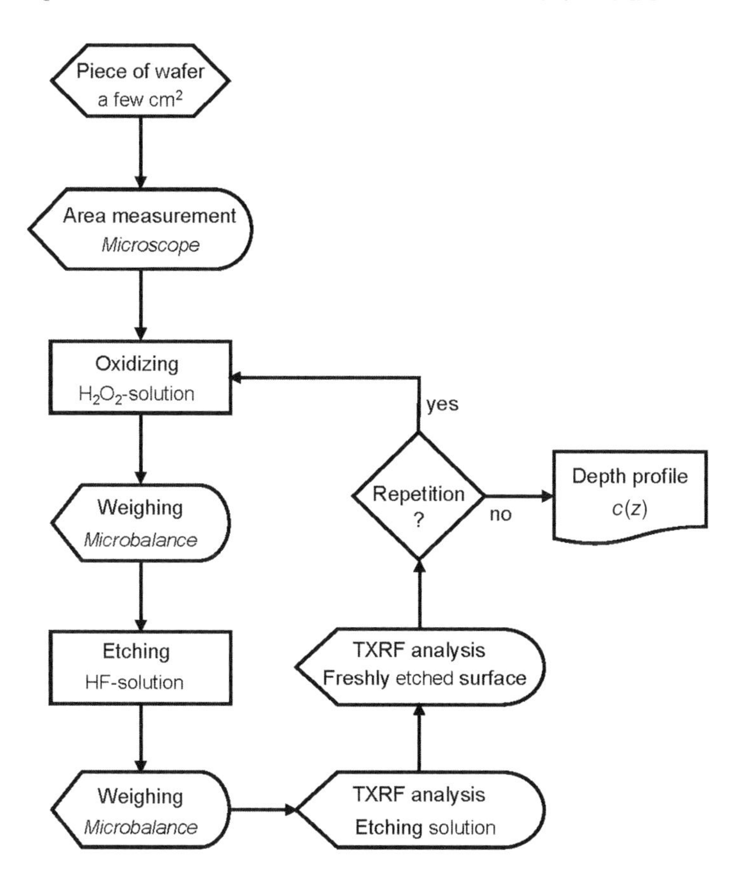

Figure 4.24. Flow chart for depth profiling by wet-chemical etching and TXRF analysis. Several steps have to be done and the loop has to be repeated for a lot of sublayers. The second weighing may be redundant if care is taken for an extremely clean working. Finally, two depth profiles can be plotted: one determined for the freshly etched surfaces, the other for the etching solutions. Figure from Ref. [72], reproduced with permission. Copyright © 1999 by Elsevier.

the oxide layer (2.27 g/cm^3) and the area of the wafer piece, the thickness of the respective sublayer can be calculated. The next step is TXRF analysis of the implanted ions (1) within the etching solution and (2) at the freshly etched surface of the sample. These measurements are carried out with an energy-dispersive Si(Li) or a Si drift detector of a TXRF spectrometer. For that purpose, (1) a small droplet (10 µl) of the etching solution is pipetted onto a clean Plexiglas carrier and analyzed by the detector and (2) a small area of the freshly etched surface (about 0.5 cm^2) is analyzed. The glancing angle is chosen to be below the critical angle of total reflection and the intensity ratio of the net counts $N_{\text{ion}}/N_{\text{Si}}$ is measured in a counting time of 100 s with an RSD of 2–3%. It leads to the mass ratio of implanted ions and atoms of the substrate, which is determined quantitatively by sensitivity factors after calibration.

The concentration of a sublayer defined by the molar ratio $c_{layer} = n_{ion}/n_{Si}$ can simply be given as a result in mole per mole and with a precision of <5% and an accuracy <8%. The sputter depth is relevant for the thickness of the *previous* sublayer (result of the etching solution) while the information depth of the primary beam is relevant for the *next* sublayer (result of the freshly etched surface). The information depth is about 3 nm due to total reflection of X-rays and both depth values should approximately be equal.

The latter five steps are repeated many times (about 20 or 30 times), until the implanted layer is completely removed (order of 10–200 nm). The terminus is controlled by a visual inspection (changing blaze of the wafer) or by TXRF (vanishing spectral peaks of implanted ions). Finally, two concentration/depth profiles are plotted showing the number density of the implanted ions versus the depth in nanometer. The first profile is based on TXRF of the etching solution and the second profile represents TXRF of the freshly etched surface. The number density of silicon in the implanted wafer is assumed to remain nearly constant during implantation (about $5.0 \times 10^{22}/cm^3$).

It has to be emphasized that all these steps necessitate extreme cleanliness. Gloves have to be worn and the wafer sample should only be touched with tweezers as little as possible and only from the side edges. Between the different steps the sample has to be kept in a Petri dish in order to prevent contamination by air dust and humidity.

An example of such a chemical etching is given in Ref. [72]. It provides the number density of implanted ions directly in a range of 10^{18} up to $10^{21}/cm^3$. The individual steps are between 0.5 and 5 nm wide, which is very favorable. However, a clear disadvantage is the troublesome chemical etching. Each sublayer needs about an hour, a total profile with 30 steps needs about 4 days. The step height cannot easily be reproduced and the requirement of a constant density of Si in the different sublayers is hardly fulfilled. Besides, this kind of chemical etching can only be applied to the class of Si wafers. The next variant is able to overcome these restrictions.

4.5.2.2 Stepwise Repeated Planar Sputter Etching

This method replaces the wet-chemical etching by a physical variant based on planar ion sputtering [73,74]. Figure 4.25 demonstrates the method of dry-physical etching by means of a flowchart [75,76]. Of course, the conditions of extremely clean working demanded for wet-chemical etching have also to be observed. The starting point again is a small rectangular piece of a wafer with an area of about $3\,cm^2$ (0.4–0.75 mm thick). Its actual dimensions are precisely determined by an optical microscope as already mentioned. Afterward, surface analysis is carried out by TXRF of the upmost layer as described earlier (Section 4.5.2.1).

The next step is sputter-etching of thin sublayers by a special ion source of the Kaufman type (RR-ISQ 40, Roth & Rau, Wüstenbrand, Germany). A sectional view and a photo of the complete device are represented in

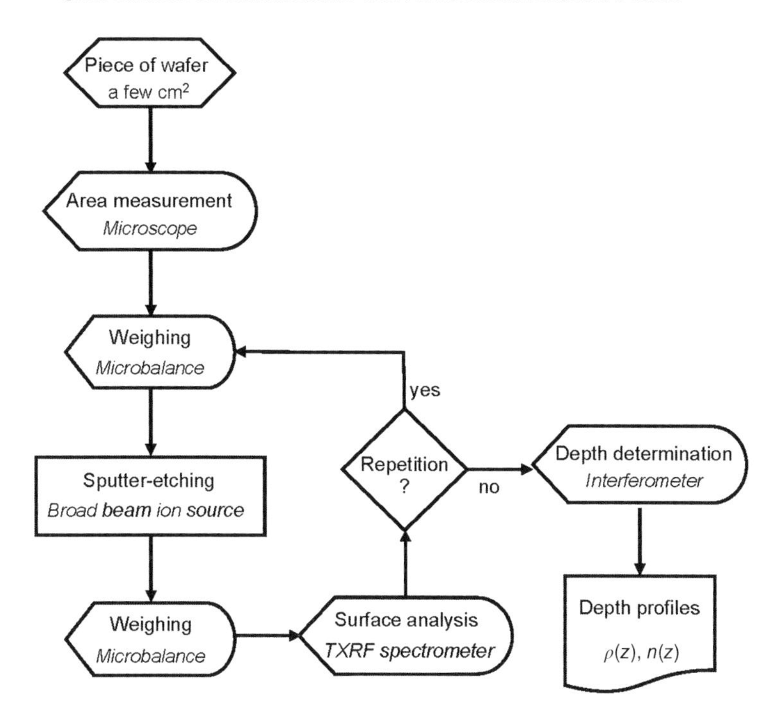

Figure 4.25. Flow chart for depth profiling by dry-physical etching or sputter etching and TXRF analysis. The loop has to be iterated sublayer by sublayer until the total layer is completely removed. Again the second weighing is redundant in case of an extreme cleanliness. Finally, a density/depth profile as well as profiles of the number density for ions or atoms in the sublayers can be plotted versus the sputter depth. Figure from Ref. [73], reproduced with permission. Copyright © 1999, John Wiley and Sons.

Figure 4.26, left and right, respectively. In contrast to the current practice of conventional ion sputtering, the Kaufman source uses a broad beam (diameter about 3 cm instead of some 5 μm) of low energy ions (about 500 eV instead of some 5 keV) at a small current of 1 mA [77]. For plasma generation, a Mo cathode is used and the ion gun is flooded with argon gas at a low flow rate of 2.5 ml/min. An Ar^+ ion beam of 2 cm in diameter can be extracted and directed normal to the sample surface. A soft and planar sputtering is carried out with a time period of 10–15 s per step. Thin sublayers of only 3–6 nm are gently removed from flat samples like pure Si or Ge wafers, and also implanted wafers. The sample surface remains flat and smooth; even after 30 repetitions the roughness is only 0.3 nm. After each step of repeated sputtering, TXRF analysis is performed at the freshly etched surface as described in the aforementioned section.

The mass of the removed sublayer is determined by differential weighing. Before and after each sputter step the sample is put on a calibrated

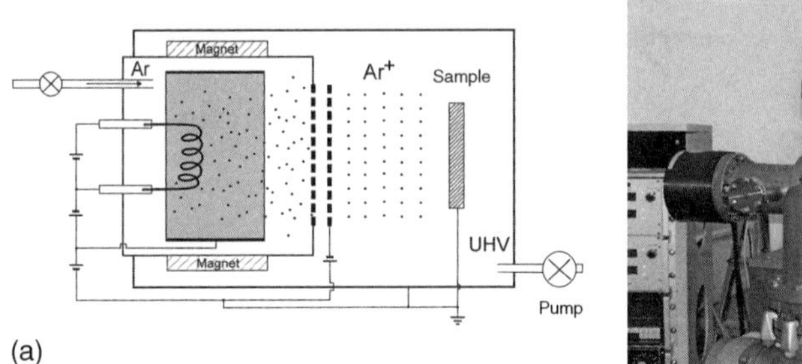

Figure 4.26. An ion source of the Kaufman type, suitable for planar sputter-etching of extended areas of a sample. (a) Sectional view; (b) the complete device with load-lock, sample chamber, ion-gun, ion-pump, and high-vacuum valve. Figure from Ref. [77], reproduced with permission from Springer-Verlag. Copyright © 2005, Springer-Verlag.

microbalance (UMT2, Mettler Toledo, Greifensee, Switzerland). Special care has to be taken: (1) the microbalance should be located in a calm, air-conditioned room free from vibrations; (2) weighing should not start before a waiting period of 5 min; and (3) weighing has to be repeated three times and averaged. The mass of about 400 mg can be determined with a precision of 0.1 μg so that the mass difference of only 2–8 μg for the sputtered sublayer can be given with an RSD of only 5–10%. The total loop is repeated sublayer by sublayer, and sputtering is reiterated until the implanted layer is completely removed from the surface.

The thickness of the total layer is determined afterward by an interferometer. For this purpose, a small droplet of a correction fluid (0.1 μl) is dripped onto the wafer piece and dried before the repetitions start. A tiny circular spot of <1 mm^2 of the sample is masked by a thin film (about 10 μm), which remains unaffected during sputtering. Such a tiny spot influences the results of repeated ion sputtering and TXRF analyses by <0.3%, consequently it can be neglected. After the layer is completely etched and the iteration process is finished, the correction fluid is removed by a solution of acetone. A truncated cone with a flat plateau is left, which is observed now, for example, by a simple Tolansky device attached to the mentioned microscope. For this purpose, the sample is coated first by a thin layer of gold with 1–2 nm thickness and illuminated then with monochromatic light, for example, of Na-D at 589 nm wavelength. Parallel fringes appear with a shift at the border of the mentioned cone. The total sputter depth can be determined in the range of 100–200 nm with an RSD < 2%. A mean thickness of a single sublayer follows by division with the number of repeated steps or sublayers.

Area measurement, differential weighing, and interferometry allow a determination of the mass density of individual sublayers [78–80]:

$$\rho_{\text{layer}} = \frac{m_{\text{layer}}}{A_{\text{sample}}\, t_{\text{layer}}} \tag{4.33}$$

Consequently, a stepped profile can be plotted for the density of all sublayers dependent on the depth below the original surface.

This information of a density profile is *quite exceptional* while a stepped profile for the molar concentration c_{layer} by TXRF is obtained as expected. It indicates the ratio of implanted ions and substrate atoms within the layers: $c_{\text{layer}} = n_{\text{ion}}/n_{\text{atom}}$. In combination with the mass density, both quantities n_{ion} and n_{atom} can be determined individually. The actual number density of implanted ions is

$$n_{\text{ion}} = \rho_{\text{layer}} \cdot \frac{c_{\text{layer}}}{M_{\text{atom}} + M_{\text{ion}}\, c_{\text{layer}}} \cdot N_{\text{A}} \tag{4.34}$$

where M_{atom} is the molar mass of substrate atoms (e.g., silicon), M_{ion} is the molar mass of the implanted ions, and N_{A} is Avogadro's number (6.022×10^{23} atoms per mole). The actual number density of substrate atoms is given by a similar formula:

$$n_{\text{Si}} = \rho_{\text{layer}} \cdot \frac{1}{M_{\text{atom}} + M_{\text{ion}}\, c_{\text{layer}}} \cdot N_{\text{A}} \tag{4.35}$$

The first factor of the latter formulas is the mass density of an individual sublayer, which is determined separately by the combined methods. It may differ significantly from the density of the crystalline material (e.g., of silicon, which is $\rho_{\text{Si}}^{0} = 4.996 \times 10^{22}/\text{cm}^3$).

An example will be given in Section 5.4.8.2 with a step height of 3–6 nm and a number density of 10^{20} up to 10^{22} As and Co ions/cm^3. The number density of the Si substrate was shown to vary by ±30% relative to the nominal value, which means a swelling or shrinking of the respective sublayers. In contrast to wet-chemical etching plus TXRF, the method of dry-physical etching with TXRF can be adapted to materials different from Si. It is less time-consuming and needs only 4 h for a total density-depth profile and another 2 h for a concentration/depth profile.

4.6 QUANTITATIVE SURFACE AND THIN-LAYER ANALYSES BY GI-XRF

The feasibility of thin-layer analysis is known from classical XRF and is based on continuous variation of the glancing angle of incidence while fluorescence intensity is recorded. The angle-dependent intensity profiles provide information on the elemental composition and thickness of surface layers by an indirect reverse fit. In contrast to the conventional method, GI-XRF is restricted to grazing incidence and occurs near the critical angle of total reflection (below and above this angle). The glancing angle is only about 0.1° instead of some 10°, and the penetration depth is in the range of nanometers instead of micrometers. In this case, standing waves appear in front of a flat sample and even within a layered structure if coherence of the incoming and reflected beam is ensured. The effect of XSW on the fluorescence intensity was dealt with in Section 2.3 and the conditions for coherence were treated in Section 4.4.3.3.

These specific features have to be considered for the recording and evaluation of angle-dependent intensity profiles of the samples. The shape and course of those curves allow a first qualitative evaluation of the samples that have to be examined. On the one hand, nanoparticles, deposits, precipitates, and contamination on or in a surface of a flat sample can be characterized and the covering or area-related mass of such fine structures can be determined. Additionally, thin near-surface layers implanted in or deposited on flat substrates can be studied. External standards of pure elements can be used for calibration, and model calculations can be carried out to fit a measured intensity/angle profile with a theoretically derived curve. By this means, an indirect quantitative determination of the elemental composition, the thickness, and even the density of near-surface layers is made possible.

Grazing incidence XRF is capable of nondestructive analysis. Problems may arise for completely unknown samples, for rough samples, and layers of some 100 nm thickness. Furthermore, problems are involved with incoherent or less coherent radiation and finally with unambiguous modeling and calculations. A helpful support can be given by the observation of the specular reflectivity of the primary X-ray beam (X-ray reflection = XRR). This can be realized easily by a simple scintillation counter positioned in the direction of the reflected primary beam some 10 cm behind the sample. XRR measurements can be carried out simultaneously with GI-XRF.

4.6.1 Recording Angle-Dependent Intensity Profiles

The basic design of an instrument suitable for surface and thin-layer analyses is shown in Figure 4.27. The primary beam is generated by a fine-focus X-ray tube, transmitted by a metal-foil filter, monochromatized by a single or double reflector with multilayer or Bragg crystal, and directed onto the flat sample. Its fluorescence intensity is recorded by a Si(Li) or silicon drift detector, while the sample is tilted. The entire instrument may be placed in a laminar flowbox in order to prevent contamination from ambient air.

A single or double monochromator can be used to select a small energy band from the primary beam. If an X-ray tube is chosen, the Kα peak of molybdenum or copper or the Lβ peak of tungsten can be used. If a synchrotron beam is available, X-ray photons with an energy between 10 and 20 keV may be selected.

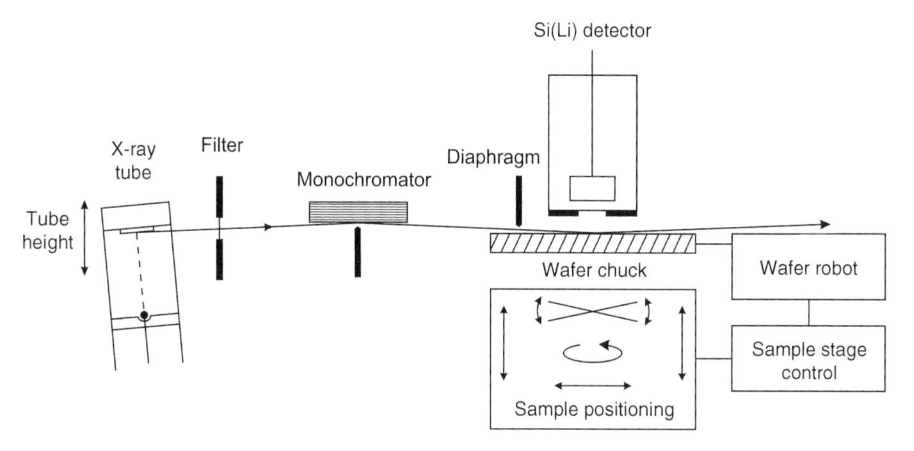

Figure 4.27. Schematic diagram of a GI-XRF instrument suitable for surface and thin-layer analyses of flat samples like wafers. The specific components are X-ray tube, monochromator, sample positioning device, wafer robot, and solid state detector. All steps for positioning and measurement may be computer-controlled. To avoid contaminations, the equipment should be placed in a clean-bench. Figure from Ref. [4], reproduced with permission. Copyright © 1996, John Wiley and Sons.

The incident beam must be monochromatized in order to produce an unambiguous dependence of the intensity and the glancing angle. X-ray photons of *different* energies would blur this relationship. Details on multilayers, Bragg crystals, and monochromators were given in Sections 3.4.2 and 3.4.3.

The wafer robot is used for loading and changing up to 50 wafers with a diameter of about 100–200 mm (4–8 in.). The wafers are placed automatically on a glass plate in rear-contact and cling to it by either an electrostatic or a vacuum chuck. Instruments are designed for 24 h unattended operation and measurement.

The sample-positioning device is the most important component (already described in Section 3.5.3). By its means, the sample or wafer is first adjusted in the observation plane. Mechanical sensors should not be employed in order to avoid any physical contact with the sample. The adjustment can simply be controlled by the fluorescence signal generated by the sample itself. Thereafter, the sample or wafer is tilted stepwise in order to run an angle scan. Starting from a setting close to zero, the incident angle is increased stepwise and fluorescence spectra are recorded during time intervals between steps. Usually, only about three to six spectral peaks or bands, which are set in so-called ROIs are recorded, rather than recording the entire spectrum. The step width may be chosen between 0.001° and 0.05°, and the counting or acquisition time between 5 and 100 s. About 50–200 steps are taken to reach an angle of about twice the critical angle of the sample material, which is generally <1°. The total measuring time is commonly between 5 min and 5 h. After that, the peak intensities of those elements that have been detected are plotted against the incident glancing angle. These angle-dependent intensity profiles are used for a qualitative and quantitative characterization of the samples.

The angle scale *can* be calibrated by taking photographs of the primary and the reflected beam. However, it is much easier to determine the critical angles for some reference samples. For this purpose, polished wafers, disks of metals, or correspondingly plated wafers are suitable [69,70]. The intensity profiles of these elements are recorded, and the inflection points of these curves (similar to that of Figure 2.17) are determined. They are set equal to the values calculated by Equation 1.68. A correction of an additive constant may still be necessary. It can be determined by a second measurement after a horizontal rotation of the sample by 180°. This calibration technique [69] allows an angle setting with an absolute accuracy of about 0.00025°. A divergence of the primary beam of about 0.01° is 10 times this value.

4.6.2 Considering the Footprint Effect

For trace analyses by conventional TXRF, the area of the small specimen on top of the sample carrier may be circular or irregular with millimeter-dimension. Usually, this area will be completely irradiated by X-rays, either of the X-ray tube or of the synchrotron beam line. Furthermore, this area is completely in view of the detector, and penetration as well as information depth remain constant during analysis. These conditions, however, change for thin-layer analysis by GI-XRF when the angle of incidence is varied by rotating the sample on its axis "*a*" (see Figure 3.13). The sample usually is a contaminated or a thin-layered substrate with centimeter-extension. The cross-section of the exciting beam and the solid angle of the detector's view remain unchanged.

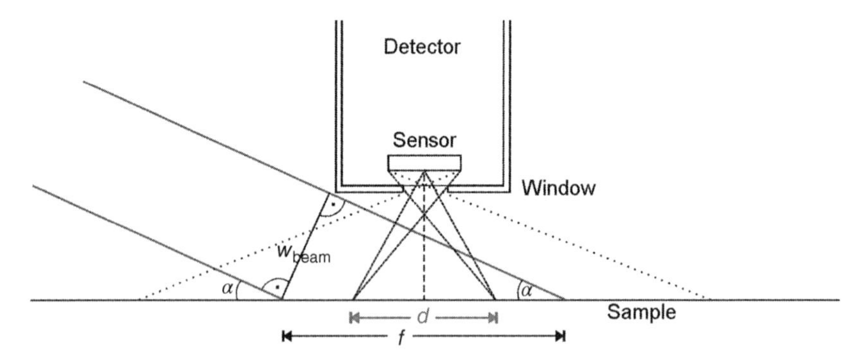

Figure 4.28. Cross-section of an energy-dispersive detector with sensor and window above a flat sample. The sensor is a silicon crystal with a circular front area, and also the window is generally circular. Both areas are parallel to each other and centered. The X-ray beam coming from the left hits the sample under the glancing angle α (<2°), and leaves a "footprint," which corresponds to the actual angle of incidence α_f. The beam width w_{beam} is about 100 µm, while the length of its "footprint" *f* varies between 3 mm and 3 m according to α_f. The detector can only observe a circular region with a diameter *d*. It is determined by the distances between sensor, window, and sample *W* and *D*, and by their diameters *s* and *w*. Figure from Ref. [47], reproduced with permission from the author.

Figure 4.28 shows the geometrical conditions. While the width of the exciting beam is constant, the X-ray beam covers an area that is stretched with a lower glancing angle in direction "b" vertical to the rotating axis. The beam leaves a "footprint" on the sample, which becomes larger and larger [47]. Perpendicular to this direction there is no change at all. Penetration and information depth, however, are reduced with smaller glancing angle in both directions.

The beam width may be called w_{beam} and the glancing angle of incidence may be called α. Then we have two special cases:

1. For a certain angle α_{d}, the irradiated area corresponds completely with the detector's field of vision with a diameter d:

$$\sin \alpha_{\text{d}} = w_{\text{beam}}/d \tag{4.36}$$

2. For smaller angles, we get a "footprint," which is larger than the detector's field of vision or even larger than the sample. For a given angle α_{f}, we can get the length f of the footprint by means of the relation

$$\sin \alpha_{\text{f}} = w_{\text{beam}}/f \tag{4.37}$$

For glancing angles $\alpha \geq \alpha_{\text{d}}$, the whole sample is excited and the fluorescence intensity is completely recorded by the detector so that a correction is not necessary. However, for a glancing angle $\alpha \leq \alpha_{\text{d}}$, a part of the irradiated area (i.e., of the footprint) cannot be observed by the detector and may even pass the sample holder. The measured intensity is reduced and has to be corrected afterward by a factor ≥ 1:

$$f_{\text{corr}} = \frac{\sin \alpha_{\text{d}}}{\sin \alpha_{\text{f}}} \approx \frac{\alpha_{\text{d}}}{\alpha_{\text{f}}} \tag{4.38}$$

This footprint correction may be called normalization and is only necessary for GI-XRF at small glancing angles. The values of d and α_{d} can be determined experimentally; d is of the order of 3–15 mm and w_{beam} may be 20–100 μm so that α_{d} will be about 0.1°–2°.

A typical geometrical arrangement places a circular window with diameter, w, in front of the sensor with diameter s in a distance W. The window should be smaller than the sensor and the distance between the sample and window D should be $wW/(s-w)$. In that case, the detector's field of vision is restricted to $d = w\ (D+W)/W$. One typical example may be given here: $s = 5$ mm, $w = 3.3$ mm, $W = 2$ mm. These specifications lead to $D \approx 4$ mm, $d \approx 10$ mm, and $\alpha_{\text{d}} \approx 0.3°$ for $w_{\text{beam}} = 50$ μm.

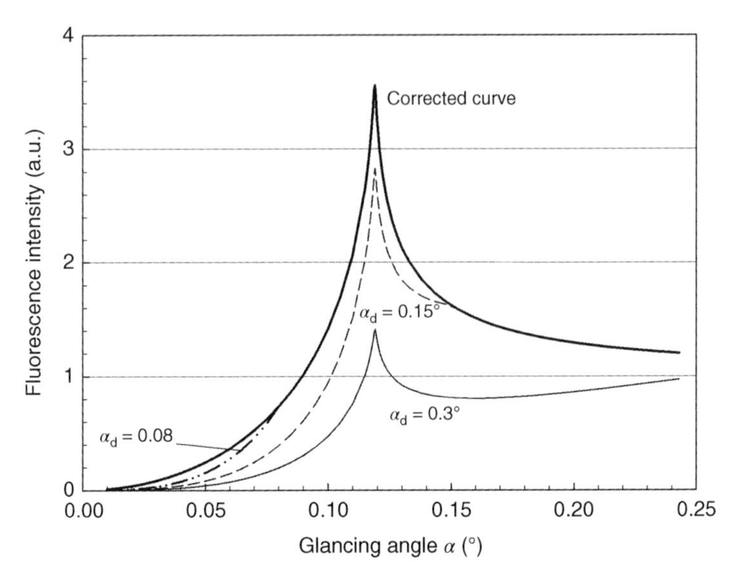

Figure 4.29. Footprint effect demonstrated for a thin layer of 10 nm on top of a Si carrier. Three different geometries are assumed with particular angles α_d (0.08°, 0.15°, and 0.3°). For $\alpha > \alpha_d$, the footprint of the primary beam is completely within the detector's field of vision, and for $\alpha < \alpha_d$, the footprint exceeds this field. Three different intensity/angle scans result, but all three curves lead to the same original curve after footprint correction.

The footprint effect is very distinct for two-dimensional thin layers but less distinct for three-dimensional particulates. Consequently, a footprint correction is absolutely necessary for GI-XRF of a thin layer, for example, the sulfur layer in a polymer, and also for all XRR measurements. GI-XRF of nanoparticles is not affected much by the footprint [81].

One example for a footprint correction may be given; for a thin layer deposited on a silicon carrier, the black solid curve in Figure 4.29 is expected, which is equivalent to curve (c) of Figure 4.22. Because of the footprint effect, a measured curve is quite different according to the particular angle α_d. Three different values were chosen for this angle and three different curves arise. After the footprint correction, all three curves are converted into the original black solid curve with a nearly fourfold intensity at the critical angle of total reflection.

4.6.3 Regarding the Coherence Length

The primary intensity used for excitation is characterized by a standing wave field above the sample carrier with nodes and antinodes at a period a. A certain fading or distortion of this XSW field with height z can be assumed because of vanishing coherence or even incoherence. Consequently, the nodes have a minimum intensity > 0 and the antinodes show a maximum intensity < 4

(relative to the intensity of the incident beam). The main reason for this fading contrast is the reduced coherence of the primary and reflected beam, reduced with the height above the flat surface. The height of coherence was found to be in the range of 1–10 μm in Section 4.4.3.3. However, a certain scattering by a rough surface of the flat carrier or layer or by the irregular shape of small particles can increase the divergence of the primary beam and thus can also reduce the coherence length and height. Up to now, the XSW field could not have been demonstrated visually.

Whatever the reason or reasons for the fading, different approaches are possible to consider it. After Brücher *et al.* [81] and Zegenhagen [82,83], the effect can be taken into account by a so-called form factor $F(z)$ when nanoparticles or thin films are deposited above the substrate. This factor can be modified by substituting m by $1/z_{\text{fade}}$ and can be extended by replacing F_1 instead of 1:

$$F(z) = (F_1 - F_0) \cdot \exp(-z/z_{\text{fade}}) + F_0 \tag{4.39}$$

This formula represents a fading of the contrast with an exponential decrease between a particular upper limit F_1 for $z = 0$ and a lower limit F_0 for large z values ($1 \geq F_1 > F_0 \geq 0$). F_1 and F_0 are fit parameters for subsequent calculations. The quantity z_{fade} can be interpreted as a height of damping or fading and is of the order of several tens of nanometers. Fading starts with F_1, decreases to $37\% F_1 + 63\% F_0$ in a height z_{fade}, and finally approximates the value F_0. Obviously, the height of fading is much smaller than the height of coherence (50 nm instead of 10 μm).

In accord with Equation 2.14 we can describe the XSW field's dependence on α and z:

$$I_{\text{XSW}}(\alpha, z) = I_0 \cdot \left\{1 + R(\alpha) + 2\sqrt{R(\alpha)} \cdot F(z) \cdot \cos[(2\pi\, z/a) - \varphi(\alpha)]\right\} \tag{4.40}$$

The form factor only influences the cosine-term for a given α value. As demonstrated in Figure 4.30, several oscillations occur with a period a around a mean value $1 + R(\alpha)$ and a damping constant z_{fade}.

A second approach for the fading field was tested by von Bohlen [84] under the condition that a first part f_0 of the incident beam (with $0 \leq f_0 \leq 1$) does not interfere with the reflected beam but is overlapped. The remaining part $(1 - f_0)$ is modulated by interference. It leads to the formula

$$I_{\text{XSW}}(\alpha, z) = I_0 \cdot \left\{[1 + R(\alpha)]f_0 + 2\sqrt{R(\alpha)} \cdot (1 - f_0) \cdot \cos[(2\pi\, z/a) - \varphi(\alpha)]\right\} \tag{4.41}$$

The parameter f_0 was determined for nanoparticles of gold. Values between 1.0 and 0.35 were found for diameters of 25 up to 250 nm.

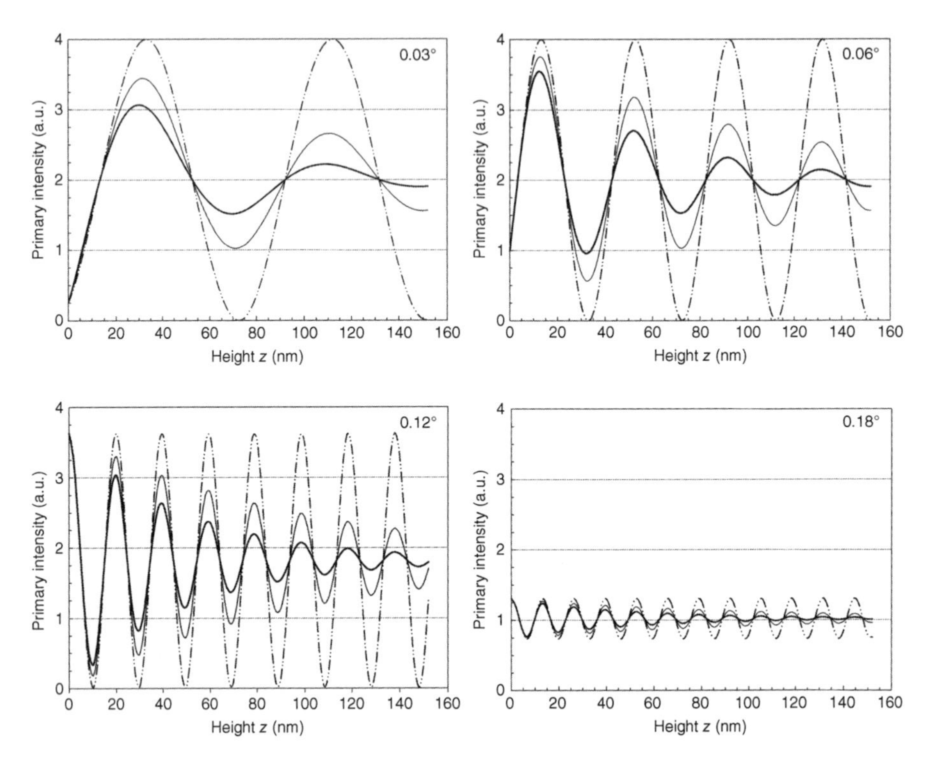

Figure 4.30. X-ray standing wave field (XSW) above a flat silicon disc. The intensity of the primary beam shows nodes and antinodes dependent on the height above the carrier ($0 < z < 160$ nm). Undisturbed oscillations between 0 (nodes) and 4 (antinodes) can be observed if the carrier is not loaded with sample material (dotted/dashed curves). The contrast, however, is reduced if small particles or thin layers are deposited on the carrier (solid curves). The dependence on the glancing angle is demonstrated by four individual angle settings. The fading of the XSW contrast is controlled by an exponential decrease with $z_{\text{fade}} = 50$ nm (narrow) and with $z_{\text{fade}} = 100$ nm (bold).

A schematic illustration, as shown in Figure 4.31, demonstrates the footprint effect and the fading contrast depending on the glancing angle. In order to make the effects obvious, total reflection was assumed to occur at angles of 10°, 20°, and even 30° instead of usual glancing angles of only about 0.1°.

4.6.4 Depth Profiling at Grazing Incidence

Thin-layered materials with a layer thickness in the range of nanometers are used as high-tech materials, especially in the semiconductor and the glass industry. The in-depth distribution of elements perpendicular to the surface of the respective samples has to be characterized by depth profiling. Several X-ray techniques are suitable for this characterization provided that grazing incidence is achieved. Glancing angles in the range of total reflection are necessary so as to be surface sensitive. GI-XRF combined with XRR is the most

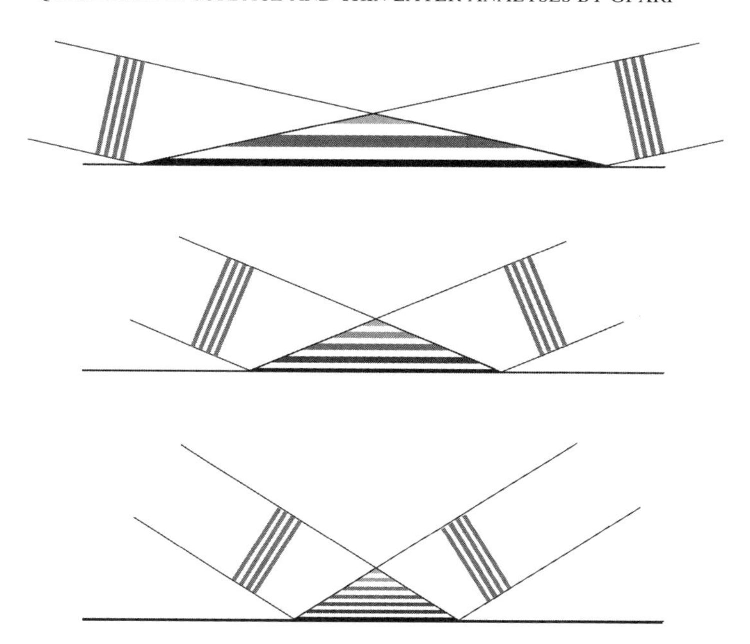

Figure 4.31. Footprint effect and fading contrast of an XSW field with nodes (dark) and antinodes (bright) above a flat sample. The glancing angles increase from top to bottom, the footprint becomes smaller, nodes and antinodes become narrower and their number increases. The contrast of nodes and antinodes fades with the height above the substrate and becomes less distinct with increasing angles. The triangle usually has a height of about 50 μm while the XSW field in this example fades within a damping height of 15 μm but this height is actually <1 μm (see also Figure 4.21).

comprehensive method of these techniques, which makes compositional depth profiling possible, even nondestructively. For that purpose, the glancing angle is increased by tilting the sample steadily. If the sample is a flat substrate, total reflection occurs and an XSW field of the primary appears above the surface with nodes and antinodes. With increasing angle, the field is compressed against the surface. If a thin layer is deposited above or below the surface or if small particles lie on top of the substrate, nodes and antinodes pass such stratified structures successively. As a result, oscillations with maxima and minima of the fluorescence intensity are induced at certain angles. This phenomenon can be used for a characterization of the thin layers or particles.

To do this, the layers and the substrate certainly must be homogeneous, flat and even, one by one.[2] In other words, the compositional profile along the height or depth should be steplike, with perfectly sharp interfaces. Layers with a roughness of only about 1 nm and a curvature of some 100 m diameter are

[2] The term *thin film* is avoided here since it should be restricted to liquid layers produced from the liquid phase.

ideally capable of total reflection. For these stratified structures, GI-XRF plus XRR can yield the element composition, the thickness, and the density of the individual layers with an accuracy of a few percent. With some mathematical effort, even a certain roughness of the interfaces can be taken into account.

The record of angle-dependent intensity profiles is the basis for the evaluation. For that purpose, the special set-up of Figure 4.27 is needed. In particular, the wafer is gradually tilted around an axis lying on its surface. In EDS with a Si(Li) or Si drift detector, the whole spectrum is recorded within an acquisition time between 5 and 100 s. The intensities are measured simultaneously for all elements of the layered sample at each angle position but only three to six spectral peaks are chosen for analysis. In WDS with a fixed crystal and an Ar gas or NaI detector, only one spectral line of one single element is observed.

The measured profile of a spectral peak or line of a particular element represents values of fluorescence intensities at different angle positions. The glancing angle is subsequently increased between about 0° and 0.3° in steps of 0.001°–0.01°. From these profiles, the respective layer parameters are evaluated by an iterative fitting procedure on the basis of modeling calculations. It was described independently by Weisbrod *et al.* [85], Schwenke *et al.* [86], and by de Boer [87], but other authors also developed such programs (e.g., Ref. [47]). In general, the procedure contains six steps.

i. A starting model is created. Several parameters of the sample have first to be estimated to lie within a certain range. By a qualitative interpretation of the element profiles, the number and sequence of the individual layers, their element composition, thickness, and density can roughly be guessed. For that purpose, a qualitative understanding of the course and the oscillation structure of the profiles is needed. A first estimation can be deduced from the form of the respective profiles as demonstrated in Figure 4.32 for nanoparticles, for a monolayer, for a thin layer, and for a thick substrate [88]. Some further hints can be found in Figure 4.33 mostly based on the findings presented in Chapter 2.
ii. The next step involves calculating the intensity of the primary beam within the layer system proposed by the starting model. This primary intensity is calculated either by a matrix formalism (as described in Section 2.4) or by a recursion formalism. The intensity I_{XSW} is determined for a given photon energy E_0 and for different glancing angles α_i dependent on the depth z normal to the surface. It reflects the inhomogeneous wave penetrating the layer system as an evanescent or standing wave with nodes and antinodes parallel to the surface. For a simple substrate and fading contrast it is quite simple and is already given by Equation 4.40 and demonstrated in Figure 4.30.
iii. The fluorescence intensity of elements excited by the primary beam is calculated next.

   If matrix effects can be neglected (for example, nanoparticles and very thin layers or thin films), the intensity is determined by an integration over

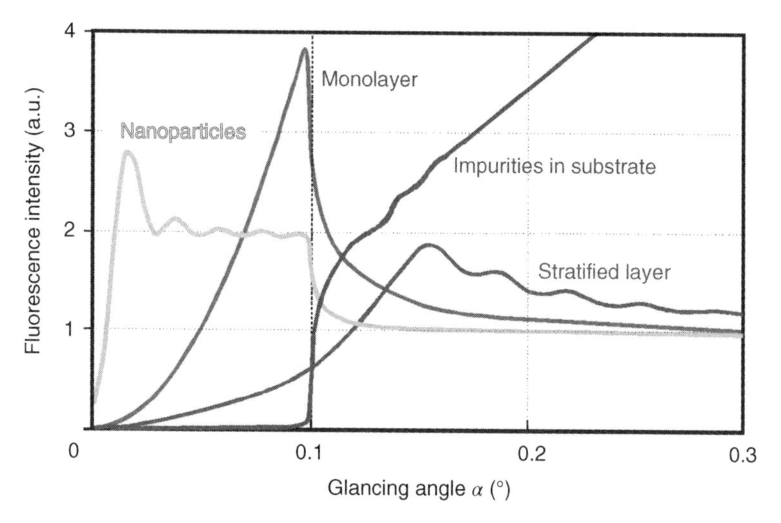

Figure 4.32. Typical profiles of an intensity/angle scan for different types of samples in a schematic representation. In contrast to Figure 4.22, oscillations of the intensity are observed: (a) for nanoparticles deposited on a flat substrate with oscillations below α_{sub}; (b) for a monolayer deposited on or implanted in this substrate with a peak at α_{sub} of the substrate; (c) for a thin layer deposited on or implanted in this substrate with oscillations above α_{impl} of the implant; (d) for homogeneously distributed impurities in the substrate, for example, in a silicon wafer, with a curve like a mountain above α_{sub}. Figure from Ref. [88], reproduced with permission. Copyright © 2005 by Elsevier.

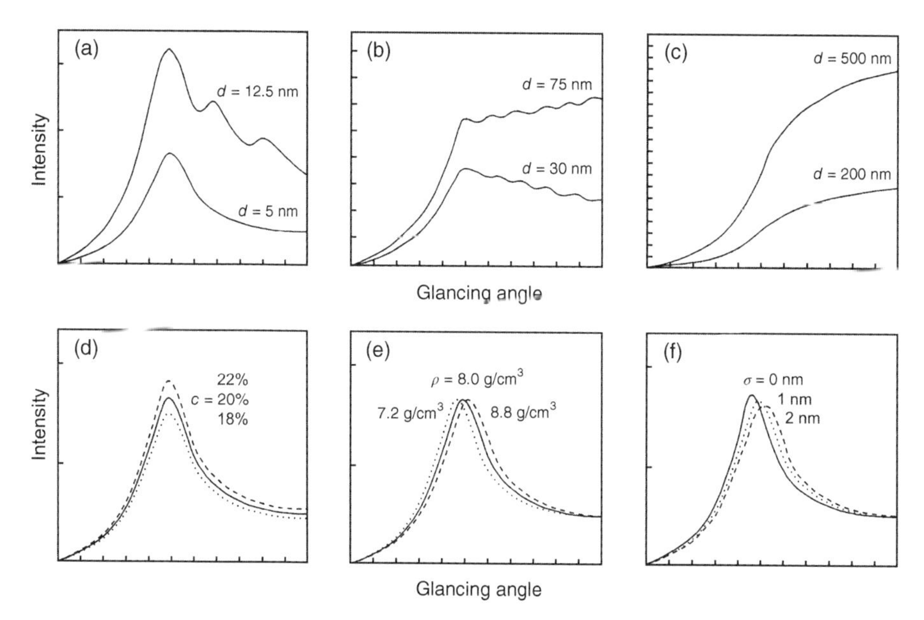

Figure 4.33. Several different types of intensity profiles distinguished by four parameters of a model layer with a weight fraction $c = 20\%$ for a given element, a thickness $d = 5$ nm, a density $\rho = 8$ g/cm^3, and a roughness $\sigma = 0$ nm (solid curves). In (a), (b), and (c), the thickness is increased by about 2.5-fold of the previous value; in (d) only c is varied by ±2%; in (e) ρ is varied while $\rho \cdot d$ is kept constant; in (f) a roughness of 1 and 2 nm is implemented. Figure from Ref. [89], reproduced with permission from Forschungszentrum Jülich.

the product of the intensity of the XSW field and the concentration profile [81,90]:

$$I_{\text{int}}(\alpha) = K \int_{o}^{z_{\max}} I_{\text{XSW}}(\alpha, z) \cdot C(z) dz \quad (4.42)$$

where K is a calibration constant, and $C(z)$ is the concentration profile of the analyte element dependent on the height z above the carrier. This profile is simple for homogeneous thin layers consisting of only one main element (plus a few light elements) and deposited on or implanted in a flat sample. If the layer is situated in a height or depth z_0 above or below the sample surface with a thickness d, it can be described by

$$C(z) = c_{\text{elem}} \quad \text{for} \quad z_0 - d/2 \leq z \leq z_0 + d/2 \quad (4.43)$$

where c_{elem} is a concentration value with $0 < c_{\text{elem}} \leq 1$. A layer with a certain concentration profile can be approximated by a stack of several successive layers with the same thickness but with different concentration. A Gaussian or Lorentzian distribution can approximate the total stepped profile of the layer.

For the deposition or residues of particulate matter three different types can be distinguished:

1. If the element is homogeneously distributed in islands with a cylindrical shape [81], the area-related mass c_{elem} represents the profile independent of the height:

$$C(z) = c_{\text{elem}} \quad (4.44)$$

2. Spherical particles with radius r meet the profile [81]:

$$C(z) = c_{\text{elem}} \cdot \pi \cdot [r^2 - (r - z)^2] \quad (4.45)$$

3. Smoothed ellipsoidal particles meet a square root law:

$$C(z) = c_{\text{elem}} \cdot 2 \cdot \sqrt{\left[\left(\frac{z}{h}\right) - \left(\frac{z}{h}\right)^2\right]} \quad (4.46)$$

iv. For particles or layers with a lateral dimension of some 10 nm or even 1 µm (see Table 4.3) a fundamental parameter approach is employed. The absorption of the fluorescence radiation by the layer system and integration over the depth z is a must and can be performed by means of the closed expression of Equation 2.56. Secondary excitation or enhancement can also be taken into account [62,87], but as a complicated second-order effect it may be neglected for layers up to 100 nm thickness. The efficiency ε of the detector (see Section 3.8.2.1) and the transmission of ambient air between the sample and the detector, however, can be included quite easily. To get intensity values really indicated by the detector, the angular divergence of the instrumentation may finally be taken into consideration. The calculated intensity of the analyte element may be written in accord with Schwenke *et al.* [86]:

$$I_{\text{elem}} = I_{\text{elem}}\left[\left(c_{\text{elem},1}, d_1, \rho_1\right), \ldots, \left(c_{\text{elem},N+1}, d_{N+1}, \rho_{N+1}\right); \alpha, \varepsilon_{\text{elem}}, T_{\text{elem}}, K\right] \tag{4.47}$$

where $c_{\text{elem},1} \ldots c_{\text{elem},N+1}$ is the concentration of element in the successive layers $1, \ldots, N+1$, $d_1, \ldots, d_{N+1}$ is the thickness, and $\rho_1, \ldots, \rho_{N+1}$ is the density of these layers, α is the glancing angle of the primary beam, and K is a calibration constant that may be determined from a bulk sample [69]. The $(N+1)$th layer usually is the substrate.

v. The intensity/angle curve $I_{\text{elem}}(\alpha)$ calculated by theory is compared with the intensity profile $I_{\text{exp}}(\alpha)$ experimentally recorded. A good fit of experimental data and theoretical values can be optimized "by eye" rather well. For fine tuning, the χ-square test is applied leading to the best fit and to the unknown parameters of the sample. The test is based on the hypothesis that I_{exp} and I_{elem} do not differ significantly. In order to execute the test, the deviations between measured values and calculated intensities for all n individual angle settings are determined and squared, divided by the calculated intensities and added. This quantity is called χ^2:

$$\chi^2 = \sum \left(I_{\text{exp}} - I_{\text{elem}}\right)^2 / I_{\text{elem}} \tag{4.48}$$

If this quantity is larger than a given value, t_{crit}, the hypothesis is rejected with a certain level of significance and the starting model will be changed. Figure 4.33 may again be used as a guide for modifications. Steps ii to v are then repeated in an iterative process until a best fit with a minimum χ^2 value is achieved—if possible below a relevant critical value. Critical values are tabulated, for example, in Table 4.4 for a level of significance, $\alpha = 5\%$ and 10%. This level is the error of the first kind, called alpha error, in the case that the hypothesis is rejected though it is correct. For $n > 20$, the values of t_{crit} can even be approximated by

$$t_{\text{crit}}(5\%) = 1.11 \cdot n + 11 \tag{4.49}$$

TABLE 4.4. Critical Values for the χ^2-Test with Level of Significance 5% and 10%. Dependent on the Number of Points in the Intensity/Angle Scan

n	10	20	30	40	50	60	70	80	90	100	110	120	130	140	150	160	170	180	190	200
$t_{crit, 5\%}$	17	30	43	55	66	78	89	101	112	123	134	146	157	168	179	189	200	211	222	233
$t_{crit, 10\%}$	15	27	39	51	62	73	84	95	107	117	128	139	150	161	172	182	193	204	214	225

and by

$$t_{crit}(10\%) = 1.09 \cdot n + 7 \tag{4.50}$$

vi. A best fit is usually reached after about five to eight iterations. The last model with parameters $c_{x1}, \ldots, c_{x,N+1}$; $d_1, \ldots, d_{N+1}$; and $\rho_1, \ldots, \rho_{N+1}$ is declared to be the final result. An averaged relative deviation can be calculated by the equation [43]

$$s_{rel,aver} = \frac{\chi^2}{\sum I_{elem}} \cdot 100\% \tag{4.51}$$

which should be just a few percent. In any case the final parameters have to be evaluated critically.

For complex layered samples, the fit procedure will become difficult. In such cases, it is advisable to use a so-called evolutionary algorithm in order to achieve results automatically [47]. This algorithm with a lot of parameters can be implemented into the simulation program.

A few software programs have been developed on the subject of matrix effects and their consideration. They require several sets of fundamental parameters, such as photoelectric mass-absorption coefficients in an energy range from 1 to 60 keV, absorption-edge jump factors, fluorescence yields, relative emission rates for the individual peaks in their series, absorption edges, peak energies, as well as densities, atomic number, and atomic mass of pure elements. Respective sets of data are available. Nevertheless, the problem is rather complex and requires a skilled operator with considerable experience in the interpretation of the profiles and modification of the models. Only thus can the iterative process converge rapidly and yield an unambiguous result.

A simple example may be given here that deals with a buried stratified transition layer of cobalt ions. They were implanted in a Si wafer with 100 keV energy and a nominal dose of 1×10^{17} ions per cm^2 [73]. An intensity/angle scan was performed with excitation by Mo-Kα at 17.44 keV. The Co-Kα peak at 6.93 keV and the Si-Kα peak at 1.74 keV were recorded with the result of Figure 4.34.

A stratified layer with a bell-shaped distribution of cobalt ions in a depth less than 0.3 μm in silicon was expected. Consequently, a model was set up with a stack of many sublayers, and fluorescence intensity values for Co-Kα were calculated. A potential fading of the XSW field was not considered because of the shallow layer of Co ions. A best fit of the calculated intensity/angle curve with the measurements was reached after four steps. Inverse modeling has led to the final result shown in Figure 4.34. The concentration/depth profile gives

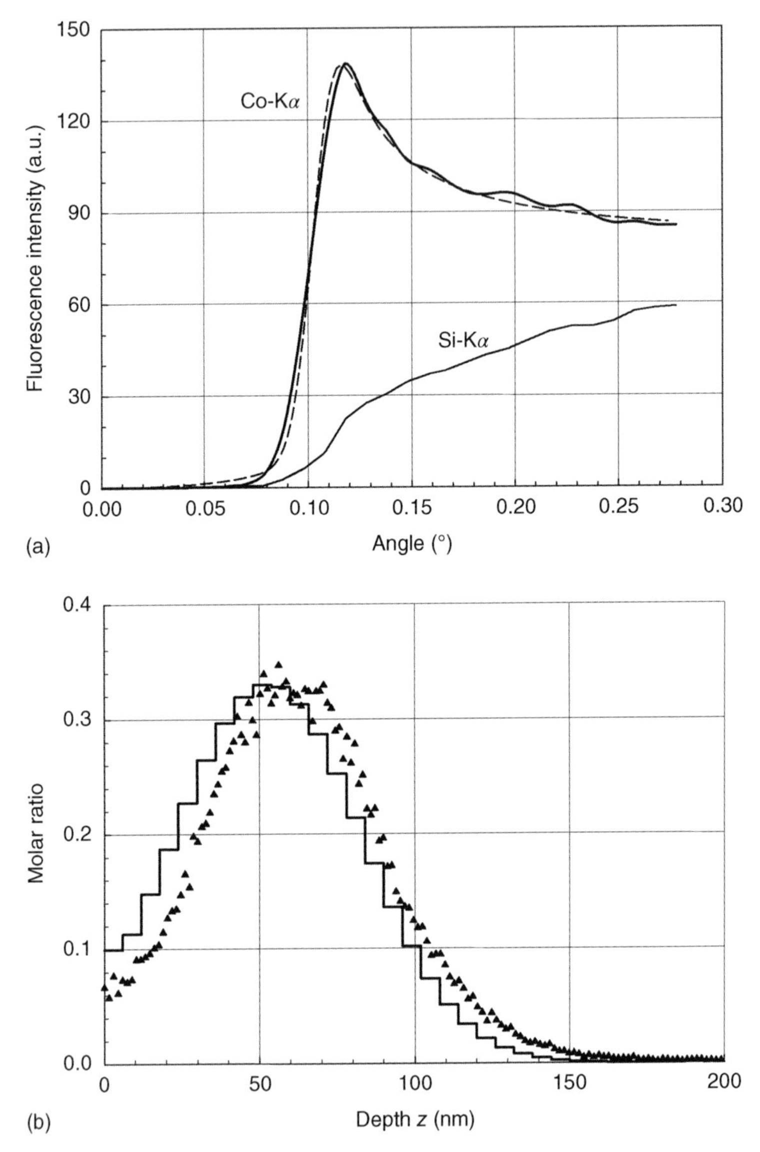

Figure 4.34. (a) Intensity/angle scan for a thin layer of cobalt ions implanted in a silicon wafer with an energy of 100 keV. The fluorescence intensity of Co-Kα and Si-Kα is plotted against the glancing angle. For the simulation, 30 sublayers were assumed with 6 nm thickness each, and the mole ratio of Co ions and Si atoms was calculated. (b) The best fit between measuring points and theoretical curve was reached for the final model represented by the stepped profile. The small triangles show the molar ratio determined by RBS. Both profiles are shifted by about 6 nm against each other, which may be caused by a difficult depth scaling.

the molar ratio of Co ions and Si atoms in every sublayer. The stepped curve represents 30 sublayers with an offset of about 10% for the surface at $z = 0$ nm and with a maximum of 33% in a depth of about 60 nm. The actual dose was determined by integration to be 1.2×10^{17} ions per cm^2, 20% above the nominal value. This profile was checked by Rutherford backscattering spectrometry (RBS), which is a well-established and reliable method for such transition layers. Both profiles match quite well with regard to their maxima and integrals (implantation dose).

In support of GI-XRF measurements, the specular reflectivity of a flat sample can simply be observed. A theoretical evaluation of XRR is possible by assuming a model of the structured sample with values for the layer thickness, concentration of different elements, density, coefficients δ and β, and the experimental parameters E, and ΔE of the primary beam, and ε of the detector. $R(\alpha)$ can be calculated according to this model and compared with the experimental values of the reflection/angle scan, again with a χ-square test.

4.6.5 Including the Surface Roughness

In practice, perfectly flat surfaces and interfaces are nonexistent. A certain nanometric roughness will be left on bulk samples even after careful polishing. Even an extremely flat and smooth wafer surface deviates from an ideal plane at the atomic scale level. For other flat samples, an average roughness of some 10 to a few 100 nm often occurs in practice.

The roughness of a surface or interface is defined by different measures. The most common quantity is the average roughness R_a, which is the average deviation of z values (height above a zero level of the xy plane) from a mean value of the respective surface or layer. Another customary parameter is the root-mean-squared roughness, R_{RMS}, which is the root of the mean of squared deviations (RMS) of z values and which is often termed σ. Generally, a stylus profilometer or laser interferometer can be used for the determination of surface roughness ranging from 10 nm to 100 μm. A scanning tunneling microscope (STM) and an atomic force microscope (AFM) can measure a roughness down to 0.2 nm or even 0.1 nm. XRR measurements can also be applied for this task, and several papers have been published with regard to roughness. Si wafers can be polished to a roughness of 0.1 nm (1 nm down to 0.1 nm). For single layers and multilayers, the roughness depends on the production process, on the substrate and on the type of materials. It increases usually with the thickness and the number of layers.

Various approaches have sought to cope with such rough surfaces [89,91–93]. Three different models were developed in order to implement roughness in the calculation of standing waves [47]:

1. The first model supposes that only the reflectivity is reduced by rough surfaces and interfaces. The beam-path, however, is assumed to be not disturbed. A "Debye–Waller" factor can be included in the coefficients of

reflection defined by Equations 2.29a, b and 2.40a, b. This model is only applicable to surfaces and interfaces with a small roughness (some nanometers).

2. For the second model it is assumed that the distribution of the height of the respective surface or interface over the *xy* plane shows a Gaussian profile. In that case, the Fresnel coefficients r_j and t_j can be modified in general. This model can be applied for a small roughness ($\sigma_j \ll$ thickness of the respective layer).
3. The third model uses so-called effective densities. It was first used by Nevot *et al.* [94,95] and seems to be rather successful. For a simple flat substrate, it describes the zone of roughness as a stack of virtual plane-parallel layers with a stepwise increasing density [89,91]. The uppermost layer may show zero density, while the lowest layer should have the bulk density. For a layered sample, the near-surface layer and also deeper layers are split into a stack of thin slices with subnanometer thickness. The total sample is treated as numerous slices with different densities but with sharp interfaces and zero-roughness between them. The density is estimated in accord with the respective volume; decrement and attenuation coefficient of the refractive index n_i are corrected accordingly.

Several examples for a characterization of nanostructured materials are described in the literature [82,83,87–92]. Thin layers or thin films and also lots of single particles with nanometer dimensions deposited on or below the surface of a flat substrate have been characterized by GI-XRF. Such thin layers were made of pure metals, metal alloys, semiconductors, oxides or nitrides, metal–organic compounds, and polymers. Different kinds of layers, such as monolayers, double layers, multilayers, Langmuir–Blodgett films, and biofilms were investigated [47,79,80,90,96,97]. Some examples will be given in Chapter 5.

REFERENCES

1. IUPAC-Nomenclature. Revision by Sandell, E.B., West, T.S., Flaschka, H., and Menis, O. (1979). Recommended nomenclature for scales of working in analysis. *Pure Appl. Chem.*, **51**, 43–49.
2. Jenkins, R. (1988). *X-Ray Fluorescence Spectrometry*, Chemical Analysis Series, Vol. **99**, Wiley: New York.
3. Tschöpel, P., Kotz, L., Schulz, W., Veber, M., and Tölg, G. (1980). Zur Ursache und Vermeidung systematischer Fehler bei Elementbestimmungen in wässrigen Lösungen im ng/ml und pg/ml-Bereich. *Fresenius Z. Anal. Chem.*, **302**, 1–14.
4. Klockenkämper, R. (1997). *Total-Reflection X-Ray Fluorescence Analysis*, 1st ed., John Wiley & Sons, Inc.: New York.
5. Prange, A. (1989). Total reflection X-ray spectrometry: method and applications. *Spectrochim. Acta*, **B44**, 437–452.

6. Prange, A. (1994). Kalibierungsfreie Multielement-Spuren- und Mikroanalyse. *Labor Praxis*, **18/2**, 30–32.
7. Knoth, J. and Schwenke, H. (1977). German Patent pend. Nr. P 2717925.5.
8. Michaelis, W., Knoth, J., Prange, A., and Schwenke, H. (1985). Trace analytical capabilities of total-reflection X-ray fluorescence analysis. *Adv. X-Ray Anal.*, **28**, 75–83.
9. von Bohlen, A., Eller, R., Klockenkämper, R., and Tölg, G. (1987). Microanalysis of solid samples by total-reflection X-ray fluorescence spectrometry. *Anal. Chem.*, **59**, 2551–2555.
10. Bilbrey, D.B., Leland, D.J., Leyden, D.E., Wobrauschek, P., and Aiginger, H. (1987). Determination of metals in oil using total reflection X-ray fluorescence spectrometry. *X-Ray Spectrom.*, **16**, 161–165.
11. Leland, D.J., Bilbrey, D.B., Leyden, D.E., Wobrauschek, P., Aiginger, H., and Puxbaum, H. (1987). Analysis of aerosols using total reflection X-ray spectrometry. *Anal. Chem.*, **59**, 1911–1914.
12. Salvà, A., von Bohlen, A., Klockenkämper, R., and Klockow, D. (1993). Multielement analysis of airborne particulate matter by total reflection X-ray fluorescence. *Quimica Analitica*, **12**, 57–62.
13. Ketelsen, P. und Knöchel, A. (1984). Multielementanalyse von Aerosolen mit Hilfe der Röntgenfluoreszenzanalyse mit totalreflektierendem Probenträger (TRFA). *Fresenius Z. Anal. Chem.*, **317**, 333–342.
14. Bethel, U., Hamm, V., und Knöchel, A. (1989). Untersuchungen zur Bestimmung von Spurenelementen in Blutserum mit Hilfe der Totalreflexions-Röntgenfluoreszenzanalyse. *Fresenius Z. Anal. Chem.*, **335**, 855–859.
15. Reus, U. (1991). Determination of trace elements in oils and greases with total reflection X-ray fluorescence: sample preparation methods. *Spectrochim. Acta*, **B46**, 1403–1411.
16. Prange, A. (1987). Totalreflexions-Röntgenfluoreszenzanalyse. *GIT Fachz. Lab.*, **6**, 513–526.
17. Prange, A., Knoth, J., Stößel, R.P., Böddeker, H., and Kramer, K. (1987). Determination of trace elements in the water cycle by total-reflection X-ray fluorescence spectrometry. *Anal. Chim. Acta*, **195**, 275–287.
18. Prange, A., Böddeker, H., and Michaelis, W. (1990). Multielement determination of trace elements in whole blood and blood serum by TXRF. *Fresenius J. Anal. Chem.*, **335**, 914–918.
19. Gerwinski, W. and Goetz, D. (1987). Multielement analysis of standard reference materials with total reflection X-ray fluorescence (TXRF). *Fresenius Z. Anal. Chem.*, **327**, 690–693.
20. Knöchel, A., Dierks, H., Hastenteufel, S., and Haurand, M. (1989). Multielement analysis of marine compartments with total reflection X-ray fluorescence analysis (TXRF). *Fresenius Z. Anal. Chem.*, **334**, 673–674.
21. Stößel, R.P. and Prange, A. (1985). Determination of trace elements in rainwater by total reflection X-ray fluorescence. *Anal. Chem.*, **57**, 2880–2885.
22. Reus, U., Freitag, K., and Fleischhauer, J. (1989). Trace analysis in ultrapure acids at levels below 1 ng/ml. *Fresenius Z. Anal. Chem.*, **334**, 674–674.

23. Reus, U. (1989). Total reflection X-ray fluorescence spectrometry: matrix removal procedures for trace analysis of high-purity silicon, quartz and sulphuric acid. *Spectrochim. Acta*, **B44**, 533–542.
24. Chen, J.S., Berndt, H., Klockenkämper, R., and Tölg, G. (1990). Trace analysis of high-purity iron by total reflection X-ray fluorescence spectrometry. *Fresenius J. Anal. Chem.*, **338**, 891–894.
25. Prange, A., Knöchel, A., and Michaelis, W. (1985). Multielement determination of dissolved heavy metal traces in sea water by total-reflection X-ray fluorescence spectrometry. *Anal. Chim. Acta*, **172**, 79–100.
26. Burba, P., Willmer, P.G., Becker, M., and Klockenkämper, R. (1989). Determination of trace elements in high-purity aluminium by total reflection X-ray fluorescence after their separation on cellulose loaded with hexamethylenedithiocarbamates. *Spectrochim. Acta*, **B44**, 525–532.
27. Gross, A. and Stosnach, H. (2013). *Ashing, pasting, plotting – Preparation of complex sample types for TXRF analysis*, 15th International Conference on TXRF and related methods, Book of Abstracts, P56, p. 136.
28. De La Calle, I., Cabaleiro, N., Romero, V., Lavilla, I., and Bendicho, C. (2013). Sample pretreatment strategies for total reflection X-ray fluorescence analysis: a tutorial review. *Spectrochim. Acta*, **B90**, 23–54.
29. Knoth, J. and Schwenke, H. (1978). An X-ray fluorescence spectrometer with totally reflecting sample support for trace analysis at the ppb level. *Fresenius Z. Anal. Chem.*, **291**, 200–204.
30. Knoth, J. and Schwenke, H. (1979). Trace element enrichment on a quartz glass surface used as a sample support of an X-ray spectrometer for the subnanogram range. *Fresenius Z. Anal. Chem.*, **294**, 273–274.
31. Miller, T.C., Sparks, C.M., Havrilla, G.J., and Beebe, M.R. (2004). Semiconductor applications of nanoliter droplet methodology with total-reflection X-ray fluorescence analysis. *Spectrochim. Acta*, **B59**, 1117–1124.
32. Miller, T.C. and Havrilla, G.J. (2004). Nanodroplets: A new method for dried spot preparation and analysis. *X-Ray Spectrom.*, **33**, 101–106.
33. Fittschen, U.E.A., Hauschild, S., Amberger, M.A., Lammel, G., Streli, C., Förster, S., Wobrauschek, P., Jokubonis, C., Pepponi, G., Falkenberg, G., and Broekaert, J.A.C. (2006). A new technique for the deposition of standard solutions in total reflection X-ray fluorescence spectrometry (TXRF) using pico-droplets generated by inkjet printers and its applicability for aerosol analysis with SR-TXRF. *Spectrochim. Acta*, **B61**, 1098–1104.
34. Fittschen, U.E.A. and Havrilla, G.J. (2010). Picoliter droplet deposition using a prototype picoliter pipette: control parameters and application in micro X-ray fluorescence. *Anal. Chem.*, **82**, 297–306.
35. Sparks, C., Fittschen, U., and Havrilla, G. (2010). Picoliter solution deposition for total reflection X-ray fluorescence analysis of semiconductor samples. *Spectrochim. Acta*, **B65**, 805–811.
36. Sparks, C.M., Fittschen, U.E.A., and Havrilla, G.J. (2013). Investigation of total reflection X-ray fluorescence calibration with picoliter deposition arrays. *Microelectron. Eng.*, **102**, 98–102.

37. Fittschen, U.E.A., Menzel, M. Scharf, O., Radke, M., Reinholz, U., Buzanich, G., Montoya, V., McIntosh, K., Hrontrich, C., Streli, C., and Havrilla, G.J. (2013). *Evaluation of total reflection X-ray fluorescence (TXRF) analysis using a color X-ray camera (CXC)*, 15th International Conference on TXRF and related methods, Book of abstracts, I9, p. 87–88.
38. Mages, M., Brauns, M., Nordhoff, E., and v. Tümpling, W. (2013). *Automatic preparation of liquid samples on quartz plate – A routine method for TXRF measurements*. The 15th International conference on total reflection X-ray fluorescence analysis and related methods (TXRF2013) and The 49th Annual conference on X-ray chemical analysis, Book of Abstracts, P6, pp. 33–34.
39. Klockenkämper, R., von Bohlen, A., Moens, L., and Devos, W. (1993). Analytical characterization of artists' pigments used in old and modern paintings by total-reflection X-ray fluorescence. *Spectrochim. Acta*, **B48**, 239–246.
40. Schneider, B. (1989). The determination of atmospheric trace metal concentrations by collection of aerosol particles on sample holders for total-reflection X-ray fluorescence. *Spectrochim. Acta*, **B44**, 519–524.
41. Bredendiek-Kämper, S., von Bohlen, A., Klockenkämper, R., Quentmeier, A., and Klockow, D. (1996). Microanalysis of solid samples by laser-ablation and total-reflection fluorescence. *J. Anal. At. Spectrom.*, **11**, 537–541.
42. Koch, J., von Bohlen, A., Hergenröder, R., and Niemax, K. (2004). Particle size distributions and compositions of aerosols produced by near-IR femto- and nano-second laser ablation of brass. *J. Anal. At. Spectrom.*, **19**, 267–272.
43. von Bohlen, A., Klockenkämper, R., Tölg, G., and Wiecken, B. (1988). Microtome sections of biomaterials for trace analyses by TXRF. *Fresenius Z. Anal. Chem.*, **331**, 454–458.
44. Knoth, J., Schneider, H., and Schwenke, H. (1994). Tunable exciting energies for total reflection X-ray fluorescence spectrometry using a tungsten anode and band-pass filtering. *X-Ray Spectrom.*, **23**, 261–266.
45. Rigaku Industrial Corporation (1992). *The Rigaku Journal* **9**, 29; Technical Note.
46. Matsushita, T. (2008) *X-ray monochromators*. http://cheiron2008.spring8.or.jp/lec_text/Sep.30/2008_T.Matsushita_1.pdf.
47. Krämer, M. (2007). Potentials of synchrotron radiation induced X-ray standing waves and X-ray reflectivity measurements in material analysis. PhD thesis, University of Dortmund, 128pages.
48. Beardon, J.A. (1964). X-ray Wavelengths, U.S. Atomic Energy Commission Report NYO-10586, 533 pp.
49. Johnson, G.G. Jr. and White, E.W. (1970) X-ray Emission Wavelengths and keV Tables for Nondiffractive Analysis, ASTM Data Series DS 46, Philadelphia.
50. Bertin, E.P. (1975). *Principles and Practice of Quantitative X-ray Fluorescence Analysis*, 2nd ed., Plenum Press: New York, p. 621–624; 661–667; 690–696.
51. Klockenkämper, R. and von Bohlen, A. (1989). Determination of the critical thickness and the sensitivity of thin-film analysis by total reflection X-ray fluorescence spectrometry. *Spectrochim. Acta*, **B44**, 461–470.
52. Welz, B. (1983). *Atomabsorptionsspektrometrie*, Verlag Chemie: 3. Auflage, Weinheim, pp. 121–123.

53. Ninomiya, T., Nomura, S., Taniguchi, K., and Ikeda, S. (1989). Quantitative analysis of arsenic element in a trace of water using total reflection X-ray fluorescence spectrometry. *Adv. X-Ray Anal.*, **32**, 197–204.
54. Yap, C.T. (1988). X-ray total reflection fluorescence analysis of iron, copper, zinc, and bromine in human serum. *Appl. Spectrosc.*, **42**, 1250–1253.
55. Klockenkämper, R., von Bohlen, A., and Wiecken, B. (1989). Quantification in total reflection X-ray fluorescence analysis of microtome sections. *Spectrochim. Acta*, **B44**, 511–518.
56. de Boer, D.K.G. (1991). X-ray standing waves and the critical sample thickness for total-reflection X-ray fluorescence analysis. *Spectrochim. Acta*, **B46**, 1433–1436.
57. Hellin, D. Fyen, W., Rip, J., Delande, T., Mertens, P. W., De Gendt, S., and Vinckier, C. (2004). Saturation effects in TXRF on micro-droplet residue samples. *J. Anal. At. Spectrom.*, **19**, 1517–1523.
58. Hellin, D., Rip, J., Geens, V., Delande, T., Conard, T., De Gendt, S., and Vinckier, C. (2005). Remediation for TXRF saturation effects on microdroplets residues from preconcentration methods on semiconductor wafers. *J. Anal. At. Spectrom.*, **20**, 652–658.
59. Segre, C. (2010) *Coherence lengths*, http://www.csrri.iit.edu/~segre/phys570/10F/lecture_04.pdf.
60. von Bohlen, A., Krämer, M., Sternemann, C., and Paulus, M. (2009). The influence of X-ray coherence length on TXRF and XSW and the characterization of nanoparticles observed under grazing incidence of X-rays. *J. Anal. At. Spectrom.*, **24**, 792–800.
61. Schwenke, H. and Knoth, J. (1995). In: Mittal, K.L. (editor) *Part. Surf. [Proc. Symp.]; Meeting Date 1992;* Dekker: New York, N.Y., pp. 311–323.
62. Weisbrod, U., Gutschke, R., Knoth, J., and Schwenke, H. (1991). X-ray induced fluorescence at grazing incidence for quantitative surface analysis. *Fresenius J. Anal. Chem.*, **341**, 83–86.
63. Hellin, D., Bearda, T., Zhaoa, C., Raskinc, G., Mertens, P.W., De Gendt, S., Heynsa, M.M., and Vinckier, C. (2003) Determination of metallic contaminants on Ge wafers using direct and droplet sandwich etch-total reflection X-ray fluorescence spectrometry. *Spectrochim. Acta*, **B58**, 2093–2104.
64. de Boer, D.K.G. and van den Hoogenhof, W.W. (1991). Total reflection X-ray fluorescence of single and multiple thin-layer samples. *Spectrochim. Acta*, **B46**, 1323–1331.
65. Hockett, R.S. (1995). A review of standardization issues for TXRF and VPD/TXRF. *Adv. X-Ray Chem. Anal. Jpn.*, **26s**, 79–84.
66. Fabry, L., Pahlke, S., Kotz, L., Adachi, Y., and Furukawa, S. (1995). Standardization of TXRF using microdroplet samples. Particulate or film type? *Adv. X-Ray Chem. Anal. Jpn.*, **26s**, 19–24.
67. Torcheux, L., Degraeve, B., Mayeux, A., and Delamar, M. (1994). Calibration procedure for quantitative surface analysis by total reflection X-ray fluorescence. *Surf. Interface Anal.*, **21**, 192–198.
68. Mori, Y., Shimanoe, K., and Sakon, T. (1995). Standard sample preparation for quantitative TXRF analysis. *Adv. X-Ray Chem. Anal. Jpn.*, **26s**, 69–72.
69. Gutschke, R. (1991). *Diploma thesis*, University of Hamburg.

70. Schwenke, H. and Knoth, J. (1995). Depth profiling in surfaces using TXRF. *Adv. X-Ray Chem. Anal. Jpn.*, **26s**, 137–144.
71. Berneike, W. (1993). Basic features of total-reflection X-ray fluorescence analysis on silicon wafers. *Spectrochim. Acta*, **B48**, 269–275.
72. Klockenkämper, R. and von Bohlen, A. (1999). A new method for depth-profiling of shallow layers in silicon wafers by repeated chemical etching and total-reflection X-ray fluorescence analysis. *Spectrochim. Acta*, **B54**, 1385–1392.
73. Klockenkämper, R., von Bohlen, A., Becker, H.W., and Palmetshofer, L. (1999) Comparison of shallow depth profiles of Co-implanted Si-wafers determined by total-reflection X-ray fluorescence analysis after repeated stratified etching and by Rutherford backscattering spectrometry. *Surf. Interface Anal.*, **27**, 1003–1008.
74. Wiener, G., Günther, R., Michaelsen, C., Knoth, J., Schwenke, H., and Bormann, R. (1997). Ion beam sputtering techniques for high resolution concentration depth profiling with glancing-incidence X-ray fluorescence spectrometry. *Spectrochim. Acta*, **B52**, 813–821.
75. Klockenkämper, R., Krzyzanowska, H., and von Bohlen, A. (2003). Density-depth profiles of an As implanted Si-wafer studied by repeated planar sputter etching and total-reflection X-ray fluorescence analysis. *Surf. Interface Anal.*, **35**, 829–834.
76. Krzyzanowska, H., von Bohlen, A., and Klockenkämper, R. (2003). Depth profiles of shallow implanted layers by soft ion sputtering and total-reflection X-ray fluorescence. *Spectrochim. Acta*, **B58**, 2059–2067.
77. Von Bohlen, A. and Klockenkämper, R. (2005). Parasitic ion-implantation produced by a Kaufman-type ion source used for planar etching of surfaces. *Anal. Bioanal. Chem.*, **382**, 1975–1980.
78. von Bohlen, A., Krzyzanowska, H., and Klockenkämper, R. (2004). A broad beam ion source used for planar sputter-etching of shallow layers from flat samples and determination of their mass density. *Nucl. Instrum. Methods Phys. Res.*, **B217**, 158–166.
79. Klockenkämper, R., Becker, M., von Bohlen, A., Becker, H.W., Krzyzanowska, H., and Palmetshofer, L. (2005). Near-surface density of ion-implantated Si studied by Rutherford backscattering and total-reflection X-ray fluorescence. *J. Appl. Phys.*, **98**, 033517, 1–5.
80. Klockenkämper, R. (2006). Challenges of total reflection X-ray fluorescence for surface and thin-layer analysis. *Spectrochim. Acta*, **B61**, 1082–1090.
81. Brücher, M., von Bohlen, A., and Hergenröder, R. (2012). The distribution of the contrast of X-ray standing waves fields in different media. *Spectrochim. Acta*, **B71–72**, 62–69.
82. Zegenhagen, J. (1993). Surface structure determination with X-ray standing waves. *Surf. Sci. Rep.*, **18**, 202–271.
83. Zegenhagen, J. and Kazimirow, A. (2013). X-ray Standing Waves in a Nutshell. In: Zegenhagen, J. and Kazimirow, A. (editors) *The X-Ray Standing Wave Technique: Principles and Applications*, Series on Synchrotron Radiation Techniques and Applications, World Scientific: Singapore, pp. 3–35.
84. von Bohlen, A. (2009). Total reflection X-ray fluorescence and grazing incidence X-ray spectrometry: a review *Spectrochim. Acta*, **B64**, 821–832.

85. Weisbrod, U., Gutschke, R., Knoth, J., and Schwenke, H. (1991). Total reflection X-ray fluorescence spectrometry for quantitative surface and layer analysis. *Appl. Phys.*, **A53**, 449–456.

86. Schwenke, H., Gutschke, R., and Knoth, J. (1992). Characterization of near surface layer by means of total reflection X-ray fluorescence spectrometry. *Adv. X-Ray Anal.*, **35B**, 941–946.

87. de Boer, D.K.G. (1991). Glancing incidence X-ray fluorescence of layered materials. *Phys. Rev.*, **B44**, 498–511.

88. Klockenkämper, R. and von Bohlen, A. (2001). Total-reflection X-ray fluorescence moving towards nanoanalysis: a survey. *Spectrochim. Acta*, **B56**, 2005–2018.

89. Hüppauf, M. (1993). Charakterisierung von dünnen Schichten und von Gläsern mit Röntgenreflexion und Röntgenfluoreszenzanalyse bei streifendem Einfall, Doctoral thesis, RWTH Aachen, and JÜL-report JÜL-2730, ISSN 0366–0885.

90. Köhnen, A., Brücher, M., Reckmann, A., Klesper, H., von Bohlen, A., Wagner, R., Herdt, A., Lützenkirchen-Hecht, D., Hergenröder, R., and Meerholz, K. (2012). Tracing a moving thin-film reaction front with nanometer resolution. *Macromolecules*, **45**, 3487–3495.

91. Schwenke, H., Gutschke, R., Knoth, J., and Kock, M. (1992). Treatment of roughness and concentration gradients in total reflection X-ray fluorescence analysis of surfaces. *Appl. Phys.*, **A54**, 460–465.

92. van den Hoogenhof, W.W. and de Boer, D.K.G. (1993). Glancing-incidence X-ray analysis. *Spectrochim. Acta*, **B48**, 277–284.

93. Kawamura, T. and Takenake, H. (1994). Interface roughness characterization using X-ray standing waves. *J. Appl. Phys.*, **75**, 3806–3809.

94. Nevot, L. and Croce, P. (1980). *Rev. Phys. Appl.*, **23**, 1675.

95. Nevot, L., Pardo, B., and Corno, J. (1988). Characterization of X-UV multilayers by grazing-incidence X-ray reflectometry. *Rev. Phys. Appl.*, **23**, 1675–1686.

96. Brücher, M., von Bohlen, A., Jacob, P., Franzke, J., Radtke, M., Reinholz, U., Müller, B.R., Scharf, O., and Hergenröder, R. (2010). The charge of solid-liquid interfaces measured by X-ray standing waves and streaming current. *ChemPhysChem*, **2010**, 2118–2123.

97. Krämer, M., von Bohlen, A., Sternemann, C., Paulus, M., and Hergenröder, R. (2006). X-ray standing waves: a method for thin-layered systems. *J. Anal. At. Spectrom.*, **21**, 1136–1142.

CHAPTER 5

DIFFERENT FIELDS OF APPLICATIONS

Total reflection X-ray fluorescence (TXRF) analysis is applicable to a great variety of sample materials. Samples of environmental origin are often analyzed for monitoring purposes when large numbers of items must be dealt with. Biological applications are related to beverages and foodstuff, and biochemical

Total-Reflection X-ray Fluorescence Analysis and Related Methods, Second Edition.
Reinhold Klockenkämper and Alex von Bohlen.

questions are concerned with metal compounds of proteins and nucleic acids. Medical, clinical, or pharmaceutical material is frequently investigated by TXRF, as quite often only small sample amounts are available. Several health problems should be revealed and their causes elucidated by the detection of essential and toxic elements in constituents of the human body. Unfortunately, the results are sometimes ambiguous with respect to cause and effect of health and disease. Nevertheless, such investigations are necessary and may help solve important and difficult problems.

Industrial and chemical applications of TXRF are aimed at high-purity materials with only little contaminations. A wide field of applications has opened up in the semiconductor industry, where TXRF and also grazing incidence XRF (GI-XRF) are used for surface and thin-layer analyses of wafers. Furthermore, both variants can be used to analyze thin films or layers in laser and X-ray optics, polymers and metal-organic compounds used for automobile production, and coatings of mass storage devices. The microcapability of TXRF is also useful in helping resolve art historical or forensic questions when precious works of art or unique pieces of evidence have to be analyzed. The analytical features of TXRF and GI-XRF have been used to promote TXRF's different kinds of applications for several types of matrices.

5.1 ENVIRONMENTAL AND GEOLOGICAL APPLICATIONS

Numerous applications concern environmental samples, where TXRF can be applied to pollution control, while geological samples are less frequently investigated. TXRF is especially suitable for ultratrace analyses of pure waters, such as rainwater, groundwater, or drinking water. The low ng/ml level is directly accessible. For river water, lake and seawater, as well as for wastewaters, a preparation is recommended to separate the suspended matter and to remove the salt content. TXRF is also suitable for the analysis of airborne particulate matter. Collection of air dust can be carried out by filtration or impaction. Furthermore, inorganic materials, such as ashes, sludge, sediments, and soils, can be analyzed after digestion and preferably after matrix removal.

Several contaminants contribute to the pollution of our environment. Aside from the absolute concentration, their bioavailability is important. In order to determine such bioavailable concentrations, so-called biomonitors are analyzed. In particular, the applicability of TXRF may be illustrated by the following examples.

5.1.1 Natural Water Samples

Pure waters, such as rainwater, groundwater, and drinking water, including tap and mineral water, can be analyzed *nearly* directly. Only a few easy preparatory

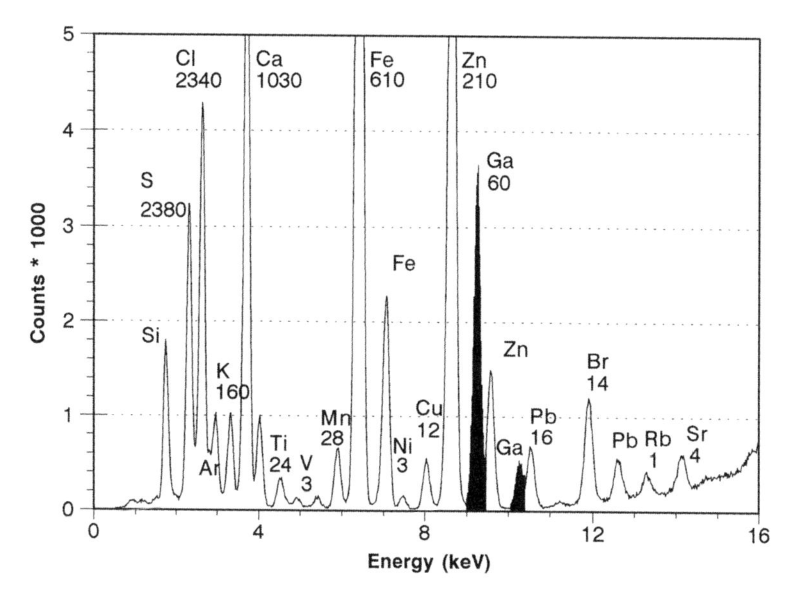

Figure 5.1. Typical TXRF spectrum of a rainwater sample. Gallium was added as internal standard with a concentration of 60 ng/ml. All values indicated in ng/ml. Figure from Ref. [3], reproduced with permission. Copyright © 1996, John Wiley and Sons.

steps have to be taken, as already shown in Figure 4.16. A volume of about 100 ml is first provided, and an aliquot of 1 or 2 ml is acidified with nitric acid (pH 2). A single-element standard, for example, Se, Co, or Ga, is added on the µg/ml level with a nitric acid base. The standard is homogeneously mixed and aliquots of 10–100 µl are pipetted onto cleaned hydrophobic sample carriers. The droplets are dried on a hot plate or under infrared light, and the residue of 0.1–10 µg is analyzed within a counting time of 100–1000 s. Quantification is carried out according to Equation 4.12.

Figure 5.1 shows a typical spectrum of rainwater. Detection limits go down to some ng/ml. Table 5.1 shows the quantitative results of TXRF applied to the reference material NIST 1643c "Water" [1,2].[1] No further preparation steps were necessary, besides the addition of yttrium as an internal standard and the evaporation of 50 µl on a clean carrier. Mo and W excitation was used with a counting time of 1000 s. Repeated determinations show a precision of 1–6% (squared mean, 4%). The relative deviations from the certified values had an accuracy of 1–12% (squared mean, 6%). For most elements, measured and reference values agree quite well—only for Cr, Sr, and Mo a relative deviation of about 8% is significant.

[1] NIST stands for the National Institute of Standards and Technology (Gaithersburg, MD).

TABLE 5.1. TXRF Results for Different Elements Determined in NIST 1643c "Water"[a]

Element	Certified Value (μg/l)		TXRF Result (μg/l; $n = 3$)		Deviation (%)	Distinction
K	2300		2280	±80	−0.9	no
Ca	36800	±1400	35300	±500	−4.1	no
V	31.4	±2.8	28.5	±2.1	−9.2	no
Cr	19.0	±0.6	17.2	±0.2	−9.5	yes
Mn	35.1	±2.2	32.6	±1.3	−7.1	no
Fe	106.9	±3.0	102.4	±4.5	−4.2	no
Co	23.5	±0.8	22.3	±1.1	−5.1	no
Ni	60.6	±7.3	60.0	±0.4	−1.0	no
Cu	22.3	±2.8	22.1	±0.6	−0.9	no
Zn	73.9	±0.9	74.4	±0.6	+0.7	no
As	82.1	±1.2	79.7	±1.4	−2.9	no
Se	12.7	±0.7	12.0	±0.2	−5.5	no
Rb	11.4	±0.2	11.5	±0.4	+0.9	no
Sr	263.6	±2.6	248.7	±3.9	−5.7	yes
Y	internal standard		100		—	—
Mo	104.9	±1.9	95.7	±3.7	−8.8	yes
Ag	2.21	±0.30	1.90	±0.3	−14.0	no
Cd	12.2	±1.0	11.3	±0.6	−7.4	no
Te	2.7		<5		—	—
Ba	49.6	±3.1	47.6	±2.4	−4.0	no
Tl	7.9		8.6	±0.5	+8.9	(no)
Pb	35.3	±0.9	34.2	±1.7	−3.1	no
Bi	12.0		13.5	±0.8	+12.5	(no)

[a] The relative deviation of these values from the certified values is given in percent. Both values are called significantly distinct (Yes) or not (No) if their regions of confidence ($x \pm s$) do not overlap or do, respectively.

Source: From Ref. [1], reproduced with permission.

The simple method just described can be improved by freeze-drying of a 10 ml volume and leaching the residue with 1 ml diluted nitric acid. Detection limits are lowered to about 10–20 pg/ml [4]. Vázquez and coworkers determined natural contaminations of groundwater in the Argentinean pampas [5] and detected several toxic elements. A concentration of 600 ng/ml was found for As, which is much higher than the maximum permissible value of 10 ng/ml for drinking water.

For river water, lake or freshwater, estuarine water, seawater, wastewater, and sewage, detection limits can be brought down to ng/ml, but some preparation is necessary. If an organic load is significant, a prior pressure filtration is recommended [6]. The filtrate of some 10 ml can be freeze-dried and the residue digested by nitric acid. If a salt matrix is troublesome, the trace elements can be separated by complexation, chromatographic adsorption, and subsequent elution—a method developed by Prange *et al.* [6–8].

A complexation of the traces with a NaDBDTC solution (sodium dibutyldithiocarbamate) leads to a coprecipitation. Subsequent adsorption of the carbamate traces by a reverse-phase column (e.g., Chromosorb, E. Merck, Darmstadt, Germany) and a final elution of the adsorbed complexes by 2 ml of subboiled chloroform gives an enriched eluate (by a factor of about 50). After the final solution is spiked with an internal standard, an aliquot of 10–100 μl is analyzed as already described. The detection limits for the aqueous solutions including the filtrate are about 0.1 ng/ml.

The suspended matter separated from the filtrate can be collected on Nuclepore filters (Nuclepore Corp., Pleasanton, California), weighed, and digested by concentrated nitric acid or a mixture of nitric and hydrofluoric acid (2:1). The determination by TXRF can be carried out as usual. Detection limits go down to 5 μg/g; precision and accuracy are characterized by a relative standard deviation of about 10% [6].

The determination of mercury in wastewater is difficult because of its high vapor pressure at room temperature, which leads to the evaporation and loss of this element. Vázquez and colleagues used APDC (ammonium pyrrolidine dithiocarbamate) or EDTA (ethylenediaminetetraacetic acid) and oxalic acid (HOOC–COOH, dicarboxylic acid) for trapping traces of mercury in industrial wastewaters [9]. The method was applied to control a cleaning process of effluents. Kregsamer and coworkers recommend a complexation of mercury with thiourea $CS(NH_2)_2$ at pH = 10. After deposition of 10 μl droplets on a quartz reflector, mercury could be determined in wastewater samples, such as industrial and municipal effluents [10].

Some national and international programs have been carried out for pollution control of rain water, river water, and seawater in the 1980s and 1990s. The river Elbe [6,11,12] was tested systematically for contaminants, and appropriate actions were taken for regeneration. Trace contaminants in the open Atlantic Ocean were studied at several deep-water stations [13], and heavy-metal traces and pollutant transfer were investigated in the North Sea [14,15]. These field experiments demonstrated the ability of TXRF to automatically handle scores of water samples with a high degree of reliability.

An impressive example stems from Prange *et al.* [11,12] who investigated the water quality of the river Elbe in an extensive international environmental research project. A specific task was the determination of arsenic in the filtrate, in suspended matter, and sediment of the river. Figure 5.2 shows the profile of arsenic in suspended matter along the river from its source to the mouth. The water samples were taken with the help of a helicopter and analyzed by TXRF after pressure filtration. Scandium was used as internal standard. Related to the dry mass of the suspended matter, the average content was found to be about 45 μg/g. In the years from 1993 to 1998, pollution from industrial districts has been reduced significantly.

Some recent examples for analyses of drinking water [16], rain water [17], river water [18], freshwater [19], and seawater [20] prove the efficiency of TXRF. Drinking water could be analyzed directly without preconcentration.

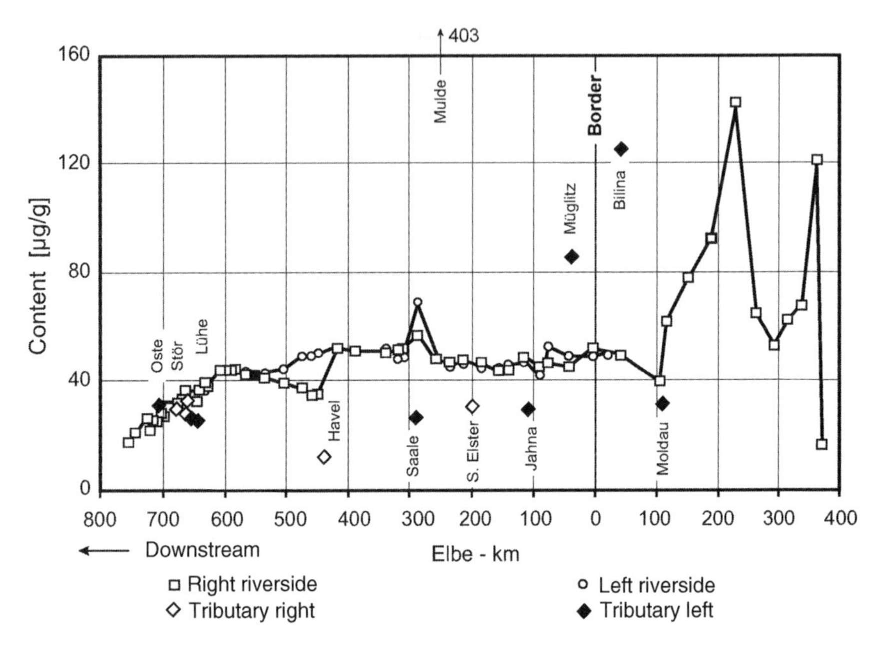

Figure 5.2. Arsenic profile of the river Elbe in October 1993. The content of arsenic is given in μg/g of suspended matter determined by TXRF. 80 checkpoints were chosen along the river from the source (right) to the mouth (left) over a length of 1200 km. The right and left riverside was controlled as well as several tributaries right and left. Figure from Ref. [12], reproduced with permission from the author.

Detection limits of 0.5 to 1.7 μg/l (ppb) were found for 13 elements (Cr, Mn, Fe, Co, Ni, Cu, Zn, As, Se, Rb, Sr, Hg, and Pb). Six further elements could be detected down to 5–11 μg/l (K, Ca, V, Cd, Sb, and Ba). A metallic pollution could be confirmed directly by TXRF, dependent on the type of plumbing inside [21]. Brine samples were diluted with ultrapure water or dissolved in a solution of the detergent Triton® X-100. The influence of self-absorption phenomena on quantification was demonstrated and data obtained by TXRF were compared to those obtained by "wet" chemistry and inductively coupled plasma mass spectrometry (ICP-MS).

Water analysis was also performed at Lake Victoria, which is the largest freshwater area in Africa [19]. Several rivers pass through agricultural and commercial centers on their way to the lake, while the Victoria Nile is the main outlet flowing on a long way of nearly 6000 km into the Mediterranean. Environmental degradation by metal pollution has important socioeconomic implications. Consequently, continuous monitoring is absolutely necessary and is supported by the International Science Programme of the Uppsala University in Sweden. Water samples were collected at several places and TXRF analyses were carried out with gallium as internal standard. K, Ca, Ti, Mn, and Fe were determined in a range of 1 μg/l (ppb) up to 10 mg/l (ppm). The concentration values were found to be partly above critical limits given by regulations for drinking water of about 25 ppb.

The quality of TXRF determinations was proved by different reference samples and by various intercomparison tests of seawater and limnetic or estuarine river water [6,22,23]. In competition with well-established methods, such as voltammetry, instrumental neutron-activation analysis (INAA), electrothermal atomic absorption spectrometry (ET-AAS), inductively coupled plasma optical emission spectrometry (ICP-OES), and ICP-MS, the micro-analytical method of TXRF showed a high performance for the certification of reference materials (see also Section 6.1.5).

5.1.2 Airborne Particulates

Atmospheric or airborne particulate matter consists of small or tiny grains or drops, either in solid or liquid form, respectively. The relevant aerosol is the mixture of these grains and drops suspended in the atmosphere and can be man-made or natural. They can influence precipitation in particular and even the climate seen as a whole; moreover, they can affect human health. Natural sources are volcanoes, sandstorms, forest and bush fires, and sea spray over the oceans. Sea salt is the largest contributor followed by mineral dust blown off from earth by wind and storm. Ninety percent of atmospheric aerosols have a natural origin. Only 10% are anthropogenic and stem from combustion of coal, wood and fossil fuel, heat and power stations, combustion engines, motor vehicles, and several industrial processes. Organic matter and elemental carbon known as soot are further components of aerosols. Smog consists mainly of SO_2, NO_x, CO, and soot.

Five different types of airborne particulate matter are distinguished: (1) suspended particulate matter (SPM) with diameters greater than 10 μm, (2) respirable suspended particles (RSP) with diameters less than 10 μm also called PM_{10} or fine dust, (3) fine particles with diameters under 2.5 μm called $PM_{2.5}$, (4) ultrafine particles with diameters less than 1 μm, and (5) soot with particles below 0.3 μm. The diagram in Figure 5.3 shows different types of atmospheric particulates and their size distribution ranging from 1 mm down to 0.1 nm [24].

Large particles (SPM) settle down to the ground by gravity within hours, whereas ultrafine particles can stay in the atmosphere for weeks and mostly fall down due to precipitations. Particulate pollution can cause serious effects on health, such as asthma, cardiovascular complaints, respiratory disease, and even lung cancer. Coarse particles are generally collected in the nose; finer inhalable particles are filtered in the bronchi. Even finer particles penetrate into the terminal bronchioles and are termed thoracic. Particles, such as PM_{10} and $PM_{2.5}$, can reach the deep parts of the lung, such as alveoli, and are called respirable. Finest particles can even pass through the lung.

The industries of most countries have to operate dust collection systems in order to reduce particulate emissions. The pollution in metropolitan areas

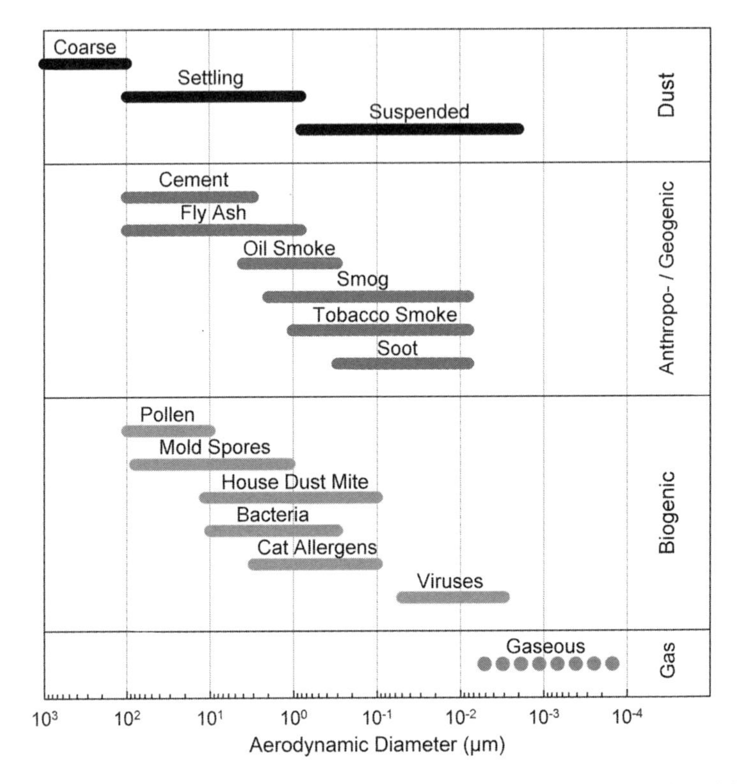

Figure 5.3. Various types of atmospheric particulates with geogenic/anthropogenic and biogenic origin and their size distribution. Figure from Ref. [24], reproduced under the Creative Commons Attribution; here with change of grouping.

of developing countries is highest. In nonpolluted rural regions the daily load of fine dust is below 20 µg/m^3; in urban and industrial regions it can be 50 µg/m^3, and extreme values are above 65 µg/m^3. Since about 2006, several states have passed regulations for a yearly and daily average of the ambient concentration of particulate matter. For PM_{10}, this limit is in the region of 100 µg/m^3; for $PM_{2.5}$ a yearly average of 15 µg/m^3 shall be kept while a daily average of 35 µg/m^3 is allowed.

Total reflection XRF can be applied to element analysis of airborne dust. The particulate matter can be collected either by filtration or after a size-fractionation with an impactor. Because of the high sensitivity of TXRF, the sampling volume and sampling time, respectively, could be reduced significantly, down to 1 m^3 ambient air and 1 h or even less [17,18,21,25–27].

When air dust is collected from air by filtration, a certain volume of air is pumped through a filter, while the aerosol droplets and particulates are deposited on the filter. If Nuclepore filters are used, the sampled material can be removed by ultrasonic treatment with a nitric acid solution. The mixture of dissolved and particulate matter can be analyzed by TXRF after internal standardization. If membrane filters are used, the loaded filter plus the internal standard can be subjected to pressure digestion with several ml of 65% nitric acid, as described in Section 4.1.2. An aliquot of 10–100 μl of the final solution is analyzed by TXRF as usual.

In the second method, air is pumped through fine nozzles and dust particles are deposited onto impaction plates due to their inertia. Different stages with nozzles of decreasing diameter are stacked on top of each other so that particles of different size classes are collected on different impaction plates. The most common impactors are Berner-, Anderson-, or Battelle-type impactors and have 2–12 stages. Membrane filters are normally used as impaction plates in a Berner impactor. They can be prepared and analyzed as just mentioned. Simple plastic plates are applied in Anderson or Battelle impactors. Consequently, Plexiglas carriers fitted for TXRF devices can be used. In order to preserve the original geometry of the impactor, the carriers are inserted in ring-shaped supports. Figure 5.4 shows a cross-section of a two-stage Battelle impactor [25].

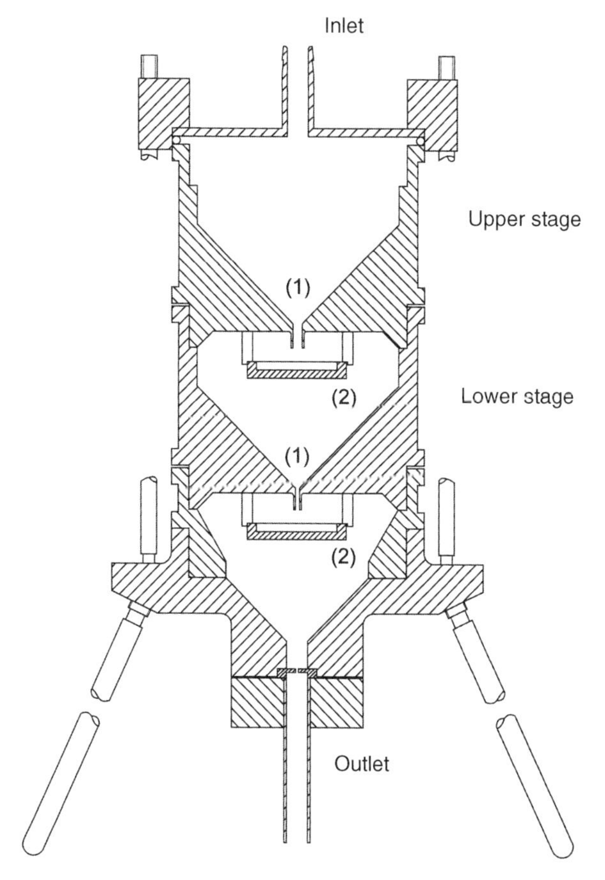

Figure 5.4. Cross-section of a two-stage Battelle-type impactor: (1) nozzles, (2) impaction plates made of Plexiglas®. Figure from Ref. [25], reproduced with permission. Copyright © 1995, Springer.

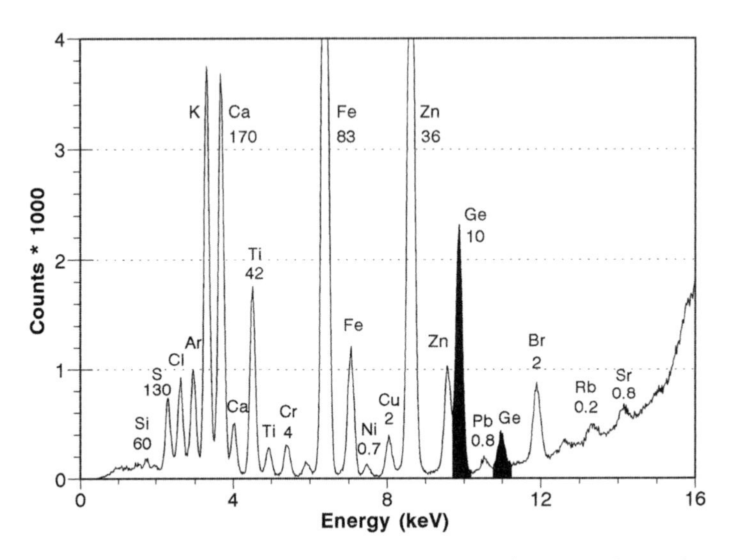

Figure 5.5. TXRF spectrum of air dust collected from ambient air. A Battelle-type impactor was used for collection of particles with diameters between 2 and 4 μm. Germanium was added as internal standard of 10 ng. The other specifications are also given in nanograms. Figure from Ref. [3], reproduced with permission. Copyright © 1996, John Wiley and Sons.

Airborne particulates of wet air or aerosols are reliably deposited and stick to the flat carriers [26,27]. Particles of dry air, however, are bounced or blown off. To avoid these effects, the Plexiglas carriers should first be siliconized with a silicone solution or even coated with a suitable grease. Medical petrolatum (e.g., Vaseline petroleum jelly) is recommended for its high purity [28]. An alkaline Ge standard can be pipetted onto the coated carrier and dried by evaporation. After a blank control, the carriers are inserted into the impactor and used for sampling. The dust-laden carriers are directly analyzed by TXRF.

Figure 5.5 shows a spectrum of dust from ambient air collected by a two-stage Battelle impactor [25]. The various elements were found in the nanogram range, with detection limits below 1 ng down to 0.01 ng [29]—three orders of magnitude less than with conventional XRF. Consequently, the collection time can be reduced to 1 h and the sampling volume to 0.5 m^3, so that pollution in the course of a day can be observed.

Due to such direct analysis by TXRF, otherwise frequent systematic errors of digestion or dissolution are avoided. However, the high sensitivity of TXRF reveals other systematic errors. In a particular study, air was sampled from a clean bench but nevertheless led to significant contamination of previously clean carriers [25,28,30]. These blank values were caused by erosion of the impactor walls made of stainless steel. The error can be avoided by use of a more suitable impactor material, for example, of an antistatic polymer.

Effective monitoring of air pollution presupposes small sampling volumes and short collection times, as is offered by TXRF. This advantage has been used to study the widespread deposition of atmospheric pollutants into forests [31,32]. Multielement analyses by TXRF were performed with great success.

In 2004, Schmeling investigated particulate air pollution in urban areas of two different cities in the USA [33]. Chicago, IL and Phoenix, AZ were chosen as big cities with local industry. Chicago is located on a flat terrain at Lake Michigan with a day-lake breeze and a night-land breeze. By contrast, Phoenix is situated near the Sonoran Desert, is surrounded by mountains, and generally has a low humidity (apart from the summer rainy season). Dependent on different topographical and meteorological conditions, but having similar high degrees of industrialization and urbanization, the particulate pollution of both cities was characterized.

Atmospheric aerosols were collected on top of a tall building in both cities. Sampling was carried out at a flow rate of about 1.1 m^3 ambient air per hour and a sampling time of 1–1.5 h. The loaded polycarbonate filters were digested with nitric acid after addition of gallium as an internal standard. TXRF analyses were carried out in about half an hour and the contents of 15 elements between 0.1 ng/m^3 and 2 $\mu g/m^3$ were determined. Time-dependent profiles show the diurnal variations of air pollution. Maxima were found in the morning and minima in the early afternoon, strongly pronounced in Phoenix. However, elements could be enriched during the day-lake breeze in Chicago with a distinct maximum in the late morning on certain days. The sum of all collections recognized five major elements: Al, S, K, Ca, and Fe. Strong differences were indicated for Al and for S. The extremely high Al contents of Phoenix (18% compared to 2%) originate from the nearby desert as a natural, crustal source (alumino-silicates of desert dust blown by wind). The significantly higher values for sulfur in Chicago (38% compared to 14%) result from steelworks and factories of the chemical industry (sulfates) as an anthropogenic source.

Boman and coworkers studied atmospheric particulate matter sampled in the urban area of Gothenburg, Sweden [34]. A particle-induced X-ray emission (PIXE) cascade impactor with nine stages was used for sampling in December 2006 and January 2007. The individual fractions with different aerodynamic diameters between 16 and 0.06 μm were collected on polished quartz carriers and directly analyzed by TXRF. Diurnal sampling was sufficient for eight elements K, Ca, Fe, Br, Cu, Zn, and Sr, whereas a 5 day sampling was necessary for the elements sulfur and chromium. The bimodal distributions of K, Cr, S, Cu, Ni, and Br indicate high-temperature processes for finest particles. Potassium and strontium likely arise from compounds used in fireworks to celebrate the New Year.

In a further study, $PM_{2.5}$ particles were investigated [35]. The samples were collected in November 2007 at the campus of the University of Gothenburg, Sweden. The $PM_{2.5}$ fractions were separated from air by means of a cyclone and were deposited on polycarbonate filters. Prior to analysis, the samples were digested with a high-pressure microwave system. TXRF and ET-AAS were used for the determination of 14 elements, including cadmium and lead.

A reflectometer was used for the analysis of soot. The large variations of the results were explained by different sources of different pollutants. Cadmium and lead were shown to meet the limits of EU regulations. The highest concentrations of sulfur were recorded on a day when air streams came from the southeast (from Poland and Romania).

Boman *et al.* also compared different analytical methods suitable for PMs [36]. Synchrotron radiation XRF (SR-XRF) and PIXE are competitors with the lowest detection limits of fg; however, both methods need a large accelerator for electrons or protons, respectively. ICP-MS is highly sensitive but needs special sample preparation since particles have to be suspended first in a solution. Furthermore, the consumption of Ar gas for the plasma is quite expensive. AAS is complicated and detection limits are worse compared to TXRF in most cases. INAA needs a neutron reactor, is time-consuming but can determine ng-traces in samples of some gram. Single particles can be detected by scanning electron microscopy with energy dispersive XRF (SEM-EDXRF).

Altogether, TXRF is suitable for many applications of aerosol analysis. A chemical speciation of selected metals in aerosols could also be carried out [37]. Particulate matter was collected on filters and subjected to sequential leaching in five stages. TXRF of the different fractions characterized five different species: elements that are mobile or exchangeable, elements bound to carbonates, oxides, and organic compounds, and elements of the insoluble residue. K, Fe, and Zn appeared as mobile elements in a church situated in an urban area, whereas Ca and Fe were shown to be extremely mobile in a church of a rural area [37].

5.1.3 Biomonitoring

Several anthropogenic contaminants contribute to the pollution of the surrounding environment, especially of biological ecosystems. The uptake, partition, and accumulation of trace elements into the trophic chain are hardly understood. The absolute concentration of the pollutants is not only important, but moreover their bioavailable concentration. In order to determine such bioavailable concentrations, so-called biomonitors are applied. They accumulate trace elements of ambient pollutants and therefore provide time-integrated measures of environmental pollutions. The behavior of toxic elements, especially of heavy metals, has been studied in several analytical investigations of biomonitoring [38–56].

Different biological objects are suitable for biomonitoring, for example, humic substances [39], aquatic biofilms and zooplankton [40,41], microcrustaceans in freshwater [42], benthic invertebrates [43], Scots pine needles [45], pollen and bee honey [46–48], fresh water mussels [49], macrozoobenthos [50], flatworms and their fish host [53], muscles and organs of freshwater fish [44] or mice [54], and also human hair [57]. Several papers report on the bioaccumulation of the platinum group metals by different aquatic organisms (e.g., Ref. [55]). Platinum, palladium, and rhodium mainly come from road dust and originally from cars with catalytic converters. A few examples of biomonitoring are given here.

Humic substances (HS) are mainly based on humus soil and are formed by microbiological processes from the remains of plants in humid soils. They can be found in meadows and marshy grounds with concentrations up to 10 or 20%, respectively, and have an amorphous inhomogeneous structure consisting of organic macromolecules with functional groups. HS can be regarded as a first stable product reconstructed by dead organic matter with high amounts of carbon, nitrogen, and oxygen. They are highly relevant to the biosphere because of their ability to form various metal complexes (about 1 mmol heavy metals per g of HS).

Metals and their macromolecular species have been determined in aquatic HS by TXRF [39]. Five reference HS originated from the German research program DFG-ROSIG. Further aquatic HS were sampled at the Venner moor near Münster, Germany. These samples were preconcentrated by preparative ultrafiltration with 1 kDa as nominal molecular mass cutoff. Afterward, 10 ml of the metal-loaded solutions were mixed with an aliquot of EDTA (ethylene diamine tetraacetic acid) in order to exchange ligands. A concentration of 10 mg/ml led to an EDTA/metal ratio of 1 : 1. After an appropriate reaction time, the samples were fractionated by ultrafiltration in order to classify their molecular mass. A multistage ultrafiltration device (MS-UF) was used with five stages equipped with ultramembranes of different nominal cutoff (1, 5, 10, 50, and 100 kDa). The sample solutions of 10 ml were pumped through this device by a six-channel peristaltic pump and filtered through the membranes, which were washed out by 10 ml of high-purity water. Afterward, five HS fractions retained by the five ultramembranes and the last fraction that passed all the filters were sampled in small reservoirs of 1 ml. These metal-loaded HS showed a concentration for dissolved organic carbon (DOC) of 0.2 to 2.6 mg/ml.

The solutions were spiked with gallium or yttrium as an internal standard and aliquots with a volume of 10 μl were deposited on quartz-glass carriers and dried. The thin "polymer" films were analyzed by TXRF with Mo excitation and a Si(Li) detector within 300 s. Detection limits for 16 elements were found between 7 and 100 ng/ml (7 and 100 ppb). Different metals (Mn, Fe, Cu, Ni, Zn, and Pb) could be determined in the humic-rich water samples with concentrations of 10 to 2000 ng/ml (ppb). It was shown that the high molecular mass-fractions prevail and natural metal species are widely EDTA inert.

Biofilms are complex, slimy, heterogeneous aggregates with a high amount of water. Several microorganisms like algae, fungi, bacteria, and protozoa are embedded in extracellular polymeric secretions. They grow on natural substrates like water plants and stones, but also bollards, pontoons, and piers in streaming water. Such biofilms were sampled from natural stones but also from ceramic plates that were exposed in the river Elbe near Magdeburg, Germany, for about 5 weeks [40]. The films were scraped off with help of a Teflon spatula in a clean bench, homogenized by stirring, evaporated to dryness (105 °C), and weighed into PTFE beakers (about 10 mg). They were digested with 2.5 ml of HNO_3 and H_2O_2 (4 : 1) within 6 h at 150 °C. The clear solutions were diluted to 5 ml with ultrapure water. Nitrous gases were evaporated and an aliquot of 1 ml was spiked with yttrium as internal standard (1 μg/ml). Finally, a portion of 20 μl was pipetted onto a clean quartz-glass carrier, polyvinyl alcohol (5 μl) was added, and the liquid drop was gently evaporated to dryness. The mass of the residue was determined by triplicate weighing on a microbalance (57 ± 3 μg).

For TXRF analysis, Mo excitation and a Si(Li) detector at 500–1000 s measuring time were chosen. The method was first checked by the application to the certified reference materials CRM 414 "Trace elements in plankton" and NIST 1643c "Trace elements in water." Parallel analyses of three independent samples showed a good precision of about 7% and an accuracy of about 9% for nine different elements. The analyses of the biofilm samples grown on exposed plates showed that K, Ca, Cr, Mn, Fe, Ni, Cu, Zn, and Pb are enriched by a "bioconcentration factor" between about 100 and 60 000 in comparison to typical values of the river Elbe water (total water phase). The bioaccumulation on natural stones is two to three times smaller.

Mages *et al.* showed that a newly developed portable TXRF spectrometer (PicoTAX, Bruker, Berlin, Germany) is even suitable for in-field trace element analyses of biological materials and applied this instrument for a further example of biomonitoring [42]. Daphnia is a genus of microcrustaceans with the colloquial name "water fleas" living in zooplankton. They were collected from a reservoir of the river Tisza, near Kisköre (Hungary) with the help of a zooplankton net and washed with lake water. Single specimens with a length of some millimeter and a mass of some milligrams were selected by a stereo-microscope, deposited on a quartz-glass carrier and digested with 5 μl HNO_3—spiked with a Ga standard—on a hot plate at 100 °C. The dried residue was analyzed by TXRF and the relative amount of several accumulated elements could be determined. This is a fast and easy possibility to select polluted sampling sites by a field study. It was also shown that a polycarbonate substrate is best suited for biofilm production [43].

In a further example of biomonitoring, Wagner and Boman used tissue samples of freshwater fish [44] to check the potential impact of a coal combustion power plant in Vietnam. For that purpose, catfish (genus *Clarias fuscus*) of similar age and size were sampled from two places: (1) from a lake 11 km southwest of the power plant situated in an agricultural area and (2) from a freshwater channel only 1 km apart and south of the power plant. During the sampling period, this channel was exposed to the emissions of the plant distributed by the monsoon wind, whereas the lake far away from the plant was considered to be unaffected.

The catfish (*C. fuscus*) were dissected and muscle and liver tissues were chosen, heated, dried, and sent to a laboratory in Sweden for analysis. Aliquots of 50 mg were digested with 2.5 ml HNO_3 in a microwave oven at 350 W for 30 s. TXRF analyses were carried out with Mo excitation, a Si(Li) detector, and 500 s lifetime. Aliquots of 5 μl were pipetted onto hydrophobic quartz-glass discs; 10 μl of a Ga solution was added as internal standard, dried on a hot plate at about 70 °C, and analyzed. The method was tested by application to NCS ZC 78005 "mussel tissue" with good precision and accuracy. Detection limits were found to be 0.3–5 μg/g dried mass. For muscle tissue, the concentration of 12 elements (P, S, K, Ca, Mn, Fe, Cu, Zn, Se, Br, Rb, Sr) for fishes of the exposed site was only slightly increased (a factor of about 1.2). All elements were far below the critical limits given by the Food and Agricultural Organization (FAO) of the United Nations and do not constitute any health risk for consumers. However, for liver tissue, the concentration of several elements (P, S, K, Ca, Mn, Fe, Cu, Zn) was significantly higher for fishes of the polluted site (a factor of about two). Obviously, these elements accumulate in fish liver. Airborne pollutants emitted from the power plant are probably the reason for the elevated concentration. Analyses of airborne particulate matter from both sites corroborated this result.

Total reflection XRF can also be used to study bioaccumulation of particular nanoparticles in vital organs [54]. For that purpose, nanorods of gold (40 nm × 10 nm) were suspended in saline at a dose of 4 μg/g, and 200 μl of this suspension was injected in mice. Tissue sections were taken from main organs, such as liver, spleen, brain, and lung, after a period of 0 h, 1 h, 1 day, 1 week, and 1 month, and digested by aqua regia and H_2O_2. Urine was sampled in addition. With As as internal standard, TXRF analysis was carried out by means of the Au-Lα peak. Tests of bovine liver CRM (certified reference material) showed a recovery rate of 99.7% and a detection limit of 110 ppb. It could be shown that liver and spleen accumulated gold particles significantly within one day while brain, lung, and urine did not take up the nanoparticles. Further studies of gold and silver nanoparticles are described in Section 5.4.8.3.

Different kinds of pollen build a further biomaterial sensitive to environmental pollution. Their use as bioindicators was tested by Pepponi *et al.* [56]. Pollen of the Common Hazel (*Corylus avellana*) was sampled at different sites in Italy with a different anthropogenic impact. The pollen was collected below bushes from ripe aglets and sieved through a mesh with 50 μm pore size. About 30 mg pollen was suspended in 1 ml of ultrapure water spiked with gallium as internal standard and stirred. Further aliquots of some 30 mg pollen were digested with concentrated HNO_3 in a microwave oven at 200 °C and 1.7 MPa for 20 min. For TXRF analysis, 5 μl of the suspension and of the solution was deposited on clean quartz-glass carriers. A Mo/W double anode tube was used for excitation. The measuring time was set to 400 s. Detection limits for iron and heavier elements were of the order of 0.3 μg/g. Seventeen elements between aluminum and lead could be determined with concentrations of several mg/g (‰) for minor elements and of 0.3 to 1 μg/g (ppm) for trace elements. The results of suspensions are comparable to those of solutions; however, the reproducibility and representativeness of the results argue in favor of the digestion.

In a further study, bee honey sampled from floral nectars in the central west of Argentina was characterized by Vázquez *et al.* [47]. The authors emphasize that honey is an environmental indicator of toxic metal contamination because it reflects more than one million interactions of bees per hive with flowers per day in an area of about 7 km^2. The concentration values of heavy metals were compared with values suggested by the Codex of Alimentarium Commission. Bee honey was also investigated by Dalipi *et al.* [48]. More than 50 samples were analyzed by TXRF as aqueous solutions and 16 elements were determined with gallium as an internal standard. By means of cluster analysis, six different groups of botanical origin could be distinguished: acacia, chestnut, eucalyptus, heather, orange, and rosmarinus. It can be generalized that TXRF is a fast and simple analytical method for safety and quality control of food and for the detection of possible toxic substances.

Aside from pollen and honey, moss and soils can be used as passive biomonitors of anthropogenic pollution and lead is the major toxic element of atmospheric deposition. Eleven sites of urban pollution were chosen in the metropolitan zone of Toluca Valley in Mexico, while natural protected areas were selected as references [50]. Special species of epiphytic moss and samples of soil were collected, prepared, and digested following reported techniques. Lead was determined with 10–110 μg/g in moss and with 30–100 μg/g in soil. Cluster and principal component analysis revealed three sites with significantly high concentration of lead due to heavy traffic and industrial activities.

For a direct checkup of environmental pollution, blood and hair samples were obtained from domestic dogs near a factory [51]. Prior to TXRF analysis they were subjected to acid digestion and 13 elements were detected. Cr, Zn, and Pb exceeded the established reference values for whole blood, while arsenic in hair showed an excessive level of 30 μg/g. It is known that these elements can cause severe toxic effects on health. Finally, human hair was tested as an indicator for food intake and environmental pollution [57]. Human hair usually grows at a rate of 1 cm/month so that a time-dependent investigation is made possible in a period of half a year. Examples are treated in Sections 5.3.4 and 5.5.4.

5.1.4 Geological Samples

Total reflection XRF can also be applied to solid samples of geological origin, such as minerals, sediments, mud, sludge, soils, ores, and rocks. Furthermore, it is just right for elemental analysis of natural waters with a low content of total dissolved solids (TDS). TXRF is especially suited if samples are disposable with only limited mass. For testing and calibration, several geological CRMs are available as powdered samples and were investigated by TXRF already in the 1980s [58]. In a recent study [59], they were prepared as suspensions where 10–50 mg of powder was mixed with 2.5 ml of aqueous solution of 1% Triton X-100 (4–20 g/l). This surfactant is a detergent that lowers surface tension. Selenium or rubidium was added as an internal standard. Aliquots of 10 μl were pipetted onto siliconized quartz-glass carriers and dried. Detection limits for several elements were found to be 1–6 μg/g (ppm) in suspensions, but 10 times lower for natural waters with low TDS. A particle size of 1–50 μm did not influence the results significantly.

For routine analyses of sediments, microwave digestion with nitric acid can be recommended [60]. Up to 20 elements were determined in sediments of the German shoals ("Wattenmeer"), that is, for the fine-grained fraction of less than 20 μm diameter. The mass fractions ranged from mg/g down to ng/g, covering five orders of magnitude. Coastal sediments were also analyzed with regard to heavy metals [61]. Metals of the lanthanides were determined in ore minerals [62]. Rock CRMs were investigated after digestion or fusion [63]. Digestion of the powdered samples was carried out with HF, HNO_3, and $HClO_4$ by heating. Fusion of the powders was applied with $LiBO_2$ (lithium metaborate) at 1100 °C followed by dissolution of the glass melt. Sample powder of 50–100 mg was decomposed with ~5 ml solutions. In each case, selenium was added as an internal standard and 10 μl of the final solutions was pipetted on quartz-glass carriers. The uncertainty of sample preparation could be neglected in comparison to that of measurements. In comparison to simple suspension, a digestion or fusion of samples deteriorates detection limits threefold. A comparison with certified values of the CRMs showed a good agreement for all three methods of sample preparation.

A further example of TXRF analysis refers to approximately 300 different soil samples that were collected at Mt. Kenya in the forest area above 1600 m [64]. The samples were dried, ground, milled, and sieved. By addition of an aqueous solution of Triton X-100 and selenium as an internal standard, slurries were prepared and 10 μl droplets with 180 μg soil were analyzed by TXRF. Elements from Al to Zn were detected; QXAS/AXIL software was used for quantification. The elemental composition could be determined between 0.01 and 100 mg/g with a standard deviation smaller than 1% and a recovery rate of 90–110%. Topsoil and subsoil samples could be distinguished and compared with Canadian soil samples.

A comprehensive review of TXRF applied to samples of geological origin was given by Revenko with 219 references [65].

5.2 BIOLOGICAL AND BIOCHEMICAL APPLICATIONS

In addition to biomonitoring, the application of TXRF in the field of biochemical and biological research has become a focus of attention in the last decade. The analysis of biological matrices can have an environmental aspect as well as a nutrimental one. The *environmental* aspect is concerned with pollution and its monitoring by appropriate plants like moss or lichen [66,67]. The *nutrimental* aspect deals with the assimilation and metabolism of various foodstuffs by an organism. The effects that certain elements and in particular some element species have on health are examined. Biochemical studies of macromolecular systems can be of medical, clinical, or pharmaceutical importance as well (Section 5.3).

A review on TXRF analyses of biological samples was presented by Marcó *et al.* [68]. Slurry sampling, *in situ* microwave digestion, *in situ* chemical modification, internal standardization, and Compton peak standardization have been evaluated to approach a routine analysis of a large number of small samples as direct as possible. A further comprehensive review on recent trends in biological applications of TXRF was given by Szoboszlai *et al.* [69]. Numerous sample matrices of animal and human origin, various analytes, several internal standards, different methods of sample preparation, and relevant references are listed in the review in two tables extending over five pages. Most of the biomatrices belong to fruits and vegetables, such as artichokes, avocados, bananas, cabbage, celeriac and kohlrabi tubers, chard, chicory, cucumber, endive, leeks, spinach leaves, lettuce, onions, radishes, and paprika.

Foodstuff generally consists of drinking water, drinks, solid food, and luxury food. Solid food contains plants, vegetables, fish, and meat, and is composed of carbohydrates, fats, and proteins. Carbohydrates are defined as a term of biochemistry and a synonym of saccharides commonly known as sugars. They have the chemical formula $C_n(H_2O)_n$. Glucose, galactose, and fructose are monosaccharides with the formula $C_6(H_2O)_6$. Starch and cellulose are polysaccharides with a long linear chain of glucose units with the formula $(C_6H_{10}O_5)_n$.

5.2.1 Beverages: Water, Tea, Coffee, Must, and Wine

Total reflection XRF is a generally suitable and well-established technique for the analysis of aqueous beverages. However, TXRF is a microanalytical method so problems may arise for a representative sampling of larger volumes of liquids. Representative sampling may usually be fulfilled for homogeneous liquids after stirring. The detection limits of TXRF are quite low and may be sufficient for a lot of analytical problems; however, competitive methods like flame atomic absorption spectrometry (FAAS), ICP-OES, and ICP-MS show even better detection limits because of a much higher consumption of sample volume commonly used for analysis. On the other hand, TXRF is a simple, fast, and economical method and will be preferred if its detection limits are sufficient. The analysis of beverages is especially suitable for TXRF because it does not require sample preparation; only a few µl of the aqueous liquid can be pipetted on a clean quartz-glass carrier and dried by evaporation, possibly with the help of an IR lamp.

Mineral waters and juices, tea and coffee, alcoholic drinks like beer, must, and wine, and spirits have been analyzed by TXRF. Tea is one of the most popular drinks throughout the world. It contains minerals and trace elements that are essential to human health. Beer mainly consists of water, and its main ingredients are malt, hop, and yeast. Its alcoholic content is about 5%. Must is a juice pressed from fruits, for example, apples, pears, or grapes. It may partly be fermented; the alcoholic content is about 7%. Wine is made of grapes pressed out and fermented. It consists of water, alcohol of about 10% (usually 9.99% ethanol and only 0.01% methanol), and residual sugar and fruit acids of about 1%.

Total reflection XRF was first applied to the detection of trace elements in tea by Xie *et al.* [70]. Infusions of 39 different kinds of tea were performed after the Chinese standard procedure for judging the flavor of tea. According to this instruction, 1 g of tea leaves was steeped in 50 ml of boiling water. In order to prevent contamination, distilled water was used instead of tap water. After 5 min, each infusion was filtered, cooled, acidified, and spiked with gallium. For analyses, aliquots of 5 µl of an infusion were pipetted onto cleaned quartz-glass carriers and dried by IR radiation. TXRF was carried out with Mo excitation and a Si(Li) detector in a live time of 200 s.

This technique was first evaluated by the analysis of a certified tea sample (GWB 08505: tea, Research Center for Eco-Environmental Sciences, Academia Sinica, Beijing, China). The detection limits were of the order of 0.1–1 µg/g. The reliability for the determination of 14 elements was determined by 12 replicate measurements. The precision was shown to be 4% for most elements, the accuracy for six elements was about 1.5%, for another seven elements it was about 9%. Apart from light elements, the concentration values for the other elements were lying within the confidence interval of the reference values. After this quality check, the 39 different infusions were analyzed and the influence of origin, type, and quality of the tea samples was studied. A particular result was found for the concentration of selenium in green tea of the Enshi district in Hubei Province, which is one of the two Se-rich regions of China. Values go up to 7.5 µg/g, being much higher than the mean selenium concentration for green Chinese tea, which is only 0.1 µg/g.

It was considered that the Keshan disease leading to serious heart failure is caused by a Se deficiency and can be prevented by the consumption of this kind of tea. The trace element selenium is a constituent of particular proteins and enzymes, which play an important role in metabolic processes. The deficiency disease was first noted and treated in 1966 in the Keshan county (Heilongjiang Province in the northeast of China). In 1975, a study on children in Sichuan Province (central part of China) showed that the symptoms of disease are reduced by supplementation of selenium. We usually receive this trace element in our diet; the daily requirement of selenium is about 50 µg. Foodstuff, such as kidney, liver, fish, seafood, and Brazil nuts with concentrations of some 10 µg/g, can provide high amounts of this essential trace element. This demand can also be met by drinking 1 l of the particular green tea per day.

Total reflection XRF was applied to coffee samples by Haswell and Walmsley in 1991 [71]. Coffee with a mass of 0.3 g was treated with 1 ml of a vanadium-standard solution, and diluted to 25 ml with ultrapure water. Aliquots of 10 µl were applied to a hydrophobic quartz-glass reflector and dried. The preparation took place in a clean bench. Afterward, up to 11 elements were determined by TXRF (Mo excitation, 500 s).

A most simple visualization of the data was made by star plots, which are a valuable tool for simple and rapid screening and classification of items [72]. This technique was applied to different coffee samples of different manufacturers as demonstrated in Figure 5.6. The concentration of 10 elements was

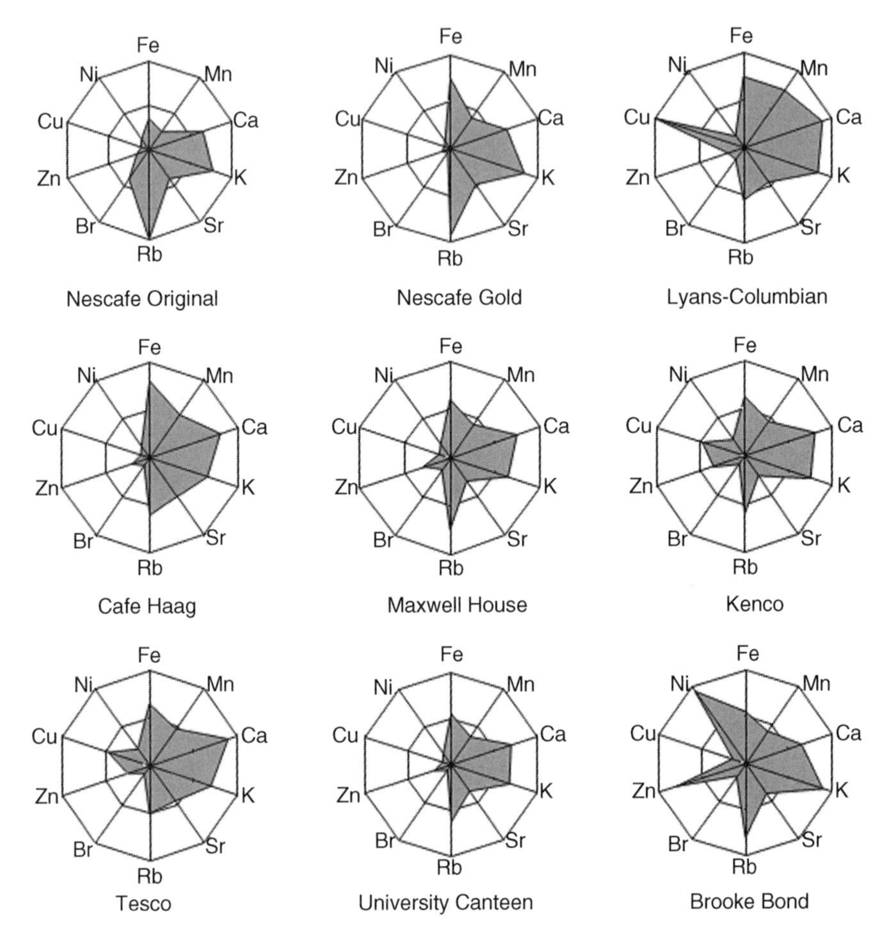

Figure 5.6. Star plots of nine different coffee samples. The element concentration was determined by TXRF for 10 elements (Fe, Mn, Ca, K, Sr, Rb, Br, Zn, Cu, and Ni) and normalized to the maximum (set to 1.0). The coffee from the canteen of the University of Hull, England, can clearly be identified as Maxwell House. Figure from Ref. [72], reproduced with permission. Copyright © 1998, Royal Society of Chemistry.

averaged, normalized to the maximum, and plotted as 10 different radii between 0.0 and 1.0 around a common origin. Each type of coffee shows its own characteristic star shape. A blind coffee sample could be identified clearly. In parallel to coffee samples, Haswell and Walmsley analyzed seven different wine samples (Cabernet Sauvignon) by TXRF in 1991 [72]. Wine was directly pipetted onto the reflector with a volume of 5 μl, followed by 5 μl of the vanadium-standard solution. The use of star plots allowed a simple and rapid classification of wine samples as was the case for coffee samples.

Analysis of wine by ICP and AAS spectrometry as well-established methods has been performed by several authors (e.g., Ref. [73]). The determination of inorganic components has been used for a characterization and classification of the wines and for a contamination control. ICP-MS methods were developed to determine trace elements in wine and compared with TXRF [73]. Thirty-one elements could be determined with detection limits between 0.5 ng/ml and 100 μg/ml. ICP-MS is superior for the detection of light elements like boron and aluminum and also of medium-heavy elements from molybdenum to lutetium ($42 \leq Z \leq 71$) especially the lanthanoids. The detection limits for these elements are one or two orders of magnitude lower, while the detection for the other elements is equally sensitive. In comparison to ICP-MS, however, TXRF analyses are easy and fast and inexpensive.

A further TXRF study dealt with several species of Madeira wine [74]. The organic matrix was removed by digestion with nitric acid. Detection limits between 1 ng/ml (ppb) and 0.1 μg/ml (0.1 ppm) were reported. The uptake of lead could be detected for wine kept in a crystal vessel with 30% lead. Recently, Austrian table wines were analyzed [75]. Eleven different wines of the same type of grape but from different years of vintage and different vineyards in lower Austria were chosen for analyses. Their elemental composition could be used as a fingerprint for their identification. A certain red wine (Blauer Portugieser 2003) matured in a barrique barrel could be distinguished from those wines matured in steel barrels by concentrations of Cr, Fe, and Ni.

Total reflection XRF was also shown to be suitable for the determination of heavy metals in must and wine from two vineyards in the Douro region in Portugal [76]. In order to assess the source of respective elements, the grapes and vine leaves were analyzed by simple EDXRF after freeze-drying and pelleting. Must and wine, however, were directly analyzed by TXRF. Selenium was added as an internal standard and aliquots of 3 μl were pipetted onto clean quartz-glass carriers. The dried residues were analyzed within 300 s. The concentration of 12 elements was determined in the range of 0.1 μg/ml up to 1 mg/ml, and the element pattern of the two different vineyards was shown to be different. In spite of the application of copper-based fungicides in the vineyards, high amounts of copper were only found in leaves, less noticeably in grapes, and not measurable at all in must and wine. All values are below the maximum allowed or regulated values except for the amount of bromine in the wine of the younger vineyard. The value 3.4 μg/ml for bromine was more than three times above the maximum allowed value. It may be explained by some ingredients, which had been added during the production process of that wine.

Finally, it should be mentioned that trace elements were also determined in soft drinks [77]. These water-based and nonalcoholic drinks are possibly carbonated, are flavored and colored, and normally contain sugar, caffeine, and possibly fruits (from 10% upward). Juices, lemonades, and cola drinks are examples. TXRF is well suitable for the analysis of such beverages. A small volume of about 5 μl can be pipetted onto a carrier and dried. Up to 15 elements can be determined simultaneously with detection limits of about 0.1 μg/ml (ppm). TXRF was also applied to the analysis of hard drinks with a high alcoholic content (above 20%), that is, of commercial or homemade spirits [78]. Vanadium was used as internal standard and Fe, Cu, and Zn could be determined as trace metals. The concentration of copper was above the allowed limit for similar beverages.

5.2.2 Vegetable and Essential Oils

The simplest preparation consists in a dilution of oil samples and has been checked for soya oil, peppermint oil, lime oil, and cumin oil by Reus and Prange [2,79]. An aliquot of 1 ml is diluted with toluene (1 : 1), and an internal standard, for example, organo-copper or -vanadium in toluene or an oil-based standard (E. Merck, Darmstadt, Germany), is added to a concentration of 10 µg/ml. Then, 20 µl of this solution is pipetted onto a quartz-glass carrier, which is heated to 100 °C on a hot plate for 5 min to remove the volatile parts of the matrix. The dry residue is analyzed by TXRF within 1000 s. Detection limits are in the range of 3–20 ng/ml.

In order to improve the detection power, a cold plasma ashing is recommended [2] in which 1 ml of an oil sample containing the internal standard is bottled in a small quartz beaker and ashed in a low-temperature oxygen plasma (1 h at 300 W and 500 Pa O_2 pressure). The residue is dissolved with 500 µl of half-concentrated nitric acid, and 50 µl of this solution is applied to TXRF. As compared to the direct method described in the previous paragraph, this method lowers detection limits by about one order of magnitude.

However, ashing requires approximately 2 h, and volatile elements like halogens or Hg, Se, Ti, and S are lost. For these elements, pressure digestion and subsequent plasma ashing are recommended but require about 10 h. For several elements, the detection limits of the aforementioned direct method are fortunately sufficient, in particular with respect to legal regulations for toxic elements.

5.2.3 Plant Materials and Extracts

Plant materials that have been investigated by TXRF are algae, tea plants, corn, hay, lichen, moss, pine, parsley, dill, poplar, beans, rye, and wheat. Analyzed samples stem from blades, bulbs, fine roots, leaves, needles, roots, rosettes, seedlings, seeds, spikes, stalks, tubers, and xylem saps. The usual preparatory technique for plant materials is nitric acid digestion of freeze-dried and pulverized components. If volatile elements have to be determined, a pressure digestion is preferable. Otherwise, an open digestion or even a cold plasma ashing can be carried out. Plant extracts like vegetable oils were analyzed after a simple dilution [67]; aqueous extracts like tea were analyzed directly [70].

The individual components of plants are thoroughly cleaned, shredded, freeze-dried, and finally pulverized in a cleaned porcelain mortar with a pestle [80,81]. Then, 100 mg of the powdered and possibly sieved plant material is doused with 3 ml HNO_3 (65%, subboiled). The mushy mixture is digested for 3 h either in a small beaker at 110 °C [1,2] or in a PTFE bomb at about 200 °C and 10 MPa [67]. After addition of an internal standard, dilution to 5–20 ml, and cooling, there is either a clear or slightly opaque solution. The latter indicates that digestion is not totally complete and that a supplementary ashing may be needed. By the way: Digestion with sulfuric acid is not suitable for TXRF because the droplets pipetted on the carriers would hardly evaporate.

The foregoing method was applied to the reference material NIST 1573 "Tomato Leaves" [1,2]. After open digestion, 20 μl of the final solution was pipetted on cleaned quartz-glass carriers and dried on a hot plate. Measurements were performed with a commercial instrument at standard settings and a counting time of 1000 s. The results for 18 elements demonstrated in Table 5.2 show a satisfactory precision (2–15%) and a sufficient accuracy (2–30%). The deviation of certified and measured mass fractions is significant only for the elements Ni, Zn, Mo, and Pb. These values, however, were determined at the detection limit of some μg/g.

The bioaccumulation of heavy metals in edible plants was studied as a risk for human health [82]. Two kinds of beans were chosen as edible legumes and put in different solutions of lead-nitrate for germination in the laboratory. TXRF was applied for screening of the beans after 12 days. Root, stem, leaves, and crops were collected, dried, and powdered; gallium was added as internal standard. The powdered samples were either suspended in water or digested in acids by microwaves, and suspensions or solutions were analyzed by TXRF. Cd, Pb, and As were determined as hazardous metals. It could be shown that lead affected the growth of the beans negatively; it was accumulated in all parts of the plants depending on the concentration of the nutrient solution (from 0 to 100 ppm).

The concentration of trace elements in tea leaves and their solubility in infusions was determined by several methods of spectral analyses like ICP-OES, ET-AAS, and INAA. TXRF was also applied to tea leaves by Xie *et al.* [70]. They investigated a variety of 39 tea samples from different provinces and districts in China, which is one of the main producers of tea. 100 mg of each sample was ground in an agate mortar, filled into a quartz tube, treated with a mixture of 1 ml suprapure HNO_3 and 0.25 ml suprapure HCl, and digested by plasma ashing in three steps (<260 °C, <45 min, <13 MPa). The digested solutions were cooled down in about 100 min, gallium was added as internal standard (4 μg/ml), and distilled water was added for dilution (1 : 4). Then, 5 μl of the digested solutions was used for TXRF analysis. The minimum concentration in tea leaves was found for selenium (about 0.1 μg/g), the maximum concentration for potassium (20 mg/g). The solubility for iron was only 2%, for selenium it is about 15%, but for K, Ni, and Rb it reached a mean level of greater than 60%.

The extracts of vegetable materials, such as lamb's lettuce and cauliflower, were investigated by Günther and von Bohlen [80,81]. Cellular extracts of these foodstuffs, called cytosols, were investigated after gel-permeation

TABLE 5.2. TXRF Results Determined for Reference Material NIST 1573 "Tomato Leaves" After Open Digestion

Element	Unit	Reference Value		TXRF Result[a] ($n = 3$)		Deviation (%)	Significant Distinction
P	mg/g	3.37	±0.22	3.44	±0.28	+2.1	No
S	mg/g	6.20	±0.40	5.99	±0.39	−3.4	No
K	mg/g	44.4	±2.4	43.8	±2.6	−1.4	No
Ca	mg/g	28.3	±2.3	23.5	±3.0	−17.0	No
Sc	mg/g	0.00017		= 2.00 int. standard		—	—
Ti	μg/g	56.0	±39.0	79.0	±2.0	+41.0	No
V	μg/g	1.20	±0.20	<4.0		—	—
Cr	μg/g	4.00	±0.50	3.10	±0.40	−23.0	No
Mn	μg/g	224	±13	231	±10	+3.1	No
Fe	μg/g	580	±110	600	±80	+3.4	No
Ni	μg/g	1.30	±0.20	3.20	±0.40	+146.0	Yes
Cu	μg/g	11.0	±2.0	12.8	±0.60	+16.4	No
Zn	μg/g	61.0	±4.0	70.1	±3.2	+14.9	Yes
As	μg/g	0.25	±0.04	<2.0		—	—
Se	μg/g	0.054	±0.006	<1.0		—	—
Rb	μg/g	17.3	±2.5	18.0	±0.9	+4.0	No
Sr	μg/g	42.0	±5.0	43.6	±3.3	+3.8	No
Zr	μg/g	–		3.10	±0.90	—	—
Mo	μg/g	0.53	±0.09	1.10	±0.40	+108.0	Yes
Cd	μg/g	2.50	±0.20	3.30	±0.90	+32.0	No
Sb	μg/g	0.036	±0.007	<7.0		—	—
Ba	μg/g	57.0	±9.0	43.0	±11.0	−25.0	No
Ce	μg/g	1.30	±0.20	<10.0		—	—
Pb	μg/g	5.90	±0.80	8.6	±1.60	+46.0	Yes
U	μg/g	0.059	±0.006	<2.0		—	—

[a] Scandium was used as internal standard; Mo and W excitation.

Source: From Ref. [1], reproduced with permission. Copyright © 1993, Springer.

chromatography [81]. About 80 fractions were separated and only 0.5 ml of each fraction was directly analyzed by TXRF. Two or three metallic species of Fe, Cu, and Zn could be distinguished, thereby elucidating the nature of metal-complexing agents.

5.2.4 Unicellular Organisms and Biomolecules

Biomineralization in unicellular organisms, such as bacteria and yeast, was investigated [83]. At first, aqueous nanofluids were prepared containing nanoparticles of maghemite (Fe_2O_3) with a diameter of 5 up to 20 nm, and traces of iron were determined by TXRF. A linear calibration was confirmed for the Fe-Kα peak and iron concentrations between 1 ppm and 1% using vanadium as an internal standard. Complementary studies were carried out for a bacterium and two types of yeast during cultivation in nutrient media of liquid and gelatinous agar. After adding Fe nanoparticles, the biomineralization process was observed in periods of half an hour. It could be shown that the concentration increased from 0.1 to 2 fg iron/cell and that this process was finished within half a day.

In addition to unicellular organisms, biological macromolecules like nucleic acids, proteins, and enzymes can be investigated by TXRF. They are based on carboxylic acids and amino acids, which are basic components of natural proteins. The human body contains 20 different amino acids; the body itself can synthesize 12 of them. The other eight amino acids are constituents of food, for example, meat, fish, and eggs, but also nuts, rice, wheat, and rye, and are called essential amino acids. Compounds of up to 100 amino acids are called peptides; compounds beyond that are called proteins. The molecular weight of proteins varies between 10 kDa and several MDa. Due to intra- and intermolecular hydrogen bonding, proteins can form pleated sheets or helices. In addition, several chains can be arranged in spheres or twisted in long fibers. Enzymes are special proteins that can catalyze biochemical reactions, such as addition, removal, or transfer of atomic groups, hydrolysis, and synthesis.

Another group of macromolecules composed of similar repeating subunits is the nucleic acids, such as RNA (ribonucleic acid) and DNA (deoxyribonucleic acid). The subunits are purine- and pyrimidine-containing compounds. These biomolecules are found in cell nuclei and are carriers of the genetic information. All these macromolecules, such as nucleic acids, proteins, and enzymes, are essential for life. A typical example of a biomolecule with biotechnological interest is shown in Figure 5.7. This complicated metalloenzyme is produced by microorganisms that synthesize hydrogen gas naturally. X-ray diffraction of the hydrogenase crystals reveals that iron-sulfide clusters are assembled stepwise [84].

Total reflection XRF is especially suitable for the detection of metals within such macromolecules. A few examples are given here.

The copper-binding of the prion protein was examined by neuropathologists [85]. Brain fractions and also membranes of mice were digested with

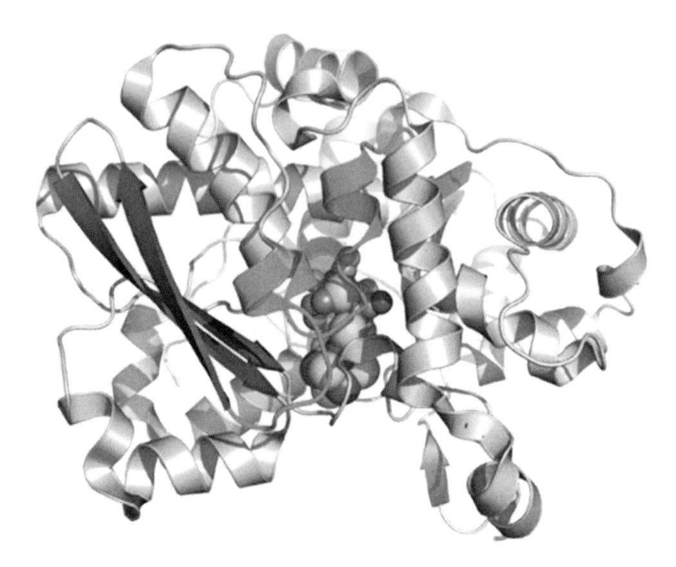

Figure 5.7. Example for the complex structure of enzymes, in this case of $HydA^{\Delta EFG}$ from *Chlamydomonas reinhardtii*. The ribbon diagram with helices and loops is based on X-ray diffractometry at the Stanford Synchrotron Radiation Light Source (SSRL) with a spatial resolution of about 0.2 nm. It shows a [4Fe-4S] cluster of the [FeFe]-hydrogenase with iron (dark gray) and sulfur (light gray). Figure from Ref. [84], reproduced with permission. Copyright © 2010, Nature Publishing Group.

concentrated HNO_3; Ga nitrate was added as an internal standard and traces of Fe, Cu, and Zn were determined by TXRF. It was found that the cellular prion-protein of the brain binds copper, while the prion-deficient protein showed significantly less copper.

A new type of enzymes was characterized by Friedrich and coworkers [86]. Sulfite dehydrogenase was isolated from the bacterium *Paracoccus pantotrophus* with strain GB17. The molecular mass was determined to be 190 kDa. Protein solutions with 12 mg/ml were digested with concentrated HNO_3 (1 : 1) and Se nitrate was added as an internal standard. Aliquots of 10 μl were chosen for TXRF analysis within 400 s and the concentration of Mn, Fe, Cu, Zn, and Mo was determined. A content of 1.3 mol molybdenum and 3.2 mol iron per mol hydrogenase was calculated.

The enzyme murine glutaminyl cyclase (mQC), which is possibly involved in osteoporosis and Alzheimer's disease, was investigated by Schilling *et al.* [87] and by Greaves *et al.* [88]. The yeast *Pichia pastoris* was used for the expression of mQC. After fermentation, the mQC samples were purified and then desalted by size-exclusion chromatography. The protein was concentrated to 3 mg/ml by ultrafiltration. Only 5 μl of the sample solution was deposited on a quartz-glass carrier and dried under IR light. For quantification, 5 μl of a selenium standard solution was added and dried. Finally, TXRF was applied to determine the concentration of zinc—revealing a stoichiometric amount of this metal-ion bound to the protein. Obviously, the depletion of zinc ions has no significant effect on the protein structure of the enzyme. The results proved the hypothesis that the catalytic activity of mQC and probably all mammalian QCs can be influenced or even reactivated by zinc.

The applications of TXRF to the analysis of biomolecules, such as proteins and nucleic acids described earlier, also have a medical aspect. The metal content of organo-metallic compounds had to be determined, which were suspected to be relevant for certain serious diseases. In Section 5.3, traces of essential and toxic elements are determined with reference to important medical, clinical, and pharmaceutical questions.

5.3 MEDICAL, CLINICAL, AND PHARMACEUTICAL APPLICATIONS

Trace elements have an important biological function and an impact on all living beings. A depletion of essential elements, such as Cr, Mn, Fe, Co, Ni, Cu, Zn, Se, or I, will lead to various human deficiency diseases. On the other side, an accumulation of elements can lead to toxic symptoms or even poisoning and is frequently caused by heavy metals like Cd, Hg, or Pb as environmental contaminants. Accordingly, there is an extensive demand for trace analytical information in the medical and clinical field. Trace analyses are carried out on different materials used as monitors for the whole organism or its constituent parts. Suitable monitors are body fluids, such as blood, plasma, serum, and urine, tissue samples from biopsies of vital organs, bone, hair, and nail.

In recent studies, most of these materials have been investigated by TXRF. The method was shown to be especially suitable for trace analyses of whole blood [57,89,90], blood plasma and serum [89–94], and amniotic fluid [95], and for micro analyses of hairs [2], organ tissue [96–98], and dental plaque [99]. Simple techniques of sample preparation have been utilized, for example, microwave digestion with nitric acid (after Ref. [89]) or freeze-cutting by a microtome (after Ref. [97]) as known from histology. Detection limits down to 20 ng/ml were reported for body fluids and down to 100 ng/g for tissue samples. The reliability of the determinations was checked using certified reference materials and found to be satisfactory. In addition to the advantages of a real multielement determination, the investigators emphasized the ease of quantification and the small sample amount required. The last feature is especially important for biopsies and generally in pediatrics.

5.3.1 Blood, Plasma, and Serum

Human blood consists of about 40% cellular components and about 60% liquid plasma, which is yellowish-clear. Cellular components are erythro-, leuco-, and thrombocytes, while the plasma contains 90% water plus proteins, glucose, urea, and coagulation factors. Blood and plasma can be separated by centrifugation. Blood serum results if not only the cellular components are separated from the whole blood but also the coagulation factors are split off.

For all the blood samples—whole blood, blood plasma, and blood serum—microwave digestion with nitric acid is recommended [2,89]. For that, 1 ml of the sample is used and 5 ml of ultrapure nitric acid is added. Digestion is speeded up in a microwave oven for 15 min at 550 W, followed by 20 min at 400 W. After cooling, cobalt or gallium is added as internal standard with a mass of 10 μg to reach a concentration of 10 μg/ml sample volume. An aliquot of 10 μl of the final solution is pipetted onto a clean quartz-glass carrier and dried by evaporation. TXRF analysis is carried out after excitation with a Mo anode and a counting time of 1000 s. Table 5.3 gives results obtained for the certified reference material NIST 909 "Human Serum." The results for the individual elements generally agree with the certified values except for calcium (relative deviation 10%). The detection limits for heavy metals are about 20–60 ng/ml (ppb).

For whole blood samples, digestion is mandatory. Plasma and serum samples, however, may be analyzed without digestion but after a simple dilution with ultrapure water (1 : 3). In this case, detection limits are three times higher but may be sufficient for all elements with concentrations above 0.2 μg/ml (ppm).

The spectra of whole-blood samples are dominated by the peaks of Fe, K, and Ca impeding the detection, for example, Mn, Ni, and Pb. In order to improve the detection limits for these elements, iron, potassium, and calcium

TABLE 5.3. TXRF Results for Different Elements Determined in NIST 909 "Human Serum"

Element	Certified Value (μg/ml)		TXRF Result (μg/ml; $n = 3$)		Deviation[a] (%)	Significant Distinction
P		—	153	±7	—	—
S		—	1111	±49	—	—
K	137.6	±4.3	133.4	±2.8	−3.1	No
Ca	120.8	±3.5	109.1	±1.8	−9.7	Yes
Cr	0.091	±0.006	0.095	±0.014	+4.4	No
Mn		—	0.099	±0.017	—	—
Fe	1.98	±0.27	2.14	±0.021	+8.1	No
Ni		—	0.085	±0.021	—	—
Cu	1.10	±0.10	1.08	±0.02	−1.8	No
Zn		—	1.21	±0.04	—	—
Se		—	0.101	±0.006	—	—
Sr		—	0.051	±0.006	—	—
Pb	0.020	±0.003	0.025	±0.012	+25.0	No

[a] The relative deviation is given with respect to the certified values. A short test was applied to decide whether the two mean values are significantly distinct (Yes) or not (No).

Source: From Ref. [89], reproduced with permission. Copyright © 1989, Springer.

are separated [89]. Initially, iron is selectively extracted from a 6 *N* HCl solution with 1 ml MIBK (methyl isobutyl ketone). Then, the salt matrix with potassium and calcium is separated according to the method developed by Prange *et al.* [7,8] for seawater analyses. One hundred microliters of the final solution is applied for the TXRF measurements. The detection limits of this procedure are in the range of 2–5 ng/ml (ppb) and are sufficient for the determination of Mn, Ni, and Pb in whole-blood samples.

Blood samples of addicts and respective pills based on sea organisms and medicinal herbs were analyzed by TXRF [100]. Blood was sampled from addicts after abuse during 5, 10, and 15 years, and for more than 16 years. After microwave digestion of the blood samples, TXRF measurements were carried out. Significant differences were observed for the blood of those addicts; P, S, and K concentration increased while Cu, Zn, and Se decreased with the period of time.

Blood plasma can also be analyzed directly without digestion as shown by Greaves *et al.* [88]. Blood samples with a volume of 0.5 or 1 or 2 ml were taken from healthy individuals and cancer patients. These samples were centrifuged to separate the blood corpuscles and the plasma. Clear plasma (3–10 μl) was pipetted onto a high-purity quartz-glass carrier, air-dried within 20 min, and analyzed in a live time of 100 to 500 s. X-ray tubes with a Mo or Ag anode were used for excitation at 35 or 50 kV. The Pt-Lα peak at 9.44 keV was used for detection, and the peak intensity was normalized by the Compton peak intensity. Four external standards were prepared by adding appropriate amounts of a standard solution (1 mg Pt/ml) to a sample of human blood plasma (Quimbiotec, Caracas, Venezuela). The standards with a platinum concentration of 0.5, 1, 5, and 10 μg/ml lead to a calibration straight line for quantification. Precision was determined to be about 4%, accuracy to about 6% for concentrations between 1 and 5 μg/ml. The detection limit for platinum is ~100 ng/ml, which is far above the normal concentration of <7 pg/ml in human blood plasma of healthy individuals.

Blood samples were taken from several patients who did undergo chemotherapy with platinum-bearing drugs. Samples were taken immediately before drug administration and 1 h after; a small volume of only 0.1–1 ml was sufficient. The time-dependent course of the Pt concentration was determined within hours and days up to 4 days after the first dose and during multiple infusions. Different doses were applied (50–300 mg) of different platinum containing drugs, for example, cisplatin, oxaliplatin, and carboplatin. The time-dependent profiles covered a range of 0.1–4 μg/ml. Such a quick, easy, and reliable analytical method is useful for pharmacokinetic studies, for example, for the determination of adequate therapeutic and subtoxic levels. It is indispensable for oncologists treating cancer patients by platinum chemotherapy under routine control. The established methods like graphite furnace AAS, ICP-MS, and voltammetry are more sensitive but time-consuming and expensive.

Khuder *et al.* investigated samples of whole blood and human head hair in parallel [57]. Direct TXRF analysis of both types of samples was carried out only for rubidium and strontium. Prior to TXRF a coprecipitation was performed in order to determine the elements Ni, Cu, Zn, and Pb. For this coprecipitation, APDC (ammonium pyrrolidine dithiocarbamate) was used instead of HMDTC as mentioned in Section 5.2.1. This method was applied for whole blood and human hair samples of the population of Damascus city (Syria) with nonoccupational exposure. The mean values were shown to be in the range of reported values for other countries.

5.3.2 Urine, Cerebrospinal, and Amniotic Fluid

The analysis of organic fluids represents a common application of TXRF. Besides whole blood, blood plasma and serum, other body fluids have been investigated, such as urine (e.g., Ref. [101]), cerebrospinal fluid (e.g., Ref. [102]), and amniotic fluid (e.g., Ref. [103]). Obviously, an investigation of perspiration, semen, spittle, and tears by TXRF is also possible.

Urine samples are not ideally suited for TXRF analyses of essential elements because of a high salt content with Na, K, Ca, and Cl. Nevertheless, Greaves *et al.* determined the concentration of platinum in urine and in serum from cancer patients treated by a chemotherapy with platinum drugs [101].

Urine and blood plasma samples from patients having undergone MRI (magnetic resonance imaging) after application of a contrast agent can be monitored by TXRF. Agents, such as Magnegita®, contain a gadolinium-based complex with Gd^{3+} ions, which are paramagnetic. Despite the toxicity of Gd^{3+}, the complex was first regarded as highly secure and stable. However, nephrogenic systemic fibrosis (NSF) has been solely observed for MRI patients with acute renal failure. Consequently, a fast and simple method for monitoring the Gd-based contrast agent in urine and blood plasma was developed by Telgmann *et al.* [104]. Blood was taken from 10 patients without kidney disease shortly after administration of Magnegita® (0.1 ml per 1 kg of body mass) and examination by MRI. The samples were centrifuged to separate the plasma. Urine was collected from two other patients A and B without kidney disease over a period of 15 to 20 h after MRI examination. Aliquots of 500 μl blood plasma and of 500 μl urine spiked with gallium as an internal standard (10 μg/ml) were stirred, 10 μl was pipetted on quartz-glass carriers, and the samples carefully dried. Finally, TXRF was carried out with Mo excitation within 1000 s each.

The detection limit for gadolinium determined by the $L\alpha$ peak at 6.06 keV was found to be 100 ng/ml urine (100 ppb), and 80 ng/ml blood plasma (80 ppb). The precision of triplicate measurements was characterized by an RSD of 1.6% for urine values, and 2.0% for plasma values. TXRF and ICP-MS results correlated well. The accuracy determined by the relative difference between TXRF and ICP-MS was about 5.7% for urine and 4.7% for plasma. The mean concentration was about 100 μg/ml plasma for the 10 patients with a high

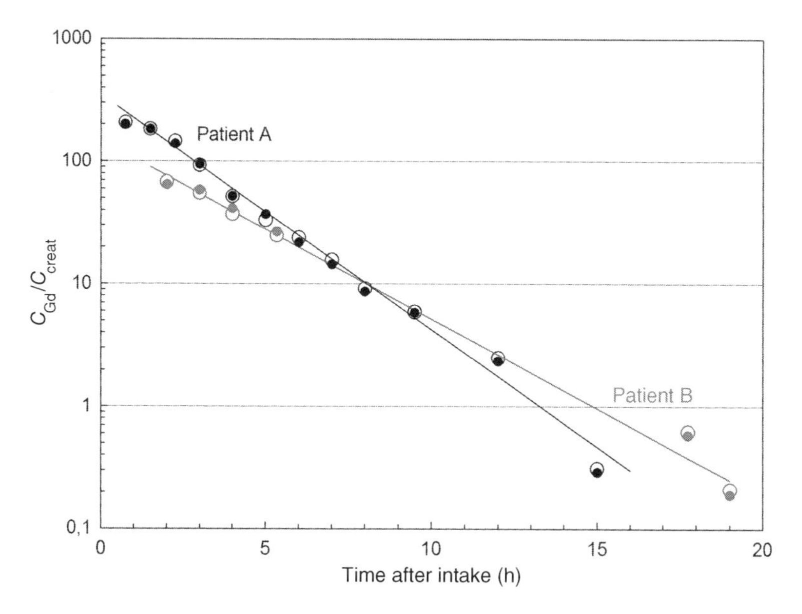

Figure 5.8. Excretion of gadolinium by urine collected from two patients after administration of a gadolinium-based contrast agent. The concentration of gadolinium was determined by TXRF (full circles) and ICP-MS (open circles) and was normalized by the concentration of creatinine determined by HPLC (high-pressure liquid chromatography). It is shown here as a semilogarithmic plot with dependence on the time after administration. The results of both methods hardly differ. The decrease follows an exponential function represented here by straight lines. Figure from Ref. [104], reproduced with permission. Copyright © 2011, Royal Society of Chemistry.

variation of 22%. The concentration of gadolinium was determined between 6500 and 5 μg/ml urine for the two patients A and B within 19 h after administration. The excretion by urine followed an exponential law as expected and demonstrated in Figure 5.8. The time constant for a reduction to $1/e$ or 37% of the initial concentration is of the order of 3 h, and a reduction to 1‰ is reached after about 20 h. Altogether, TXRF was shown to be an attractive alternative for fast and simple Gd determinations in human urine and blood plasma in clinical laboratories during daily routine. TXRF is as precise and accurate as ICP-MS; however, it is easier and faster than ICP-MS because it does not need digestion or dilution of samples.

Cerebrospinal fluid is a clear colorless liquid filling the ventricles of the brain and the spinal cord. It is produced in the choroid plexus of the brain and provides basic protection and neutral buoyancy of the brain. This fluid was sampled from patients with amyotrophic lateral sclerosis (ALS) and a control group [102]. ALS is a motor-neuron disease mainly characterized by muscular and spasmodic atrophy. Altogether, the disease occurs globally but very rarely (5 : 100 000), is incurable so far, and its cause is unknown. Cerebrospinal fluids were taken during routine diagnosis, some μl was deposited on quartz-glass carriers, and concentrated nitric acid was added to the samples. In this way, the organic matrix was decomposed. Scandium and rubidium were used as internal standards and up to eight elements were determined. Lower values for Na, Mg, and Zn and higher values for calcium were found in the ALS group compared to the control group. In parallel to these measurements, serum samples from both groups were taken and investigated with a raised concentration of bromine.

Amniotic fluid is a clear slightly yellowish liquid in the amniotic sac of pregnant females protecting the fetus. In a first approach, the cellular content and suspended particles were separated by centrifugation [103]. In the second approach, samples were treated with nitric acid and subsequently digested by oxygen plasma. After that, the metal composition could be determined quantitatively [105].

5.3.3 Tissue Samples

Organ tissues can be analyzed after ashing and/or digestion. It may, however, be preferable to simply cut a tissue sample in thin sections as known from histology, and directly place these sections on TXRF carriers [96–98]. Besides simplicity, this method offers the advantage of preventing contamination and losses caused by chemical preparation.

5.3.3.1 Freeze-Cutting of Organs by a Microtome

A small piece of tissue with a dimension of 4–8 mm, a volume of less than 0.5 cm^3, and a mass below 500 mg is frozen at a temperature of −15 to −25 °C. The frozen sample is cut by a microtome (Reichert-Jung, Nußloch, Germany) in thin sections, each about 5–15 μm thick. Every section is slid onto the scalpel with the aid of a device called a stretcher. After about 30 s, the section can be placed in the center of a glass carrier by slightly touching it with the surface of a cleaned carrier. Quartz-glass carriers must not be siliconized because hydrophobic carriers lead to tears and droplets. Furthermore, the carriers must have a moderate temperature. If a carrier is too cold the section may be repelled, and if it is too warm it may be attracted too fast and may develop waves. Figure 5.9 shows a microtome and a thin section of mussel tissue on a quartz-glass carrier [98]. After positioning, the carrier is warmed up to room temperature. The section dries within a few minutes and adheres evenly to the surface. Then, a first spectrum can be recorded in order to choose a suitable internal standard element. Ga, Se, or Y are usually absent in the sample so that they can serve as the internal standard. For that purpose, the sections are spiked with a droplet of this standard, for example, 10 μl of a gallium-standard solution with a 1 μg/ml concentration. The droplet is soaked up by the section and the section is thoroughly dried by evaporation on a hot plate or in a small clean oven (about 30 min at 50 °–80 °C depending on the kind of tissue). At the same time, the section shrinks to a thickness of 5-10 μm. After that, a rest time of about 1 h is

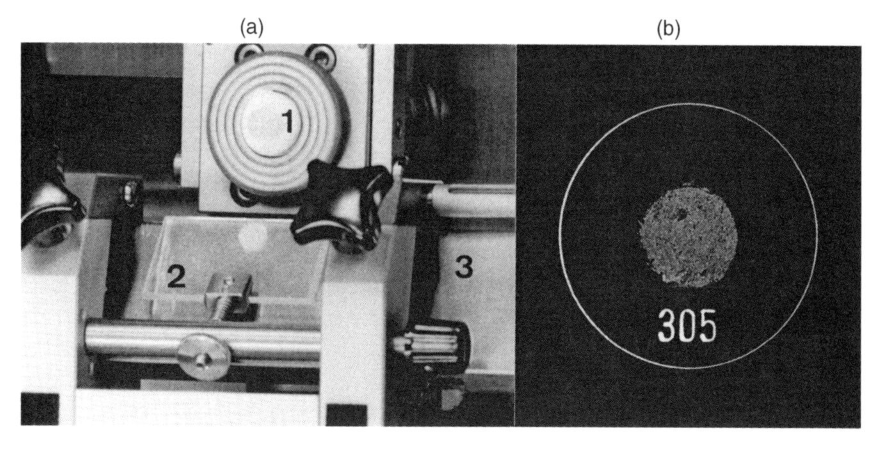

Figure 5.9. (a) Freezing microtome: (1) tissue sample; (2) stretcher with a thin section; (3) scalpel. (b) A thin section of mussel tissue is placed on a quartz-glass carrier. Figure from Ref. [98], reproduced with permission. Copyright © 1989, Elsevier.

absolutely recommended for cooling down to room temperature and conditioning of the section.

Finally, the individual sections are analyzed by TXRF within a counting time of 100–1000 s. Afterward, the dry mass of the section has to be determined. For this purpose, a difference weighing is carried out by means of a microbalance with a precision of 0.1 μg. Every carrier is weighed twice, first *with* the deposited section and then *without* the section, after it was scraped off with a scalpel. The carriers must not be touched by hand, and gloves must be used in order to precisely determine dry masses of only 100 μg or so. The mass fractions are finally determined by Equation 4.12.

The reliability of the method was checked by the analysis of different reference materials [98]: NIST 1573 "Tomato Leaves"; NIST 1566 "Oyster Tissue"; NIST 1577 "Bovine Liver"; and NIES[2] No. 6 "Mussel Tissue". Because these materials are only available as powders, 100 mg was first mixed with several μl of ultrapure water and then this powder mash was freeze-cut and analyzed by TXRF. The detection limits were found to be about 1 μg/g. The relative deviations between certified and measured values were about 10% for elements with a mass fraction >10 μg/g. A possible Fe contamination by the iron scalpel was precluded by control tests.

The freeze-cut technique can be applied to the analysis of human tissues, for example, of kidney, liver, and lung. Within a project on occupational hazards, Wiecken and colleagues studied the dust stress caused by the inhalation of heavy-metal particles [96]. A first group of 12 persons with exposure to normal dust stress was examined. The spectra obtained from their lung tissue differed

[2] NIES stands for the National Institute for Environmental Studies (Tsukuba, Japan).

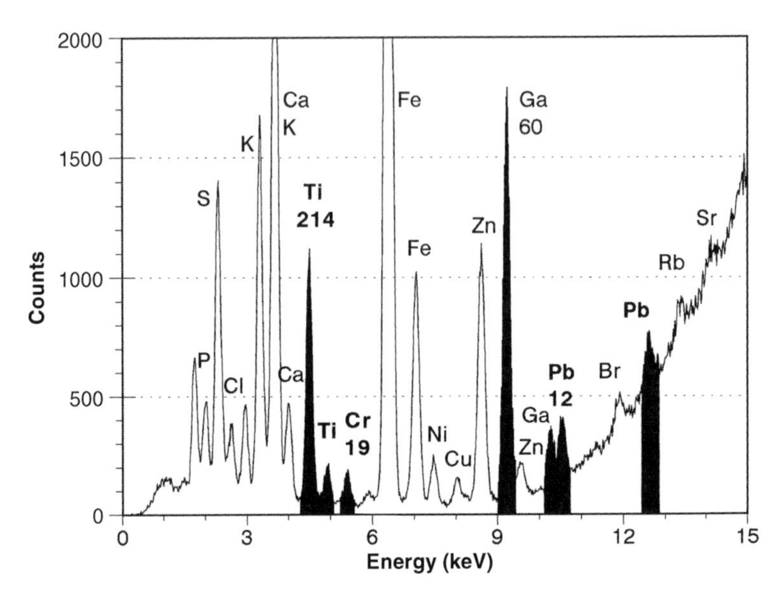

Figure 5.10. TXRF spectrum of lung tissue from a foundryman. Dust deposits show a typical enrichment of Pb, Cr, and Ti with 12, 19, and 214 µg/g, respectively. Gallium was used as internal standard. Figure from Ref. [3], reproduced with permission. Copyright © 1996, John Wiley and Sons.

only slightly and lead to an averaged "normal spectrum." Significant deviations, however, could be observed for the spectra of people who suffered from an increased dust stress due to their occupation. Figure 5.10 shows a spectrum of lung tissue obtained from a foundry worker. The quantitative results show heavy-metal contaminations of Ti, Cr, and Pb in the upper µg/g range, which is obviously caused by an occupational exposure to metalliferrous dusts. The spectra of painters showed titanium and lead, the spectra of steel workers displayed Ti, Cr, Mn, and Ni. For goldsmiths, gold and silver were additionally found in the spectra of lung tissue. Of course, there were many people for whom no significant deviations could be established despite their occupational dust stress.

The foregoing method is not restricted to investigations of organ tissues but can also be applied to the analysis of plant and animal foodstuffs, for example, nuts, mushrooms, shrimps, hen eggs, rice, paprika, peas, cheese, sausages, and in general to any compact or liquid biomaterial. The method can be carried out simply and rapidly and is therefore suitable for screening or monitoring in the clinical, nutritional, and environmental fields.

5.3.3.2 Healthy and Cancerous Tissue Samples

Trace elements in human tissue can be essential or toxic for biochemical processes, which generally depends on the concentration. Many elements are suspected to affect certain diseases. For instance, specific mechanisms of trace elements seem to be responsible for carcinogenesis. Free radicals, especially activated oxygen, can be produced in diversified processes and can induce biologically harmful processes, such as damage to cell membranes [106]. Some elements like copper and zinc can react as cofactors and influence the activity of antioxidant enzymes. A growing tumor may accumulate several elements serving as electrolytes. A selenium deficiency of cancer patients was shown a few times, but this observation is not the only one that was not proved beyond doubt and can also be regarded as a symptom but not as cause of the disease [106].

Breast cancer is the most common kind of cancer among women. It is associated with extreme estrogen levels and with mutation of the DNA leading to tumor formation. Several carcinogens, including some trace elements, induce the formation of reactive oxygen species (ROS) and hereby cause mutation and damage of the DNA. Copper and zinc particularly can assist the function of antioxidant enzymes. Furthermore, the need for increased blood supply of a growing tumor may lead to an accumulation of several elements, such as iron and potassium. Potassium bromate ($KBrO_3$) has been classified as a carcinogen based on experimental evidence. It is a food additive found in flour, bread, fermented beverages, and fish paste. It is also used in cold-wave hair lotion.

In the last decade several studies have focused on the content of minor and trace elements in tumor tissues in comparison with healthy tissues of the same individuals. The samples were usually digested and analyzed by different analytical methods, for example, by ET-AAS, ICP-OES, ICP-MS, EDXRF, synchrotron-radiation-induced X-ray emission (SRIXE), PIXE, and TXRF [107]. Carvalho and coworkers [105,106] studied the behavior of trace elements in cancerous and healthy tissues by TXRF after freeze-cutting with a microtome, while Denkhaus and coworkers performed TXRF analyses of cancerous tissue samples after digestion [108,109]. Unfortunately, the number of investigated individuals is mostly insufficient, and a lot of results are not significant but ambiguous. Furthermore, a special correlation can be a particular result of certain circumstances rather than an unambiguous reason for a disease.

Usually, a multitude of results are determined with different concentration values of several elements for each individual sample simply represented by a star plot (e.g., Ref. [71]). In order to arrange and classify such a variety of variables, multivariate statistics can be applied. A *first* simple and suitable method is cluster analysis representing each sample as a single point in a multidimensional space. Outliers can easily be recognized and the distances of these points can be determined. The aim is to detect particular scatter plots or clusters with neighboring points and to identify only a few clusters with a small variance of the distance within the clusters. The so-called Ward method is suitable to generate such "homogeneous" groups by drawing up dendograms. The *second* more complicated method is the main-component analysis. It produces a few so-called main components by correlating all the variables with each other and building special linear combinations between them. The individual coefficients can be interpreted as weighing factors. Altogether, 10–20 variables per sample may be reduced to two to four clusters or main components. A meaning of components has to be found, which sometimes may be difficult and the results may be ambiguous and doubtful.

As already described earlier, tissue samples can be freeze-cut and analyzed directly by TXRF [97,98]. The method is simple and fast and provides the simultaneous detection of about 10–15 elements within a few minutes. Carvalho and coworkers applied this technique for breast, colon, rectum, stomach, lung, uterus, and prostate tissues of several patients during surgery [74,110]. Healthy and cancerous tumor tissue was taken from the same individual during surgery. After a representative excision of about $0.1\,cm^3$ tissue each, the samples were kept separately in clean formalin at room temperature for several weeks until freeze-cutting.[3] Many sections with a diameter of about 4–6 mm and a thickness of about 5–10 μm were cut by a freezing microtome. Contiguous but nonconsecutive sections were taken, that is, only each sixth section was chosen, and about five sections of each tissue were analyzed by TXRF.

Each section was placed on a clean quartz-glass carrier and dried at room temperature. After recording a first spectrum, the best internal standard was chosen between Ga, Y, and Se and added to the section. After drying at 50 °C for roughly 15 min, the final X-ray spectrum was recorded within 180 s. A Mo tube adjusted to 50 kV and 38 mA was used for excitation, and a Si(Li) detector was applied for spectral registration. After a rest time of 1 h, the mass of each section was determined by a microbalance as mentioned in Section 5.3.3.1 with typical masses of about 100 μg.

A few but significant results may be given here. The concentration values determined by TXRF are roughly between 0.1 μg/g and 3000 μg/g. It was shown that the concentration of several elements and their behavior is tissue dependent. Phosphorus and potassium show increased concentrations in most cancerous tissues, copper and zinc are increased for cancerous breast, lung, and prostate tissue, selenium is decreased in cancerous colon tissue but increased for cancerous breast and lung tissue, bromine is strongly increased for cancerous breast tissue, whereas it is decreased for all other cancerous samples. Copper and zinc show increased levels of concentration not only for the tissue but also for the serum of cancer patients. Consequently, the Cu concentration and moreover the Zn/Cu ratio with a better precision and a higher significance can be a useful screening tool for a preliminary and fast cancer diagnosis as already pointed out by Marco-Parra *et al.* [111]. Three groups of elements with

[3] Formalin is an aqueous solution of formaldehyde (CH_2O) used for the germination and immobilization of tissue samples. For TXRF investigations it should be extremely clean. Nevertheless, there is a risk of introducing and/or washing out certain elements. Such a possible effect should be controlled by TXRF of the formalin bath before and after use.

identical or similar behavior could be built: (1) K and Br; (2) Ca, Fe, and Ni; (3) P and S, Cu and Zn.

Tissue samples of 15 patients with breast cancer were investigated and eight elements detected (P, S, K, Ca, Fe, Ni, Cu, Zn, and Br). Absolute precision and accuracy are strongly improved with higher concentrations; however, the *relative* reliability is nearly constant for concentrations significantly above the detection limits. Consequently, a logarithmic plot for all net intensities was chosen so that a constant relative reliability is demonstrated by a constant scatter. Furthermore, the logarithmic plot reflects the log-normal distribution of elements in the earth's crust. For cluster analysis, three-dimensional plots for the combinations of three elements were demonstrated as two-dimensional figures [110]. Altogether, 83 different combinations are feasible; four of them with specific results are shown in Figure 5.11. Each of them is representing data of nearly 75 healthy and 75 cancerous tissue samples from 15 patients. Two or three clusters can be distinguished depending on the individual element combination. Figure 5.11a and b for Br, Fe, and S, and for Br, Zn, and S show two clusters. Apart from some outliers, they can be separated by the presence or absence of bromine. However, the clusters with high Br values can be split into two subgroups with cancerous and healthy tissue. Figure 5.11c and d for S, P, and Fe and for S, P, and Zn represent two clusters apart from some outliers: one cluster with higher Fe or Zn concentrations for cancerous tissues and another cluster with lower Fe or Zn values for healthy tissues. Further studies on cancerous tissues carried out by XANES are presented in Section 7.3.3.

5.3.4 Medicines and Remedies

Several investigations of TXRF applied to medicines and remedies have been carried out. The method is suitable for the determination of inorganic impurities, especially heavy metals in drugs and also identifying a poisoning due to an improper use of remedies. Two examples are given here.

The quality control of drugs with regard to inorganic impurities was proposed by Wagner *et al.* [112,113]. Arsenic, zinc, and heavy metals are suspected to cause toxic effects. Of course, the toxicity depends on the daily dose and the duration of treatment. The main spectrometric methods for testing drugs are AAS, ICP-OES, and conventional XRF. However, TXRF is a method especially recommended in cases where only a small mass is available. For that purpose, aliquots of only ~10 mg could be dissolved in nitric acid and rubidium added as internal standard (1 mg/l). A droplet less than 50 μl has to be pipetted on a clean quartz-glass disc and dried. Insoluble substances should be decomposed by high-pressure digestion. Several substances were investigated: lecithin used as an emulsifier in pharmaceutical products and also in foodstuffs, human insulin as an essential drug for people suffering from diabetes, procaine used as an anesthetic and neural therapeutic, and tryptophan from barbiturates or antidepressants. Human insulin could be analyzed

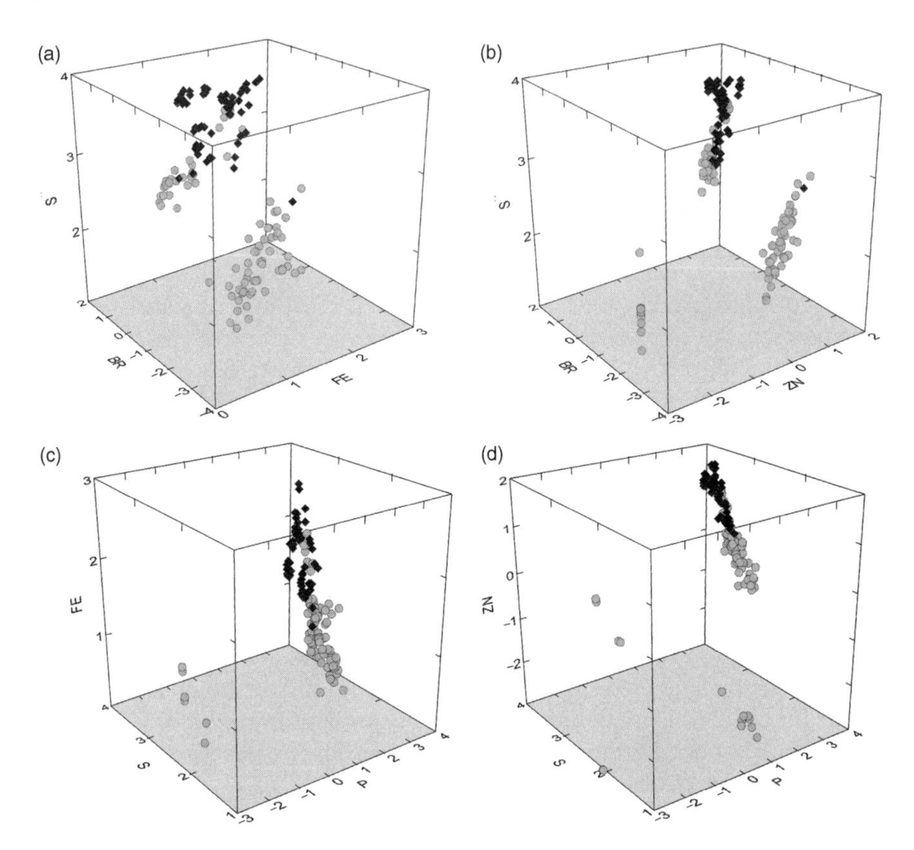

Figure 5.11. View of three-dimensional scatter plots for tumor (black) and healthy (grey) tissue samples. The logarithm of the concentration of three elements is represented: (a) Br, Fe, S; (b) Br, Zn, S; (c) S, P, Fe; and (d) S, P, Zn. Each of the four diagrams contains nearly 150 dots of 15 patients building clusters and possibly subclusters. Figure from Ref. [110], reproduced with permission. Copyright © 2009, Elsevier.

after dissolution, but the other substances had first to be digested. The elements Ti, Mn, Ni, As, and Se could be determined with 0.1–10 µg/g in insulin. Cr, Fe, Ni, Cu, Zn, and Ba could be found with concentrations of some µg/ml in lecithin. Ratios of P, S, K, Fe were determined in procaine. It was shown that TXRF is suitable for a discrimination of different batches and different production or purification processes of pharmaceuticals by a simple fingerprint analysis.

A traditional Ayurvedic remedy was applied to treat urological problems [114]. An Indian patient showed clinical signs of heavy-metal poisoning after a 6 month use of a mineral whitish powder and a daily intake of about 5 g. TXRF analyses were carried out on the medicine and on the patient's hair. For that purpose, 1 cm of hair was cut, deposited on a quartz-glass carrier, and 10 µl of nitric acid (65%) was added. The dried residue was analyzed by TXRF.

The drug powder was ground (<10 μm), mixed with an aqueous solution of Triton X-100, and homogenized. Gallium was added as an internal standard (20 μg/ml). Finally, 10 μl of the slurry was deposited on quartz glass, simply dried in air, and analyzed by TXRF. The method was tested by a quantitative analysis of the standard reference material IAEA-085 "Human hair" and found to be reliable for several elements; lead could not be detected. The content of lead in the powder was about 17 mg/g (29 mg/g zinc and 23 mg/g tin). Consequently, the patient ingested 80 mg lead per day and about 15 g during 6 months.

An intake of 60 μg lead per day during a period of 1 month will cause chronic poisoning with life-threatening symptoms (kidney, lung, bones). However, the patient was in a very poor but not yet critical condition. Of course, the crystalline compound of PbO can hardly be leached in a toxic form in the gastrointestinal tract. In the patient's hair, the element lead was found with a concentration of 17.2 mg/g. One hundred and forty persons in the region of Brescia, Italy were checked, but only 22 persons showed lead in hair with a low average of 1.6 mg/g.

All the examples demonstrate the suitability and versatility of TXRF for a rapid screening of biosamples in the medical, clinical, and pharmaceutical practice. Different samples, such as blood, serum, and other body fluids, but also human tissue, nails, and hair as well as drugs can be analyzed in a rapid and simple manner. Only small or even tiny samples are necessary, offering the advantage of noninvasive and cheap clinical screening and monitoring of patients and the possibility of time-dependent investigations.

5.4 INDUSTRIAL OR CHEMICAL APPLICATIONS

Because of its simplicity, multielement capability, and detection power, TXRF is highly suitable for the analysis of industrial and mainly chemical products. TXRF is the method of choice for quality control of ultrapure reagents needed in the electronics, cosmetics, and pharmaceutical industry. Some high-purity-grade acids, bases, and solvents, including high-purity water, can be analyzed nearly directly down to the level of 0.1 ng/ml [115]. Further applications are concerned with high-purity metals like aluminum [116] or iron [117] and with ultrapure nonmetals like silicon [118,119] or silica [118]. These solid products can be analyzed in the sub-μg/g region after an appropriate matrix separation.

The TXRF analysis of hydrocarbons and their polymers, and of mineral and synthetic oils, is another field of industrial interest, for example, for the petroleum refining industry and for engineering. It is especially useful for the determination of trace metals in light and heavy crude oils [120], lubricating oils [121–123], motor oils [124], and diesel fuels [125,126]. Light oils can be applied to the carrier as drops but should first be diluted with chloroform [121] or toluene [122] and then heated. Such a preparation can be highly convenient if detection limits of 1 μg/ml are sufficient. If, however, lower detection limits of some ng/ml are necessary, a cold plasma-ashing with oxygen may be preferable, as already described in Section 5.2.2.

Finally, the TXRF analysis of alloys for major and minor constituents at the percent level is possible after an appropriate digestion, but it is not suitable for production control. In that respect, conventional wavelength-dispersive XRF is more convenient and faster and has the highest reproducibility. However, TXRF can be used for the control of vaporization and sputter processes by which metal or metal oxide layers are prepared. Contamination can easily be detected [127].

5.4.1 Ultrapure Reagents

Strong acids, bases, or solvents that have a volatile matrix can be analyzed by TXRF nearly directly. Only a few simple preparation steps are necessary, as already shown in Figure 4.16. A volume of 1 ml is spiked with an internal standard, and then a 100 μl volume is pipetted onto a glass carrier in a few steps and evaporated to dryness within 10 min. Even highly concentrated acids like HNO_3, HCl, or HF and bases like NH_3 solutions can be analyzed without dilution. Element impurities can be detected down to or even below 0.1 ng/ml with good accuracy, sufficient for certification [115].

Some useful hints should be noted here: (i) quartz-glass carriers must be used for almost all acids, except that pure silicon carriers are needed for hydrofluoric acid; (ii) commercial stock solutions applied for standardization should first be diluted to about 100 ng/ml; (iii) such solutions should be acidified (pH 2) in order to prevent wall-adsorption effects and to avoid hydrolysis.

In order to further improve the detection power, a simple preconcentration step has been recommended [115]. In this procedure, 5 ml of acid is bottled in a Teflon vessel and heated on a hot plate. At the same time, a gentle nitrogen stream removes the acid steam into a water trap. Evaporation down to 1 ml gives detection limits below 20 pg/ml (ppt) for most elements.

The aforementioned method was applied to some commercial high-purity-grade acids: HNO_3, HCl, and HF. In order to further reduce impurities, an additional purification of acids was carried out afterward by means of a subboiling distillation. The result was checked by TXRF and showed a significant reduction of the impurities. Individual results for the purification of hydrofluoric acid are listed in Table 5.4. The geometric mean of the reduction factors gives a value of five as a rough figure of merit.

Reagents like H_2SO_4 or $(NH_4)F$ are not nearly as volatile and need somewhat more complicated preparation. The matrix with a volume of, for example, 200 μl, can be removed by a low-temperature oxygen plasma, as already proposed for vegetable oils. After about 1 h, the residue can be taken up with some 100 μl of diluted HNO_3 and transferred to a sample carrier for TXRF analysis.

TABLE 5.4. Concentration of Elements in Commercial Suprapure Hydrofluoric Acid Before and After Purification by Subboiling Distillation[a]

Element	Suprapure (ng/ml; $n = 4$)		Subboiled (ng/ml; $n = 6$)		Reduction Factor
P	<3		<2		
S	73.4	±9.9	3.0	±1.00	24
K	2.33	±0.23	1.11	±0.19	2
Ca	3.5	±1.4	2.3	±1.4	1.5
Ti	0.41	±0.11	0.35	±0.10	1.2
V	<0.06		<0.06		
Cr	0.73	±0.02	0.18	±0.08	4
Mn	0.11	±0.01	<0.03		≥4
Fe	9.35	±0.31	0.40	±0.13	23
Co	<0.04		<0.02		2
Ni	0.13	±0.05	0.09	±0.02	1.4
Cu	0.04	±0.01	0.04	±0.01	1
Zn	0.38	±0.03	0.04	±0.02	10
Se	<0.02		<0.02		
As	0.23	±0.04	<0.02		≥11
Sr	0.06	±0.01	<0.03		≥2
Zr	<0.10		—		
Sn	0.3	±0.2	—		
Ba	1.25	±0.10	<0.15		≥8
Pb	4.47	±0.05	0.06	±0.02	75

[a] Rubidium was used as internal standard.
Source: From Ref. [115], reproduced with permission. Copyright © 1991, Elsevier.

5.4.2 High-Purity Silicon and Silica

High-purity Si and SiO_2 are extremely important products for the semiconductor industry. The determination of impurities on the sub-μg/g level by TXRF is made possible by a dissolution of the materials and a separation of Si as silicon fluoride, which is volatile [117,118].

Fragments of about 100 mg silicon can be dissolved with a mixture of 0.4 ml HNO_3 (65%, suprapure, subboiled) and 2 ml HF (40%, p.a. or subboiled) in an open PTFE vessel. The chemical reaction can be described by

$$3Si + 4HNO_3 + 18HF \leftrightharpoons 3H_2SiF_6 + 4NO + 8H_2O \qquad (5.1)$$

For silica, a similar sort of dissolution can be effected by 1 ml HF. The reaction is

$$SiO_2 + 6HF \leftrightharpoons H_2SiF_6 + 2H_2O \qquad (5.2)$$

Dissolution occurs in the presence of an excess of HF and by heating the vessel to about 130 °C for nearly 30 min.

After either reaction 5.1 or 5.2, the hexafluorosilicic acid is decomposed according to

$$H_2SiF_6 \leftrightharpoons SiF_4 + 2HF \qquad (5.3)$$

The volatile fluoride is evaporated by further heating the PTFE vessel at 100 °C for 2 h, thereby reducing the volume of the solution. After evaporation to dryness and cooling, the residue is taken up with 1 ml diluted HNO_3 (10%, suprapure) and spiked with an internal standard (Rb or Ga). A portion of 10–50 µl is transferred to a cleaned but not siliconized quartz-glass carrier, dried, and analyzed by TXRF.

If impurities of TiO_2 or Al_2O_3 are present, the dissolution will be incomplete. But at any rate, it will lead to a usable fine suspension (grain size <1 µm). Table 5.5 shows the results found for the certified reference material BCS-CRM[4] 313/1 "High-Purity Silica" [118]. The TXRF values agree quite well with the reference values. Beyond that, about 20 additional elements could be detected. Detection limits were estimated from the blanks and reached a level of 0.1 µg/g. Only about 1 µg silicon of the original matrix of 100 mg was found to be left. Thus, trace elements were enriched by a factor of about 10^5.

5.4.3 Ultrapure Aluminum

For the technical evaluation of ultrapure aluminum, about 20 trace elements needed to be determined in the range of ng/g to µg/g. To this end, a multistage procedure had to be applied consisting of the digestion of the solid, the adsorption of trace metals as their hexamethylene-dithiocarbamates (HMDTC) on a reversed-phase cellulose, and the elution of the collected trace metals [116]. Detection limits of some 10 ng/g could be reached by subsequent TXRF.

In this method, 2 g of aluminum is first dissolved in 25 ml HCl (30%, suprapure) by heating at 100 °C for 4 h. The resultant solution is diluted with ultrapure water to a volume of 75 ml. The pH value is adjusted to 2.5–3 by addition of 5 ml NaOH in order to prevent hydroxide precipitation. This solution is mixed with 5 mg HMDTC salt (E. Merck, Darmstadt, Germany) and dissolved in 100 µl methanol p.a. The final solution is pumped through a small glass column (inner diameter 5 mm) filled with a suspension of 0.1 g acetylated cellulose in 10 ml ultrapure water. A peristaltic pump is used at a rate of 2 ml/min, so that leaching requires 40 min.

Afterward, the trace metals loaded on the column are slowly eluted by 0.1 ml methanol, by 1 ml nitric acid (25%), and lastly by 0.9 ml ultrapure water. The eluate of 2 ml is spiked with an internal standard, for example, gallium at

[4] BCS stands for British Chemical Standard - Certified Reference Material.

TABLE 5.5. Concentration of Impurities Found in the Certified Reference Material BCS-CRM 313/1 "High-Purity Silica" by TXRF After Matrix Removal with HF

Element	Certified Values (μg/g)		TXRF Values (μg/g); $n=5$)		Deviation (%)	Significant Distinction
Al	190	±21	170	±15	−10.5	No
P		—	7.5	±2.5	—	—
S	24[a]		20	±2	−17	(No)[b]
K	42	±17	49.0	±2.0	+17	No
Ca	43	±7	48.2	±0.8	+12	No
Ti	100	±20	95.3	±3.0	−4.7	No
V		—	0.2	±0.10	—	—
Cr	1[a]		0.7	±0.09	+30	(Yes)[b]
Mn	1	±0.2	1.03	±0.10	+3.0	No
Fe	84	±7	83.4	±2.5	−0.7	No
Ni		—	0.43	±0.06	—	—
Cu		—	0.58	±0.04	—	—
Zn		—	0.34	±0.04	—	—
Ga		—	0.02	±0.01	—	—
As		—	0.15	±0.02	—	—
Rb		—	0.18	±0.02	—	—
Sr		—	1.32	±0.07	—	—
Y		—	0.61	±0.03	—	—
Zr	15[a]		16.0	±1.5	+6.7	(No)[b]
Nb		—	0.31	±0.07	—	—
Mo		—	0.07	±0.03	—	—
Sn		—	0.10	±0.05	—	—
Ba		—	7.9	±1.0	—	—
Ce		—	1.6	±0.40	—	—
Hf		—	0.44	±0.04	—	—
W		—	0.08	±0.04	—	—
Pb		—	0.31	±0.04	—	—
Th		—	0.26	±0.06	—	—

[a]Value not certified, but only recommended.
[b]Parentheses indicate the test for significant distinction described for Table 5.1 was somewhat changed. Both values are called significantly distinct or not if the recommended value is inside the region of confidence ($x \pm 2s$) of the TXRF value or outside, respectively.
Source: From Ref. [118], reproduced with permission. Copyright © 1989, Elsevier.

200 ng/ml, and aliquots of 20 μl are used for TXRF analysis within a counting time of 200 s.

Precision and accuracy of the foregoing procedure were examined by means of an industrial reference material. Parallel to TXRF, the trace concentrate was analyzed by Burba *et al.* [116] using FAAS and applying the injection technique. Figure 5.12 shows the results to be in a good agreement [128]. Detection

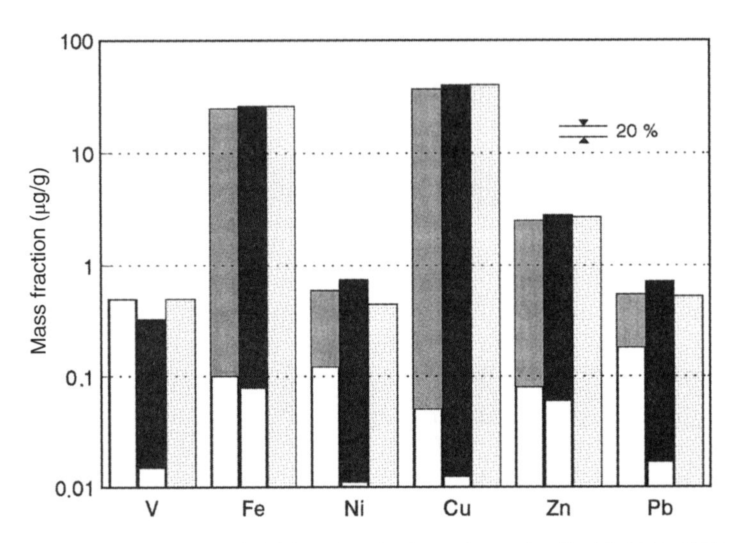

Figure 5.12. Trace metal contents in high-purity aluminum determined by TXRF (solid black) and FAAS (medium gray) in comparison with certified values (light gray). Detection limits are indicated by a white area. The reference material VAW R 14 [128] was chosen for analysis (VAW = Vereinigte Aluminium Werke, Bonn, Germany). Figure from Ref. [116], reproduced with permission. Copyright © 1989, Elsevier.

limits mainly determined by the blanks of the total procedure are on the order of 10–20 ng/g. The remaining mass of aluminum is about 80 µg, so that an enrichment factor of 2.5×10^4 was attained.

5.4.4 High-Purity Ceramic Powders

A ceramic is an inorganic solid material most commonly with a crystalline structure. Ceramics older than 25 000 years were made of clay and hardened in fire. They were produced all over the world and have a glazed and smooth surface, possibly with colored ornaments. Ceramics were used as floor, wall, and roof tiles, bricks and pipes, linings of kilns, crucibles, tableware, cookware, and sanitary ware. Earthenware, stoneware, and porcelain mostly made of kaolin ($Al_2O_3{\cdot}2SiO_2{\cdot}2H_2O$) are known as whiteware ceramics. Today, ceramics are used for special technical devices, such as burner nozzles, nuclear pellets, coatings of jet and rocket engines, tiles of spacecraft, and cones of missiles, knife blades, brake discs, medical implants, and dentures.

Technical ceramics can be divided into three groups. The first group includes oxides, such as alumina (Al_2O_3), beryllia (BeO), ceria (CeO_2), and zirconia (ZrO_2). These materials are usually white or pale, yellow-white powders, odorless, and insoluble in water. Because of their high melting point (between 2072 and 2715 °C) and their high electrical resistance, most of these ceramics are used as refractory materials and electrical insulators. Alumina is the ceramic oxide with the highest economic importance. It is used as rubies or yttrium-alumina-garnets for lasers, and used as heat-resistant combustion chambers in high-pressure sodium lamps.

The second group of technical ceramics consists of a great variety of nonoxides, such as carbides (Be_2C, Al_4C_3, B_4C, WC, and SiC also called carborundum), nitrides (e.g., BN, TiN, Si_3N_4), silicides (e.g., $MoSi_2$, $NbSi_2$, WSi_2), and borides (e.g., SiB_3, LaB_6). Some of them are applied as thin metal coatings.

The third group of ceramics includes composite materials, including combinations of oxides and nonoxides possibly reinforced by particulates or fibers. Most of these products are available as high purity powders with impurities of less than 0.1%.

The first application of TXRF on the analysis of high-purity ceramic powders was reported by Graule *et al.* in 1989 [129]. Several powders of Al_2O_3, AlN, Si_3N_4, and SiC were analyzed by ICP-OES with slurry atomization using a special nebulizer. TXRF was proposed and introduced as a competitive method and was shown to have comparable capabilities for routine analysis and production control. More than 15 years later, TXRF was used for the determination of nine elements in Al_2O_3 ceramic powders [130]. Slurries containing 10 mg of powder per ml of pure water were prepared with cobalt as an internal standard (200 µg/g). Aliquots of 5 µl were deposited on quartz carriers, evaporated, and excited by synchrotron radiation. Systematic errors are reduced for fine powders with a grain size of less than 10 µm and reliable results were obtained. The detection limits for Ca, V, Cr, Mn, Fe, Ni, Cu, Zn, and Ga were found to be in the range of 0.3–7 µg/g (ppm).

In another paper, boron nitride powders were studied by TXRF [131]. Slurries of BN with a good stability were obtained by the addition of PEI (polyethylenimine) at a concentration of 1 mg/g and a pH value of 4. This time, 5 mg ceramic powder was suspended in 1 ml pure water and gallium was added as an internal standard. Six elements (Ca, Ti, Cr, Mn, Fe, and Cu) could be determined in four different ceramic powders. Detection limits for the elements calcium and copper determined by TXRF were found to be 0.06–1.6 µg/g, and herewith are higher than those obtained for GF-AAS with slurry sampling and for ICP-OES with electrothermal evaporation. However, detection limits of TXRF are lower than those of GF-AAS and ICP-OES after a previous chemical digestion of the boron nitride powders.

In addition to the mentioned ceramic powders, TXRF was applied to the analysis of the certified reference materials SiC and Al_2O_3. Three elements Ca, Ti, and Fe, were determined by calibration straight lines in a concentration range of 2.5 µg/g to 1.5 mg/g and correlation coefficients greater than 0.991 were achieved. The advantage of a simple calibration by an internal standard valid for all matrices was emphasized.

A similar study was carried out for boron carbide powders after stabilization [132]. Eight elements (Ca, Ti, Cr, Mn, Fe, Co, Ni, and Cu) could be determined in seven different ceramic powders. Again, gallium served as the internal standard. Detection limits ranging from 0.1 to 1.2 µg/g (ppm) were found to be comparable to those of ICP-OES after wet-chemical digestion of the ceramics. The average relative standard deviation for all elements was below 10%, the correlation coefficient was 0.997—similar to ICP-OES values.

5.4.5 Impurities in Nuclear Materials

Nuclear materials used for power production in a reactor have to be analyzed with the special condition of small sample size. The analysis has to be consumption-free so that a loss of radioactive and precious materials is avoided. Consequently, XRF and especially TXRF is the method of choice. TXRF is simple and fast, needs only a minute sample amount, and is accurate even if it needs only one single internal standard for quantification. Apparently, there is no alternative.

Traces of several elements at ppb levels in UO_2, ThO_2, and zircalloy as well as chlorine traces in U_3O_8, and PuO_2 and In traces in heavy water were determined by Misra and coworkers [133,134]. The determination of chlorine traces in nuclear fuel materials is very important because chlorine is extremely corrosive. It can be incorporated into nuclear materials during the various stages of fabrication and processing. Even in low concentrations, chlorine can lead to a depassivation of the oxide layer and corrosion of cladding materials. Consequently, the specification limits for chlorine in nuclear materials are very low. Depending on the different materials, these limits are between 5 and 50 μg/ml (ppm).

Chlorine was determined by TXRF after its separation from the solid matrices by pyrohydrolysis [133]. For that purpose, 500 mg of a solid sample was put into quartz glass boats, and heated at 900 °C. Volatile impurities, such as B, F, S, and Cl, were set free, and led into a cooled condenser by a stream of Ar/O_2 and steam. Chlorine was collected as HCl in an aqueous solution of 5 mM NaOH. Afterward, cobalt was added as an internal standard to the distillate and aliquots of 30 μl were analyzed by TXRF with a W tube used for excitation.

Several materials were investigated, such as U_3O_8, (U,Pu)C, PuO_2, and Pu alloys. Concentrations down to about 0.1 μg/ml (ppm) were determined. The precision could be improved from about 27 to 8% by a continuous helium flow of 800 cm^3/min instead of atmospheric air. The results were compared with those obtained from ion chromatography and found to deviate by about 14% in the range of 1 to 70 μg/ml (ppm).

Total reflection XRF was shown to be simple, fast, and accurate. It requires only a small sample volume of some 10 μl and consequently produces less analytical liquid waste. A complete and cumbersome dissolution of the nuclear material is avoided by pyrohydrolysis at 900 °C. Matrix matched standards are not required for quantification but only one single internal standard is needed.

5.4.6 Hydrocarbons and Their Polymers

In addition to inorganic materials, aliphatic, cyclic, and aromatic compounds of the organic chemistry can also be analyzed and distinguished by TXRF. About

9 million such carbon compounds are known and they play an important role in biology, medicine, pharmaceutics, and engineering.

The large group of hydrocarbons consists of alkanes (C_nH_{2n+2}), alkenes (C_nH_{2n}), and alkynes (C_nH_{2n-2}). Paraffins, benzene, ethers, and esters are representatives of this group, mainly produced by the petrochemical industry. The quality of such products can be affected by elemental impurities and can lead to malfunctions (e.g., corrosion). For quality control, a fast and simple method like TXRF is desirable [135]. Aliquots of different liquid petrochemical products were deposited on quartz-glass carriers. The organic matrix was simply evaporated and organic gallium was added as an internal standard solution. Fifteen elements could be determined by TXRF. Detection limits were lower than 50 ng/g and in most cases below 5 ng/g. The different hydrocarbons showed typical metal contents of less than 1 μg/g. The advantages of TXRF were elucidated with respect to the reference method, ICP-MS. TXRF was found to be best suited for monitoring the quality of such petrochemical products.

Polymers are macromolecular substances with a high molar mass (1000–800 000 g/mol) and high viscosity. They have great technological uses in industrial manufacturing, for biomedical and pharmaceutical applications, and play an important role in areas, such as enhanced oil recovery, drug controlled release, and biological corrosion protection. A well-known polymer is polyethylene oxide (PEO), which is a synthetic polyether containing a linear chain of $-(CH_2-O-CH_2)_n-$. A natural substance similar to this polymer is scleroglucan excreted by a specific fungus.

PEO can have a molar mass greater than 100 000 g/mol. It differs from polyethylene glycol (PEG), which is a polyalcohol with the chain $-(OH-CH_2-CH_2-OH)_n-$ and a molar mass less than 100 000 g/mol. Depending on the molar mass, the polymers vary from liquid to solid phase; light PEGs with $M < 1000$ are colorless, viscous fluids while PEGs with $M > 10\,000$ g/mol are white, waxy solids. These polymers are inexpensive, readily available, and approved to be nontoxic. For that reason, they build the basis of plastics, excipients, ointments, creams, drugs, roughage, toothpastes, suppositories, and gels. They are used for chemical, biological, medical, pharmaceutical, and cosmetic applications. PEO is water-soluble, which is noteworthy for organic polymers and can produce aqueous polymeric solutions and gels with a concentration less than 1%.

The high viscosity of polymer matrices prevents the application of ICP methods to spectral analysis of such samples. Traditional radiotracer methods need safety measures and security checks, whereas TXRF is a safe option applicable for micro amounts of a sample. It was applied to the characterization of trace elements in highly viscous polymers. A paper presenting the capabilities of TXRF for polymer analysis was given as a review by Vázquez [136]. Particularly, the diffusion of metal ions in aqueous polymeric solutions, such as PEO and scleroglucan, was studied [137]. For that purpose, several capillaries (0.5 mm diameter, 50 mm length, sealed at the bottom) were filled with a polymeric solution containing Mn or Ba nitrate with a concentration of 50 μg/l

(50 ppb). A set of 10 capillaries was completely immersed in a water bath of 25 °C placed on a stirrer. Every 30 min, one of the capillaries was removed from the bath and its content of 9.8 μl was placed onto a clean carrier and dried in a microwave oven. The dried residues were analyzed by TXRF with a Mo X-ray tube and a lifetime of 500 s. The Kα peak of manganese and Lα peak of barium were used for analysis and the Compton peak of molybdenum served as an internal standard.

Because of the concentration gradient between solutions and water, the nitrate salts diffused out of the upper opening of the capillaries while the viscous polymeric macromolecules remained inside. The manganese and barium concentration was determined for 10 periods in steps of 0.5 h and for 20 different polymer solutions between 0.01 and 1%. The diffusion coefficients were determined in accord with Fick's law. The authors could show that these diffusion coefficients decrease in the diluted regime $<0.2\%$ as expected. However, they surprisingly increase in the semidiluted regime of 0.2% up to 0.7% and again decrease for higher concentrations in the "gel regime." This behavior could be explained by structural characteristics of the system.

Marking of polymer macromolecules with heavy atoms was proposed in a further study [138]. Iodine was used for labeling polysaccharides, such as scleroglucan, guar gum, galactomannan, glucomannan, gelatine, and starch. The OH groups of these hydrogels were substituted by iodine atoms in a chemical reaction guided by a variable temperature, time, concentration, and pH value. TXRF with synchrotron excitation was used for quantitative analysis of the labeled macromolecules.

5.4.7 Contamination-Free Wafer Surfaces

Wafers are the main product of the semiconductor industry, which started in the 1960s in the Silicon Valley of California and spread over the whole world. Today large manufacturers have worldwide facilities and top manufacturers, such as IBM, Intel, Samsung, Toshiba, Infineon, Taiwan Semiconductor Manufacturing, STMicroelectronics, Texas Instruments, and GlobalFoundries share the billion dollar market. They mainly produce integrated circuits (ICs), which are basic components of almost all electrical and moreover electronic devices today. Above all, silicon is the source material, which is grown in large monocrystalline ingots with a diameter of several inches (8 inches ≈ 200 mm; 12 inches ≈ 300 mm, or 18 inches ≈ 450 mm). These extremely pure ingots are sliced into wafers less than 0.75 mm thick by means of long, thin steel wires coated with abrasive slurry. Several wafers are sliced by a saw with several wires simultaneously. Afterward, the front sides of the raw wafers are polished to a roughness of only 0.1 nm.

The subsequent fabrication process of highly sophisticated ICs covers a number of different complex steps of deposition, removal, patterning of the deposited material by lithography, modification by doping, implantation, or thermal treatment, and finally packaging. Recleaning steps may be added in between the other steps. Within the last 30 years, the fine pattern or structure of a wafer chip has been miniaturized from about 1 μm to about 20 nm (a human hair is nearly 70 μm thick). The "nodes" of an older IC have a 500 nm distance down to 130 nm. Most of today's wafers are processed with a node of 90, 65, 45, 32, 22, down to 14 nm. According to the International Technology Roadmap for Semiconductors (ITRS), the nanotechnological process shall aim at nodes of 11 nm in the year 2015. The total fabrication process for a wafer can take several weeks and is performed in cleanrooms of highly specialized plants referred to as fabs.

All of these steps are susceptible to infiltration of contaminants as explained by Zaitz [139]. They may stem from the raw material itself, slurries of slicing, polishing and etching materials, gases, tools, masks or walls of an implanter, a furnace, a vaporizer, from holders, and even from cleaning agents or ambient air. Large particles can be accumulated in so-called nooks and crannies. Such contaminations can be the cause of malfunctions of the finished ICs, especially in VLSI (very-large-scale integration) technology and even more so in the ULSI (ultra-large-scale integration) technology. Alkali elements like sodium or potassium can reduce the requisite threshold voltage, transition elements like Cr, Fe, Co, Ni, Cu, and Zn may induce a leakage current, short circuits or phase-to-phase faults, and a failure of the whole device. The actinides uranium and thorium can cause a malfunction by α emission. To avoid such failure modes, wafers must have extremely low impurities and extremely clean surfaces. Current guidelines or "roadmaps" of the semiconductor industry desire an area density of less than 2.5×10^9 atoms/cm^2 for nearly 10 different elements (K, Ca, Cr, Mn, Fe, Co, Ni, Cu, Zn), which is equivalent to 4.3×10^{-14} g of copper per cm^2 (43 fg/cm^2).

For that reason, contamination control is indispensable [139]. TXRF and ICP-MS are highly suitable for direct examination. If both methods are coupled with VPD (shown later) of 300 mm wafers, they are capable of detection of 10^7 atoms/cm^2, which is highly sufficient for this task. A new development of ICP-MS with a ShieldTorch System and Micro Flow nebulizer even reduces detection limits by a factor of 25 down to about 3.5×10^5 atoms/cm^2 for 450 mm wafers.

Several commercial TXRF instruments have been developed especially for wafer applications, and today about 300 are estimated to be utilized in the semiconductor industry worldwide, especially in the United States and in Japan. Numerous reports have appeared and overviews on the different applications have been given by Hockett [140,141]. Most studies have been concerned with contamination of transition metals on silicon wafers [140–152], but germanium wafers [153] and gallium-arsenide wafers [154] have also been evaluated by TXRF. Monitoring contamination of transition metals could be improved by monochromatic synchrotron excitation [155]. Compound semiconductors, such as InP and $In_xGa_{1-x}As$, are essential components of photonic devices, such as solid-state lasers and photon detectors, and can also be searched for contaminants.

The detection of low-level metallic contamination on wafers may lead to the sources of contamination brought in by the different steps. The aim is to reduce the contaminants to a minimum, to remove the remainder by effective cleaning steps, and finally to improve the manufacturing process. TXRF can be applied to monitor the entire process automatically and to serve as a means of production control. Two modes of operation can be distinguished. The first is direct TXRF, which can be applied quickly and nondestructively to determine trace-metal contamination above 10^9 atoms/cm^2 and even to make a complete map of the total wafer surface. For each point of the surface, only a single measurement at one fixed angle position is necessary. It is even possible to distinguish between particulate and thin-layer-type contaminations. For this purpose, however, at least two measurements at two distinct angle positions are required.

The second mode of operation involves contamination down to the level of 10^7 atoms/cm^2. They can be determined if they are first collected from the entire surface of a wafer and then concentrated on a small spot. This technique is based on etching of the wafer surface—and consequently requires that the usual goal of nondestructiveness be abandoned in this case. This mode of operation is called vapor-phase decomposition (VPD) TXRF.

5.4.7.1 Wafers Controlled by Direct TXRF

The so-called direct TXRF is the commonly used nondestructive method applied to a quick contamination control of wafers. Usually, five to nine different spots of a wafer with a diameter of about 0.8 cm are analyzed directly, that is, without any preconcentration. A triple beam excitation from a twin X-ray source is advisable in order to detect all elements from sodium to uranium simultaneously. The measuring time is chosen to be 50 to 500 s per spot. Detection limits for the critical transition metals, such as Cr, Mn, Fe, or Ni are on the level of 10^9 atoms per cm^2, which is equivalent to ~0.1 pg.

A device is needed for wafer positioning, especially for shifting and fine-angle adjustment, as already demonstrated in Figure 4.27. A W tube is preferably chosen and the W-Lβ instead of the Lα peak is selected by a multilayer monochromator in order to excite the transition metals, including zinc, and to avoid the excitation of germanium in case of Ge wafers or gallium and arsenic in case of GaAs wafers. A single measurement is needed for the determination of contaminants at each spot of the surface chosen for the control. For this purpose, a fixed glancing angle has to be adjusted that should amount to about 70% of the critical angle of the wafer material according to Equation 4.28. The angle adjustment can be controlled by the fluorescence intensity of the wafer material itself (e.g., silicon), as described in Section 4.6.1. Calibration can be carried out by an external standard (e.g., a nickel-plated wafer), as described in Section 4.5.1.

Figure 5.13 shows the results of a wafer mapping for different elements [149]. The complete mapping with about 225 spots needs about 12 h measuring time.

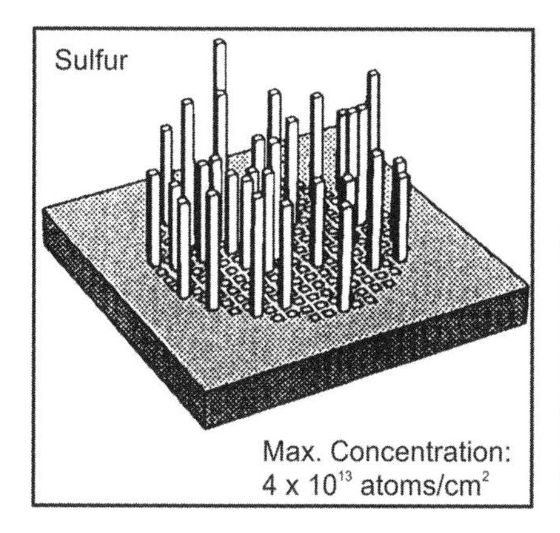

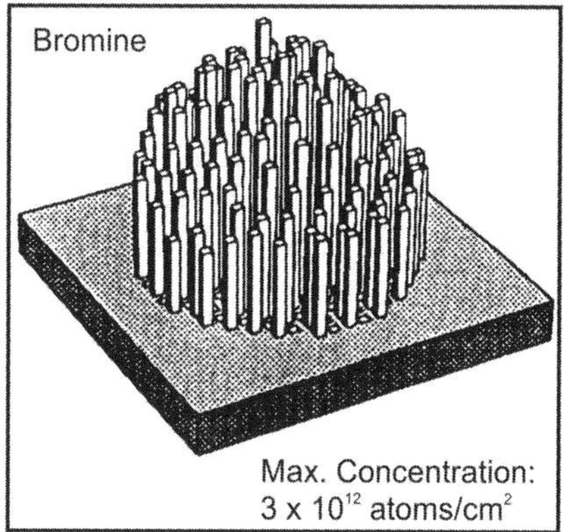

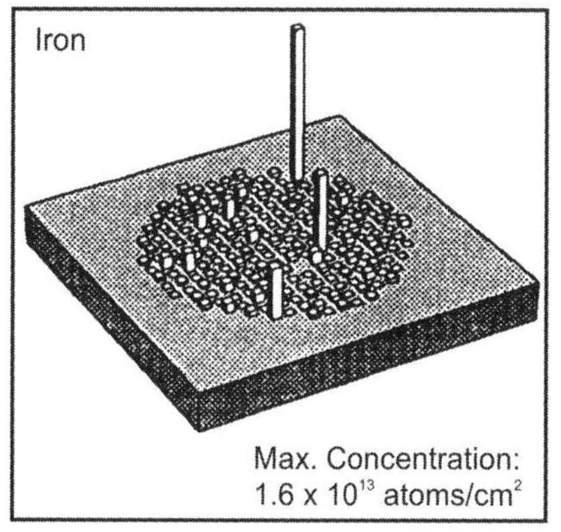

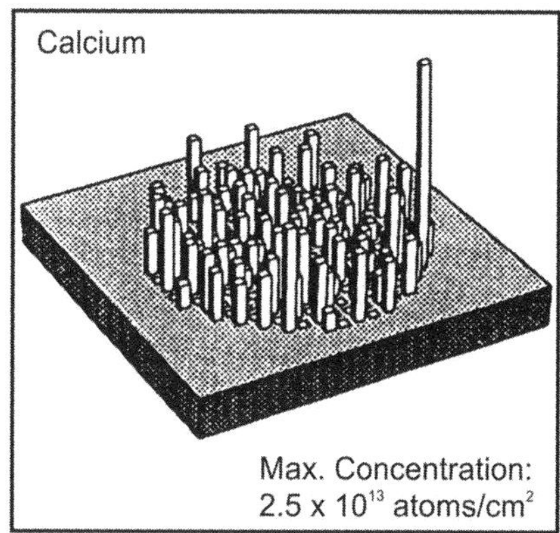

Figure 5.13. TXRF results of a wafer mapping for the elements S, Br, Fe, and Ca. About 225 different spots with an area of 0.5 cm^2 were investigated on an 8 inch wafer. Figure from Ref. [149], reproduced with permission. Copyright © 1993, Elsevier.

The lateral resolution is determined by the observation area of the detector, which is on the order of 0.5 cm^2. Contaminants are in the range of 10^{11} to some 10^{13} atoms/cm^2. The nonuniform distribution differs for the individual elements. Such results can give a first hint of existing contaminations. The information they provide may give valuable clues as to the source of the contamination and can help reduce or avoid them.

Sweeping TXRF is a new approach to directly check a whole wafer for contamination within a reasonable period of time [156]. To this end, the wafer is moved step-by-step automatically in a helical way and a mapping is carried out. Figure 5.14 shows a map of four transition metals represented by 221 colored spots on the wafer. The level of concentration equivalent to the detection limits is 10^{10} atoms/cm^2. Only about 20 min was needed for the control of this 12 inch silicon wafer. By the addition of all spectra, a mean value and a standard deviation can also be evaluated for each contaminant. This technique can also

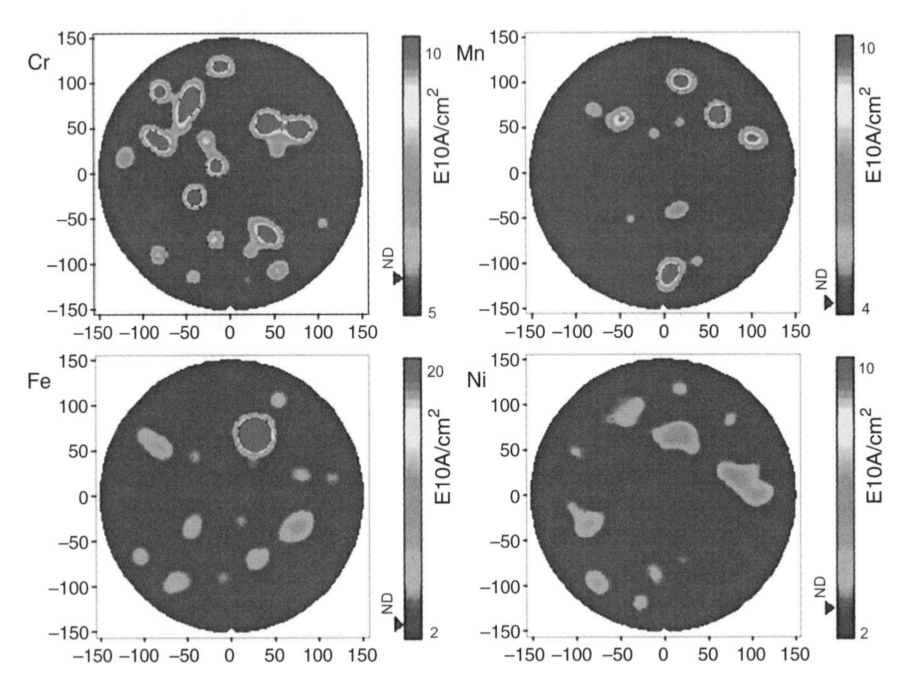

Figure 5.14. Area map of four elements Cr, Mn, Fe, and Ni on a 300 mm Si wafer. The wafer was moved step-by-step in a spiral fashion by means of a fast-positioning stage. Altogether, 221 colored spots are shown representing an area with a diameter of 0.8 cm each and concentrations between 2×10^{10} and 20×10^{10} atoms/cm^2. The measuring time was set to 5 s per spot so that only 20 min was needed per wafer. Figure from M.A. Zaitz of IBM, NY, reprinted with permission from the author. (See colour plate section)

be used for the backside of wafers, for patterned wafers, and thin films on wafers (M.A. Zaitz, personal communication). As far as is known, no other method is capable of a visualized localization of such low contamination levels.

Furthermore, a distinction between two kinds of contamination—particulate and thin-layer type—is possible. As described in Section 4.5.1.4, two angle measurements are sufficient, although a recording of a total intensity profile with 20–30 different positions is preferable. Concrete examples of this approach were first given by Prange and Schwenke [157] and Schwenke and Knoth [158]. As demonstrated in Figure 5.15, the elements zinc and bromine are detected as contaminations of 6.2×10^{11} and 2.7×10^{11} atoms/cm^2, respectively. Zinc is mainly deposited in particulates (85%), whereas bromine is deposited nearly exclusively in a thin layer (97%). The analytical error is about 5%.

5.4.7.2 Contaminations Determined by VPD-TXRF

The detection limits for critical transition metals are on the order of 10^9 atoms/cm^2 if determined by direct nondestructive TXRF. They can be improved

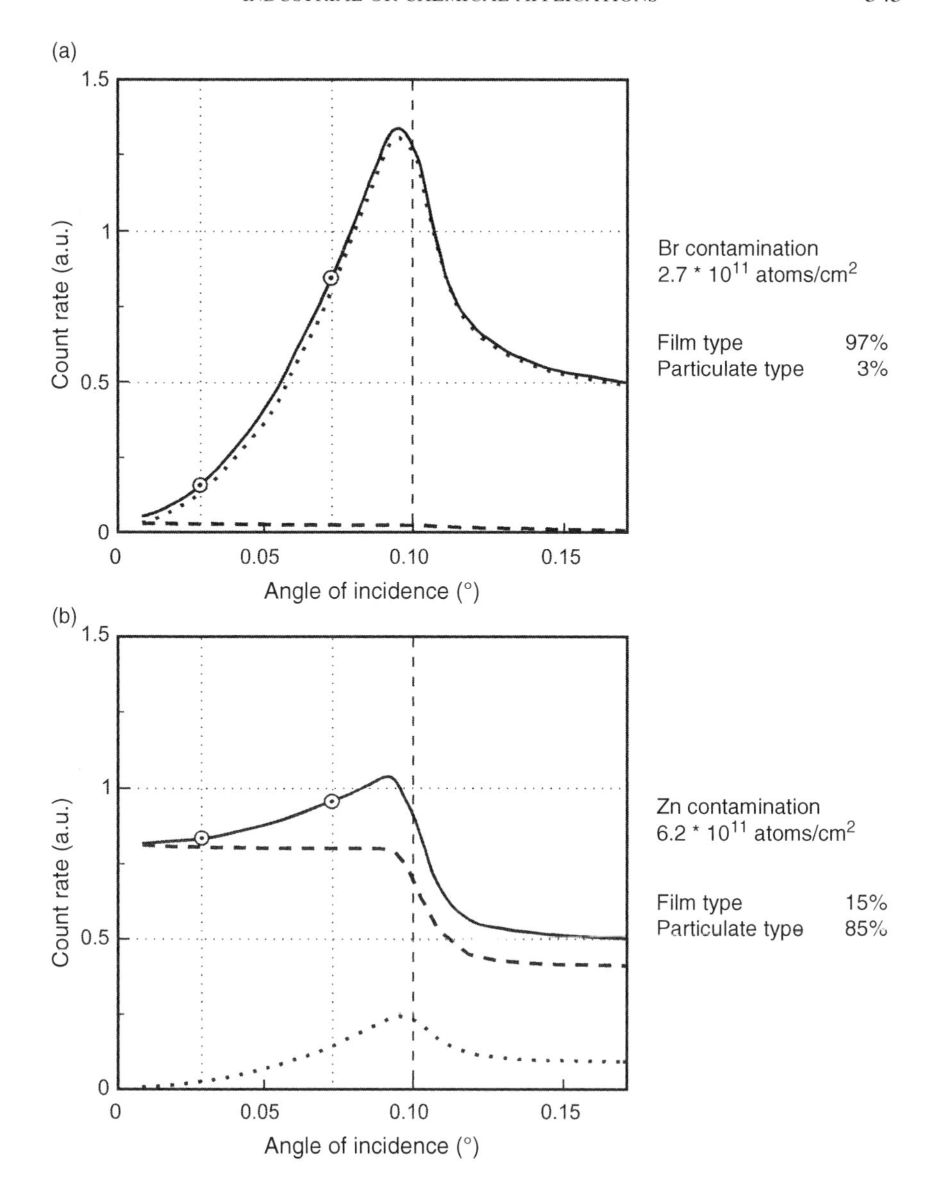

Figure 5.15. Angle-dependent intensity profiles for (a) bromine and (b) zinc. The measured profiles (solid lines) were composed of a particulate curve (dashed) and a thin-layer curve (dotted). Thereby, the total contamination was distributed to a particulate and a thin-layer type with a respective percentage. Figure from Ref. [157], reproduced with permission. Copyright © 1999, ICDD.

by more than two orders of magnitude, reaching the level of 10^7 atoms/cm^2 if the impurities of the entire surface of the wafer (4–8 inches in diameter) are collected and preconcentrated prior to TXRF analysis. The improvement is given by the ratio of the entire area of the wafer and the observation area of the detector.

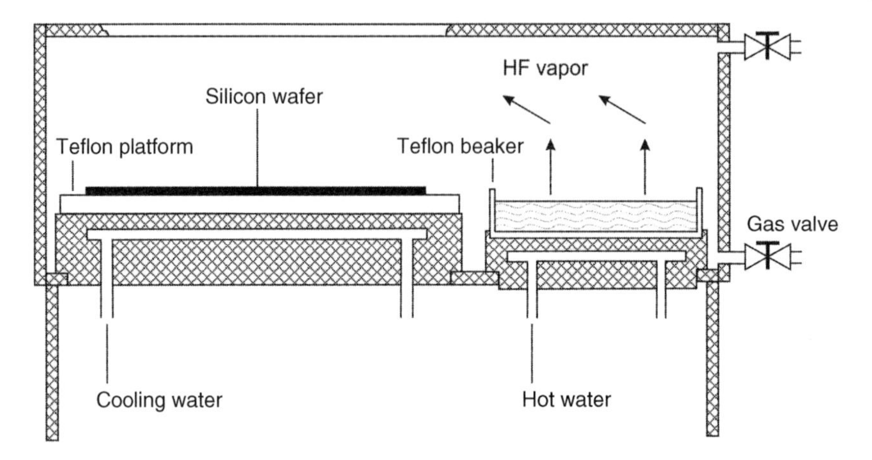

Figure 5.16. Diagram of a VPD reactor suitable for vapor-phase decomposition of the oxidized surface of a silicon wafer. Figure from Ref. [161], reproduced with permission. Copyright © 1991, Elsevier.

A common preconcentration technique is VPD. Originally developed for AAS measurements, it was later adapted to TXRF determinations [159–161]. ET-AAS and ICP-MS also make use of it [162,163] but in contrast to both methods TXRF is nonconsumptive so that a repetition of the actual X-ray measurement is possible. The combination of VPD and TXRF is a destructive method but highly sensitive for element traces *on* and *in* the uppermost layers of silicon wafers.

A special VPD reactor, as described by Neumann and Eichinger [161], is schematically illustrated in Figure 5.16. Some 10 ml of a 20% HF solution is poured onto a heated Teflon dish and evaporated. The wafer, placed on a Teflon platform and cooled down, is exposed to the HF vapor in the closed reactor. As a result, the uppermost layer of Si oxide is etched within about 30 min. The oxide layer, either native and only about 3 nm thick or thermally grown and some 10–100 nm thick, is dissolved according to Equation 5.2 already given in Section 5.4.2:

$$SiO_2 + 6HF \leftrightharpoons H_2SiF_6 + 2H_2O$$

Lots of fine water droplets appear and build a moisture film on the hydrophobic surface. These droplets contain most of the contaminants or impurities previously present on or in the oxide layer of the wafer. The etching stops automatically just at the interface of Si oxide and silicon.

After opening the gas valves and the reactor itself, the wafer is taken out. One single drop of 10 or 100 μl ultrapure deionized water is pipetted onto the surface and moved spirally in order to collect all the fine droplets that include the contaminations. Finally, the collecting drop which was moved to the center is dried by evaporation. The complete procedure is carried out in a clean bench.

Afterward, TXRF is applied for the analysis of the granular dry residue at a fixed-angle adjustment.

Several wafers (up to 10) can be stacked on top of one another and all the droplets may be drained by rocking motions and collected in a glass tub at the bottom [164]. From there, a larger drop can be taken for analysis by TXRF.

Care must be taken to get a residue within an area of less than 1 mm diameter and to place it under the center of the detector [165]. For quantification, an external multielement standard can be prepared as a similar residue and measured under the same conditions. Subsequent addition of an internal standard is not recommended because a homogeneous distribution would not be guaranteed. In any case, the residue of metal impurities should have a mass less than 1 μg. Correspondingly, the original contamination of the entire surface must not exceed the level of 10^{13} atoms/cm^2; otherwise, negative effects mentioned in Section 4.4.3 will become evident.

Some problems arise by drying the accumulated μl drops [163]. For example, the dried residues are usually distributed in ring-shaped walls so that absorption or saturation effects occur and the dynamic range of linearity is exceeded. New approaches, for example, the nanoliter deposition technique or the inkjet printing technique mentioned in Section 4.1.3, overcome these problems.

A further requirement for accurate results is high collection efficiency for all the different metal impurities. Most of them show an efficiency >70% with the exception of copper. The copper atoms are obviously adsorbed on the silicon surface after the oxide layer is stripped away by the HF vapor. To remove the adsorbed copper, some hydrogen peroxide H_2O_2 should be added to the deionized water used for collection [166].

Figure 5.17 shows an automated VPD system applicable to Si wafers with a diameter from 100 to 300 mm [163,167]. In the left chamber the vapor of

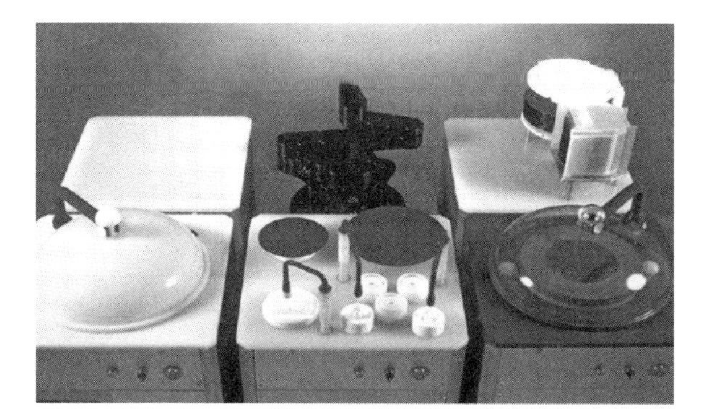

Figure 5.17. Automatic VPD reactor applicable to Si wafers of 100–300 mm diameter. The left chamber is used for the generation of HF vapor, the middle station is for the collection of small droplets by scanning a 50 μl drop of high-purity water, and the right chamber serves for drying this drop and analyzing the dried residues. A cassette system is behind, a robot in the background. The system was developed in a cooperation of GeMeTec Munich and Wacker Siltronic, Burghausen, Germany. Figure from Ref. [167], reproduced with permission. Copyright © 2006, Elsevier.

hydrofluoric acid is generated by isothermal distillation and the Si wafer is exposed to this vapor. Lots of fine water droplets are loaded with the contamination of the wafer surface. In the middle station, all these fine droplets are collected by a single drop, which is scanned on a spiral over the whole wafer surface. In the right chamber, the accumulated drop is dried and the dry residue is analyzed by TXRF. The transport of the wafer is done by a robot in the background. The enrichment factor is between 157 and 1414 for a wafer with 100 mm and 300 mm diameter, respectively. Detection limits go down to 10^7 atoms/cm^2. Such low limits are demanded for the production of ultraclean wafers in the semiconductor industry.

The VPD technique described earlier can also be applied to germanium wafers because it can be based on a chemical reaction for GeO_2 similar to that for SiO_2:

$$GeO_2 + 6HF \rightleftharpoons H_2GeF_6 + 2H_2O \tag{5.4}$$

Detection limits are about 10^8 atoms/cm^2; however, the recovery rate is only about 60% due to matrix effects that are produced by heavy Ge etch products [153].

5.4.8 Characterization of Nanostructured Samples

Structured materials with nanometer dimensions represent an ambitious task for today's methods of analysis. Thin layers or thin films, deposited or implanted with nanometer thickness, junction layers, electric double layers, but also biofilms, adsorbates, and small nanoparticles have to be characterized. The thickness of a single layer, the position and number density of a buried layer, the thickness of organic and biofilms, the composition and thickness of periodic multilayers, the ion distribution in a solution, the composition and thickness of nanoparticles, and the thickness and coverage of clusters can be determined by GI-XRF. A standard method for stratified layers is RBS (Rutherford backscattering); another one is SIMS (secondary ion-mass spectrometry). However, GI-XRF based on X-ray standing waves is also suitable for the mentioned examples. Repeated sputter etching plus usual TXRF of the remaining surface can only be applied to thin stratified layers.

5.4.8.1 Shallow Layers by Sputter Etching and TXRF

Depth profiling of implanted layers in Si wafers was already described in Section 4.5.2.2. One remarkable result may be given here—the depth profiles of arsenic and cobalt ions implanted in Si wafers and characterized by repeated sputter etching and TXRF of the surface being left after sputtering [168,169]. Two implanted samples could be examined: (1) arsenic ions implanted in silicon at 100 kV and with a nominal dose or fluence of 10^{17} ions/cm^2, and (2) cobalt ions implanted in silicon at 25 kV and with a nominal fluence of

10^{16} ions/cm^2. The molar fraction of implanted ions and Si atoms was determined on the nanometer scale. The corresponding stepped profiles are demonstrated in Figure 5.18a.

Both profiles represent a bell-shaped distribution of ions in a depth normal to the surface plane. The step height is 6 nm for the arsenic and 3 nm for the cobalt sample. Both profiles show a maximum in a certain depth of 60 nm (for As) and 30 nm (for Co) and with a certain width. They have an offset or surface value at $z = 0$ and are skewed to the right. Both profiles were compared with RBS measurements carried out independently at Bochum and Munich University. TXRF and RBS profiles match quite well, with differences only occurring at the surface of both implantations and at the maximum of the As implantation. Detection limits for the molar fraction are about 0.01. The depth resolution is 10–15 nm for RBS and only 3 nm for TXRF (four times better).

The differences of TXRF and RBS profiles are caused by an assumption simply and solely made for RBS—that the original number density of crystalline silicon remains unchanged during implantation. This assumption, however, is not permitted. During implantation, the original crystalline structure is damaged and the crystal is amorphized. A swelling (expansion) of the crystal can generally be observed at a depth of some 10 nm and a shrinking (compression) usually occurs at a depth of 10–50 nm. For even deeper layers, the crystalline structure of silicon remains unchanged. In contrast to RBS, the application of TXRF allows an individual determination of both quantities n_{ion} and n_{Si} (see Section 4.5.2.2), in accord with Equations 4.34 and 4.35. The respective results are shown in Figure 5.18b. The values of n_{ion} for RBS are directly proportional to c_{layer} of Figure 5.18a and show a similar profile; however, respective values for TXRF lead to changed profiles that are less smoothed because of uncertainties of mechanical measurements. Both implantations show a good match of RBS and TXRF; differences only appear on their increasing flank down to the maximum and can be explained by the different depth resolution of RBS and TXRF [168,169]. Both curves correspond very closely on their decreasing flank, that is, in a depth where the crystalline structure is nearly undestroyed.

5.4.8.2 Thin-Layer Structures by Direct GI-XRF

X-ray fluorescence at grazing incidence can be employed to characterize thin-layer structures in the near-surface range of 1–500 nm with respect to their composition and thickness *nondestructively*. This requires, however, an angle scan of the layered wafer in order to record and interpret angle-dependent intensity profiles.

A variety of thin layers deposited on flat surfaces (mainly silicon wafers) have been analyzed by GI-XRF [170–184]. These single- and double-layer systems consisted of pure metals, metal alloys, metal oxides, or nitrides. The mass fraction of the individual elements covered the total percentage range, the

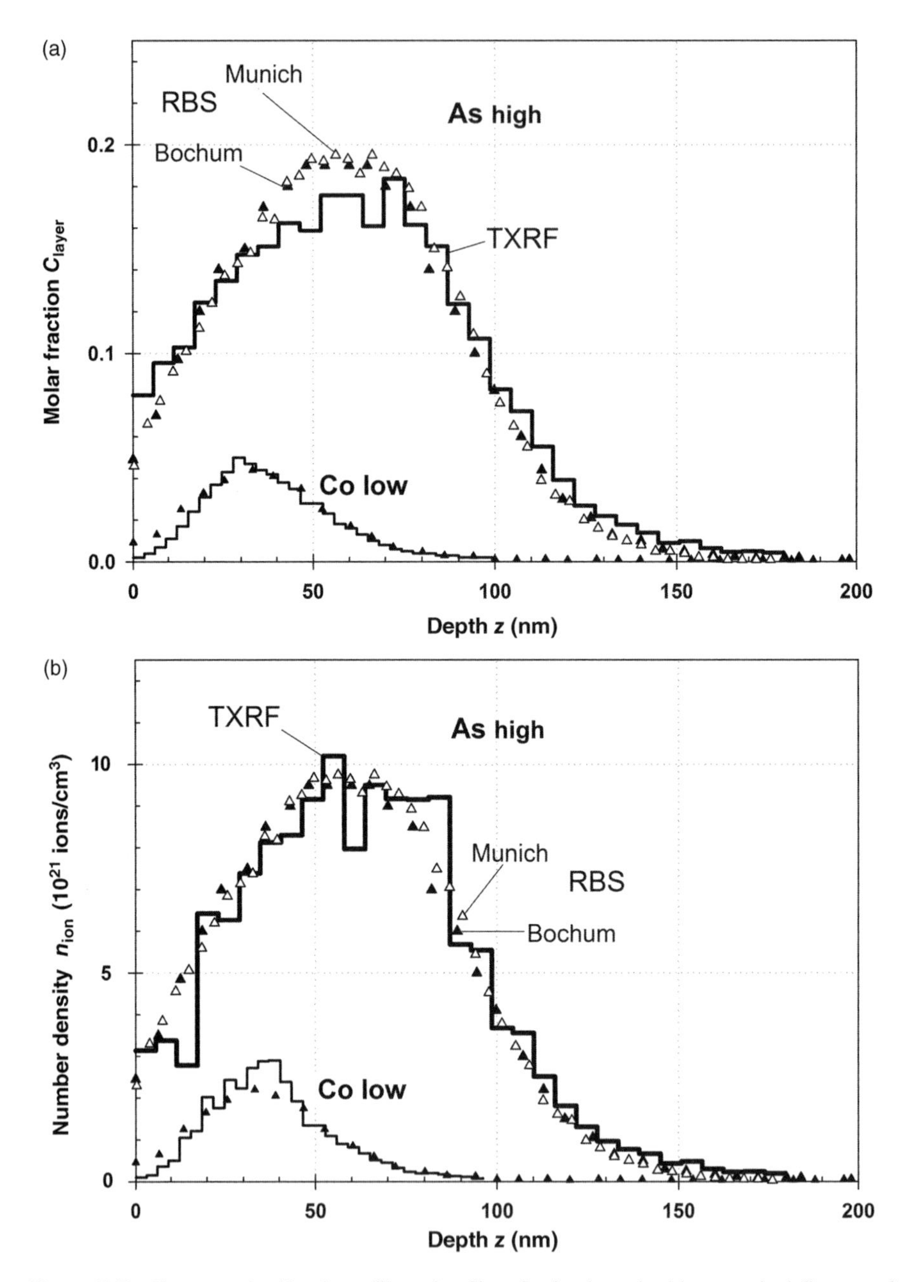

Figure 5.18. Concentration/depth profiles of a Si wafer implanted with a nominal fluence of 10^{17} arsenic ions/cm^3 or with 10^{16} cobalt ions/cm^3. TXRF was performed after sputter etching in steps of 6 nm for the As-implanted wafer and of 3 nm for the Co implantation. RBS measurements were carried out independently at Bochum and Munich University (full and open triangles). (a) Molar fraction and (b) number density. Figure from Ref. [169], reproduced with permission. Copyright © 2005, American Institute of Physics.

thickness ranged from about 1 to 500 nm, and the density was between 3 and 12 g/cm^3. Angle-dependent intensity profiles were recorded for these determinations, and the profiles were interpreted qualitatively and calculated quantitatively. The performance, already described in Section 4.6.4, can be demonstrated by some examples.

The first simple example is the investigation of semiconductor layers, such as germanium deposited on silicon wafers by GI-XRF. Intensity/angle curves were calculated by a recursive algorithm after Krämer [183] regarding the standing wave field of the primary beam. Calculations after the integrated intensity of Equation 2.47 lead to the same results. Excitation by a primary radiation of 15.2 keV and detection of the Ge-Kα peak at 9.89 keV were supposed and the layer thickness was chosen between 1 and 500 nm. The results are shown as Ge graphs in Figure 5.19, which are similar to the Co curves in Figure 2.22.

For glancing angles <0.17° (critical angle of germanium), the primary beam is totally reflected at all Ge layers. It is evanescent with a penetration depth of about 2.2 nm, and the induced fluorescence is generally infinitesimal. However, for ultrathin layers of about 1 or 2 nm, the evanescent primary beam can reach the Si substrate and is totally reflected at 0.12° (critical angle of silicon). It excites the Ge layer leading to a peak because of constructive interference (Figure 5.19b). This peak represents a maximum that is three times the limit approximated for larger angles. A Ge layer of 10 nm absorbs the primary beam so that the Si substrate is not reached and an intensity peak could not be observed. At 0.17°, the primary beam penetrates in this layer and excites it to fluorescence. The intensity shows an inflection point, reaches a maximum of 1.5 times the large-angle limit, and shows some smaller oscillations. Layers with a thickness of 20 nm up to 100 nm show less distinct oscillations (Figure 5.19a) behind the inflection point. For angles above 1° (not demonstrated in this figure), the primary beam penetrates in the Ge layers with a μm depth of $\approx 0.060 \cdot (\alpha/\alpha_{crit})$ or $0.364 \cdot \alpha°$ in μm. Layers of 1 μm are excited completely under angles >5°; at vertical incidence layers up to 10 μm can even be reached. The fluorescence intensity increases linearly with the thickness; however, absorption effects lead to an asymptotical approximation.

Germanium layers with different thickness are commercially available (Jenoptik, Jena, Germany). Excitation with synchrotron radiation of 15.2 keV photon energy and detection of the Ge-Kα peak by an SDD were performed [183]. An intensity/angle scan was recorded for Ge layers of 29 nm of nominal thickness. This thickness could be ascertained with an accuracy of 2 nm.

Theoretical calculations for Co layers deposited on silicon wafers were proven true by de Boer and van den Hoogenhof [177]. Experiments were carried out with a Mo tube and a diffractometer equipped with an energy-dispersive detector and show all characteristic features of the intensity/angle scan for layers with a thickness of 1, 30, and 100 nm. The accuracy for the determination of a layer with 30 nm thickness was estimated to be 1 nm.

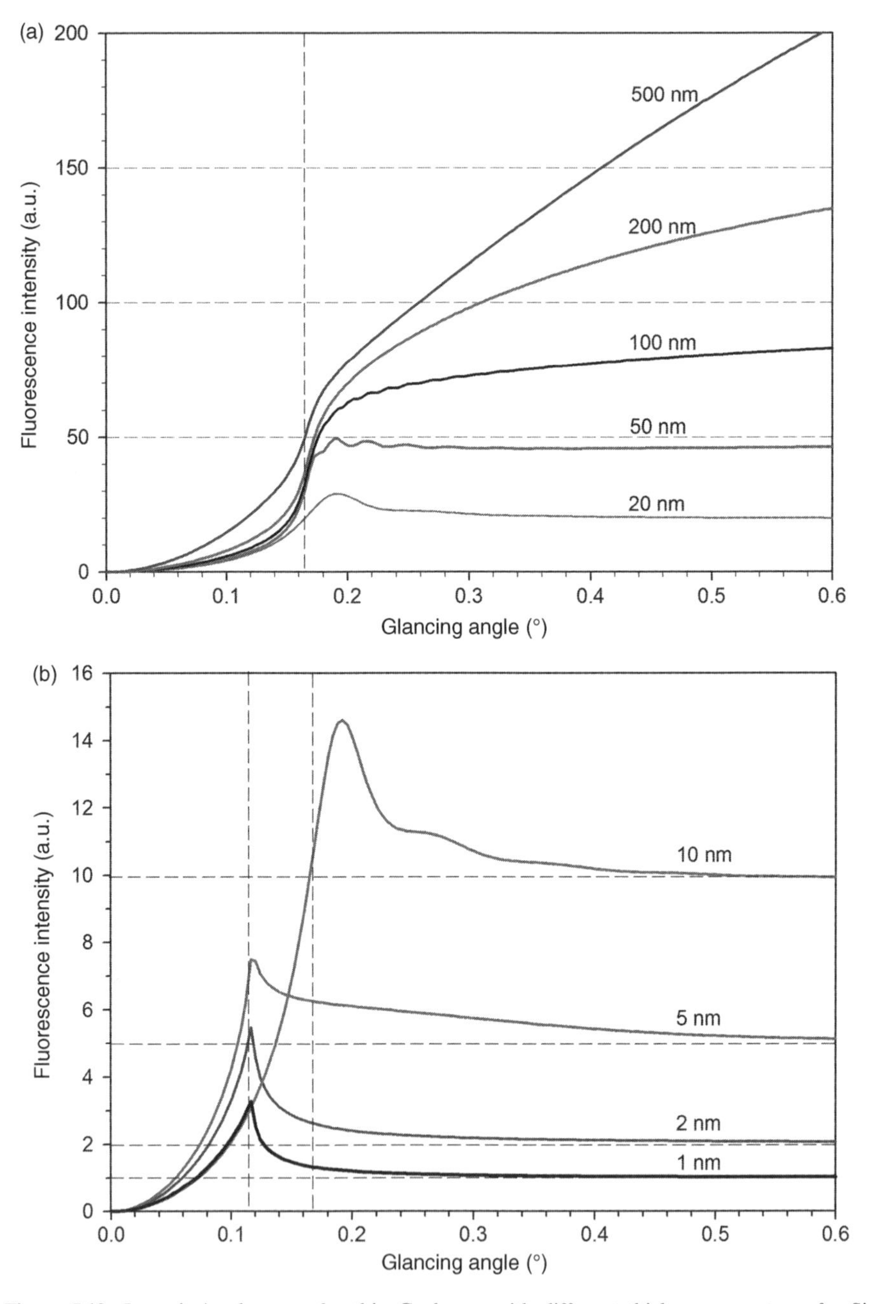

Figure 5.19. Intensity/angle scans for thin Ge layers with different thicknesses on top of a Si wafer [182]. It has been assumed that excitation was carried out by monochromatic X-rays of 15.2 keV at a glancing angle α_0 and fluorescence of Ge-Kα at 9.89 keV was detected vertically to the sample surface. (a) Curves between 20 and 500 nm. (b) Curves for a thickness ≤10 nm. Sample and detector were assumed not to be limited so that a footprint effect could not occur. The authors are indebted to M. Krämer from AXO, Dresden, for the simulation of these profiles.

Krämer also studied thin films of a specific protein and a lipid. Cytochrome is a colored chromo-protein and phosphor lipid was prepared as a bilayer by the Langmuir–Blodgett technique [183]. Both films were shown to have only about 5 nm thickness.

A further example of GI-XRF deals with a one-element layer deposited on a Si substrate and covered by a thin multielement layer [175,182]. Figure 5.20 represents intensity profiles of five elements Cr, Fe, Ni, Pd, and Si measured for a layered sample at glancing angles up to 0.7°. From the similar course of the Cr, Fe, and Ni curves, it can be concluded that there is a first layer consisting of a Cr/Fe/Ni alloy. The course of these curves suggests a thin layer of about 10 nm thickness. The course of the Pd curve refers to a second layer of pure palladium, which in contrast to the first layer is thicker than 100 nm. The Si curve indicates a pure silicon substrate.

This first qualitative impression of the curves is a rough estimate, as shown in the inset of Figure 5.20. It serves as a starting point for the quantitative calculations described in Section 4.6.4. The solid curves of Figure 5.20 represent the best fit after several iterations. They are based on a model presented in Table 5.6 as the final result. It contains the element composition, the thickness, and the density of the two layers and the substrate, thereby making the first

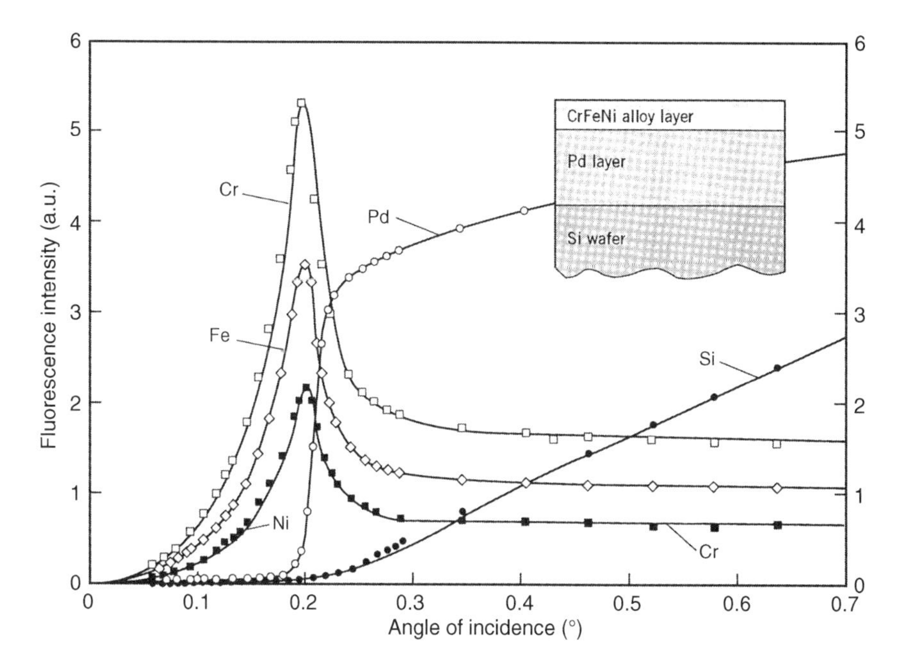

Figure 5.20. Fluorescence intensities of five elements measured on a layered sample by an angle scan. Excitation was performed by the Kα peak of a Mo X-ray tube. The solid curves do not represent a regression line of the measuring points but are a best fit of calculations based on a model as demonstrated in the inset on top, right. Figure from Ref. [182], reprinted with permission. Copyright © 1992, American Chemical Society.

TABLE 5.6. Final Results of the Fitting Procedure Applied to a Stratified Structure on a Silicon Wafer

Structure	Element Composition (%)					Thickness (nm)	Density (g/cm^3)
	Cr	Fe	Ni	Pd	Si		
Upper layer	46.7	29.5	23.8	—	—	5.9	7.0
Lower layer	—	—	—	100	—	257.0	11.2
Substrate	—	—	—	—	100	∞	2.3

Source: Table from Ref. [3], reproduced with permission.

rough estimate more sophisticated [175]. The relative standard deviation of the individual data is about 1–2%.

Ultrashallow junction layers with boron and arsenic ions implanted in Si wafers can serve as acceptors and donors, respectively. Such layers are important for downsizing the components in the silicon ULSI technology. The ions were implanted with energies of 0.2 to 3 keV in a depth of 3 to 20 nm and a total dose or fluence of 1×10^{14} to 5×10^{15} ions/cm^2. Hönicke *et al.* [185] determined the boron and arsenic concentrations as a function of depth. GI-XRF was carried out with synchrotron radiation at BESSY II using low-excitation energy, a UHV chamber, and a windowless SDD. Relevant concentration/depth profiles could also be calculated *ab initio* with a reference-free fundamental parameter method. The total implanted fluence or dose was determined by integration of these profiles. Deviations from the nominal dose were below 10%. In contrast to GI-XRF, the well-established and approved method SIMS (secondary ion mass spectrometry) showed differing profiles that were shifted toward the surface. Such significant deviations are caused by matrix and transient effects, especially for low-energy implantations and the near-surface region. The profiles calculated by the theoretical TRIM program (Transport and Range of Ions in Matter) fit even much better.

Silicon dioxide has been used as a gate oxide material in the last decades. The thickness of such dielectric layers could be permanently reduced; however, below 2 nm the leakage current due to tunneling increased and the reliability suffered severely. For that reason, the silica gate was replaced by a material with a high dielectric constant (k value > 3.9). Hafnium or zirconium can be implanted as a shallow layer of some nanometer thickness in a silicon wafer showing a double capacity but much smaller leakage current (0.01-fold in comparison to a thin silica layer). Ingerle *et al.* investigated shallow implanted layers of arsenic in silicon and additionally of Hf-based high-k layers with 2 and 5 nm thickness, called ultrashallow junctions (USJ). GI-XRF was shown to be a powerful technique for the determination of the composition of the different layers and the whole implantation dose or fluence [186,187]. XRR was applied to get additional information on the thickness of mono- and multilayers, directly and nondestructively. A software package was developed for combined measurements [187]. The conventional technique SIMS and the well-known simulation program TRIM were quoted for comparison.

Thin layers of TiN are used as a metal gate in CMOS devices and as a barrier against copper diffusion in microelectronics. Such layers of 50 nm thickness were deposited by physical vapor deposition (PVD) and chemical vapor deposition with metal organic substances (MOCVD). They could be characterized by GI-XRF profiles of Ti, N, O, and C layers in a reference-free mode combined with XRR [188]. Another example dealt with thin marker layers of titanium buried in a carbon layer of nearly 100 nm thickness. These layers with only 0.2 nm thickness were buried at a depth of 33, 49, and 66 nm. A standard Cr X-ray tube was used for excitation and a Si(Li) detector chosen for spectral recording. The marker layers could be localized reliably and nondestructively [189].

The last complex example is related to thin layers of ions included in a polymer layer of a few nanometers, which is deposited on a Si substrate and covered by a thin suspension. Such insoluble functional films can be used for multilayer organic light-emitting diodes. This somewhat difficult example stems from Brücher *et al.* [190] and Köhnen *et al.* [191]. A flat and clean Si substrate was first spin-coated with a suspension of PEDOT:PSS (poly(3, 4-ethylenedioxythiophene:polystyrenesulfonate)) and dried. This basic layer of about 10 nm thickness was covered by a polymer solution of toluene by spin-coating and was heat-treated at 200 °C. During the polymerization process the PEDOT:PSS layer decomposed into the anions HSO_4^- or SO_4^{2-}, which migrated through the polymer layer in the form of a polymerization front upward to the surface. Finally, nonpolymerized material was removed by rinsing it with tetrahydrofuran.

This sample was excited to X-ray fluorescence by a synchrotron beam with a photon energy of 10 keV. The Kα peaks of sulfur at 2.31 keV and of silicon at 1.74 keV were used for an intensity/angle scan with a step-width of 0.001°. Simultaneously, the intensity of the reflected beam was recorded by XRR. Figure 5.21 shows the results. The sulfur intensity emerges at an angle of 0.12°, which is the critical angle of total reflection of the interface polymer/Si. It rises up and oscillates with three maxima up to an angle of 0.17°, which is the critical angle for the interface air/polymer. These oscillations occur when the antinodes of an XSW field successively pass through a thin layer of sulfur. A model was set up in accord with Equation 4.47 with a basic layer of S-containing molecules, which is about 10 nm thick. A much smaller layer was assumed to be included in the polymer at a height of 102 nm with a thickness less than 10 nm but a threefold concentration. A fading of the contrast was taken into account with $z_{\text{fade}} = 100$ nm. Reflectivity measurements confirmed that the upper S layer is less than 5 nm while the overall thickness of the polymer layer is about 115 nm.

Further examples can be found in the literature. For instance, simple boroncarbide layers of 1–5 nm thickness buried under a silica layer of 2.5 nm were investigated by Unterumsberger *et al.* [192]. A further example from Brücher *et al.* [193] dealt with the distribution of ions in thin water films on

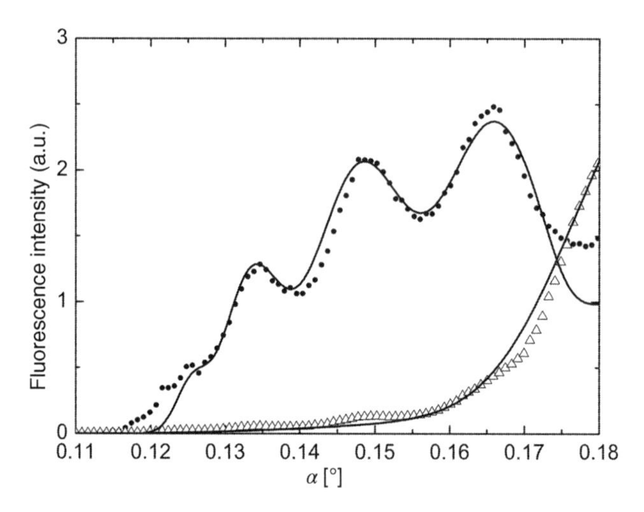

Figure 5.21. Profiles of the sulfur intensity taken by GI-XRF. The X-ray fluorescence intensity of S-Kα (dots) and of Si-Kα (triangles) is plotted vs. the glancing angle. Three maxima occur between 0.13° and 0.17° when the antinodes of the XSW field with a period of about 25 nm successively pass through the sulfur containing layer. This profile was reproducibly measured with high accuracy. The simulated intensity was calculated from a model with a 12 nm layer on the Si substrate ($z = 0$ nm) and a 6 nm layer just below the surface of the polymer layer ($z = 102$ nm). Figure from Ref. [191], reprinted with permission. Copyright © 2012, American Chemical Society.

functionalized silicon surfaces. Br^{1-} was chosen as anionic, Fe^{3+} as cationic markers with a concentration of 10 µg/ml at two different pH values. The electric double layer at the solid–liquid interface could be characterized by diffuse layers with a Debye length of 2 or 4 nm for the bromine ions.

5.4.8.3 Nanoparticles by TXRF and GI-XRF

Nanoparticles are of increasing interest for environmental studies, pollution control, and industrial applications. For example, gold nanoparticles can be used for heterogeneous catalysis and silver nanoparticles are utilized for coating fabrics as a resistance against microorganisms, such as fungi, bacteria, and viruses [194]. However, such particles can be mobilized by aerosols and can cause toxic effects being harmful to human health. For monitoring, aerosols were collected from a test control unit of the factory with a low-pressure Berner impactor. They were deposited on carriers made of silicon wafers and analyzed by TXRF directly. A nitrogen atmosphere was provided in order to eliminate the argon peak of ambient air and to avoid coincidence with the Ag-Lα peak. The detection limit for silver was improved down to about 0.2 ng.

Apart from simple TXRF analysis of nanoparticles, GI-XRF was first used for their investigation by von Bohlen [195]. Fine gold particles were produced by means of a fs laser pointed at a Au foil under a thin film of water and a suspension of 20 µl was pipetted on a silicon carrier. In order to reduce the high

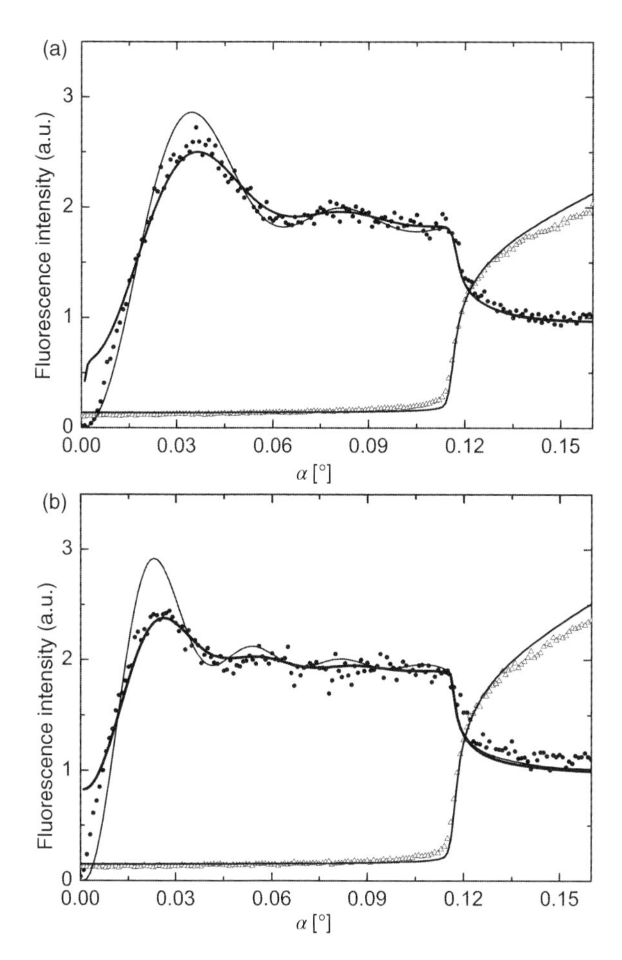

Figure 5.22. GIXRF analysis of tiny spherical gold particles with a nominal diameter of 47 nm (a, above) and 73 nm (b, below). The Au-Lα signals at 9.71 keV are indicated by small dots; the Rayleigh scattering signals at 15 keV are marked by small triangles. The simulation without fading of the XSW field is plotted by a narrow line that deviates from the measuring points. It is corrected by a fading contrast with a height of 120 nm and leads to the well-fitted bold lines. Figure from Ref. [190], reproduced with permission. Copyright © 2012, Elsevier.

occupation of particles, the residue was diluted by a droplet of alcohol, spread over the surface, and dried again. Nevertheless, SEM pictures showed conglomerates of particles with nonuniform grain size between 15 and 120 nm and a density greater than 100 particles per 4 μm^2. The intensity/angle scans taken for the deposits of such particles are typical for particulate-type residues represented by curve (b) in Figure 4.32.

Several suspensions with nanoparticles of gold are also commercially available in stock solutions of about 10^8 particles per ml. The nominal diameters lie between 25 and 250 nm (Nanopartz, USA or Microspheres–Nanospheres, Germany). For their investigation, aliquots of 5 μl were pipetted on a Si wafer, diluted with 20 μl ethanol, and dried in ambient air. Pictures of the dried residues were taken by a SEM demonstrating a very thin distribution of the spherical particles. Up to 20 agglomerates per 4 μm^2 were found containing some 10 to 30 spherical particles with uniform grain size. Areas with only 20–40 single particles per 4 μm^2 also exist. The resolution of a SEM, however, is restricted to some nanometers so that particles with a diameter in the lower nanometer region cannot be depicted reliably [195,196].

For GI-XRF investigations in the nanometer region, a synchrotron beam line is necessary with a high brilliance and monochromaticity. For that purpose, the primary beam of a synchrotron was monochromatized by a DCM with Si (311) and adjusted to 15 keV so that the critical angle of total reflection is 0.12°. The samples were fixed on a goniometer stage and tilted at grazing incidence. This angle scan was recorded between 0.000° and 0.200° in 200 steps of 0.001° with an acquisition time of 20 s per spectrum [196]. Depending on the glancing angle, the fluorescence intensity of Au-Lα at 9.71 keV and of Si-Kα at 1.74 keV was recorded by an SDD.

Figure 5.22 shows two profiles of the intensity/angle scans for two samples of gold nanoparticles. The curves start at zero intensity and rise to a maximum of nearly three, then oscillate with several smaller maxima up to the critical angle of total reflection. At this angle the intensity falls down to a constant value, which is set to be 1.0. A second signal, which is plotted in the figure, represents the intensity of the elastically scattered radiation, the so-called Rayleigh peak in the spectrum. For small angles it is nearly zero but rises abruptly at 0.12° because the primary beam penetrates into the Si wafer below the critical angle.

A first calculation of the fluorescence intensity of the gold particles in the primary XSW field was carried out in accord with Equation 4.40. It was taken for granted that the particles are spherical so that the concentration profile described by Equation 4.45 could be applied. This first simulation with the nominal diameter of the particles did not fit the oscillations of Figure 5.22 very well. For that reason, a fading of the XSW field was assumed and taken into account by a factor $F(z)$ of Equation 4.39. With a parameter z_{fade} of 100 nm the theoretical curves match quite well with the experimental values. Deviations at small angles can be explained by a substrate that is not ideally flat and even. Deviations of the scattering signal can be attributed to X-ray absorption in the substrate.

Further good results were also produced for several other gold particles with diameters down to 30 nm and up to 250 nm. Even mixtures of particles with five different sizes could be characterized and gold and silver clusters deposited on a polymer layer on top of a silicon carrier [183]. Consequently, it can be concluded that nanoparticles of metals can simply be characterized by GI-XRF.

5.5 ART HISTORICAL AND FORENSIC APPLICATIONS

Valuable works of fine arts have long been investigated by chemical analyses. Such an analytical characterization can obviously be helpful for the purposes of art history and for answering questions of restoration and conservation. In cases with favorable outcomes, dating of the work of art is made possible and forgeries can be detected.

Various analytical techniques have been applied, and the results obtained are often excellent. But most of the techniques require a small amount of sample material for analysis. Although test portions of some mg are usually sufficient, such sampling obviously would still damage the work of art. A method that can work nondestructively is conventional XRF. However, this method requires a particular bulky apparatus for objects with a dimension of 30 cm or more. In contrast to conventional XRF, the TXRF method needs only μg portions and is therefore virtually nondestructive. A very gentle method of microsampling has been developed that is especially applicable to oil paintings under restoration [197].

In the field of forensic science, the analyst frequently must do with mg quantities (microsamples) or even μg quantities (ultramicrosamples) of the sample in question. Consequently, TXRF as a microanalytical tool is highly suitable for such forensic questions. In recent years, several specific problems have been investigated and appropriate solutions have been developed. These investigations have involved ultramicroanalyses of hair [2,157,198], glass fragments [2,199], blood, paints, and other samples of evidence [200–202], for example, tape fragments, drug powders, semen traces, and gunshot residues. Generally, in such cases, two similar samples have to be compared and their identical nature must be either confirmed or rejected. For this purpose, minute samples are placed on a glass carrier and normally analyzed by TXRF without any further preparation. The spectral patterns are used as fingerprints of the samples and compared as described in Section 4.3.3. For TXRF there is no consumption at all; the minute samples on the carriers can be preserved as pieces of evidence, which may be highly important.

5.5.1 Pigments, Inks, and Varnishes

Pigments and inks once used for oil paintings and illuminated manuscripts are objects of valuable cultural heritage. Today they allow insights of historical and archaeological relevance. Oil paints generally are a mixture of pigment powders with an appropriate oil or resin that serves as bonding agent. A lot of pigments are inorganic, consisting of fine grains of some 0.1–1 μm. The variety of these pigments is not very large and amounts to only about a hundred different types. Most of them are metal oxides, hydroxides, sulfides, and sulfates or consist of mixed compounds. Each pigment can be characterized by one to five major elements, which can be detected by simple TXRF, besides the light elements H, C, and O. The detectable elements may be called "key

Figure 5.23. Soft sampling of an oil painting by a small cotton-wool bud (Q-tip). It is demonstrated here for the Flemish painting "The annunciation" created in 1646 by the past master Antoon Van Den Heuvel and to be seen as an altarpiece of Saint Nicholas in Ghent, Belgium. The TXRF spectrum (inset above) indicates vermilion and white lead as pigments with key elements mercury and lead. Figure from Ref. [203], reproduced with permission. Copyright © 2000, John Wiley and Sons. (See colour plate section)

elements" and are usually sufficient for the identification of a particular pigment or paint.

A sampling technique recently proposed for TXRF can be applied to oil paintings either *without* a clear varnish or *with* a varnish that must be removed anyway because of an intended restoration. A minute amount of paint can be wiped off the surface by means of a dry cotton-wool bud ("Q-tip") as is demonstrated in Figure 5.23. Different Q-tips are applied for different spots of the painting [203,204]. This technique is nearly nondestructive since only about 1 μg of paint is rubbed off in each case. The Q-tips are locked in bottle caps made of Plexiglas for safekeeping and transportation. For analysis at an experienced TXRF laboratory, they are dabbed onto a sample carrier by a single tip. An amount of less than 100 ng is transmitted, which is enough for identifying the pigments by TXRF.

The minute sample mass cannot be determined exactly, so an internal standardization is not practicable. Nevertheless, a quantification is made possible by normalization to 100% according to Equation 4.13. This method was extended and applied to the analysis of pigment mixtures [197]. The accuracy was checked by a synthetic mixture of three different pigments:

titanium white, zinc white, and strontium yellow. The nominal proportion of 1 : 1 : 1 was approximately confirmed by the ratio of 0.94 : 1.04 : 1.03. The precision was characterized by a relative standard deviation of only 4%.

The repeatability of the method was examined by a multiple sampling at 14 different spots of the modern oil paint cobalt green or Rinman's green (CoO·5 ZnO with key elements cobalt and zinc). A relative standard deviation of only 4% demonstrates a good repeatability. A quite different result was obtained for 14 different spots of a certain color on a real painting. The old Flemish painting of Figure 5.23 was chosen and the "red" color of the Madonna's dress was identified as a mixture of vermilion (key element mercury) and white lead (key element lead). A high scattering of the results indicates that the artist (Van Den Heuvel in Ghent, about 1646) has applied the "red" color in different mixing proportions. Obviously, sampling is only representative for a small restricted area of an oil paint.

The foregoing method is fast and convenient and has already been applied to a systematic screening of numerous oil paintings under restoration, as well as to murals, painted sculptures, and book illuminations [204]. The samples came from 12 museums in the whole of Europe, and many questions of restoration, conservation, and dating had been answered [203–205]. Some examples will demonstrate the capabilities of TXRF.

The mentioned painting of Figure 5.23 had to be restored because it suffered from the so-called "ultramarine disease," which is feared in all museums around the whole world. It has discolored the brilliant blue into a pale greenish-gray in the dark folds of the blue coat. Ultramarine could not be identified, since the key elements sodium and aluminum could hardly be detected by TXRF without vacuum. Among blue pigments of the seventeenth century, ultramarine might be recognized by the absence of copper and cobalt. Indeed, both elements are absent in all blue parts of the coat, but the key element cobalt was found in the discolored parts. Cobalt is characteristic for the blue pigment smalt—a cobalt-bearing glass. Arsenic was determined as an impurity because cobaltite (CoAsS) and smaltite ($CoAs_2$) are sources of cobalt [203,205]. Consequently, it was inferred that the ultramarine disease is actually a smalt disease. The less valuable and cheap smalt pigment had been used in previous restorations. The deterioration of smalt is based on the migration of cobalt and cannot be reversed. For the future, only the precious and expensive pigment ultramarine has to be applied for restoration.

Another Flemish oil painting was also examined before restoration. It is called "Piet Hein goes ashore" and was painted by Reintjens in 1846. A white cloud on a blue sky was sampled and the white color was identified as a mixture of permanent white (key element barium), white lead (key element lead), and titanium white (key element Ti) with a mixing portion of 1 : 3 : 1. The appearance of TiO_2 on a picture of 1846 arouses suspicion because titanium white was not discovered until 1908 and was generally not applied on paintings before World War II. However, the painting had been restored in the 1950s [197].

In general, the use of most pigments is known in chronological order [204,205] so that an approximate dating of artifacts or restored parts of a painting is made possible. On the condition that a pigment on the painting appeared after a certain time (*post quem*) and another one disappeared before that time (*ante quem*), a rough dating is obvious. Another way to detect forgeries was demonstrated by the examination of a picture allegedly painted by Modigliani (1884–1920). It showed pigments that do not correspond with the well-known palette of this artist. In particular, the pigments emerald green and cerulean blue appeared in the spectra of this picture while cadmium yellow and zinc white were absent. Consequently, the painting was strongly suspected of being a forgery [203].

Vázquez *et al.* combined TXRF with Fourier transform IR (FT-IR) and gas chromatography mass spectrometry (GC-MS) for the investigation of rock art samples at an archaeological site near the town of Susques, Argentina [206]. Inorganic and organic pigments could be identified. Manganese and iron were detected by TXRF as oxides of a black pigment. FT-IR showed the presence of lipids, while GC-MS identified fatty acids pointing to an animal source.

Several iconographic illuminations were also investigated by TXRF after Q-tip sampling. It was confirmed that at least three artists are responsible for the illumination of the "Breviarium Mayer van den Bergh" (about 1510, Antwerp). An example is demonstrated in Figure 5.24 [203,207]. By a characterization of specific pigment mixtures it could be confirmed that

Figure 5.24. Two illuminations in the Flemish breviarium "Mayer van den Bergh": (a) f° 536v, Birth of Blessed Virgin Mary; (b) f° 552v, St. Michael. TXRF analyses of three different pigment mixtures (1–3) corroborated the hypothesis that both pictures had been produced by the same artist Jan Provoost. (1) light-green = azurite, white lead, and lead–tin yellow type I; (2) violet = azurite and vermilion; (3) dark green=organic paint with the same contamination (Ca, Cu, Zn, and As) in both figures. Figure from Ref. [203], reproduced with permission. Copyright © 2000, John Wiley and Sons. (See colour plate section)

both pictures were painted by the same artist and that the art historical suspicion was right. Furthermore, TXRF analysis of inks was carried out in order to study manuscripts from a collection of Raphael de Mercatellis (1437–1508) at the University Library, Ghent. The ink was identified as a ferro-gallus ink [203]; additives are very similar throughout the whole manuscript but with one significant exception—a colophon indicating the acquisition date, 1505.

Usually, oil paintings are protected against humidity and visible light by thin layers of varnishes. Simultaneously, they are finished according to the spirit of the times. Varnishes are also applied to wooden surfaces, for example, of furniture and musical instruments like violins. Historical varnishes generally consist of a resin dissolved in a volatile thinner like oil or spirit with the addition of nonvolatile components like wax and pigments. They should yield a thin homogeneous and nonsticky film that is scratch-resistant, shows a certain luster, and is light-fast. Three main groups of historical varnishes can be distinguished: "drying" oil-based varnishes, essential-oil-based varnishes, and spirit varnishes. For a characterization of such complex artifacts, different analytical methods have been applied. TXRF is capable of a quantitative determination of the inorganic additives as shown by von Bohlen and Meyer [208,209].

First, the surface of the object under restoration should be cleaned carefully. After that, small flakes of the varnish can be taken by means of a clean scalpel. The tiny samples with a mass of 50–100 μg are accurately weighed with a microbalance and digested in 90 μl of nitric acid plus 10 μl of hydrogen peroxide at moderate temperature (<95 °C). The solution is spiked with a 10 μl droplet of a standard solution containing 50 ng of gallium or selenium serving as an internal standard. Finally, aliquots of 10 μl are dried on quartz-glass carriers and analyzed by simple TXRF within 100 s. Results for historical varnishes of a 200-year-old wooden banister, a 100-year-old piece of furniture, and a Bohemian contrabass are presented in Ref. [208]. They show 14 minor and trace elements with mass fractions between 14 mg/g and 10 μg/g. The sum of these elements is below 3% of the total mass of the examined flakes representing only a minute amount of inorganic ingredients.

In a further study, varnishes from old violins and cellos of Amati, Guarneri, and Stradivari were examined (1560–1730). It was shown that only the finest fractions of yellow and red pigments, for example, auripigmentum, massicot, realgar, and red lead, were used to preserve the transparency of the varnish [209]. Stradivari used iron oxide and vermilion as red pigments [210].

5.5.2 Metals and Alloys

Metals and their alloys can simply be touched by a hard quartz-glass carrier or vice versa. This way of sampling by a single and slight touch or by a gentle streak is similar to the old and famous technique of a touchstone already used

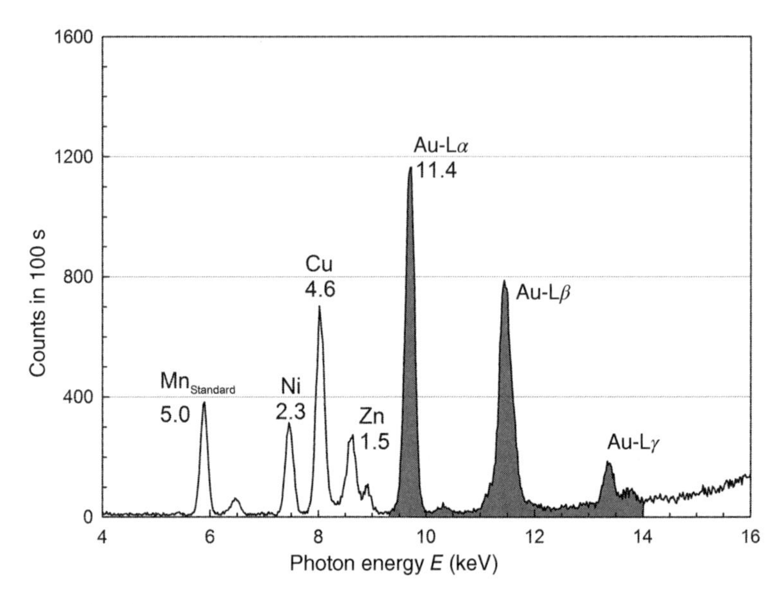

Figure 5.25. TXRF spectrum of a golden ring rubbed on a quartz-glass carrier by a single stroke. 5 ng of manganese was added as an internal standard; the low-intensity spectrum was taken in 100 s. Absolute mass and relative content of four main elements were determined by relative sensitivity values. Figure from Ref. [211], reproduced with permission. Copyright © 2001, Elsevier.

in ancient Greece.[5] Actual covering and thickness of the material smeared on the quartz-glass carrier usually fulfill the conditions for quantitative analysis established in Section 4.4.3 with m less than 8 μg/cm^2 and t below 10 nm.

A characteristic example is the fast and very simple analysis of an old gold ring of medium hardness [211]. The ring was softly rubbed on a quartz-glass carrier and only about 20 ng of this alloy was transferred, which is sufficient for TXRF analysis. A droplet of a standard solution with 5 ng of manganese was added as an internal standard and dried. The spectrum taken within a single minute is shown in Figure 5.25. With help of sensitivity values, the mass of the four major and minor components, Au, Cu, Ni, and Zn, was originally determined in nanograms. In relation to the total amount, the content of the four elements could be given in mg/g. The gold value of 574 mg/g corresponds quite well to the hallmark of 585 with a deficiency of −1.9%. The accuracy for the other elements was better than 5%. The method is fast and nearly nondestructive, and obviously preferential abrasion does not occur.

Rubbing an ancient precious object on a hard quartz-glass carrier is generally not allowed. However, a nonvisible amount of a few micrograms of the sample may be removed by touching it with a clean cotton bud ("Q-tip"). This method of a very gentle sampling was applied to several small beads of two

[5] A touchstone is a flat black stone, such as lydite, which was used for assaying gold alloys because different alloys leave differently colored streaks on the stone.

ancient bracelets made of silver and gold [212]. The bracelets were investigated in tombs at El-Mahasna, Egypt, stem from the fourth millennium BC, and are the only gold artifacts known for the Naqada I period. After sampling the loaded cotton buds were placed in clean beakers and transported to the laboratory. TXRF analyses were performed within a few minutes and semi-quantitative results were obtained by relative sensitivities. The main constituents were silver (430–990 mg/g) and gold (<300 mg/g), determined with an accuracy of about 5%. Secondary components were iron, copper, and zinc. It could be concluded that most of the beads probably consist of a rather strange alloy, that is, natural aurian silver alloy. Such beads could be smelted from naturally occurring ores of Egyptian or Nubian "gold" mines.

Unfortunately, this method of sampling cannot be applied to hard materials, such as alloys of stainless steel and alloys of refractory metals. For metallic alloys with phases of different hardness, the troublesome problem of preferential abrasion arises. Consequently, bronze and brass can only be sampled in the mentioned method for a first qualitative survey. Quantitative results show a surplus of soft lead components, which are preferably rubbed off. As shown by the analysis of relevant standard reference materials, repeatability and accuracy get worse and the result for the soft components or phases can be incorrect by a factor of >2.

Steel alloys used for the core components in nuclear reactors were investigated after dissolution by Pepponi *et al.* [213]. Several elements, including rare earth elements, can cause a much higher activity than the main components of steel alloys, such as iron or titanium. In order to limit nuclear radioactive waste, only small amounts of the material were taken and dissolved in concentrated acids. The main components, iron and titanium, were separated first by means of ion-exchange chromatography. Rubidium and zirconium were added as internal standards, and 2 μl droplets of the acidic solution with 20 μg steel were deposited on a reflector for TXRF analysis. In order to excite the K shell of niobium, photon energies of 19.7 keV were used. For the rare earth elements, the white synchrotron radiation with high photon energies was applied. A side-looking geometry was chosen for detection and several alloy samples were investigated. The detection limit for niobium was about 50 ng/g, for rare earth elements, for example, terbium, 500 ng/g. Similar values were also found for simple EDXRF when larger droplets of 25 μl were deposited on 150 nm thick AP1™ films.

5.5.3 Textile Fibers and Glass Splinters

A special method has been developed for the analytical characterization of single textile fibers [214]. A variety of 35 different types and models of uncolored textile fibers (e.g., polyester, viscose, or wool) was first analyzed in order to get the respective element patterns.

A sample of 500 μg of every kind of fiber was placed on a carrier; then, 20 μl of nitric acid with 20 ng of an internal standard (gallium) were added and dried by evaporation. The spectra were recorded directly in order to first determine volatile elements like sulfur. Afterward, a cold oxygen plasma ashing was applied in order to determine the other elements with a higher sensitivity. Mass fractions were found between 10 μg/g and 10 mg/g. The elements K, Ca, Fe, and Zn were regarded as contaminants and not taken into account. However, six further elements (P, S, Ti, Mn, Sb, and Ba) were used to establish an appropriate database for the total set of 35 chosen fibers.

Subsequently, individual pieces of the set with a length of 2 mm were placed on carriers and analyzed after standard addition. The mass of these small pieces was roughly estimated to be 0.7 μg (350 μg/m) in order to determine the mass fractions of the six elements chosen. By application of the 2σ criterion, 65% of the fibers could be identified unambiguously. For the remaining 35%, the number of candidates could be restricted to less than three of all the 35 fibers chosen for this investigation.

Glass splinters or fragments from windshields or headlights may come from a hit-and-run car crash, from burglaries, or vandalisms. These fragments can serve as pieces of evidence that can be discriminated by their refractive index and their chemical composition. However, due to an improved quality control of manufacturers, the refractive index and also the main components of different glass samples hardly differ from each other. For that reason, the pattern of trace elements may be used for discrimination. Relevant elements can be identified by a microanalytical method like TXRF; their composition can be determined as a fingerprint, which can be compared with that of another glass or with several known glasses [199]. When window glass is shot through, lots of tiny glass splinters are ejected from the window and several of them should be found in the clothes of a suspect. Gunshot residues usually stick on a suspect's hand. Such pieces of circumstantial evidence like splinters and fragments can be identified by a powerful method of microanalysis like TXRF.

Nishiwaki *et al.* [199] developed a technique for TXRF of very small glass fragments. Samples with a volume of about or less than 0.2 mm^3 and a mass less than 0.5 mg were decomposed with 50 μl of HF and 50 μl HNO_3 by ultrasonification for 2 h. This mixture was condensed to a droplet of 10 μl in a water bath at 80 °C, dropped onto a Si wafer and dried. The residue with a lateral dimension of less than 1 cm was analyzed by TXRF with Mo excitation and a Si (Li) detector of 80 mm^2 active area within 1000 s. The precision of this method was checked by the analysis of NIST SRM 612 "Trace elements in glass." The peak area of 13 elements was normalized by that of strontium with the highest sensitivity and showed a mean relative standard deviation of about 5%.

Fragments of 23 figured sheet glasses produced by five different manufacturers could generally be characterized and discriminated by their refractive index in the range between 1.5118 and 1.5256 (relative difference 0.9%). Two samples are said to be different if their indices differ by >0.0002. Two groups could be built, which could roughly be distinguished. Four glass samples within the first group could not be discriminated from one another (difference ≤0.0001) and seven samples within the second group could hardly be

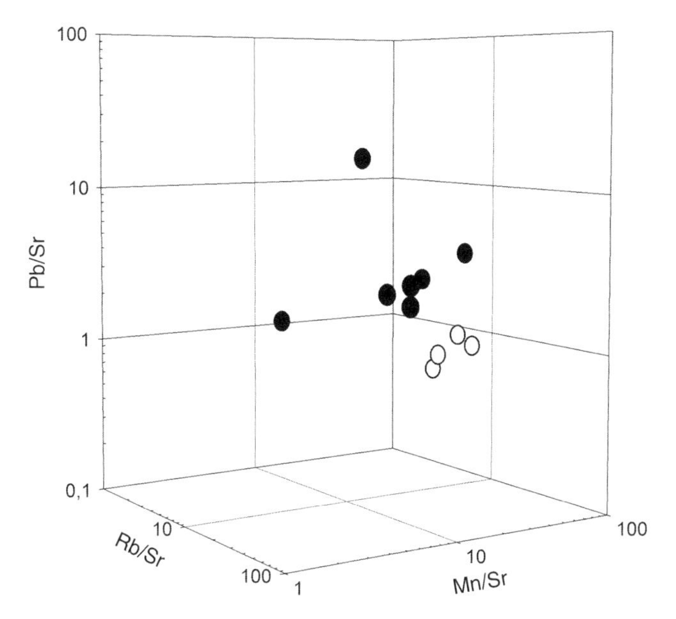

Figure 5.26. View of a three-dimensional scatter plot for different glass fragments. The logarithm of the concentration ratio of three elements, Mn, Rb, and Pb, with respect to strontium is represented. Group 1 with four glass fragments (○) is building a first cluster. It can clearly be distinguished from group 2 with seven fragments (●) building a second cluster. From Ref. [198], permission not required.

distinguished (difference <0.0003). However, TXRF could significantly discriminate the different glass samples of these two groups. It could be shown that the three intensity ratios of Mn/Sr, Rb/Sr, and Pb/Sr are sufficient for distinguishing small glass fragments. Figure 5.26 demonstrates a three-dimensional plot of these three elemental ratios.

Another example for forensic examinations is the identification and discrimination of white paint fragments from automobiles. It could be shown by Nishiwaki *et al.* [215] that 10 trace elements in TiO_2 pigments (rutile) unambiguously reflect the origin of such fragments. The detection of six heavy elements, Zr, Nb, and Sn and even Hf, Ta, and W, was especially helpful for the discrimination. These elements need high-energy excitation by a synchrotron in order to identify their Kα peaks in the spectra (>70 keV). Normal XRF with synchrotron excitation at 116 keV was applied to small fragments with dimensions of approximately 0.4 mm, but TXRF is also suitable for this task.

5.5.4 Drug Abuse and Poisoning

Certain drugs or remedies can be used as laxatives or may help to treat urological problems. The intake of such preparations with high concentration of arsenic or heavy metals like mercury or lead can be a health hazard and lead to severe chronic and life-threatening problems, such as osteomalacia, kidney malfunction, or obstructive lung disease. Drugs and remedies can be analyzed by TXRF as shown in Section 5.3.4, and blood and urine of a patient can be investigated after Sections 5.3.1 and 5.3.2. The hair of a patient can serve as a biomonitor and can also be analyzed by TXRF. The analysis of the hair of a dead person can reveal the cause of death even years or centuries later—the death may be induced by a deliberate attempt or by an unintentional accident.

Two pieces of hair of the violin virtuoso Nicolo Paganini (1782–1840) were cut with 2 cm length, placed on a quartz-glass carrier, dried, weighed, and directly analyzed by TXRF. Mercury was determined with a concentration of about 20 μg/g dried hair mass. This result, significantly above the reference value for human hair, was confirmed by ICP-MS. The authors explain the high mercury value by an abuse of mercury (II) chloride ($HgCl_2$) or mercury (I) chloride (Hg_2Cl_2), respectively "calomel." It could possibly be taken as a laxative and as a therapy against syphilis [216].

Narcotic drugs are generally prohibited worldwide with the exception of a few countries and for medical use. The production, selling, and abuse of narcotics are usually prosecuted as a crime. Muratsu *et al.* applied TXRF with synchrotron excitation (SR-TXRF) to narcotics as evidence [202]. A leaflet of marijuana was directly placed on a Si wafer and a soft lump of opium was smeared onto a wafer and analyzed. Furthermore, they dissolved drugs, such as cocaine, heroin, and amphetamines in pure water or acidic solutions (10 μg/μl) and analyzed the dried residues. Picograms of a contamination in a residue of 10 μg (i.e., 0.1 ppm) could be detected so that the spectral pattern of several trace elements could be determined and used as a fingerprint in order to identify producers, dealers, and abusers.

Besides narcotics, TXRF or XRF in general can identify poison and help to find out their origin in crimes. In July 1998, a summer festival was held in Wakayama city, 60 km southeast of Osaka, Japan. During this festival, curry was prepared by volunteers in different pots and served to visitors. By the next morning, four people died and another 63 fell seriously ill [217,218]. The police detected a toxic substance in the curry powder, which was proved to be white arsenic. They arrested an insurance saleswoman suspected of having committed the crime. The police were searching for substantial evidence. In the years after, a wave of copycat poisonings occurred in Japan.

Scientists at SPring-8 and KEK-PF analyzed the curry and found traces of arsenious acid. They compared the results with an arsenic sample from the suspect's husband who worked as an exterminator. By means of XRF analysis with synchrotron radiation they found out that the spectra of both samples were similar. Specific impurities of both samples were shown to be the same by pattern recognition. Consequently, the origin of the poison was assumed with no doubt. In the Japanese TV and also in the daily newspapers the result was celebrated. In 2002, the woman was condemned to death. In 2009, this death sentence was confirmed by the Supreme Court of Japan. Meanwhile, the Japanese government decided that X-ray spectrometers—mainly from

Rigaku—were installed in several police prefectures and hospitals all over the country. By this, X-ray methods rose to fame in Japan in one go [217].

The woman petitioned for a retrial—her counsels were arguing that only circumstantial evidence existed. In 2013, Kawai [219] reviewed two previous inquiry reports of the Wakayama crime. The first report was said to be a qualitative analysis by "pattern recognition" of two spectra, while the second report was confirmed to be the first quantitative report of SR-XRF analysis. Besides arsenic, the first report was based on three minor elements, Sn, Sb, and Bi, while the second report additionally used the elements selenium and lead. The peaks of these elements, however, were useless because of interferences. A recent analysis showed another four impurity elements in the raw spectra of the first report, particularly Fe, Zn, Mo, and Ba. In contrast to previous investigations, this result showed that the arsenic powder found in the flat of the condemned woman was significantly different than the powder in the paper cup used for poisoning.

On the one hand, this example indicates the generally high expectations placed on spectral analysis in criminology. On the other hand, it demonstrates considerable difficulties and limitations in complicated analytical situations. An exceptional problem is that lawyers and scientists talk in different languages and use difficult terms with different meanings, such as error, uncertainty, significance, and representativeness. Such terms are exactly defined and strictly used in science but have a vague border in colloquial language and can even have a different value judgment.

REFERENCES

1. Reus, U., Markert, B., Hoffmeister, C., Spott, D., and Guhr, H. (1993). Determination of trace metals in river water and suspended solids by TXRF spectrometry. A methodical study on analytical performance and sample homogeneity. *Fresenius J. Anal. Chem.*, **347**, 430–435.
2. Reus, U. and Prange, A. (1993). *Application Notes*, Atomika Instruments GmbH.
3. Klockenkämper, R. (1997). *Total-Reflection X-Ray Fluorescence Analysis*, 1st ed., John Wiley & Sons, Inc.: New York.
4. Stößel, R.-P. and Prange, A. (1985). Determination of trace elements in rainwater by total reflection X-ray fluorescence. *Anal. Chem.*, **57**, 2880–2885.
5. Farías, S.S., Casa, V.A., Vázquez, C., Ferpozzi, L., Pucci, G.N., and Cohen, I.M. (2003). Natural contamination with arsenic and other trace elements in ground waters of Argentine Pampean Plain. *Sci. Total Environ.*, **309**, 187–199.
6. Prange, A., Böddeker, H., and Kramer, K. (1993). Determination of trace elements in riverwater using total-reflection X-ray fluorescence. *Spectrochim. Acta*, **48B**, 207–215.
7. Prange, A., Knöchel, A., and Michaelis, W. (1985). Multielement determination of dissolved heavy metal traces in sea water by total-reflection X-ray fluorescence spectrometry. *Anal. Chim. Acta*, **172**, 79–100.

8. Prange, A. (1983). Entwicklung eines spurenanalytischen Verfahrens zur Bestimmung von gelösten Schwermetallen in Meerwasser mit Hilfe der Totalreflexions-Röntgenfluoreszenzanalyse, PhD Thesis. University of Hamburg.
9. Custo, G., Litter, M.I., Rodríguez, D., and Vázquez, C. (2006). Total reflection X-ray fluorescence trace mercury determination by trapping complexation: application in advanced oxidation technologies. *Spectrochimica Acta*, **B61**, 1119–1123.
10. Marguí, E., Kregsamer, P., Hidalgo, M., Tapias, J., Queralt, I., and Streli, C. (2010). Analytical approaches for Hg determination in wastewater samples by means of total reflection X-ray fluorescence spectrometry. *Talanta*, **82**, 821–827.
11. Prange, A., von Tümpling Jr., W., Niedergesäß, R., and Jantzen, E. (1995). Die gesamte Elbe auf einen Blick – Elementverteilungsmuster der Elbe von der Quelle bis zur Mündung. *Wasserwirtschaft - Wassertechnik*, **7**, 22–33.
12. Prange, A. and coworkers (2001). Erfassung und Beurteilung der Belastung der Elbe mit Schadstoffen, Abschlussbericht Band 2/3, Anhang Band 1/3, GKSS-Forschungszentrum, Geesthacht.
13. Schmidt, D., Gerwinski, W., and Radke, I. (1993). Trace metal determinations by total-reflection X-ray fluorescence analysis in the open Atlantic Ocean. *Spectrochim. Acta*, **48B**, 171–181.
14. Freimann, P. and Schmidt, D. (1988). Application of total reflection X-ray fluorescence analysis for the determination of trace metals in the North Sea. *Spectrochim. Acta*, **44B**, 505–511.
15. Haarich, M., Schmidt, D., Freimann, P., and Jakobsen, A. (1993). North Sea research projects ZISCH and PRISMA: application of total reflection X-ray fluorescence in sea-water analysis. *Spectrochim. Acta*, **48B**, 183–192.
16. Barreiros, M.A., Carvalho, M.L., Costa, M.M., Marques, M.I., and Ramos, M.T. (1997). Application of total-reflection XRF to elemental studies of drinking water. *X-Ray Spectrom.*, **26**, 165–168.
17. Muia, L.M., Razafindramisa, F.L., and Van Grieken, R.E. (1991). Total reflection X-ray fluorescence analysis using an extended focus tube for the determination of dissolved elements in rain water. *Spectrochim. Acta*, **B46**, 1421–1427.
18. Miesbauer, H. (1997). Multielement determination in sediments, pore water and river water of Upper Austrian rivers by total-reflection X-ray fluorescence. *Spectrochim. Acta*, **B52**, 1003–1007.
19. Gatari, M.J., Maina, D.M., Wagner, A., Boman, J., Gaita, S.M., and Bartilol, S. (2013). Trace metals in waters of Lake Victoria basin, 15th International Conference on TXRF and related methods, Book of Abstracts, P15, p. 48.
20. Gerwinski, W. and Schmidt, D. (1998). Automated solid-phase extraction for tracemetal analysis of seawater: sample preparation for total-reflection X-ray fluorescence measurements. *Spectrochim. Acta*, **B53**, 1355–1364.
21. Pashkova, G.V., Revenkov, A.G., and Finkelshtein, A.L. (2013). Study of factors affecting the results of natural water analyses by total reflection X-ray fluorescence. *X-Ray Spectrom.*, **42**, 524–530.
22. Prange, A., Knoth, J., Stößel, R.-P., Böddeker, H., and Kramer, K. (1987). Determination of trace elements in the water cycle by total-reflection X-ray fluorescence spectrometry. *Anal. Chim. Acta*, **195**, 275–287.

23. Freimann, P., Schmidt, D., and Neubauer-Ziebarth, A. (1993). Reference materials for quality assurance in sea-water analysis: performance of total-reflection X-ray fluorescence in the intercomparison and certification stages. *Spectrochim. Acta*, **48B**, 193–198.
24. https://en.wikipedia.org/wiki/Particulates; Wikimedia Foundation, Inc., headquarter in the United States of America, 09-Sept-2014
25. Bayer, H., von Bohlen, A., Klockenkämper, R., and Klockow, D. (1995). Choice of a suitable material for construction of a Battelle type impactor to minimize systematic errors in sampling of airborne dust. *Mikrochim. Acta*, **119**, 167–176.
26. Schneider, B. (1989). The determination of atmospheric trace metal concentrations by collection of aerosol particles on sample holders for total-reflection X-ray fluorescence. *Spectrochim. Acta*, **44B**, 519–524.
27. Injuk, J. and Van Grieken, R. (1995). Optimization of total-reflection X-ray fluorescence for aerosol analysis. *Spectrochim. Acta*, **50B**, 1787–1803.
28. Salvà, A., von Bohlen, A., Klockenkämper, R., and Klockow, D. (1993). Multi-element analysis of airborne particulate matter by total reflection X-ray fluorescence. *Quimica Analitica*, **12**, 57–62.
29. Injuk, J., Van Grieken, R., Klockenkämper, R., von Bohlen, A., and Kump, P. (1997). Performance and characteristics of two total reflection X-ray fluorescence and a particle induced X-ray emission setup for aerosol analysis. *Spectrochim. Acta*, **52B**, 977–984.
30. Klockenkämper, R., Bayer, H., and von Bohlen, A. (1995). Total reflection X-ray fluorescence analysis of airborne particulate matter. *Adv. X-ray Chem. Anal. Jpn.*, **26s**, 41–46.
31. Michaelis, W., Schönburg, M., and Stößel, R.-P. (1989). *Mechanisms and Effects of Pollutant-Transfer into Forests*, Kluwer Academic Publishers: Dordrecht, p. 3.
32. Pepelnik, R., Erbslöh, B., Michaelis, W., and Prange, A. (1993). Determination of trace element deposition into a forest ecosystem using total-reflection X-ray fluorescence. *Spectrochim. Acta*, **48B**, 223–229.
33. Schmeling, M. (2004). Characterization of urban air pollution by total reflection X-ray fluorescence. *Spectrochim. Acta*, **59B**, 1165–1171.
34. Wagner, A., Boman, J., and Gatari, M.J. (2008). Elemental analysis of size-fractionated particulate matter sampled in Göteborg, Sweden. *Spectrochim. Acta*, **63B**, 1426–1431.
35. Boman, J., Wagner, A., and Gatari, M.J. (2010). Trace elements in $PM_{2.5}$ in Gothenburg, Sweden. *Spectrochim. Acta*, **65B**, 478–482.
36. Boman, J., Gaita, S., Petterson, J., Hallquist, M., Gatari, M.J., and Maina, D. (2013). Is TXRF suitable for analysis of ambient aerosols? 15th International Conference on TXRF and related methods, Book of Abstracts, I14, p. 105.
37. Samek, L., Ostachowicz, B., Worobiec, A., Spolnik, Z., and Van Grieken, R. (2006). Speciation of selected metals in aerosol samples by TXRF after sequential leaching. *X-Ray Spectrom.*, **35**, 226–231.
38. Osan, J., Török, S., Alfoldy, B., Alsecz, A., Falkenberg, G., Baik, S.Y., and Van Grieken, R. (2007). Comparison of sediment pollution in the rivers of the Hungarian Upper Tisza Region using non-destructive analytical techniques. *Spectrochim. Acta*, **62B**, 123–136.

39. Aster, B., von Bohlen, A., and Burba, P. (1997). Determination of metals and their species in aquatic humic substances by using total-reflection X-ray fluorescence spectrometry. *Spectrochim. Acta*, **52B**, 1009–1018.
40. Friese, K., Mages, M., Wendt-Potthof, K., and Neu, T.R. (1997). Determination of heavy metals in biofilms from River Elbe by total-reflection X-ray fluorescence spectrometry. *Spectrochim. Acta*, **52B**, 1019–1025.
41. Mages, M. (2013). Elementbestimmungen in aquatischen Biofilmen und Zooplankton mittels Totalreflexions-Röntgenfluoreszenzanalytik, Dissertation, Universität Lüneburg, Institut für Ökologie und Umweltchemie, 127 Seiten.
42. Mages, M., Woelfl, S., Ovary, M., von Tümpling, W., and Encina, F. (2004). The use of a portable total reflection X-ray fluorescence spectrometer for trace element determinations in freshwater microcrustaceans (Daphnia). *Spectrochim. Acta*, **59B**, 1265–1272.
43. Kröpfl, K., Vladar, P., Szabo, K., Acs, E., Borsodi, A.K., Szikora, S., Caroli, S., and Zaray, G. (2006). Chemical and biological characterisation of biofilms formed on different substrata in Tisza river (Hungary). *Environ. Pollut.*, **144**, 626–631.
44. Wagner, A. and Boman, J. (2003). Biomonitoring of trace elements in muscle and liver tissue of freshwater fish. *Spectrochim. Acta*, **58B**, 2215–2226.
45. Viksna, A., Znotina, V., and Boman, J. (1999). Concentrations of some elements in and on Scots pine needles. *X-Ray Spectrom.*, **28**, 275–281.
46. Kump, P., Necemer, N., and Snaider, J. (1996). Determination of trace elements in bee honey, pollen and tissue by total reflection and radioisotope X-ray fluorescence spectrometry. *Spectrochim. Acta*, **51B**, 499–507.
47. Enrich, C., Boeykens, S., Caracciolo, N., Custo, G., and Vázquez C. (2007). Honey characterization by total reflection X-ray fluorescence: evaluation of environmental quality and risk for the human health. *X-Ray Spectrom.*, **36**, 215–220.
48. Dalipi, R., Borgese, L., Zacco, A., Fenotti, M., Piro, R., Bontempi, E., and Depero, L.E. (2013). TXRF as a tool for determination of food safety and traceability, 15th International Conference on TXRF and related methods, Book of Abstracts, P55, p. 134.
49. Wagner, A. and Boman, J. (2004). Biomonitoring of trace elements in Vietnamese freshwater mussels. *Spectrochim. Acta*, **59B**, 1125–1132.
50. Martinez, T., Zarazua, G., Fernandez, A., Marquez, C., Tejeda, S., Avilla-Perez, P., Poblano-Bata, J., and Macedo, G. (2013). Total reflection X-ray fluorescence determination of lead in environmental samples from the metropolitan zone of Toluca valley, 15th International Conference on TXRF and related methods, Book of Abstracts, P53, p. 130.
51. Vazquez, C., Rodriguez Castro, M.C., Andreano, V., and Marco, L. (2013). Measuring exposition to different contaminants in tissues by means of total reflection X-ray fluorescence, 15th International Conference on TXRF and related methods, Book of Abstracts, I15, p. 107.
52. Miesbauer, H., Köck, G., and Püreder, L. (2001). Determination of trace elements in macrozoobenthos samples by total reflection X-ray fluorescence analysis. *Spectrochim. Acta*, **56B**, 2203–2207.
53. Woelfl, S., Mages, M., and Torres, P. (2008). Trace metal concentrations in single specimens of the intestinal broad platworm (Diphyllobothrium latum), compared

to their fish host (Onchorhynchus mykiss) measured by total reflection X-ray fluorescence spectrometry. *Spectrochim. Acta*, **63B**, 1450–1454.

54. Fernandez-Ruiz, R., Redrejo, M.J., Friedrich, E.J., Ramos, M., and Fernandez, T. (2013). Pioneer application of the TXRF spectrometry to the evaluation of the bioaccumulation kinetic of gold nanorods by several vital organs on mice, 15th International Conference on TXRF and related methods, Book of Abstracts, P48, p. 121.
55. Zimmermann, S., Messerschmidt, J., von Bohlen, A., and Sures, B. (2005). Uptake and bioaccumulation of platinum group metals (Pd, Pt, Rh) from automobile catalytic converter materials by the Zebra mussel (Dreissena polymorpha). *Environ. Res.*, **98**, 203–209.
56. Pepponi, G., Lazzeri, P., Coghe, N., Bersani, M., Gottardini, E., Cristofolini, F., Clauser, G., and Torboli, A. (2004). Total reflection X-ray fluorescence analysis of pollen as indicator for atmospheric pollution. *Spectrochim. Acta*, **59B**, 1205–1209.
57. Khuder, A., Bakir, M.A., Karjou, J., and Sawan, M.K. (2007). XRF and TXRF techniques for multi-element determination of trace elements in whole blood and human hair samples. *J. Radioanal. Nucl. Chem.*, **273**, 435–442.
58. Gerwinski, W. and Goetz, D. (1987). Multielement analysis of standard reference materials with total reflection X-ray fluorescence (TXRF). *Fresenius Z. Anal. Chem.*, **327**, 690–693.
59. Cherkashina, T.Y., Panteeva, S.V., and Pashkova, G.V. (2013). Applications of total reflection X-ray fluorescence spectrometry for geological problem solving, 15th International Conference on TXRF and related methods, Book of Abstracts, I12, p. 95.
60. Koopmann, C. and Prange, A. (1991). Multielement determination in sediments from the German Wadden Sea - investigations on sample preparation techniques. *Spectrochim. Acta*, **46B**, 1395–1402.
61. Battiston, G.A., Gerbasi, R., Degetto, S., and Sbrignadello, G. (1993). Heavy metal speciation in coastal sediments using total-reflection X-ray fluorescence spectrometry. *Spectrochim. Acta*, **48B**, 217–221.
62. Freiburg, C., Krumpen, W., and Troppenz, U. (1993). Determinations of cerium, europium and terbium in the electroluminescent materials gadolinium oxysulfide (Gd_2O_2S) and lanthanum oxysulfide (La_2O_2S) by total-reflection X-ray spectrometry. *Spectrochim. Acta*, **48B**, 263–267.
63. Panteeva, S.V. and Cherkashina, T.Y. (2013). Multielement analysis of rocks by total reflection X-ray fluorescence: sample preparation methods, 15th International Conference on TXRF and related methods, Book of Abstracts, P18, p. 53.
64. Gatari, M.J., Mugoh, F., Njenga, L.W., Shepherd, K.D., Sila, A., Kamau, M.N., and Maina, D.M. (2013). Prediction of soil properties: an experiment using Mt Kenya forest soils, 15th International Conference on TXRF and related methods, Book of Abstracts, O7, p. 23.
65. Revenko, A.G. (2010). The special features of analytical techniques for geological samples using TXRF spectrometers. *Analytics and Control*, **V14**, 42–64.
66. International Atomic Energy Agency, Austria. (1993). Intercomparison run on trace and minor elements in candidate lichen research material (IAEA-336) and AQCS cabbage material (IAEA-359), Progress report No. 1.

67. Schmeling, M., Alt, F., Klockenkämper, R., and Klockow, D. (1997). Multielement analysis by total reflection X-ray fluorescence spectrometry for the certification of lichen research material. *Fresenius J. Anal. Chem.*, **357**, 1042–1044.

68. Marcó, P., Lué, M., and Hernández-Caraballo, E.A. (2004). Direct analysis of biological samples by total reflection X-ray fluorescence. *Spectrochim. Acta*, **59B**, 1077–1090.

69. Szoboszlai, N., Polgary, Z., Mihucz, V.G., and Zaray, G. (2009). Recent trends in total reflection X-ray fluorescence spectrometry for biological applications. *Anal. Chim. Acta*, **633**, 1–18.

70. Xie, M.Y., von Bohlen, A., Günther, K., Jian, X.H., and Klockenkämper, R. (1998). Multielement analysis of Chinese tea (Camellia Sinensis) by total-reflection X-ray fluorescence. *Z. Lebensm. Unters. Forsch. A.*, **207**, 31–38.

71. Haswell, S.J. and Walmsley, A.D. (1991). Chemometrics–the key to sensor array development. *Anal. Proc.*, **28**, 115–117.

72. Haswell, S.J. and Walmsley, A.D. (1998). Multivariate data visualisation methods based on multi-elemental analysis of wines and coffees using total reflection X-ray fluorescence analysis. *J. Anal. At. Spectrom.*, **13**, 131–134.

73. Castineira, M.M., Brandt, R., von Bohlen, A., and Jakubowski, N. (2001). Development of a procedure for the multielement determination of trace elements in wine by ICP-MS. *Fresenius J. Anal. Chem.*, **370**, 553–558.

74. Carvalho, M.L., Barreiros, M.A., Costa, M.M., Ramos, M.T., and Marques, M.I. (1996). Study of heavy metals in Madeira wine by total reflection X-ray fluorescence analysis. *X-Ray Spectrom.*, **25**, 29–32.

75. Gruber, X., Kregsamer, P., Wobrauschek, P., and Streli, C. (2006). Total-reflection X-ray fluorescence analysis of Austrian wine. *Spectrochim. Acta*, **61B**, 1214–1218.

76. Pessanha, S., Carvalho, M.L., Becker, M., and von Bohlen, A. (2010). Quantitative determination on heavy metals in different stages of wine production by total reflection X-ray fluorescence and energy dispersive X-ray fluorescence: comparison on two vineyards. *Spectrochim. Acta*, **65B**, 504–507.

77. Yap, C.T. and Gunawardena, K.V.R. (1989). Analysis of trace elements in soft drinks using total reflection X-ray fluorescence spectrometry. *Int. J. Environ. Stud.*, **32**, 297–302.

78. Capote, T., Marcó, L.M., Alvarado, J., and Greaves, E.D. (1999). Determination of copper, iron and zinc in spirituous beverages by total reflection X-ray fluorescence spectrometry. *Spectrochim. Acta*, **54B**, 1463–1468.

79. Reus, U. (1991). Determination of trace elements in oils and greases with total reflection X-ray fluorescence: sample preparation methods. *Spectrochim. Acta*, **46B**, 1403–1411.

80. Günther, K. and von Bohlen, A. (1990). Simultaneous multielement determination in vegetable foodstuffs and their respective cell fractions by total-reflection X-ray fluorescence (TXRF). *Z. Lebensm. Unters. Forsch.*, **190**, 331–335.

81. Günther, K. and von Bohlen, A. (1991). Multielement speciation in vegetable foodstuffs by gel permeation chromatography (GPC) and total-reflection X-ray fluorescence (TXRF). *Spectrochim. Acta*, **46B**, 1413–1419.

82. Bilo, F., Borgese, L., Zacco, A., Lazo, P., Zoani, C., Bontempi, E., and Depero, L.E. (2013). TXRF as a screening technique to assess environmental pollution,

15th International Conference on TXRF and related methods, Book of Abstracts, O11, p. 97.

83. Kulesh, N.A., Muela, A., Fdez-Gubieda, L., Novoselova, Y.P., Denisova, T.P., Safranov, A.P., Zarubina, K., and Kurlyandskaya, G.V. (2013). Study of trace iron concentration in biological samples by total reflection X-ray fluorescence spectrometry, 15th International Conference on TXRF and related methods, Book of Abstracts, O12, p. 99.
84. Mulder, D.W., Boyd, E.S., Sarma, R., Lange, R.K., Endrizzi, J.A., Broderick, J.B., and Peters, J.W. (2010). Stepwise [FeFe]-hydrogenase H-cluster assembly revealed in the structure of $HydA^{\Delta EFG}$. *Nature*, **465**, 248–252.
85. Brown, D.R., Qin, K., Herms, J.W., Madlung, A., Manson, J., Strome, R., Fraser, P.E., Kruck, T., von Bohlen, A., Schultz-Schaeffer, W., Griese, A., Westaway, D., and Kretzschmar, H. (1997). The cellular prion protein binds copper in vivo. *Nature*, **390**, 684–687.
86. Quentmeier, A., Kraft, R., Kostka, S., Klockenkämper, R., and Friedrich, C.G. (2000). Characterization of a new type of sulfite dehydrogenase from Paracoccus pantotrophus GB17. *Arch. Microbiol.*, **173**, 117–125.
87. Schilling, S., Cynis, H., von Bohlen, A., Hoffmann, T., Wermann, M., Heiser, U., Buchholz, M., Zunkel, K., and Demuth, H.U. (2005). Isolation, catalytic properties, and competitive inhibitors of the zinc-dependent murine glutaminyl cyclase. *Biochemistry*, **44**, 13415–13424.
88. Greaves, E.D., Angeli-Greaves, M., Jaehde, U., Drescher, A., and von Bohlen, A. (2006). Rapid determination of platinum plasma concentrations of chemotherapy patients using total reflection X-ray fluorescence. *Spectrochim. Acta*, **61B**, 1194–1200.
89. Prange, A., Böddeker, H., and Michaelis, W. (1989). Multielement determination of trace elements in whole blood and blood serum by TXRF. *Fresenius J. Anal. Chem.*, **335**, 914–918.
90. Knöchel, A., Bethel, U., and Hamm, V. (1989). Multielement analysis of trace elements in blood serum by TXFA. *Fresenius Z. Anal. Chem.*, **334**, 673.
91. Knoth, J., Schwenke, H., Marten, R., and Glauer, J. (1977). Total-reflection X-ray fluorescence spectrometric determination of elements in nanogram amounts. *J. Clin. Chem. Clin. Biochem.*, **15**, 557–560.
92. Yap, C.T. (1988). X-ray total reflection fluorescence analysis of iron, copper, zinc, and bromine in human serum. *Appl. Spectrosc.*, **42**, 1250–1253.
93. Ayala, R.E., Alvarez, E.M., and Wobrauschek, P. (1991). Direct determination of lead in whole human blood by total reflection X-ray fluorescence spectrometry. *Spectrochim. Acta*, **46B**, 1429–1432.
94. Dogan, P., Dogan, M., and Klockenkämper, R. (1993). Determination of trace elements in blood serum of patients with Behcet disease by total reflection X-ray fluorescence analysis. *Clin. Chem. (Washington, D.C.)*, **39**, 1039–1041.
95. Greaves, E.D., Meitín, J., Sajo-Bohus, L., Castelli, C., Liendo, J., and Borgerg, C. (1995). Trace element determination in amniotic fluid by total reflection X-ray fluorescence. *Adv. X-Ray Chem. Anal. Jpn.*, **26s**, 47–52.
96. von Bohlen, A., Klockenkämper, R., Otto, H., Tölg, G., and Wiecken, B. (1987). Qualitative survey analysis of thin layers of tissue samples-Heavy metal traces in human lung tissue. *Int. Arch. Occup. Environ. Health*, **59**, 403–411.

97. von Bohlen, A., Klockenkämper, R., Tölg, G., and Wiecken, B. (1988). Microtome sections of biomaterials for trace analyses by TXRF. *Fresenius Z. Anal. Chem.*, **331**, 454–458.
98. Klockenkämper, R., von Bohlen, A., and Wiecken, B. (1989). Quantification in total reflection X-ray fluorescence analysis of microtome sections. *Spectrochim. Acta*, **44B**, 511–518.
99. von Bohlen, A., Rechmann, P., Tourmann, J.L., and Klockenkämper, R. (1994). Ultramicroanalysis of dental plaque films by total reflection X-ray fluorescence. *J. Trace Elem. Electrolytes Health Dis.*, **8**, 37–42.
100. Polyakova, N.V. (2013). Determination of elemental composition in biological samples by TXRF method, 15th International Conference on TXRF and related methods, Book of Abstracts, P1, p. 25.
101. Greaves, E.D., Marco-Parra, L.M., Rojas, A., and Sajo-Bohus, L. (2000). Determination of platinum levels in serum and urine samples from pediatric cancer patients by TXRF. *X-Ray Spectrom.*, **29**, 349–353.
102. Ostachowitz, B., Lankosz, M., Tomik, B., Adamek, D., Wobrauschek, P., Streli, C., and Kregsamer, P. (2006). Analysis of some chosen elements of cerebrospinal fluid and serum in amyotrophic lateral sclerosis patients by total reflection X-ray fluorescence. *Spectrochim. Acta*, **B61**, 1210–1213.
103. Liendo, J.A., Gonzalez, A.C., Castelli, C., Gomez, J., Jimenez, J., Marco, L., Sajo-Bohus, L., Greaves, E.D., Fletcher, N.R., and Bauman, S. (1999). Comparison between proton-induced X-ray emission (PIXE) and total reflection X-ray fluorescence (TXRF) spectrometry for the elemental analysis of human amniotic fluid. *X-Ray Spectrom.*, **28**, 3–8.
104. Telgmann, L., Holtkamp, M., Künnemeyer, J., Gelhard, C., Hartmann, M., Klose, A., Sperling, M., and Karst, U. (2011). Simple and rapid quantification of gadolinium in urine and blood plasma samples by means of total reflection X-ray fluorescence (TXRF). *Metallomics*, **3**, 1035–1040.
105. Carvalho, M.L., Custodio, P.J., Reus, U., and Prange, A. (2001). Elemental analysis of human amniotic fluid and placenta by total-reflection X-ray fluorescence and energy-dispersive X-ray fluorescence: child weight and maternal age dependence. *Spectrochim. Acta*, **B56**, 2175–2180.
106. Magalhães, T., Carvalho, M.L., von Bohlen, A., and Becker, M. (2010). Study on trace elements behaviour in cancerous and healthy tissues of colon, breast and stomach: total reflection X-ray fluorescence applications. *Spectrochim. Acta*, **65B**, 493–498.
107. Carvalho, M.L., Magalhães, T., Becker, M., and von Bohlen, A. (2007). Trace elements in human cancerous and healthy tissues: a comparative study by EDXRF, TXRF, synchrotron radiation and PIXE. *Spectrochim. Acta*, **62B**, 1004–1011.
108. Benninghoff, L., von Czarnowski, D., Denkhaus, E., and Lemke, K. (1997). Analysis of human tissues by total reflection X-ray fluorescence. Application of chemometrics for diagnostic cancer recognition. *Spectrochim. Acta*, **52B**, 1039–1046.
109. von Czarnowski, D., Denkhaus, E., and Lemke, K. (1997). Determination of trace element distribution in cancerous and normal human tissues by total reflection X-ray fluorescence analysis. *Spectrochim. Acta*, **52B**, 1047–1052.

110. Magalhães, T., Becker, M., Carvalho, M.L., and von Bohlen, A. (2008). Study of Br, Zn, Cu and Fe concentrations in healthy and cancer breast tissues by TXRF. *Spectrochim. Acta*, **63B**, 1473–1479.

111. Marco-Parra, L.M., Jimenez, E., Hernandez, E.A., Rojas, C.A., Greaves, E.D. (2001). Determination of Zn/Cu ratio and oligoelements on serum samples by total reflection X-ray fluorescence spectrometry for cancer diagnosis. *Spectrochim. Acta*, **56B**, 2195–2201.

112. Wagner, M., Rostam-Khani, P., Wittershagen, A., Rittmeyer, C., Kolbesen, B.O., and Hoffmann, H. (1997). Trace element determination in drugs by total-reflection X-ray fluorescence spectrometry. *Spectrochim. Acta*, **52B**, 961–965.

113. Wagner, M., Rostam-Khani, P., Wittershagen, A., Rittmeyer, C., Hoffmann, H., and Kolbesen, B.O. (1996). Application of total reflection X-ray fluorescence spectrometry (TXRF) to trace element determination in pharmaceutical substances. *Pharmazie*, **51**, 865–868.

114. Borgese, L., Zacco, A., Bontempi, E., Pellegatta, M., Vigna, L., Patrini, L., Riboldi, L., Rubino, F.M., and Depero, L.E. (2010). Use of total reflection X-ray fluorescence for the evaluation of heavy metal poisoning due to improper use of a traditional ayurvedic drug. *J. Pharm. Biomed. Anal.*, **52**, 787–790.

115. Prange, A., Kramer, K., and Reus, U. (1991). Determination of trace element impurities in ultrapure reagents by total reflection X-ray spectrometry. *Spectrochim. Acta*, **46B**, 1385–1393.

116. Burba, P., Willmer, P.G., Becker, M., and Klockenkämper, R. (1989). Determination of trace elements in high-purity aluminium by total reflection X-ray fluorescence after their separation on cellulose loaded with hexamethylenedithiocarbamates. *Spectrochim. Acta*, **44B**, 525–532.

117. Chen, J.S., Berndt, H., Klockenkämper, R., and Tölg, G. (1990). Trace analysis of high-purity iron by total reflection X-ray fluorescence spectrometry. *Fresenius J. Anal. Chem.*, **338**, 891–894.

118. Reus, U. (1989). Total reflection X-ray fluorescence spectrometry: matrix removal procedures for trace analysis of high-purity silicon, quartz and sulphuric acid. *Spectrochim. Acta*, **44B**, 533–542.

119. Klockenkämper, R., Becker, M., and Bubert, H. (1991). Determination of the heavy-metal ion-dose after implantation in silicon-wafers by total-reflection X-ray fluorescence analysis. *Spectrochim. Acta*, **46B**, 1379–1383.

120. Ojeda, N., Greaves, E.D., Alvarado, J., and Sajo-Bohus, L. (1993). Determination of vanadium, iron, nickel and sulphur in petroleum crude oil by total-reflection X-ray fluorescence. *Spectrochim. Acta*, **48B**, 247–253.

121. Bilbrey, D.B., Leland, D.J., Leyden, D.E., Wobrauschek, P., and Aiginger, H. (1987). Determination of metals in oil using total reflection X-ray fluorescence spectrometry. *X-Ray Spectrom.*, **16**, 161–165.

122. Freitag, K., Reus, U., and Fleischhauer, J. (1989). The application of TXRF spectrometry for the determinations of trace metals in lubricating oils. *Fresenius Z. Anal. Chem.*, 334, 675-??

123. Hahn, J.U. and Jaschke, M. (1993). Determination of metals in cooling lubricants by use of total-reflection X-ray fluorescence analysis. *Staub-Reinhalt. Luft*, **53**, 109–113.

124. Freitag, K., Reus, U., and Fleischhauer, J. (1989). Extension of the analytical range of total reflection X-ray fluorescence spectrometry to lighter elements (11 < Z < 16) and increase in sensitivity by excitation with tungsten La radiation. *Spectrochim. Acta*, **44B**, 499–504.

125. Yap, C.T., Ayala, R.E., and Wobrauschek, P. (1988). Quantitative trace element determination in thin film samples by TXRF using the scattered radiation method. *X-Ray Spectrom.*, **17**, 171–174.

126. Schirrmacher, M., Freimann, P., Schmidt, D., and Dahlmann, G. (1993). Trace elemental determination by total-reflection X-ray fluorescence (TXRF) for the differentiation between pure fuel oil (bunker oil) and waste oil (sludge) in maritime shipping legal cases. *Spectrochim. Acta*, **48B**, 199–205.

127. Hoffmann, P., Lieser, K.H., Hein, M., and Flakowski, M. (1989). Analysis of thin layers by total-reflection X-ray fluorescence spectrometry. *Spectrochim. Acta*, **44B**, 471–476.

128. Kudermann, G. (1988). Characterization of high-purity aluminium. *Fresenius Z. Anal. Chem.*, **331**, 697–706.

129. Graule, T., von Bohlen, A., Broekaert, J.A.C., Grallath, E., Klockenkämper, R., Tschöpel, P., and Tölg, G. (1989). Atomic emission and atomic absorption spectrometric analysis of high-purity powders for the production of ceramics. *Fresenius' Z. Anal. Chem.*, **335**, 637–642.

130. Peschel, B.U., Fittschen, U.E.A., Pepponi, G., et al. (2005). Direct analysis of Al_2O_3 powders by total reflection X-ray fluorescence spectrometry. *Anal. Bioanal. Chem.*, **382**, 1958–1964.

131. Amberger, M.A., Höltig, M., and Broekaert, J.A.C. (2010). Direct determination of trace elements in boron nitride powders by slurry sampling total reflection X-ray fluorescence spectrometry. *Spectrochim. Acta*, **65B**, 152–157.

132. Amberger, M.A. and Broekaert, J.A.C. (2009). Direct multielement determination of trace elements in boron carbide powders by total reflection X-ray fluorescence spectrometry and analysis of the powders by ICP atomic emission. *J. Anal. At. Spectrom.*, **24**, 1517–1523.

133. Dhara, S., Misra, N.L., Thakur, U.K., Shah, D., Sawant, R.M., Ramakumar, K.L., and Aggarwal, S.K. (2012). A total reflection X-ray fluorescence method for the determination of chlorine at trace levels in nuclear materials without sample dissolution. *X-Ray Spectrom.*, **41**, 316–320.

134. Misra, N.L. (2013). Advanced X-ray spectrometric techniques for characterization of nuclear materials, The 15th International conference on total reflection X-ray fluorescence analysis and related methods (TXRF2013) and The 49th Annual conference on X-ray chemical analysis, Book of Abstracts, I11, p. 94.

135. Cinosi, A., Andriollo, N., Pepponi, G., and Monticelli, D. (2011). A novel total reflection X-ray fluorescence procedure for the direct determination of trace elements in petrochemical products. *Anal. Bioanal. Chem.*, **399**, 927–933.

136. Vázquez, C. (2004). The capabilities of total reflection X-ray fluorescence in the polymeric analytical field. *Spectrochim Acta*, **59B**, 1215–1219.

137. Boeykens, S., Caracciolo, N., D'Angelo, M.V., and Vazquez, C. (2006). Metal ions diffusion through polymeric matrices: a total reflection X-ray fluorescence study. *Spectrochim. Acta*, **61B**, 1236–1239.

138. Boeykens, S., Vazquez, C., and Temprano, N. (2003). Macromolecules by total reflection X-ray fluorescence: marking techniques. *Spectrochim. Acta*, **58B**, 2169–2175.

139. Zaitz, M.A. (2013). Developments with TXRF analysis for semiconductor metallic contamination for both new materials and processes, 15th International Conference on TXRF and related methods, Book of Abstracts, I05, p. 11.

140. Hockett, R.S. (1994). TXRF semiconductors applications. *Adv. X-Ray Anal.*, **37**, 565–575.

141. Hockett, R.S. (1995). A review of standardization issues for TXRF and VPD/TXRF. *Adv. X-Ray Chem. Anal. Jpn.*, **26s**, 79–84.

142. Iida, A., Sakurai, K., Yoshinaga, A., and Gohshi, Y. (1986). Grazing incidence X-ray fluorescence analysis. *Nucl. Instrum. Methods*, **A246**, 736–738.

143. Iida, A., Sakurai, K., and Gohshi, Y. (1988). Near surface analysis of semiconductor using grazing incidence X-ray fluorescence. *Adv. X-Ray Anal.*, **31**, 487–494.

144. Hockett, R.S., Baumann, S.M., and Schemmel, E. (1988). *In* T.J. Shaffner and D.K. Schroder (Eds.) *Diagnostic Techniques for Semiconductor Materials and Devices*, ECS Proceedings **88–20**, 113.

145. Hockett, R.S. and Katz, W. (1989). Comparison of wafer cleaning processes using total reflection X-ray fluorescence (TXRF). *J. Electrochem. Soc.*, **136**, 3481–3486.

146. Penka, V. and Hub, W. (1989). Application of total reflection X-ray fluorescence in semiconductor surface analysis. *Spectrochim Acta*, **44B**, 483–490.

147. Eichinger, P., Rath, J., and Schwenke, H. (1989). *In* D.C. Gupta (Ed.) "*Semiconductor Fabrication: Technology and Metrology*", ASTM STP 990, Application of Total Reflection X-Ray Fluorescence Analysis for Metallic Trace Impurities on Silicon Wafer Surfaces. American Society for Testing and Materials: Philadelphia, 305–313.

148. Nishihagi, K., Yamashita, N., Fujino, N., Taniguchi, K., and Ikeda, S. (1991). Impurity analysis on silicon wafer using Monochro-TREX. *Adv. X-Ray Chem. Anal. Jpn.*, **22**, 121.

149. Berncike, W. (1993). Basic features of total-reflection X-ray fluorescence analysis on silicon wafers. *Spectrochim. Acta*, **48B**, 269–275.

150. Torcheux, L., Degraeve, B., Mayeux, A., and Delamar, M. (1994). Calibration procedure for quantitative surface analysis by total reflection X-ray fluorescence. *Surf. Interface Anal.*, **21**, 192–198.

151. Gambino, V., Moccia, G., Girolami, E., and Alfonsetti, R. (1995). Study of metal contamination induced by ion implantation process using TRXRF and SIMS techniques. *Adv. X-Ray Chem. Anal. Jpn.*, **26s**, 35–40.

152. Mori, Y., Shimanoe, K., and Sakon, T. (1995). Standard sample preparation for quantitative TXRF analysis. *Adv. X-Ray Chem. Anal. Jpn.*, **26s**, 69–72.

153. Hellin, D., Bearda, T., Zhaoa, C., Raskinc, G., Mertens, P.W., De Gendt, S., Heynsa, M.M., and Vinckier, C. (2003). Determination of metallic contaminants on Ge wafers using direct and droplet sandwich etch-total reflection X-ray fluorescence spectrometry. *Spectrochimica Acta*, **58B**, 2093–2104.

154. Kamakura, T., Sugamoto, H., Tsuchiya, N., and Matsushita, Y. (1995). The application of TXRF for the adsorbed impurities on the GaAs wafers. *Adv. X-Ray Chem. Anal. Jpn.*, **26s**, 169–174.

155. Baur, C., Brennan, S., Pianetta, P., and Opila, R. (2002). Looking at trace impurities on silicon wafers with synchrotron radiation. *Anal. Chem.*, **74** 609A–616A.

156. Mori, Y., Uemura, K., and Iizuka, Y. (2002). Whole surface analysis of semiconductor wafers by accumulating short-time mapping data of total reflection X-ray fluorescence spectrometry. *Anal. Chem.*, **74**, 1104–1110.

157. Prange, A. and Schwenke, H. (1992). Trace element analysis using total-reflection X-ray fluorescence spectrometry. *Adv. X-Ray Anal.*, **35B**, 899–923.

158. Schwenke, H. and Knoth, J. (1995). *In* Mittal, K.L. (Ed.) *Part. Surf. [Proc. Symp.]*; Meeting Date 1992; Dekker: New York, N.Y., 311–323.

159. Huber, A., Rath, H.J., Eichinger, P., Bauer, Th., Kotz, L., and Staudigl, R. (1988). Sub-ppb monitoring of transition metal contamination of silicon wafer surfaces by VPD-TXR. In: Shaffner, T.J. and Schroder, D.K. (editors) *Diagnostic Techniques for Semiconductor Materials and Devices.*

160. Huber, A., Rath, H.J., Eichinger, P., Bauer, Th., Kotz, L., and Staudigl, R. (1988). Sub-ppb monitoring of transition metal contamination of silicon wafer surfaces by VPD-TXRF. *Proc. Electrochem. Soc.*, **88–20**, 109–112.

161. Neumann, C. and Eichinger, P. (1991). Ultra-trace analysis of metallic contaminations on silicon-wafer surfaces by vapour phase decomposition/total reflection X-ray fluorescence analysis. *Spectrochim. Acta*, **46B**, 1369–1377.

162. Neumann, C. and Eichinger, P. (1991). Ultra trace analysis of metallic contaminations on silicon wafer surfaces by VPD-TXRF. *Spectrochim. Acta*, **46B**, 1369–1377.

163. Pahlke, S. (2002). Quo vadis total reflection X-ray fluorescence? *Spectrochim. Acta*, **58B**, 2025–2038.

164. Arai, T. (1994). Lecture on *Wafer Surface Analysis*, held at Rigaku, Osaka, Japan.

165. Yakushiji, K., Ohkawa, S., and Yoshinaga, A. (1993). The fundamental consideration for the droplet analysis on the wafer by the total reflection X-ray fluorescence. *Adv. X-Ray Chem. Anal. Jpn.*, **24**, 87–95.

166. Shimono, T. and Tsuji, M. (1989). Proc. 1st Workshop on ULSI Ultra Clean Technology, pp. 49–72; Tokyo.

167. Klockenkämper, R. (2006). Challenges of total reflection X-ray fluorescence for surface- and thin-layer analysis. *Spectrochim. Acta*, **61B**, 1082–1090.

168. Klockenkämper, R., Krzyzanowska, H., and von Bohlen, A. (2003). Density-depth profiles of an As implanted Si-wafer studied by repeated planar sputter etching and total-reflection X-ray fluorescence analysis. *Surf. Interface Anal.*, **35**, 829–834.

169. Klockenkämper, R., Becker, M., von Bohlen, A., Becker, H.W., Krzyzanowska, H., and Palmetshofer, L. (2005). Near-surface density of ion-implantated Si studied by Rutherford backscattering and total-reflection X-ray fluorescence. *J. Appl. Phys.*, 98, 033517-1–033517-5.

170. Schwenke, H., Knoth, J., and Weisbrod, U. (1990). Current work on total reflection X-ray fluorescence spectrometry at the GKSS research centre. *X-Ray Spectrom.*, **20**, 277–281.

171. Lengeler, B. (1990). X-ray techniques using synchrotron radiation in materials analysis. *Adv. Mater.*, **2**, 123–131.

172. Gutschke, R. (1991). Diploma thesis. Universität Hamburg.

173. de Boer, D.K.G. (1991). Glancing incidence X-ray fluorescence of layered materials. *Phys. Rev.*, **B44**, 498–511.

174. Weisbrod, U., Gutschke, R., Knoth, J., and Schwenke, H. (1991). X-ray induced fluorescence at grazing incidence for quantitative surface analysis. *Fresenius J. Anal. Chem.*, **341**, 83–86.

175. Weisbrod, U., Gutschke, R., Knoth, J., and Schwenke, H. (1991). Total reflection X-ray fluorescence spectrometry for quantitative surface and layer analysis. *Appl. Phys.*, **A53**, 449–456.

176. de Boer, D.K.G. and van den Hoogenhof, W.W. (1991). Total-reflection X-ray fluorescence of thin layers on and in solids. *Adv. X-Ray Anal.*, **34**, 35–40.

177. de Boer, D.K.G. and van den Hoogenhof, W.W. (1991). Total reflection X-ray fluorescence of single and multiple thin-layer samples. *Spectrochim. Acta*, **46B**, 1323–1331.

178. Iida, A. (1992). Grazing incidence X-ray fluorescence analysis using synchrotron radiation. *Adv. X-Ray Anal.*, **35**, 795–806.

179. Schwenke, H., Gutschke, R., and Knoth, J. (1992). Characterization of near surface layer by means of total reflection X-ray fluorescence spectrometry. *Adv. X-Ray Anal.*, **35B**, 941–946.

180. Huang, T.C. and Lee, W.Y. (1995). Characterization of multiple-layer thin films by X-ray fluorescence and reflectivity techniques. *Adv. X-Ray Chem. Anal. Jpn.*, **26s**, 129–136.

181. de Boer, D.K.G., Leenaers, A.J.G., and van den Hoogenhof, W.W. (1995). Glancing-incidence X-ray analysis of thin-layered materials: a review. *X-Ray Spectrom.*, **24**, 91–102.

182. Klockenkämper, R., Knoth, J., Prange, A., and Schwenke, H. (1992). Total reflection X-ray fluorescence. *Anal. Chem.*, **64**, 1115A–1123A.

183. Krämer, M. (2007). Potentials of synchrotron radiation induced X-ray standing waves and X-ray reflectivity measurements in material analysis. PhD thesis, University of Dortmund, 128 pages.

184. Kregsamer, P., Streli, C., and Wobrauschek, P. (2002). Total-reflection X-ray fluorescence. In: Van Grieken, R. and Markowicz, A. (editors) *Handbook of X-Ray Spectrometry*, 2nd ed., Marcel Dekker: pp. 559–602.

185. Hönicke, P., Beckhoff, B., Kolbe, M., Giubertoni, D., van den Berg, J., and Pepponi, G. (2010). Depth profile characterization of ultra shallow junction implants. *Anal. Bioanal. Chem.*, **396**, 2825–2832.

186. Ingerle, D., Meirer, F., Zoeger, N., Pepponi, G., Giubertoni, D., Steinhauser, G., Wobrauschek, P., and Streli, C. (2010). A new spectrometer for grazing incidence X-ray fluorescence for the characterization of Arsenic implants and Hf based high-k layers. *Spectrochimica Acta*, **65B**, 429–433.

187. Ingerle, D., Pepponi, G., Meirer, F., Giubertoni, D., Demenev, E., Wobrauschek, P., and Streli, C. (2013). New approach for characterization of ultra-shallow implants by simultaneous evaluation of GI-XRF and XRR, 15th International Conference on TXRF and related methods, Book of Abstracts, O08, p. 90.

188. Detlefs, B., Hönicke, P., Müller, M., Nolot, E., Veillerot, M., and Beckhoff, B. (2013). Traceable characterization of nanostructured microelectronic devices by

GI-XRF, 15th International Conference on TXRF and related methods, Book of Abstracts, O09, p. 91.

189. Di Fonzo, S., Jark, W., Lagomarsino, S., Cedola, A., Müller, B.R., and Pelka, J.B. (1996). Electromagnetic field resonance in thin amorphous films: a root for non-destructive localization of thin marker layers by use of a standard X-ray tube. *Thin Solid Films*, **287**, 288–292.

190. Brücher, M., von Bohlen, A., and Hergenröder, R. (2012). The distribution of the contrast of X-ray standing waves fields in different media. *Spectrochim. Acta Part*, **B71-72**, 62–69.

191. Köhnen, A., Brücher, M., Reckmann, A., Klesper, H., von Bohlen, A., Wagner, R., Herdt, A., Lützenkirchen-Hecht, D., Hergenröder, R., and Meerholz, K. (2012). Tracing a moving thin-film reaction front with nanometer resolution. *Macromolecules*, **45**, 3487–3495.

192. Unterumsberger, R., Pollakowski, B., Müller, M., and Beckhoff, B. (2011). Complementary characterization of buried layers by quantitative X-ray fluorescence spectrometry under conventional and grazing incidence conditions. *Anal. Chem.*, **83**, 8623–8628.

193. Brücher, M., Jacob, P., von Bohlen, A., Franzke, J., Sternemann, C., Paulus, M., and Hergenröder, R. (2009). Analysis of the ion distribution at a charged solid–liquid interface using X-ray standing waves. *Langmuir*, **26**, 959–966.

194. Menzel, M. and Fittschen, U.E.A. (2013). TXRF analysis of sliver nanoparticles in fabrics, 15th International Conference on TXRF and related methods, Book of Abstracts, P49, p. 123.

195. von Bohlen, A. (2010). Totalreflexions–Röntgenfluoreszenz an partikelförmigen Feststoffen, Dissertation, Fachbereich Physik, Universität Dortmund.

196. Krämer, M., von Bohlen, A., Sternemann, C., Paulus, M., and Hergenröder, R. (2006). X-ray standing waves: a method for thin-layered systems. *J. Anal. At. Spectrom.*, **21**, 1136–1142.

197. Klockenkämper, R., von Bohlen, A., Moens, L., and Devos, W. (1993). Analytical characterization of artists' pigments used in old and modern paintings by total-reflection X-ray fluorescence. *Spectrochim. Acta*, **48B**, 239–246.

198. Prange, A. (1989). Total reflection X-ray spectrometry: method and applications. *Spectrochim. Acta*, **44B**, 437–452.

199. Nishiwaki, Y., Shimoyama, M., Nakanishi, T., Ninomiya, T., and Nakai, I. (2006). Application of total-reflection X-ray fluorescence spectrometry to small glass fragments. *Anal. Sci.*, **22**, 1297–1300.

200. Nomura, S., Ninomiya, T., and Taniguchi, K. (1988). Trace elemental analysis of titanium oxide pigments using total reflection X-ray fluorescence analysis. *Adv. X-Ray Anal. Jpn.*, **19**, 217–226.

201. Ninomiya, T., Nomura, S., Taniguchi, K., and Ikeda, S. (1995). Application of GIXF to forensic samples. *Adv. X-Ray Chem. Anal. Jpn.*, **26s**, 9–18.

202. Muratsu, S., Ninomiya, T., Kagoshima, Y., and Matsui, J. (2002). Trace elements of drugs of abuse using synchrotron radiation total reflection X-ray fluorescence analysis (SR-TXRF). *J. Forensic Sci.*, **47**, 944–949.

203. Klockenkämper, R., von Bohlen, A., and Moens, L. (2000). Analysis of pigments and inks on oil paintings and historical manuscripts using total reflection X-ray fluorescence spectrometry. *X-Ray Spectrom.*, **29**, 119–129.

204. Moens, L., Devos, W., Klockenkämper, R., and von Bohlen, A. (1995). Application of TXRF for the ultra micro analysis of artists' pigments. *J. Trace and Microprobe Techniques*, **13**, 119–139.

205. Moens, L., Devos, W., Klockenkämper, R., and von Bohlen, A. (1994). Total reflection X-ray fluorescence in the ultramicro analysis of artists' pigments. *Trends in Anal. Chem.*, **13**, 198–205.

206. Vázquez, C., Maier, M.S., Parera, S.D., Yacobaccio, H., and Solá, P. (2008). Combining TXRF, FT-IR and GC-MS information for identification of inorganic and organic components in black pigments of rock art from Alero Hornillos 2 (Jujuy, Argentina). *Anal. Bioanal. Chem.*, **391**, 1381–1387.

207. von Bohlen, A., Vandenabeele, P., De Reu, M., Moens, L., Klockenkämper, R., Dekeyzer, B., and Cardon, B. (2003). Pigmente und Tinten in mittelalterlichen Handschriften. *Restauro*, **2/2003**, 118–122.

208. von Bohlen, A. (2004). Quantitative analysis of minor and trace elements in historical varnishes using total reflection X-ray fluorescence. *Anal. Lett.*, **37**, 491–498.

209. von Bohlen, A. and Meyer, F. (1997). Microanalysis of old violin varnishes by total-reflection X-ray fluorescence. *Spectrochim. Acta*, **52B**, 1053–1056.

210. Echard, J.P., Bertrand, L., von Bohlen, A., Le Ho, A.S., Paris, C., Bellot-Gurlet, L., Derieux, A., Thao, S., Robinet, L., Lavedrine, B., and Vaiedelich, S. (2010). The nature of the extraordinary finish of Stradivari's instruments. *Angew. Chem.*, **49**, 197–210.

211. Klockenkämper, R. and von Bohlen, A. (2001). Total-reflection X-ray fluorescence moving towards nanoanalysis: a survey. *Spectrochim. Acta*, **B56**, 2005–2018.

212. Hauptmann, A. and von Bohlen, A. (2011). Aurian silver and silver beads from tombs at El-Mahasna, Egypt, in: Proceedings of the third international conference "Origin of The State. Predynastic and Early Dynastic Egypt", London 2008, ed. by Friedman, R.F., and Fiske, P.N.

213. Pepponi, G., Wobrauschek, P., Hegedus, F., Streli, C., Zoger, N., Jokubonis, C., Falkenberg, G., and Grimmer, H. (2001). Synchrotron radiation total reflection X-ray fluorescence and energy dispersive X-ray fluorescence analysis on AP1^{TM} films applied to the analysis of trace elements in metal alloys for the construction of nuclear reactor core components: a comparison. *Spectrochim. Acta*, **56B**, 2063–2071.

214. Prange, A., Reus, U., Böddeker, H., Fischer, R., and Adolf, F.-P. (1995). Microanalysis in forensic science: characterization of single textile fibers by TXRF. *Adv. X-Ray Chem. Anal. Jpn.*, **26s**, 1–8.

215. Nishiwaki, Y., Watanabe, S., Shimoda, O., Saito, Y., Nakanishi, T., Terada, Y., Ninomiya, T., and Nakai, I. (2009). Trace elemental analysis of titanium dioxide pigments and automotive white paint fragments for forensic examination using high-energy synchrotron radiation X-ray fluorescence spectrometry. *J. Forensic Sci.*, **54**, 564–570.

216. Kijewski, H., Beck, J., and Reus, U. (2012). Krankheit und Tod des Violin-Virtuosen Nicolo Paganini – Interpretation auf Basis neuer Haaruntersuchungen. *Arch. Kriminol.*, **229**, 11–24.

217. Kawai, J. and Gohshi, Y. (1999). Guest editorial. *X-Ray Spectrom.*, **28**, 419–420.

218. https://en.wikipedia.org/wiki/Masumi_Hayashi_(poisoner)

219. Kawai, J. (2014). Forensic analysis of arsenic poisoning in Japan by synchrotron radiation X-ray fluorescence. *X-Ray Spectrom.*, **43**, 2–12.

CHAPTER

6

EFFICIENCY AND EVALUATION

With regard to its efficiency, the special analytical methods of total reflection X-ray fluorescence (TXRF) and grazing incidence X-ray fluorescence (GI-XRF) analyses may be characterized as being quite economical with respect to costs of instrumentation and maintenance, as well as low time consumption. Simultaneous multielement detection is possible, with detection limits in the low picogram range. Quantitative determinations can be carried out by the simple and reliable method of internal standardization. As already shown in Chapter 5, TXRF and GI-XRF are applicable to a great variety of sample materials. Their advantages and limitations will be pointed out in this chapter, and their utility for micro- and trace analyses as well as for surface and thin-layer analyses will be emphasized. The competitiveness with other efficient and well-established methods of atomic spectroscopy will be scrutinized. Because of their efficiency and competitiveness, both variants of X-ray fluorescence analysis no doubt play an increasingly important role within the family of atomic spectrometric methods. Reports on the progress in X-ray spectrometry, including TXRF and GI-XRF, have been published periodically (e.g., Refs [1–4]).

Total-Reflection X-ray Fluorescence Analysis and Related Methods, Second Edition.
Reinhold Klockenkämper and Alex von Bohlen.

6.1 ANALYTICAL CONSIDERATIONS

The efficiency of TXRF is mainly a result of the fact that it is an energy-dispersive method of X-ray fluorescence analysis. Specimens need be deposited on totally reflecting carriers in only small amounts. This capability for microanalyses is the second feature that determines the efficiency of TXRF. The benefits of this efficiency affect cost, detection power, reliability, and applicability.

6.1.1 General Costs of Installation and Upkeep

The simplicity of the TXRF instrumentation described in Chapter 3 ought to be reemphasized. Complete instruments as well as individual supplements or components are commercially available. At present five different manufacturers offer a certain selection of devices. The basic purchase costs amount to about $180 000 for a complete instrument including a high-voltage (HV) generator, X-ray tubes, reflectors, sample changer, detector, multichannel analyzer, computer, and software.[1] An instrument additionally offering an angle-scan of the sample and therefore suited for surface and thin-layer analyses (GI-XRF) may cost approximately $20 000 more. The complete instruments are already adjusted, mechanically stable, and nearly maintenance-free. Operational problems are very infrequent, but sometimes a leakage of the thin detector window does occur, which then has to be replaced by the manufacturer.

For installation, a space of only about 4 m^2 is required and the room should be air-conditioned. A power supply of 3 kW is needed, and a coolant system with a flow rate of 5 l water/min must be connected. Low-power X-ray tubes do not need water cooling because they are air-cooled but they are less effective. The older detectors, such as Si(Li) and Ge(Li) detectors, need to be cooled with liquid nitrogen, and an accessory Dewar of 10 l volume has to be filled at least once but not more than twice a week. However, the modern silicon drift detectors (SDDs) run without liquid nitrogen because they are thermoelectrically cooled (Peltier cooling). In general, vacuum will not be needed if one does not intend to detect light elements with low atomic numbers ($Z < 11$). For light-element detection a fine vacuum (1 hPa) is sufficient; furthermore, a detector with a special diamond-like window or a thin polypropylene foil or even a windowless detector will have to be utilized, the latter being more susceptible to trouble.

Sample preparation before analysis takes the bulk of the time and effort, especially when samples cannot be analyzed directly but have to be digested prior to analysis. However, a lot of preparation techniques are well-tried and tested for approved methods like electrothermal atomic absorption spectrometry (ET-AAS) or inductively coupled plasma optical emission

[1] All costs are given in US currency here.

spectrometry (ICP-OES) and can also be applied for TXRF. For analysis, a small specimen of about 10 μl or 10 μg of a sample has to be deposited on a flat glass carrier and dried by evaporation. Sample preparation and presentation necessitate an extremely clean working space, preferably in a clean bench of class 100.[2] The glass carriers chosen for deposition have first to be checked for their cleanliness. The expensive quartz-glass carriers ($40 each) can be cleaned and reused; the cheap Plexiglas carriers (5 cents each) are usually applied only once.

The next step for TXRF analysis is the recording of an energy-dispersive spectrum, which is rather straightforward. A total spectrum or a smaller part of it can be recorded in some seconds, but usually a counting time of 1 or 2 min is chosen, and for extreme traces 20 min is preset. The total spectrum is recorded simultaneously, so that even an element that may be in the sample or specimen unexpectedly will be detected and no element will be overlooked. The processing of the spectra is likewise simple and rapid, and is usually done via a software program. Quantification is performed by internal standardization, either by addition of the analyte element in a few concentration steps or by addition of one other element (i.e., previously not present in the sample) in a single concentration level. The latter method, which is normally applied for multielement determinations, saves some more time and effort (see Section 4.4.2.2).

For micro- and trace analyses, only one angle position is used. For surface and (even more) for thin-layer analyses (GI-XRF), the sample must be tilted in several steps for an angle scan. The time needed for angle-dependent intensity profiles can extend to about 1 h or more. The evaluation requires the adaptation of a model with several parameters by an iterative process until calculations and measurements are in an acceptable correspondence. A correction of absorption-enhancement effects by a fundamental parameter method is necessary. Appropriate software programs for the complex calculations are available.

6.1.2 Detection Power for Elements

Total reflection XRF is especially valuable because of its high detection power. All elements with atomic numbers $Z \geq 11$ (sodium) can be detected without a vacuum. If windowless detectors are used and a vacuum is applied, even the lighter elements down to $Z = 6$ (carbon) may be detectable and the other elements with atomic numbers $Z \leq 20$ (calcium) can be determined at a higher sensitivity. Elements can be detected simultaneously if their spectral peaks do not strongly overlap, but strong interference is rather an exception. Generally, some 10–15 (but anyhow less than 20) different elements can be determined in

[2] Class 100 for a cleanroom means: <100 particles/ft^3 of <3500/m^3 with diameters ≤0.5 μm; no particles ≥5 μm (after US Federal Standard 209 D).

a single run. In order to detect all potential elements of a sample, two or even three runs are needed at different excitation modes, as mentioned in Section 4.2. Consequently, two or even three spectra have to be recorded and evaluated if the number of elements in the same specimen is not restricted.

Detection limits can be determined according to the International Union of Pure and Applied Chemistry (IUPAC) rules [5]. The minimum detectable amount or mass of an element can be calculated from the formula

$$m_{\text{min}} = k \frac{s_{\text{blank}}}{B} \tag{6.1}$$

where k is a factor for which a value of 3 is strongly recommended (not 2). s_{blank} is the standard deviation of the blank measurements, and B is the absolute sensitivity for the element of interest. For X-ray spectroscopical methods, s_{blank} can widely be influenced by photon counting of the spectral background. It is limited by Poisson statistics:

$$s_{\text{blank}} \geq \sqrt{\overline{N_{\text{back}}}} \tag{6.2}$$

where $\overline{N_{\text{back}}}$ is the averaged background. If the blank or background correction is carried out separately for each single spectrum and not for the averaged blank spectrum, $\sqrt{(2N_{\text{back}})}$ is the decisive quantity. The sensitivity can be measured by the net-counts N_{net} of the analyte element related to the quantity m of this element coming from the specimen. Consequently, an *ideal* value of the detection limit can be determined from

$$m_{\text{min}} = 3 \frac{m}{N_{\text{net}}} \sqrt{(2\,N_{\text{back}})} \tag{6.3}$$

Detection limits were measured with the residues of aqueous standard solutions [6,7] and are presented in Figure 6.1. The three excitation modes already described in Section 4.2 were applied. The live-time was chosen to be 1000 s, so that the actual elapsed time was about 1200 s (20 min) due to a dead time portion of 20%. The Si(Li) detector used for these measurements had a frontal area of 80 mm^2 and a spectral resolution of 150 eV for the Mn Kα peak.

Figure 6.1 shows the familiar relationship between the detection limit and the atomic number with a vertex or minimum for a particular element. All analytes with atomic numbers greater than 11 can be detected either by their K or L peaks. The detection limits go down to the low-picogram level. By means of an appropriate excitation, *nearly all* elements are detectable at a level of 1–10 pg. Consequently, the ng/l region would be reached if 100 μl of a high-purity water or acid was applied, and the ng/g range would be reached if 100 μg of an organic matrix was used. Metal contaminants on wafer surfaces can thus be determined down to or even below 10^{10} atoms/cm^2.

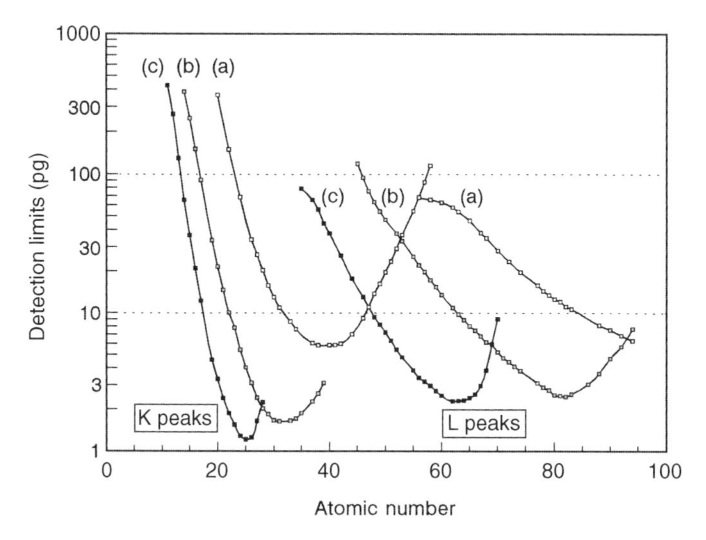

Figure 6.1. Detection limits of TXRF for the residues of aqueous solutions as a function of the atomic number of the analyte element. Three excitation modes were used: (a) W tube, 50 kV; Ni filter, cutoff 35 keV; (b) Mo tube, 50 kV; Mo filter, cutoff 20 keV; (c) W tube, 25 kV; Cu filter. The three curves on the left were determined by the detection of K peaks, the three curves on the right by that of L peaks. All six curves show a minimum for a particular element. Figure from Ref. [7], reproduced with permission. Copyright © 1999, ICDD.

Such ideal values, however, can only be realized for minute residues of aqueous solutions. It is a further prerequisite that the main spectral peaks of the analytes be free from interferences. Furthermore, only photon fluctuations should be significant for the background noise, that is, that all mechanical and electrical units of the spectrometer should be stable on a level $<10^{-4}$, especially the power unit of X-ray tubes with current and voltage. Finally, special blanks must not disturb analysis. In practical examples, these conditions are often only partly fulfilled.

Relative detection limits have actually been determined in many practical applications. These real values collected from the literature are represented in Figure 6.2. The actual detection limits are dependent on the original sample matrix and the kind of sample preparation or matrix separation. For high-purity waters or acids, detection limits go down to the low µg/l level (ppb) after simple drying of µl specimens. After freeze-drying of ml volumes, detection limits are lowered to the ng/l level (ppt). For natural waters, burdened with dissolved or suspended matter, digestion and matrix separation are necessary to achieve detection limits below µg/l. For direct analysis of light samples with a bio-medical or environmental origin, detection limits are at about 0.1 µg/g but can be improved to 10 ng/g if the matrix is first digested. Inorganic solid samples, for example, high-purity metals, can be analyzed down to about 0.1 µg/g after digestion and matrix separation.

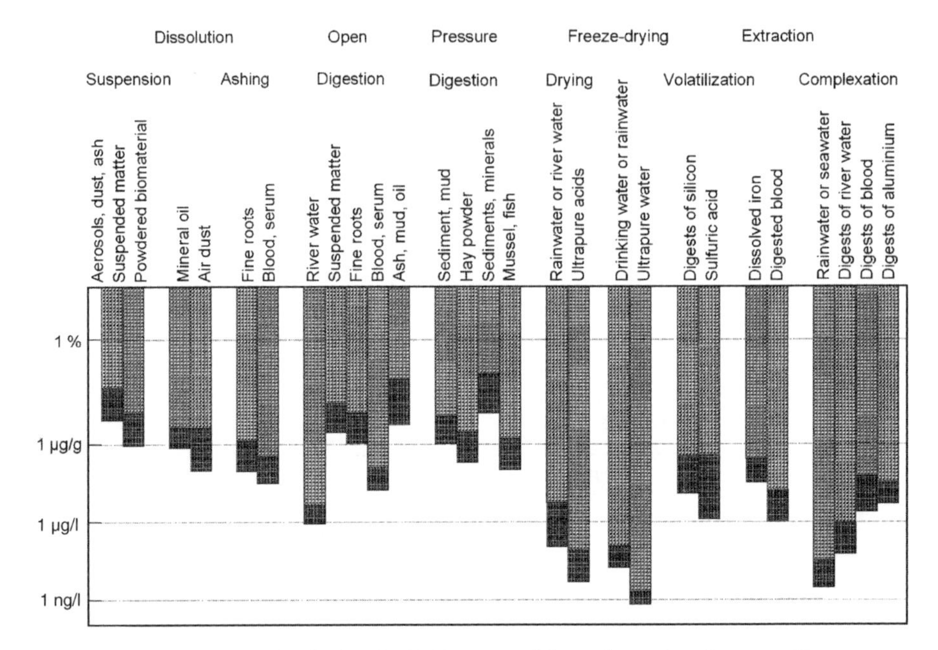

Figure 6.2. Relative detection limits of TXRF measured for real samples after a specific preparation. Because of various analyte elements, the detection limits have a range of uncertainty that spans one order of magnitude. Figure from Ref. [8], reproduced with permission. Copyright © 1996, John Wiley and Sons.

Excitation by synchrotron radiation, use of monochromators, vacuum sample chambers, and novel silicon drift detectors with ultrathin windows have improved detection limits within the last decade. In particular light and transition elements—determined by their K and L peaks, respectively—have profited by such instrumental developments. Details can be found in Section 7.2.1.

6.1.3 Reliability of Determinations

Because the spectrum is recorded as a whole, no detectable element may be overlooked, but because of a relatively low spectral resolution, some peak interferences can occur. For example, the Kα peaks of the transition metals ${}_{21}$Sc to ${}_{30}$Zn are partly overlapped by the Kβ peaks of the neighboring elements with lower atomic number. The Lα and Lβ peaks of the lanthanoids overlap in a similar way. Furthermore, additional peaks, like escape or sum peaks, can occur that have to be noticed carefully (Section 4.3.1). In general, however, the identification of the low-numbered peaks in an X-ray spectrum is easy, the detection of elements is unambiguous, and a qualitative analysis can be carried out rapidly and reliably.

For a quantitative analysis, the determination of the net intensity of peaks is relevant, ranging from about 10^2 to some 10^6 counts. For that reason, the determination of the content of *one* element is restricted to four orders of magnitude. This dynamic range can be spanned for one definite instrumental setting of parameters. However, up to six orders of magnitude can be reached simply if the tube current and acquisition time are changed.

Precision and accuracy are the decisive figures of merit for quantitative analyses. The *precision* can be characterized by the relative standard deviation of repeated quantitative determinations. The *accuracy* is defined by the mean relative deviation of the actual determination from the nominal or true value.

According to Equation 4.8, the concentration of an analyte is determined by the ratio of net intensities. Consequently, *multiplicative* errors are compensated and instrumental fluctuations, for example, of current or voltage, do not influence the results. Only the photon counting is decisive, which is controlled by Poisson statistics. The net intensities of both the analyte peak and the internal standard peak are obtained by subtracting the spectral background from the respective gross intensities. Because of this subtraction, nonspecific *additive* errors are eliminated that are caused, for example, by different carriers. Such fluctuations do not affect the result, which again is only influenced by the photon noise.

The relative standard deviation of the concentration c_x follows from Equation 4.12 and is given by

$$s_{\mathrm{rel}}(c_x) = s_{\mathrm{rel}}\left(\frac{N_x}{N_{\mathrm{is}}}\right) \tag{6.4}$$

where N is the net intensity of either the analyte or the internal standard with index x or is, respectively. Due to the counting statistical fluctuations, it follows that

$$s_{\mathrm{rel}}(c_x) = \sqrt{\frac{s^2(N_x)}{N_x^2} + \frac{s^2(N_{\mathrm{is}})}{N_{\mathrm{is}}^2}} \tag{6.5}$$

where $s(N_x)$ and $s(N_{\mathrm{is}})$ are the absolute standard deviations of the respective net intensities. These quantities are determined by the square root law:

$$s(N_x) = \sqrt{N_{x+b} + N_{b,x}} \tag{6.6}$$

$$s(N_{\mathrm{is}}) = \sqrt{N_{\mathrm{is}+b} + N_{b,\mathrm{is}}} \tag{6.7}$$

where N_{x+b} and $N_{\mathrm{is}+b}$ are gross intensities, and $N_{b,x}$ and $N_{b,\mathrm{is}}$ are the background corrections of the analyte and the internal standard, respectively. If we

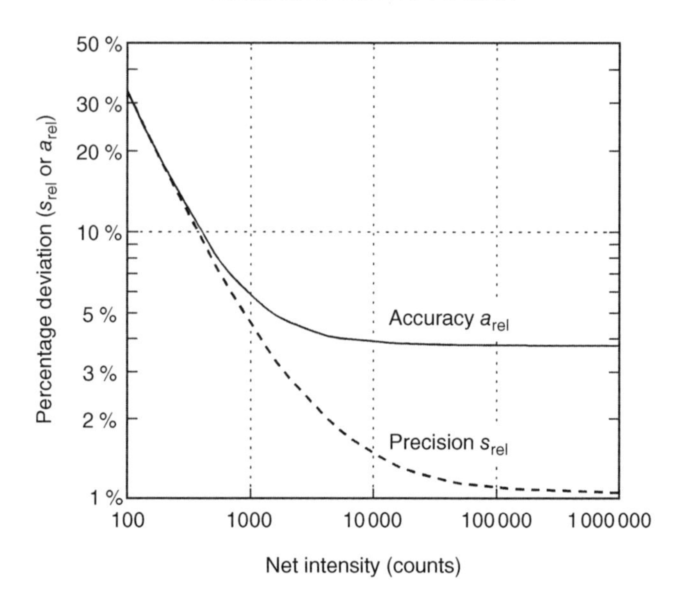

Figure 6.3. Dependence of precision and accuracy represented by a percentage deviation, on the net intensity measured for the analyte peak. The net intensity for the internal standard was assumed to be 10 000 counts and the spectral background was estimated to 500 counts. Figure from Ref. [8], reproduced with permission. Copyright © 1996, John Wiley and Sons.

take these relations into account, it follows that

$$s_{\rm rel}(c_x) = \sqrt{\frac{N_{x+b} + N_{b,x}}{\left(N_{x+b} - N_{b,x}\right)^2} + \frac{N_{{\rm is}+b} + N_{b,{\rm is}}}{\left(N_{{\rm is}+b} - N_{b,{\rm is}}\right)^2}} \tag{6.8}$$

This relationship is illustrated in Figure 6.3 for a typical set of intensity values. The background intensities are assumed to be 500 for both the analyte and internal standard; the net intensity of the internal standard is set to 10 000, and that of the analyte is varied between 100 and 1 000 000. The concentration of the analyte does not appear explicitly but is proportional to the associated net intensity N_x. Determinations significantly above the detection limit can be carried out with a precision of 1–5%.

The deviations between the actual determinations and the nominal values, which characterize the accuracy of an analytical method, will surpass these values. From Equation 4.12, they can be estimated according to

$$a_{\rm rel}(c_x) = \sqrt{s_{\rm rel}^2\left(\frac{N_x}{N_{\rm is}}\right) + a_{\rm rel}^2\left(\frac{S_x}{S_{\rm is}}\right) + a_{\rm rel}^2(c_{\rm is})} \tag{6.9}$$

where $s_{\rm rel}(N_x/N_{\rm is})$ is the aforementioned relative standard deviation, $a_{\rm rel}(S_x/S_{\rm is})$ is the relative uncertainty in the determination of the relative sensitivities, and

$a_{rel}(c_{is})$ is that of the internal standard concentration. These quantities are estimated to be 3% and 2%, respectively, so the relative deviation $a_{rel}(c_x)$ can be calculated. In addition to s_{rel}, the quantity a_{rel} is represented in Figure 6.3. From there, we can see that an accuracy between 3% and 6% can be expected for TXRF analyses, significantly above the detection limit.

In the most favorable cases, for example, for the analysis of pure waters or fine powders, such precision and accuracy can be realized, as was shown in Section 5.1.1. Larger systematic errors can occur if a chemical pretreatment of the sample is necessary, for example, a digestion or matrix separation. Several intercomparison tests, however, demonstrate a high reliability of TXRF and confirm its competitiveness with the well-established methods of ICP-OES, INAA (instrumental neutron activation analysis), and atomic absorption spectrometry (AAS) (e.g., Refs [9–11]).

In cases where the results found for a small specimen need to be transferred to a larger sample, that sample has to be homogeneous or must first be homogenized. Only in that case can microanalysis by TXRF give representative results for the total sample. The sampling error for TXRF will be negligible if the sample material is homogeneous below the micrometric scale. This condition can easily be met for liquids or complete solutions but is difficult to fulfill for suspensions, aerosols, or powdered materials. Solid samples generally show coarse-grained inhomogeneities of some 10 μm and must be thoroughly ground and/or digested prior to analysis. This requirement can be dropped for a true microanalysis where the results must not be transferred to a larger sample. But in any case, the analyte and internal standard should be homogeneously distributed within the small specimen in order to provide reliable quantitative results.

6.1.4 The Great Variety of Suitable Samples

Generally, all liquids and solids will be suitable for analysis if they can be placed onto a sample carrier in a small or even minute amount. Individual particles like grains, fibers, shavings, or filings can simply be put down, powders can be dusted, foils or sections laid down, and layers deposited on a sample carrier. Liquids, solutions, or suspensions can easily be pipetted on a hydrophobic carrier but have to be dried prior to analysis.

Sample materials of many varieties have been analyzed by TXRF as already shown in Chapter 5. A diversity of waters [12–36], but also acids [37,38], oils [39–43], and body fluids [44–58], are just part of the *liquids* that have been investigated. The analyzed *solid* materials are either inorganic—and mostly have an industrial origin or geogeneous [59–70]—or they are organic and mostly biogeneous [71–77]. Table 6.1 gives an overview of various sample materials. The samples can be analyzed directly, but mostly a suitable preparation step is carried out previously. To get a representative result, the sample should first be homogenized. To reach the smallest possible detection limits, the sample matrix should first be separated.

TABLE 6.1. Sample Material Already Analyzed by TXRF

Liquids	Solids (Anorganic)	Solids (Biogeneous)
• *Waters*: drinking and mineral water rain-, river-, seawater wastewater • *Body fluids:* blood, plasma and serum urine, amniotic and cerebrospinal fluid • *Pure chemicals:* acids, bases, salts solvents, water hydrocarbons, polymers • *Oils and greases:* crude and fuel oil essential and vegetable oil fat and grease	• *Suspended matter:* aerosols, dusts, fly ash metallic nano particles • *Soils:* historical and contaminated mud, sediments, sewage sludge • *Minerals and Glasses:* ores, rocks, silica, silicium glass fragments • *Metals:* aluminum, iron, steel gold and silver platinum group metals • *Pigments:* oil paints, inks, varnishes powder and creams • *Thin deposits:* contaminations, films, foils, coatings, implanted layers	• *Plant materials:* algae, hay, leaves, lichen moss, needles, roots, wood xylem sap humic substances • *Biomaterial:* proteins, enzymes nucleic acids pollen, daphnia • *Foodstuff:* flour, fruits, honey, nuts fish, crab-, mussel meat mushrooms, vegetables • *Tissue samples:* human hair and nails kidney, liver, lung breast tissue, healthy and cancerous tissue

Source: From Ref. [8], reproduced with permission. Copyright © 1996, John Wiley and Sons.

Total reflection XRF is a microanalytical method. About 50 μl of a liquid, 10 to 100 μg of an inorganic solid sample depending on the density, and about 50 μg of an organic biotic sample can be used at the maximum. Only a low level of totally dissolved solids can be used. When a maximum of 1% is exceeded, the detection limits get distinctly worse and are just comparable to those of conventional XRF [78].

Total reflection XRF can be applied to micro-, trace-, and surface or near-surface-layer analyses. Because only a small sample amount is required, TXRF is extremely suitable for microanalyses. These will be necessary if the sample material is valuable, unique, or if only a small sample amount is available. Because of the low detection limits, TXRF can also be employed for trace or ultratrace analyses. For that purpose, a larger sample amount of several ml or mg is first homogenized and then one or several small specimens are analyzed. For optically flat samples like wafers, surface and near-surface-layer analyses are performed (GI-XRF). A special device for fine-angle setting is needed to record angle-dependent intensity profiles. Furthermore, a modeling program based on fundamental parameters is needed to evaluate these profiles.

It should be emphasized that the instrumental equipment meets all safety regulations and is rather easy to use. The development of a suitable method for resolving a particular problem, especially that of the sample preparation, initially needs the experience of a specialist. But relatively unskilled personnel can operate the instruments—which partly work under computer control—and can use the method for routine analyses. An automated design with a sample changer can facilitate unattended operation, and even overnight operation becomes possible. The sample throughput is typically 10 samples per hour for the whole set of elements. Extreme trace analyses can take a longer time.

Because of the necessary sampling, TXRF in general is not applied non-destructively. However, TXRF is a *nonconsumptive* method because the specimens are not used up. After an analysis, the specimen is still available for further repeat determinations. This possibility may be important, for example, for the analysis of forensic pieces of evidence and in general for reference analyses.

6.1.5 Round-Robin Tests

A round-robin test or proficiency test is a method of external quality measurement (QM) organized by special laboratories with accreditation, for example, after EN ISO/IEC 17043. Each test is performed by several laboratories and therefore it is also an interlaboratory test. In general, identical samples are investigated, either by different methods and equipment or by the same method with different equipment. Participation is not free of charge (some hundred US$ per test). The aim is either the accreditation of a laboratory, for example, after DIN EN ISO/IEC 17025, the validation of a new method, or the certification of a new standard reference sample distributed by an accredited organization. After a period of some months the organizer usually publishes the results and hands out the certificates. Round-robin tests have been performed with participation of laboratories applying TXRF or GI-XRF. Three characteristic examples will be given here.

Lichen was proposed as a reference material for biomonitors by the International Atomic Energy Agency (IAEA). The species *Evernia prunastry* was collected in unpolluted regions of Portugal, cleaned, dried, ground, and sieved. Material with a grain size less than 125 µm was homogenized and stored in a freezer. The resulting powder, called lichen-336, was distributed by the IAEA for analysis by different analytical methods for an intercomparison test. Twenty-seven different laboratories participated in the test with six different methods (OES, AAS, NAA, MS, XRF, and voltammetry). A first report was given in 1993 [79].

Schmeling *et al.* applied TXRF to the analysis of this reference material three years later [80]. Six samples of about 500 mg were dried in three replicates, 250 mg portions were weighed, mixed with 2 ml of concentrated nitric acid, and digested by high-pressure ashing at 260 °C and 13 MPa for 3 h. After that, each solution was replenished with deionized water to 10 ml, diluted 1 : 1 with nitric acid, and spiked with 1 µg/ml of gallium serving as internal standard. Aliquots of 10 µl were pipetted on cleaned quartz-glass carriers and dried by IR light. The residues were analyzed by TXRF (Mo excitation, counting time 200 s) and eight elements could be determined in a concentration range between 1.8 µg/g for rubidium and 2.8 mg/g for calcium with a relative standard deviation of about 5%. All TXRF values were in the range of accepted values. The average relative deviation from the certified values was about 8%. A typical example comparing all submitted results for copper is demonstrated in Figure 6.4. The TXRF result of 3.68 µg/g matches very well with the certified value of 3.57 µg/g. The relative deviation is only 3%, thus demonstrating the ability of TXRF for the certification of reference materials.

For contamination control of silicon wafers, round-robin tests were initiated by individual researchers (Gohshi, 1996) and industrial companies (Philips, 1996) and were organized by IMEC in 1996, and by ISO and SEMI in 1997. A European round-robin test organized by Rink *et al.* [81] started in 1998 with 12 companies or institutes and 21 different TXRF systems. Silicon wafers intentionally contaminated with a solution of NH_4OH/H_2O_2, which were spiked with Ni and Fe standard solutions, respectively, served as test samples. Five Ni wafers were prepared by spin-coating with standard solutions of five different concentrations (1 ppb, 5 ppb, 10 ppb; 1 ppm and 5 ppm), and two Fe wafers were prepared by immersing the wafers in standard solutions of two different concentrations (0.2 ppb and 1 ppb). Ten different TXRF tools were used and Mo-Kα or W-Lβ was applied for excitation. All participants analyzed the wafers at three predefined positions; however, they used different external standards for calibration: a droplet standard, a spin-coated standard, an immersion standard, and a bulk standard of a pure metal. The so-called iso-angle for particulate and thin-layer-type contaminations was generally chosen for Mo excitation (0.074°). For W excitation the iso-angle is 0.132°; the glancing

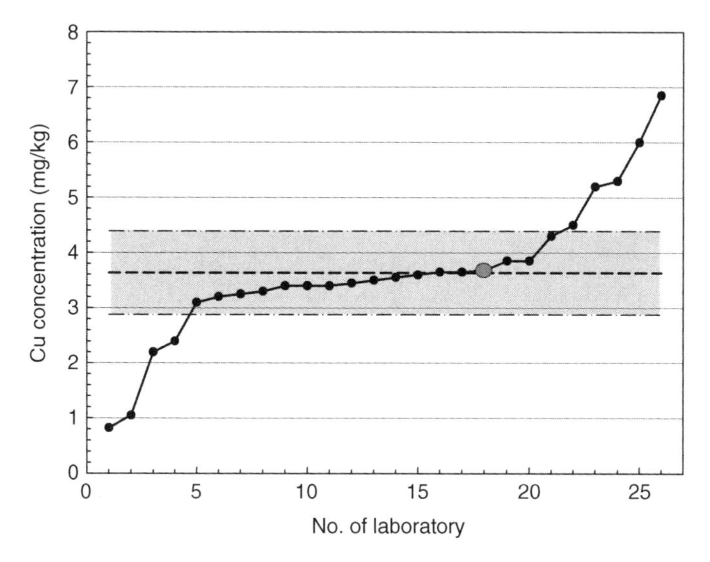

Figure 6.4. Results of an intercomparison test for certification of the IAEA reference material lichen-336. Copper was determined by 26 laboratories and 6 different analytical methods. The mean value is 3.57 mg/kg and the standard deviation is 0.26 mg/kg. The 99% confidence interval is shown as a shaded band. The additional TXRF value (●) is very near the mean value. Figure from Ref. [79], reproduced with permission. Copyright © 1997, Springer-Verlag Berlin Heidelberg.

angle, however, was set to 0.083°, 0.100°, 0.132°, or 0.135° for the different tools. Most measurements were carried out within half a year while the last results were obtained after 2 years.

In addition to the straight TXRF measurements, one specific laboratory carried out Rutherford backscattering spectrometry (RBS) as an absolute reference method and accordingly all TXRF results were related with respect to the RBS values (set to 1.0). Excepting one outlier, three groups of participants or laboratories could be classified: the first group with five participants showed relative deviations below 5% from the RBS values; the second group with 11 participants showed averaged deviations between 20% and 50%; the third group with the remaining five participants deviated by 50% up to 90%. An impact due to the equipment seems not to be obvious; differences may be caused by the type of standard used for calibration. The samples themselves were not ideally layer-like but showed nanocrystals of about 5 nm.

A proficiency test was carried out for a sediment sample analyzed by three different methods of trace analysis. The test called IMEP-14 (international measurement program) was organized by the Institute for Reference Materials and Measurements (IRMM) in Geel, Belgium, and focused on the determination of 12 elements in sediment, mainly on the level of mmol/kg (ppb). The fine sediment powder was first analyzed by seven experienced reference laboratories in a certification campaign, mostly by isotope dilution mass spectrometry (IDMS) and neutron activation analysis (NAA) as primary methods of measurement (PMM). Aliquots of 40 g were forwarded to interested laboratories for a charge of €300, with the so-called reference values undisclosed. Two hundred thirty-nine laboratories from 43 countries took part in this extensive proficiency test. Half of them were accredited, and 13 different methods of analysis were applied. The results had to be delivered before a deadline of six months. The final report was given by IRMM after eight further months [82].

Klockenkämper *et al.* participated in the test with the established technique of flame atomic absorption spectrometry (FAAS) and with the younger methods of TXRF and ICP-MS [83]. For TXRF, 500 mg of the sediment sample were poured into 4 ml of nitric acid (65%) and 0.5 ml of hydrochloric acid (30%). This mixture was digested by high-pressure ashing at a maximum temperature of 320 °C and a pressure of 13 MPa. The insoluble residues of siliceous components were treated with 2 ml of hydrofluoric acid (40%) and dissolved. After evaporation almost to dryness, the residue was taken up with 10 ml of deionized water. In this way several different sediment samples and also acid blanks were prepared. For TXRF, aliquots of 1 ml were diluted 1 : 10 and spiked with selenium as an internal standard of 5 mg/l (5 ppm). Droplets of 10 µl of each solution were pipetted onto siliconized quartz-glass carriers, dried under IR light, and the dry residues of some micrograms were analyzed by TXRF with a counting time of 300 s. Altogether, six samples were analyzed via sensitivity values, which had been determined by aqueous standard solutions sometime beforehand. Six elements could be determined quantitatively (Cr, Fe, Cu, Ni, Zn, and Pb) with concentrations between 0.4 mmol/kg (ppm) for lead and 460 mmol/kg for iron.

The results are demonstrated in Figure 6.5 as a horizontal-bar plot [83]. The colored error bars more or less overlap with the reference bars, demonstrating a high accuracy of the three techniques. There is no evidence for a systematic error. The relative deviations between measured and reference values range from −10% to about +11% for all three methods. The arithmetic mean is +0.6% for TXRF, +1.0% for FAAS, and +3% for ICP-MS. The root mean square of the deviations is 6.4% for TXRF, 3.5% for

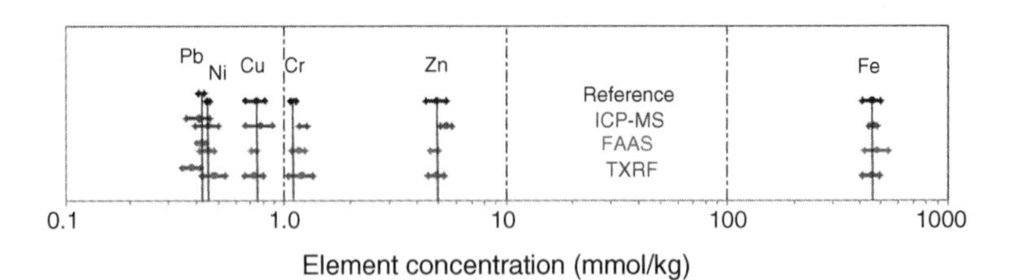

Figure 6.5. Horizontal-bar plot for six different elements of a sediment sample (Cr, Cu, Fe, Pb, Ni, and Zn) determined by FAAS (light gray), ICP-MS (dark gray) and TXRF (gray). Mean values and confidence intervals are shown by points and bars with a length of 2σ. The reference values are plotted in black. Figure from Ref. [83], reproduced with permission. Copyright © 2001, Royal Society of Chemistry.

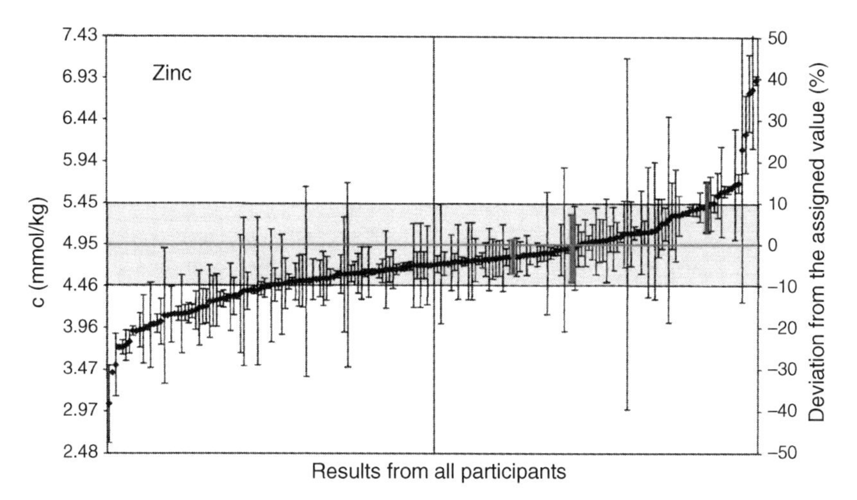

Figure 6.6. Results from all participants of IMEP-14 (sediment) related to the content of zinc [82]. The fitted black curve shows the form of a deckchair, which is typical for proficiency tests with a large number of participants. The shaded band contains the reference value 4.95 mmol/kg (yellow straight line) with a confidence interval of ±10%. Individual values were determined by FAAS (green), by ICP-MS (blue), and by TXRF (red). Figure from Ref. [83], reproduced with permission. Copyright © 2001, Royal Society of Chemistry. (See colour plate section)

FAAS, and 5.6% for ICP-MS. These values show a rather equal ranking of the three techniques.

The final report of IRMM represents an overview of the results of all participants [82]. For any element, the results are ordered and plotted in diagrams with increasing values. Figure 6.6 gives a typical example for the element zinc [83]. The three bars of TXRF, FAAS, and ICP-MS are colored red, green, and blue, respectively, and are lying within the uncertainty band of the reference value almost completely. The test demonstrates the high accuracy and multielement capability of the three techniques, even for a complex matrix. FAAS seems to be slightly superior to TXRF and ICP-MS, although it is more laborious and time consuming. TXRF shows somewhat poorer detection limits; however, it requires a much smaller amount of sample volume and it is the most simple and rapid technique.

6.2 UTILITY AND COMPETITIVENESS OF TXRF AND GI-XRF

Total reflection XRF is a universal and economic method of multielement analysis. It is a microanalytical tool suitable for cases when only small sample amounts are available. The minute specimens have only to be placed on flat, totally reflecting carriers. In addition, TXRF is effectively applied to trace analyses of the elements when larger sample amounts are available. Aqueous solutions, high-purity acids, and body fluids are analyzed down to the pg/ml region. Only a few droplets need to be pipetted on carriers and dried by simple evaporation. If a more complex sample is to be analyzed, a decomposition and separation of the matrix becomes necessary. Simple quantification is made possible, since matrix effects do not appear because of the small sample amounts used for analysis. Furthermore, contaminations on flat surfaces, such as wafers, can be quickly determined by TXRF. Depth profiles of biogeneous materials can be recorded after sectioning of the material with a freezing microtome. Stratified near-surface layers can be characterized by a depth profile after planar sputter etching. This technique combined with TXRF of the remaining surface is especially suitable for thin layers of nanometer thickness deposited on wafers even though it is destructive and a bit time-consuming.

In contrast to TXRF after sputter etching, GI-XRF is nondestructive and moreover is a fast method for depth profiling of deposited or implanted layers in the lower nanometer range. However, GI-XRF makes higher demands on instrumentation and evaluation considering matrix effects. New applications are concerned with semiconducting layers and even thin polymeric films.

With respect to its capabilities, TXRF has far surpassed conventional XRF. Indeed, TXRF has attained a leading position in atomic spectroscopy. The outstanding features compete very well with those of instrumental neutron activation analysis (INAA) and ICP-MS. For several applications, TXRF even has distinct advantages over these methods because of its simplicity and rapidity. The detection power may be inferior, but for many applications is sufficient. The detection of light and transition elements, however, needs special and further efforts.

In the field of surface analyses, TXRF is highly effective in the contamination control of wafers and therefore is a widespread analytical tool in the semiconductor industry. For analyses of thin stratified layers, GI-XRF is able to compete with the reference methods RBS and secondary ion mass spectrometry (SIMS).

6.2.1 Advantages and Limitations

Several strengths of TXRF and GI-XRF can be listed on a short inspection: (i) generally nonconsumptive and also nondestructive investigations; (ii) a versatility of various sample types; (iii) no sample preparation or only an easy and fast handling; and (iv) mature instrumentation being commercially available. The weaknesses of TXRF and GI-XRF are: (i) a minor sensitivity for light and transition elements; (ii) poorly representative results for macrosamples; and (iii) a lack of on-line investigations. Standardization by ISO, ASTM, or DIN is just being developed.

Several features of TXRF are especially worth noting. Most of them are advantageous, although some present limitations. Some drawbacks are

TABLE 6.2. Benefits and Drawbacks of TXRF Applied to Element Analyses

Benefits	Drawbacks or Limitations
Unique microanalytical capability, generally nonconsumptive	Certain sample preparation is mostly necessary
Great variety of samples and applications	Limitation for nonvolatile liquids
Simultaneous multielement determination	
Low detection limits	Restriction for low-Z elements Limitation for transition elements Limitation by high matrix contents
Simple quantification by internal standardization	
No matrix or memory effects	Limitation to a low covering and thickness of samples
Wide dynamic range	
Nondestructive thin-layer analysis	Restriction to flat or polished samples
Simple automated operations	
Low running costs and maintenance	

Source: From Ref. [8], reproduced with permission. Copyright © 1996, John Wiley and Sons.

accepted as the price that has to be paid for the benefits. In Table 6.2, the benefits and drawbacks of TXRF are compared.

Total reflection XRF's unique capability for micro- or even ultramicroanalyses of small sample volumes should first be mentioned. No additional device is needed for that purpose as is the case for ET-AAS and ICP-MS. It may be mentioned in passing that TXRF is a nonconsumptive method like all X-ray techniques. On the other hand, an entirely nondestructive analysis is mostly impossible because of the need for minute specimens. A great variety of samples can be analyzed and a wide scope of applications can be outlined. Solutions, however, are usually evaporated prior to analysis with the result of improved detection limits. Nonvolatile liquids can be investigated *directly* but with lower detection power. For improved detection limits, their matrix has to be digested.

Next, TXRF's usefulness for multielement determinations should be emphasized. Up to 30 elements can be determined simultaneously and low detection limits below 10 pg can be reached for about 70 elements, with the exception of low-Z elements. Detection limits worsen if heavy elements are predominant in the sample matrix and cannot be separated. Also, detection will be impeded if a neighbor in the periodic table is present with a higher concentration (more than 100-fold).

Total reflection XRF generally provides a simple means of quantification via an internal one-element standard. No labor-intensive calibration is required. Because of the small sample volume needed for analysis, no matrix effects occur. The prerequisite is a minute specimen, which has to be deposited as a

TABLE 6.3. Advantages of TXRF and Necessary Prerequisites

Advantages	Prerequisites
Capability for microanalyses	Handling of micrograms or microliters
Large variety of samples	Decomposition of several kinds of samples
Low detection limits	Matrix separation for non-aqueous samples
	Clean-bench working methods
Reliable quantification	Careful homogenization
Surface and thin-layer analyses	Exact fine angle control
Simple unattended operation	Stable equipment and automation

Source: From Ref. [8], reproduced with permission. Copyright © 1996, John Wiley and Sons.

single particle, as a powder, or as a thin layer in the center of a flat and clean carrier. Mostly, a separate quartz-glass or Plexiglas carrier is used for each sample so that no memory effects occur. As a result of these measures, quantification becomes easy and reliable and can be performed in a large dynamic range of five orders of magnitude.

A new field of applications has been opened up by the GI-XRF technique that examines a sample under variation of the angle of incidence. It enables nondestructive surface and thin-layer analyses. The technique, however, is restricted to optically flat or polished samples, like wafers and glass disks, uncoated, coated, or implanted. Samples of a certain roughness (above 0.1 μm) blur the effect of total reflection used by this technique and so have to be treated particularly by special calculation programs.

Finally, TXRF's simplicity of operation should be mentioned, supported by automated sample changing, measurement, and evaluation. The running costs are quite low (e.g., $13 per 10 l liquid nitrogen per week), and the maintenance of the instrumentation is rather easy. An X-ray tube that needs to be replaced after a few years costs about $4000 and a broken window for the detector can be replaced for $1200.

Some prerequisites have to be met in order to reap the advantages of TXRF. They are compiled in Table 6.3. The capability for microanalysis requires skillful handling of microgram masses or microliter volumes, for example, by micropipettes, inkjet printers, or microdispensing systems. A great variety of samples, especially solid samples, can only be analyzed if decomposition is carried out prior to analysis, for example, by high-pressure digestion. Low detection limits necessitate chemical matrix separation. A clean-bench working methodology is a matter of course in ultratrace analysis, not only for TXRF analysis. Naturally, a reliable quantification gives representative results only when a macrosample is carefully homogenized prior to sampling.

6.2.2 Comparison of TXRF with Competitors

A great variety of instrumental methods can be applied to element analysis in different fields of investigation. Usually, many elements, with different atomic numbers, with different concentrations down to extreme traces, and in quite different matrices, have to be determined. Consequently, methods are required that are capable of a multielement detection, have a high detection power, and are applicable to a multitude of sample materials. Moreover, the methods should give accurate results and work economically. With regard to all these features, TXRF plays an important role in atomic spectroscopy [84].

Total reflection XRF is a powerful method of microanalysis and is also suitable for trace element detection. Masses of only a few μg of a solid or volumes of few μL of a liquid sample are sufficient for analysis. Specimens have to be deposited in the center of a flat and ultraclean carrier, for example, on a quartz-glass carrier or on a cheap Plexiglas support. The time necessary for recording a spectrum is of the order of minutes.

All elements can be detected simultaneously with the exception of light elements, that is, TXRF is a real multielement method. Detection limits are usually on the picogram level, but for low-Z and transition elements the detection limits decrease by one or two orders of magnitude. Concentrations can be determined down to 1 ng/g for solid samples or 1 pg/ml for liquid samples. For a simple quantification, one prerequisite has to be met absolutely; samples have to be deposited with a restricted covering and thickness. Small samples with a covering of $\mu g/cm^2$ and a thickness of μm, do not show annoying matrix effects and do not destroy the coherence of the incoming and totally reflected beam. In this case, quantification is made possible by a simple addition of a single internal standard element.

Competing methods of TXRF for laboratory analyses are ET-AAS (electrothermal atomic absorption spectrometry, also called GF-AAS or graphite furnace AAS), ICP-OES, ICP-MS, and INAA. The group of classical methods including gravimetry, titrimetry, voltammetry, and chromatography cannot compete in any way with one of the universal and effective methods of modern spectrometry like TXRF. Most of the nonspectrometric methods are only suitable for single-element detection and have only mediocre detection limits—at μg or ng levels. However, the simplicity of their equipment and the accuracy of their results assure them firm places in the analytical laboratory.

Atomic absorption spectrometry with flames (FAAS) was the most common method of atomic spectroscopy several years ago. However, only the technique using graphite furnaces (ET-AAS) is suitable for trace analysis. It is a microanalytical method like TXRF and shows low detection limits (at picogram levels) even for light elements. On the other side, it is mostly applied as a single-element method so that it is time-consuming especially for the successive determination of many elements. Furthermore, the standard-addition method used for calibration is rather laborious. A simultaneous multielement detection has been envisaged for laser AAS. At present, only an oligoelement technique has been developed, using four hollow-cathode lamps and an Echelle spectrometer.

A very common excitation source for optical emission spectrometry is ICP-OES. It is a multielement method for macrosamples, applicable to solutions of several milliliters. However, the detection limits are at the nanogram level (1–100 ng), so that ICP-OES cannot compete with more effective methods like TXRF. The variant using a microwave-induced plasma (MIP-OES) seems to be superior but is not yet widespread.

Several techniques of laser spectroscopy have recently been developed, based on the absorption, fluorescence, or ionization of a cloud of atomic vapor by means of a strong laser. Some of these techniques permit determinations at femtogram levels and are very promising for ultratrace analysis. However, any comparison would be premature. At present, the higher costs of laser systems applicable to the UV spectral region hinder further progress and spread of these approaches.

There are, however, two powerful multielement methods of trace analysis that are highly developed and well established and consequently are strong competitors: INAA and ICP-MS. They profit from their macroanalytical capability and are more sensitive with lower detection limits; however, they usually require much more sample material. Furthermore, INAA needs a large neutron reactor for irradiation of the samples (about 10 h), the method is inherently slow (typically some days or weeks) and not time saving as is TXRF. Nevertheless, the ICP methods have become more popular in the laboratory practice of macroanalytical analyses. TXRF as a microanalytical method is less established; however, it is often needed for a first overview of new and totally unknown samples and for a confirmation of the results obtained by other methods.

Some comparisons of TXRF with competing methods have already been made, and some examples have been presented in Chapter 5. They show alternate priorities of the different methods, depending on the matrices to be analyzed and the elements to be detected. Intercomparison tests with respect to a variety of environmental samples were first performed by Michaelis [85]. A multielement characterization of tobacco smoke condensates was carried out by Krivan *et al.* [86], and comparative analyses of spinach, cabbage, and domestic sludge were performed by Pepelnik *et al.* [87].

Results found for several elements in the standard reference material NIST 1570 "Spinach" are demonstrated in Figure 6.7. The element mass fractions range from about 36 mg/g for K down to 30 ng/g for Hg. The results of ICP-MS, INAA, and TXRF agree quite well, and most of them are also in agreement with the certified values. The deviations are on the order of 2–30% but are generally less than 10% and prove the fairly good accuracy of all three methods.

The detection power of the three methods and additionally of ET-AAS is illustrated in Figure 6.8. It represents relative detection limits of the different methods applied to trace analysis of aqueous solutions [84,88,89]. The figure confirms ICP-MS to have the higher detection power for most elements. However, a more detailed evaluation of the competitive methods should consider additional characteristics. Moreover, the analytical problems to be solved have to be taken into account in order to make a careful choice of the

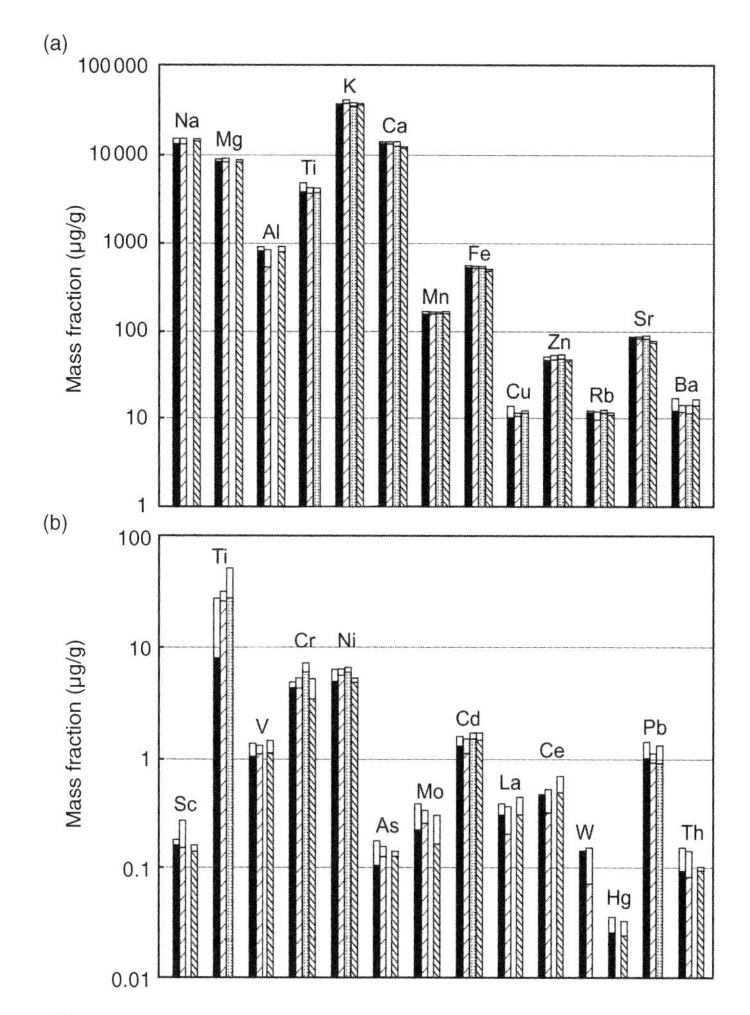

Figure 6.7. Element concentrations found in NIST 1570 "Spinach" by ICP-MS (left oblique lines), TXRF (mottled grey), and INAA (right oblique lines) in comparison to the certified values (solid black). The bars represent mean values ± standard deviations at $n = 6$ determinations. Elements with mass fractions (a) above 10 µg/g and (b) below 100 µg/g. Figure from Ref. [87], reproduced with permission. Copyright © 1994, Royal Society of Chemistry.

most suitable method. Some characteristics, advantages, and disadvantages that are of help in such an evaluation are listed in Table 6.4.

As demonstrated by Lieser *et al.* [90], TXRF can have distinct advantages over INAA if extremely small volumes (10 µl) are to be analyzed on traces. On the other hand, INAA is a nondestructive technique in principle. It is suited for direct analysis of *solids* without any sample preparation. In most reactors, however, it is forbidden to irradiate *solutions* as a safety precaution but only on principle; it is actually possible to irradiate solutions.

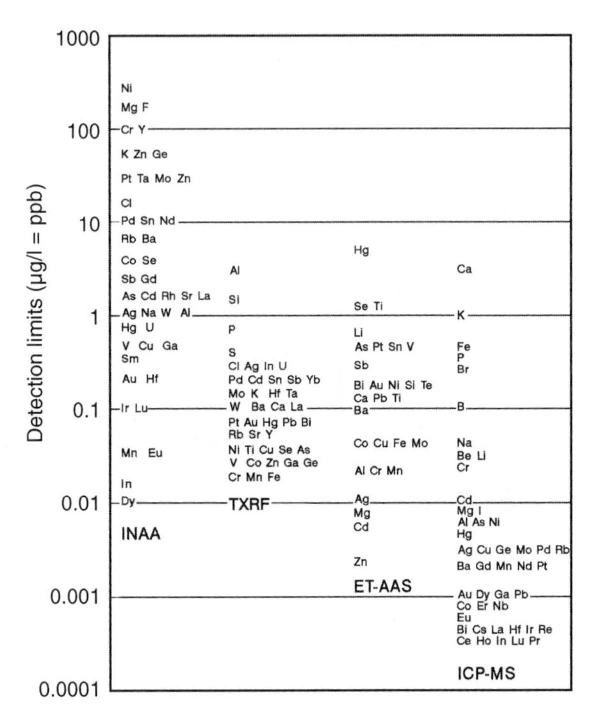

Figure 6.8. Relative detection limits of INAA, TXRF, ICP-MS, and ET-AAS applied to trace analysis of aqueous solutions or their residues after evaporation. A 50 µl specimen was used for TXRF and ET-AAS; 3 ml was needed for INAA and ICP-MS. The individual values should be considered approximate at best. Data from Ref. [84], reproduced with permission. Copyright © 1993, Elsevier.

In contrast to INAA, TXRF and ICP-MS are well suited for liquids or solutions. Solids first need to be digested or at least dissolved as a suspension. As a result, losses and contaminations can occur. Furthermore, a limitation of the dissolved portion, that is, a low salt concentration, has to be observed for both ICP-MS and TXRF. Highly concentrated acids or bases must be diluted only for ICP-MS but can be directly analyzed by TXRF. No such restrictions exist for INAA. The total sample volume applied for analysis is consumed by ICP-MS, whereas TXRF and INAA are nonconsumptive.

Total reflection XRF and INAA permit detection on the low-pg level; ICP-MS has detection limits that can even be lower by one or two orders of magnitude. Limitations exist for certain sets of elements, which are different for the three methods. While TXRF is limited in the detection of light elements with $Z < 13$ and less effective for some transition elements, ICP-MS is hampered in the determination of some elements introduced by air, for example, nitrogen and oxygen. For some other elements like H, F, P, and S, the ionization probability in the plasma is rather low. A determination of the halogens chlorine and bromine in liquid samples is effective for TXRF; however, it is difficult for ICP-OES and ICP-MS [91].

TABLE 6.4. Comparison of Important Analytical Features of Three Competitive Methods: ICP-MS, TXRF, and INAA

Analytical Feature	ICP-MS	TXRF	INAA
Samples:			
Volume or mass	2–5 ml (10 μl)	5–50 μl	10–200 mg
Preparation of solids	Digestion or suspension	Digestion or suspension	None
Dissolvation portion	<0.4%	<1%	Any
Dilution of acids	1 : 100	None	None
Consumption	Yes	No	No
Detection:			
Detection limits	Excellent	Very good	Very good
Element limitations	H, C, N, O, F, P, S	$Z < 13$ and $41 < Z < 45$	$Z < 9$; Tl, Pb, Bi
Spectral interferences	Several	Few	Few
Isotope + species detection	Yes	No	No
Quantification:			
Calibration	Several external + internal standards	One internal standard	Some pure element foils
Matrix effects	Severe	None	None
Memory effects	Yes	No	No
Time consumption	<3 min	<20 min	20 min – 30 days
Expenditure:			
Equipment	Ar plasma + Quadrupol MS	EDS with SSD	Nuclear reactor + γ- spectrometer
Capital costs	Medium	Medium	Very high
Running costs	High	Low	High
Maintenance	Frequently	Seldom	Seldom

Source: From Ref. [8], reproduced with permission. Copyright © 1996, John Wiley and Sons.

A majority of elements (about 70) can be determined by INAA, with high sensitivity for rare earth elements. But again there are some exceptions; light elements with $Z < 9$ and some heavier elements, for example, Ti, Nb, Tl, Pb, and Bi, are not detectable without an additional device or a chemical procedure. Spectral interferences mostly occur for ICP-MS. Molecular ions built by the solvent or the carrier gas, for example, argides by argon, can lead to numerous overlaps. On the other hand, ICP-MS is capable of isotope and species analysis. A high degree of reliability in quantitative analysis can be achieved by isotope dilution for INAA.

For quantification, a highly elaborate calibration must be carried out in ICP-MS. Several external standards have to be used, and some internal standards may have to be added as well. In order to reduce the severe matrix effects, the external standards have to be adapted to the sample solution with respect to concentration and even acidity. Matrix effects can also be corrected for by standard addition and isotope dilution. But even then, memory effects and the already mentioned spectral overlaps can restrict the accuracy of ICP-MS. In contrast to ICP-MS, the quantification for TXRF is much easier and more reliable. An internal standardization with a single element is possible. In general, neither matrix effects nor memory effects obstruct TXRF analysis, and the same is true for INAA. On the other hand, ICP-MS is a very fast method like TXRF, while INAA is generally time-consuming because of the long irradiation and decay times in addition to the measuring time.

Total reflection XRF and ICP-MS are strong competitors in the contamination control of wafers with the lowest detection limits. The variant of VPD-TXRF and VPD-ICP-MS show the highest sensitivity (about 10^7 atoms/cm^2) and are particularly suitable to meet the high-purity demands in this field. A direct *local* analysis of contaminations and its area distribution, however, is only possible by means of TXRF with a microspot. A unique feature of TXRF with sputter etching is the determination of the density of near-surface layers in addition to their composition and thickness.

Regarding costs, INAA requires a high neutron flux for a high detection power. Such flux is only available at a nuclear reactor that exceeds the laboratory standard. ICP-MS and TXRF only need customary equipment of a medium size and price. However, the capital expenditure for ICP-MS will double if a focusing mass spectrometer is used instead of a simple quadrupole MS. Such an instrument with high spectral resolution is recommended in order to avoid troublesome interferences. TXRF, in contrast, has moderate capital expenditures and also requires quite low running costs and little maintenance. By contrast, the argon consumption of ICP-MS is a serious item (some \$10 000 per year). Nebulizer, skimmer, pumps, and gas supply need frequent maintenance. For TXRF, liquid nitrogen is needed only for Si(Li) detectors (about \$600 per year), while modern silicon drift detectors are maintenance-free. Flat carriers are needed but may be reused after cleaning.

The spread or rank of an analytical method may be read from sales figures. Such information is collected and published biannually by Strategic Directions

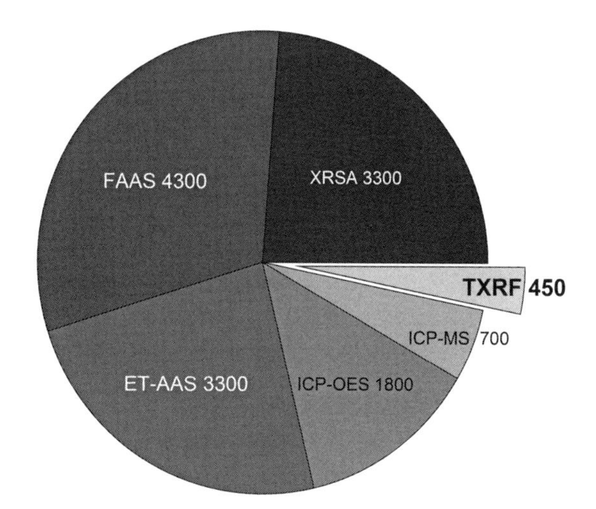

Figure 6.9. Sales figures of devices provided in 2003 for six different methods of spectral analysis. FAAS = flame atomic absorption spectrometry; ET-AAS = electrothermal –AAS; ICP-OES = inductively coupled plasma optical emission spectrometry; ICP-MS = ICP-mass spectrometry; XRSA = X-ray spectral analysis; TXRF = total reflection X-ray fluorescence. Data from Ref. [92] and figure from Ref. [93], reproduced with permission. Copyright © 2003/2006, Elsevier.

International, Inc. (SDi) covering a period of 5 years. Figure 6.9 illustrates sales figures of six different instrumental methods of element analysis demonstrated as a pie chart [92,93]. Altogether 13 850 devices were sold in 2003. The classical and very cheap method of FAAS is most widespread with 31%, followed by conventional XRSA with 24% (excluding TXRF). The microanalytical method ET-AAS covers 24%, while ICP-OES and ICP-MS account for 13% and 5%, respectively. TXRF as a variant of XRS is sold with 450 devices or 3%. The whole market represents a total sum of more than \$1 billion (10^9).

Another hallmark or rank can be inferred from the number of publications in a certain period. Scientific papers published within 43 years between 1971 and 2014 were recorded by Chemical Abstracts Service (CAS), a division of the American Chemical Society; the numbers were retrievable online.[3] Attention was paid to seven main methods of element analytical spectroscopy [94]. The bar-plot of Figure 6.10a, shows the rate of four common methods, FAAS, ICP-OES, XRSA (without TXRF), and ICP-MS. These mainly macroanalytical methods show strongly increasing publication rates. The oldest method, FAAS, with only 160 publications per year is the tail-light while the youngest, ICP-MS, with about 2300 publication per year is the front-runner. The less common special methods like ET-AAS, INAA, and TXRF have a much lower publication rate. As demonstrated in Figure 6.10b,

[3] Such information is also available from Thomson-Reuters Corp. on ISI Web of Science (formerly from Thomson Scientific on ISI Web of Knowledge).

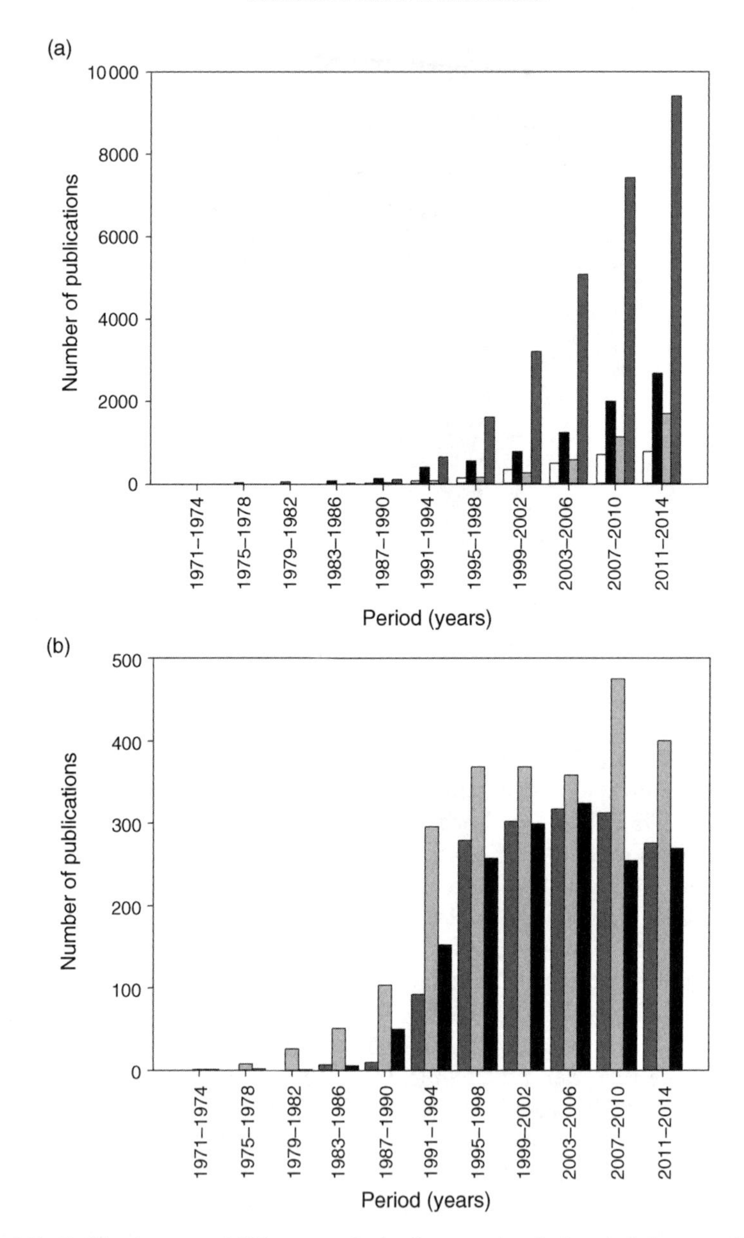

Figure 6.10. Publication rate of different methods of spectrochemical analysis between 1971 and 2014. Eleven periods of 4 years each were set up. (a) The rate for the common methods FAAS, XRSA (without TXRF), for ICP-OES, and ICP-MS. (b) The rate for ET-AAS, INAA, and TXRF. Data from Ref. [94]. Copyright © 2005, American Chemical Society.

their rate is nearly constant within the last 20 years. INAA with about 100 publications per year is followed by ET-AAS and TXRF, both with about 75 papers per year. Obviously, INAA with about 3%, ET-AAS and TXRF both with 2% of all scientific papers of analytical spectrometry seem to be less popular but nevertheless are powerful techniques.

6.2.3 GI-XRF and Competing Methods

Total reflection XRF and moreover GI-XRF can be applied to surface and near-surface layer analyses. Both methods are suitable for the investigation of flat surfaces, especially for the characterization of wafer contaminations and of nanostructured materials like ultrathin stratified layers or films, and nanoparticles deposited on wafers [93,95–97].

Surface and thin-layer analyses by GI-XRF are based on an angle scan of the sample and its excitation by a monochromatized X-ray beam. It presupposes additional instrumental devices, in particular a crystal monochromator and a six-stage goniometer. Generally, a synchrotron beamline with a double crystal monochromator is used. The sample is tilted by a goniometer up to approximately 0.2° in about 200 steps of 0.001° so that an intensity/angle scan of a single sample takes about 1 h. Fine-angle control with a precision of 0.0001° is indispensable but is state of the art. In any case, a simple and unattended operation is only possible if a stable and compact piece of equipment is employed that is automated and under computer control.

Grazing incidence XRF and TXRF are nonconsumptive, can be nondestructive, need no vacuum (only for the detection of light elements), and have no charging problems. The instrumentation for both methods is comparably simple apart from devices at synchrotron beamlines. The depth resolution amounts to several nm. Applications are mainly restricted to thin layers on optically flat samples such as wafers, but successful efforts have been made to analyze samples with a certain degree of roughness up to some 10 or 100 nm (see Section 4.6.5).

Grazing incidence XRF can compete quite well with conventional methods like XPS (X-ray photoelectron spectrometry), AES (Auger electron spectrometry), SIMS, GD-OES (graphite-furnace OES), and RBS. Only a few comparisons have as yet been published [98]. All methods have their own pros and cons. Depth, thickness, and concentration of a layer are the quantities to be determined usually. GI-XRF is the simplest and least expensive method. For thin-layer analysis, XPS, AES, SIMS, and GD-OES have to sputter-etch the samples and consequently destroy it. Nonconductive samples suffer from charging effects and precautions have to be taken. All these competitors need high vacuum or even ultra-high vacuum like RBS and SIMS. RBS is a reference method with high accuracy, while all the other methods suffer from different errors in quantification. Systematic errors can occur, which can reach one order of magnitude for the mentioned quantities [99].

6.3 PERCEPTION AND PROPAGATION OF TXRF METHODS

A brief overview of previous TXRF instrumentation and its worldwide application has already been given in Section 1.2.1. A more detailed description of today's commercial instruments and their worldwide distribution shall demonstrate the perception and propagation of TXRF and related methods. Furthermore, efforts concerning standardization, international activities, and cooperation shall be enumerated.

6.3.1 Commercially Available Instruments

In the last decade, the TXRF market changed considerably. Two major companies no longer exist (Technos and Atomica), another manufacturer no longer supplies TXRF instruments (Cameca/FEI as successor of Atomica), and Seifert has stopped the production of TXRF instruments. The company G.N.R. (initials of the owners) supplies the instruments formerly distributed by ItalStructures. On the other side, the ATI (Atominstitut in Vienna) constructs a spectrometer designed for academic use only (like the previous Wobi module) and a small company, OurStex in Japan, manufactures a new portable instrument.

At first, an overview of today's commercially available instruments will be given before dealing with particular instrumental developments. Table 6.5 summarizes five current manufacturers or suppliers of complete TXRF spectrometers (ATI, Bruker, G.N.R., OurStex, and Rigaku). The different suppliers with sites, instruments, excitation, monochromators, detectors, weight, and particularities are listed here. Half of the instruments have high-power X-ray tubes so they need water-cooling; the other half have low-power tubes, which are air-cooled. Most instruments use a multilayer for monochromatizing, only one uses an X-ray waveguide, and all these new instruments offer an SDD detector instead of a Si(Li) detector.

Many instruments are benchtop devices like the Wobistrax from ATI, the Picofox from Bruker, and the Nanohunter from Rigaku. Half of all spectrometers have a sample changer and allow vacuum measurements. The Wobistrax is produced by the nonprofit ATI and is distributed to scientific institutes only as mentioned earlier. The only portable spectrometer is the 200TX of OurStex [100,101]. It is equipped with a low-wattage X-ray tube of 5 W [102] and the sample chamber can be flushed with nitrogen gas [103]. The company G.N.R. supplies the TX2000, which allows a choice between TXRF with (0°/90°) geometry and conventional XRF with (45°/45°) geometry. The Nanohunter of Rigaku is suitable for GI-XRF measurements with an angle scan up to 2°. Several further ingenious devices of Rigaku are available for TXRF analysis in the wafer industry. They need a room of a whole lab possibly with an ultraclean environment and are applicable for wafers up to 300 mm and even 450 mm in diameter. Rigaku is headquartered in Tokyo, Japan, but other affiliated companies are located in Japan, in the USA, and in Europe (in Kent, UK, and Berlin, Germany).

TABLE 6.5. Different Manufacturers of Complete Spectrometers Applicable to TXRF and Related Methods

Company	Site	Instrument	Excitation	Monochrom.	Detector	Weight (Volume)	Special
ATI wobi@ati.ac.at	Vienna, Austria www.ati.ac.at	WOBISTRAX	(Mo tube) 2 kW	Multilayer Mo/Si	SDD Ketek	about 50 kg	Benchtop Rough vacuum Sample changer
Bruker Nano info@bruker-nano.de	Berlin, Germany www.bruker.com	S2 PICOFOX	Mo/W tube 30 or 40 W	Multilayer 17.5 keV	SDD Bruker	37 kg	Benchtop Sample cassette automatic
G.N.R. gnrtech@gnr.it	Agrate Conturbia Italy	TX 2000	Mo/W tube 2.2 kW	Multilayer Si/W	SDD	185 kg	Sample changer Variable geometry
OurStex info@ourstex.co.jp	Osaka, Japan www.ourstex.co.jp	200TX	W tube 5 W	X-ray waveguide	SDD	8 kg	Portable device Rough vacuum
Rigaku www.rigaku.com	Tokyo, Japan www.rigaku.co.jp	NANOHUNTER	Mo/Cu tube 50 W	Multilayer Mo/Si	SDD	70 kg	Benchtop Sample changer Angle scan <2°
Rigaku	Tokyo, Japan	TXRF 3760, 3800	High-power rotating anode	Multilayer	SDD	10–20 m^3	200 mm wafer
www.rigaku.com	www.rigaku.co.jp	TXRF 310, V310	”	”	”	”	300 mm wafer
		TXRF 450, V450	”	”	”	”	450 mm wafer

Several manufacturers produce and supply energy-dispersive detectors only, in recent years mainly silicon-drift detectors. Such companies are Amptec, Bruker, Canberra, EDAX, KETEK, Kevex, Link, Ortec, Oxford Instruments, Princeton Gamma Tech, SGX Sensortech, Thermo Scientific, and Vortex. The XR100 detector of Amptek already collected spectra on the surface of Mars in 1997. Today, KETEK has a market share of 75%. The main advantage of SDDs in comparison to SiLis is the thermoelectric cooling instead of liquid nitrogen cooling. Detectors can be cooled down to –40 °C or even –60 °C with an electric energy of only 2 W up to 7 W. They are available with an active area of 10 mm^2 up to 150 mm^2 and a thickness of 500 μm. The spectral resolution measured by the FWHM of the Mn-Kα peak of iron-55 is about 125 eV at a shaping time of 8 μs. By reducing this time to only 0.2 μs, the input count rate can reach the level of 10^6 cps. A new SDD can simply be adapted to a complete TXRF system with only little mechanical and electrical effort and leads to an easy upgrade [104]. For particular signal processing, application-specific integrated circuits (ASICs) are developed that can contain several millions of logic gates for digital electronics.

Multilayers are provided by other small companies such as AXO, Dresden, and incoatec, Geesthacht, Germany. Such layers are produced by physical vapor deposition (VPD) of heavy metals and light elements in alternation. Between ~10 and 500 different double layers are deposited on Si substrates or quartz glass. Prominent examples are W/C, Mo/Si, Ta/Si, Ni/C, Pt/SiC, La/B_4C, and WSi_2/C. Such multilayered coatings can be used (1) as beam adapters for X-ray sources, (2) as monochromators for energy tuning, and (3) as reference samples for thin-layer analysis. In comparison to natural crystals, synthetic multilayers can be fabricated as plane or even bent mirrors. Kirkpatrick-Baez and Göbel mirrors are well-known examples of X-ray optics with multilayered coatings of substrates that are elliptically or parabolically bent or prefigured. They can collimate and monochromatize a divergent beam simultaneously and increase the intensity of X-radiation. In order to focus the X-ray beam, two mirrors can be cross-coupled orthogonal to each other. Such devices that weigh several tons and cost $100 000 and more are usually installed at synchrotron beamlines. Apart from such big devices, simple monolayers can be produced with a thickness smaller than 1 nm by means of VPD. Such layers of a single element show a good lateral homogeneity down to the pm range, that is, with a density of submonolayers [105].

Thin windows suitable for X-ray tubes, detectors, and vacuum chambers can be purchased from special companies, such as Moxtec Inc., Orem, UT. The windows mainly consist of beryllium or polymer foils, and their transmittance depends on the thickness and the energy of X-ray photons. Ultrathin windows have a thickness of about 0.3 μm and are coated with a thin Al layer or a multilayer in order to absorb visible, UV, or IR light and to prevent charging. They are supported by a silicon or carbon grid with about 75% transmittance.

6.3.2 Support by the International Atomic Energy Agency

The International Atomic Energy Agency (IAEA) belongs to the United Nations Organization (UNO), promoting the peaceful use of nuclear energy and inhibiting its military use by nuclear weapons. It was created in 1957 after a proposal of the former president of the United States, Dwight D. Eisenhower. The IAEA is an autonomous organization with 159 member states pursuing the "safe, secure, and peaceful uses of nuclear sciences and technology." Its headquarters are located in Vienna, Austria, and two "Reginal Safeguard Offices" are in Toronto, Canada, and in Tokyo, Japan. In 2005, the IAEA and its former director, Mohamed ElBaradei of Egypt, were awarded the Nobel Prize for peace.

The IAEA runs three laboratories: in Monaco, in Vienna, and in Seibersdorf, Austria. The latter publishes a monthly newsletter reporting on the activities of the IAEA laboratories and the member states. The main objectives are activities in the field of nuclear analytical techniques, of XRF and TXRF analysis, and environmental monitoring. Relevant meetings, regional and interregional training courses, and workshops are announced. Dedicated programs and activities are organized with the aim of international collaborations and interdisciplinary applications.

The IAEA supports Technical Cooperation or TC projects worldwide, especially for developing member states [106]. The Wobrauschek module, for instance, is a low-cost attachment for existing X-ray equipment, available since 1986 from ATI in Vienna, Austria. This module has been distributed to about 40 developing countries around the world by a TC project. Recently, it was modified with a vacuum chamber and an uplooking SDD [107]. Also, the first WOBISTRAX was installed under the auspices of the IAEA at the University of the West Indies in Kingston, Jamaica [108]. It may be stressed again that these devices are nonprofit systems for academic use only.

All the instruments need software as an essential tool for data evaluation. It was a consequent action that software programs were developed, which allow deconvolution of the X-ray spectra and quantification after calibration. Two software packages are available called AXIL-QXAS and WINQXAS. They are widespread in the TXRF community and not only applied for ATI instruments.

6.3.3 Worldwide Distribution of TXRF and Related Methods

Three groups of TXRF users can be distinguished: (i) users from universities and scientific institutes; (ii) users at synchrotron beamlines, and (iii) users from the semiconductor industry. The first group was represented quite well by the attendees of the TXRF 2013 conference in Osaka, Japan. A survey was carried out with help of these attendees in order to demonstrate the worldwide distribution of TXRF equipment and the different fields of applications. Additional support was given by the four small manufacturers.

Total reflection XRF instrumentation exists on six continents and in more than 50 countries at about 200 institutes and laboratories. The number of running spectrometers applied for TXRF and related methods amounts to nearly 300. Most of them are desktop instruments. Forty-three percent are localized in Europe, 23% in Asia, 16% in North and Central America, and 11% in South America. Countries with a high percentage are Germany with 17%, USA with 9%, and Japan with 6%. Most instruments were produced by the five companies enumerated in Section 6.3.1.

Several small and developing countries also have such devices, mostly low-priced attachment modules. Nearly 50 ATI instruments were distributed with the help of the IAEA to countries, such as Costa Rica, Guatemala, and Panama in Central America; the Dominican Republic and Jamaica in the Caribbean; Mongolia and Singapore in Asia; and Bolivia, Paraguay, and Uruguay in South America. A very few instruments are home-made.

Figure 6.11 shows the distribution of today's running TXRF equipment in a world map with six continents. The Robinson projection depicts the whole globe as a flat image with a good area and angular conformity aside from severe distortions near the poles. Figure 6.11a represents TXRF equipment of the first group of users (universities and scientific institutes) as red bars in 51 different countries. Figure 6.11b depicts an accumulation of TXRF instruments in 24 subregions, largely delineated after the United Nations. All data are minimum values and give a picture of TXRF's distribution, which is fairly realistic though not absolutely exact.

In addition to the first group of TXRF users, a second group is considered working at several synchrotron beamlines at a work place exclusively dedicated to TXRF and related methods. Instead of a separate X-ray tube, they can use the synchrotron beam for excitation preferably with a monochromator. In addition, a six-axis sample stage and an energy-dispersive detector are necessary. Unfortunately, precise numbers of all particular working stations for TXRF in the different subregions are not available. For 2014, the authors of this book estimate their total number at about 55 stations in 25 countries. Several beamlines have even two or three work places used for TXRF applications. They are represented in Figure 6.11b as blue bars in 15 subregions. Recently, a new work station was established at the Shanghai Synchrotron Radiation Facility (SSRF), dedicated to XANES and XRR measurements at grazing incidence [109]. In a first study, Ti/Ni/Ti layers were investigated with different nm-thick central Ni layers and interfacial diffusion could be observed.

The third group of TXRF users is represented by chemical laboratories in industry, especially in the semiconductor industry with particular interest in wafer control. Some figures illustrate the use of TXRF in the wafer industry [110]: in 2013 more than 25 companies with more than 150 semiconductor fabrication plants—known as fabs—produced and supplied silicon wafers worldwide. They are distributed in more than 15 countries. The United States with 36% leads the group (23% are concentrated in six states alone: AZ, ME, MN, NY, OR, and TX). It is followed by Taiwan with 19%, China with 8%, and

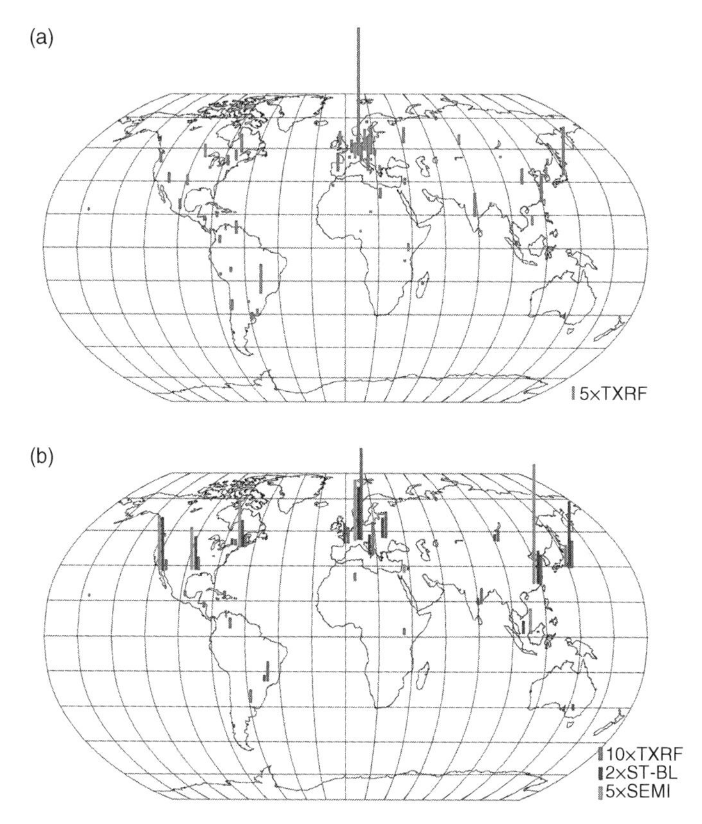

Figure 6.11. Worldwide distribution of TXRF instrumentation after a survey of conference attendees in 2013. The maps are shown in Robinson projection. (a) The red bars represent TXRF instrumentation in 54 countries. (b) TXRF equipment as shown in Figure 6.11a with red bars, but accumulated in 24 subregions according to the United Nations. The blue bars show synchrotron beamlines with TXRF facility, and the green bars stand for semiconductor fabs that probably apply TXRF for contamination control. The height of the bars is a measure for the number of devices. Red, blue, and green bars are scaled so that their total sum is equalized. Maximum for TXRF instruments is 34 in Europe, for the beamlines it is 10 in the USA, and for the semiconductor fabs it is 28 in Taiwan. (See colour plate section)

Germany, Japan, France, and Singapore with about 5% each. A single plant costs about US $1 billion ($10^9$ US$). Most of these fabs have been working since about 2000 and have a capacity between 10 000 and 80 000 wafers per month. If 30 000 wafers of 300 mm diameter were produced per month, each appropriate for 1000 chips then US$10 per chip would lead to US $3.6 billion sales per year and fab. A 1 h stop of a manufacturing plant means a loss of US $400 000.

This group of potential TXRF users is shown in Figure 6.11b as green bars. The number of fabs was taken from the Internet showing a list of 152 plants in 18 countries or 8 subregions (http://en.wikipedia.org/wiki/List_of_semiconductor_fabrication_plants). The fabs were assumed to have two

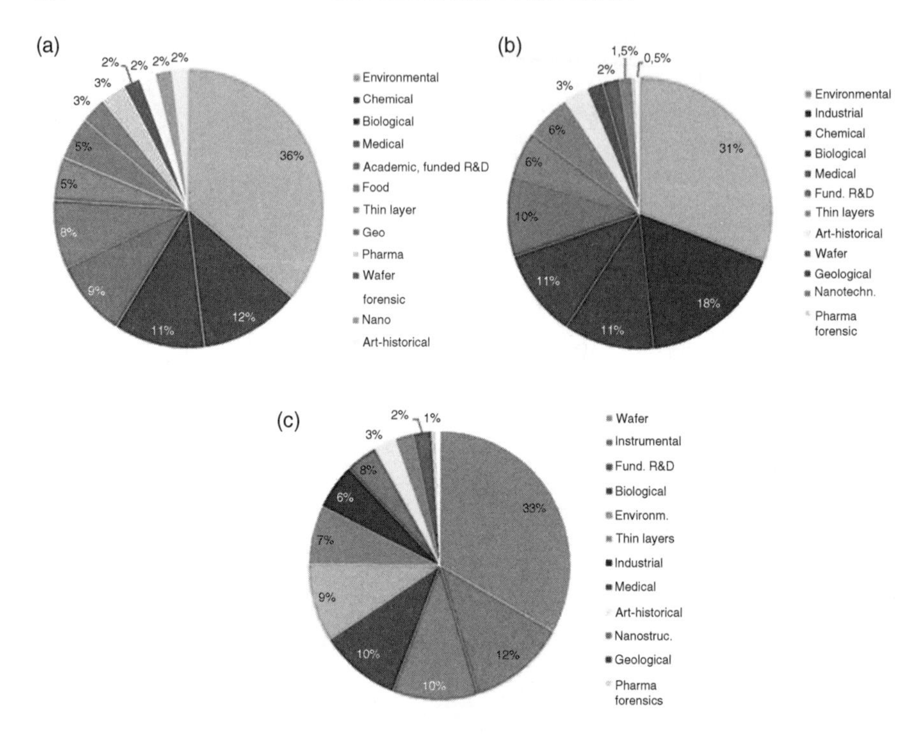

Figure 6.12. Pie charts with percentage of different fields of investigations or applications carried out by TXRF and related methods. (a) After customers of Bruker devices; reproduced by courtesy of A. Gross from Bruker. (b) After a survey of conference attendees in 2013. (c) After book of abstracts compiled for the conference TXRF 2013. It was taken for granted that 150 semiconductor fabs use TXRF for wafer control. (See colour plate section)

TXRF sets on average for wafer control. Consequently, about 300 devices can be estimated to be still working in the semiconductor industry as of 2013.[4] Approximately 600 sets were installed for about US $700 000 per set between 1989 and 2013. Most of them are floor-mounted large devices instead of desktop instruments mentioned earlier.

Figure 6.11a and b demonstrates that TXRF is mainly distributed in institutions of the northern hemisphere, especially in Europe, North America, and East Asia. The same is valid for TXRF at synchrotron beamlines. The semiconductor industry with TXRF equipment is concentrated in 51 fabs in the USA, with 40 fabs in China and Taiwan, and 23 fabs in central Europe.

Apart from wafer control, TXRF and related methods work on different fields of investigations or applications, especially at universities and scientific institutes. The pie charts of Figure 6.12 demonstrate the percentage of such fields. In order to meet different aspects, three different presentations are

[4] If the average number per fab is 2 because of a necessary substitute for claim and if the life-span is 8 years, the total number should be 600 within 25 years.

shown: (a) after the TXRF market according to Bruker installations; (b) after a survey of scientists attending the conference TXRF 2013 in Osaka; (c) after papers within the book of abstracts of TXRF 2013 in addition to 150 fabs performing wafer control in the semiconductor industry.

Bruker divided the TXRF market according to applications of their own installations [111]. The pie chart of Figure 6.12a shows a main emphasis of 36% on environmental applications, especially for the analysis of sewage. Chemical and biological applications with 12% and 11%, respectively, are next to it. Semiconductor applications have a small percentage of only 5% for thin layers and 4% for wafers, which can be explained by the fact that wafer analysis is mainly carried out with Rigaku spectrometers in the semiconductor industry.

The survey among attendees of TXRF 2013 gives a similar picture as demonstrated in Figure 6.12b. It reflects nearly 200 applications in 13 different fields. The main part of nearly 31% deals with environmental problems (different kinds of water, aerosols, biomonitors, nutrients), followed by industrial applications with 18% (ultrapure water, waste water, fuels, high-tech materials). Applications with chemical aspects (high-purity fluids and solids, ceramics, polymers), with biological background (biomolecules, cells, plants, beverage, food), or medical origin (urine, blood, tissue) have a part of about 11% each. As could be expected for a scientific conference, nearly all investigations refer to R&D (research and development), but only 6% exclusively. Thin layers and wafers together take about 9%, while art-historical investigations cover only 3%. Instrumental developments seem not to be planned according to the survey of conference attendees. Besides, it can be established that both presentations of Figure 6.12a and b correspond with an intersection of 80%.

Figure 6.12c grants 33% roughly to wafer control because of the high proportion of instruments routinely used in about 150 fabs of the semiconductor industry. The remaining part corresponds to 67% devices at institutes and work places at beamlines. According to 90 contributions of TXRF in the book of abstracts, the different fields show a high percentage of 12% for instrumental development followed by fundamental, biological, and environmental applications with nearly 10% each. Industrial and medical analyses cover about 5% each, while art-historical, geological, pharmaceutical, and forensic applications have small parts of only a few percent. The intersection of all three presentations of Figure 6.12 is nearly 50%.

All these results demonstrate the distribution of TXRF and related methods, its perception, and strength in several numerical details.

6.3.4 Standardization by ISO and DIN

ISO is the international organization for standardization, which published nearly 20 000 international standards available in the ISO store. A standard is a document that describes specifications, guidelines, or characteristics of materials, products, processes, and services—hereby ensuring that they are fit for purpose, that is, they are safe, reliable, and of good quality. International standards benefit technology, the economy, society, business, and government. Popular standards are ISO 9000—Quality management; and ISO 14000—Environmental management.

An ISO standard is developed by a technical committee (TC) with several experts in the field being considered. If a need for a standard is established, the relevant panel of experts discusses and draws up a draft that is distributed among ISO's members. They are asked to give comments and to suggest changes. The draft is revised and finally put to a vote.

For a great many of routine applications in analytical chemistry, standardized methods are required or even mandatory. This is the case for elemental analysis in the wide field of industrial, clinical, pharmaceutical, environmental, and biological applications. For the distribution of TXRF and related methods and their instrumentation, the development of specific standards is *essential*. A strong international effort and cooperation is necessary.

ISO/TC 201 is a technical committee for surface chemical analysis. It includes analytical techniques that direct a beam of atomic particles (atoms, ions, protons, and electrons) or photons toward a specimen of the material to be analyzed and detect atomic particles or photons emitted or scattered from the specimen. Analytical information is obtained from near-surface regions within 20 nm thickness. Relevant standards may be found from the ISO website and summaries are published in the journal "Surface and Interface Analysis." In the last decade, 28 standards were developed for ISO/TC 201; seven standards were published for XPS (X-ray photoelectron spectrometry), four standards for AES and XPS each, three standards for both AES and XPS, and two standards for TXRF and GD-OES (glow-discharge OES) each.

By request of the semiconductor industry, both TXRF standards were worked out for the measurement of elemental contamination on silicon wafer surfaces. After 12 years of work, the first standard was accepted [112]. It is called ISO 14706: 2000—Surface Chemical Analysis—Determination of surface elemental contamination on silicon wafers by total reflection X-ray fluorescence (TXRF) spectroscopy. Chemo-mechanically polished or epitaxial silicon wafer surfaces are the objects of investigation. The method is applicable to elements with atomic numbers between 17 and 92, that is, from sulfur to uranium. For direct analysis by TXRF, the surface density of atoms ranges between 10^{10} and 10^{14} atoms/cm^2. By means of vapor-phase decomposition this range can be lowered to 1×10^8 atoms/cm^2.

This standard is based on three predecessors existing since about 1995, namely ASTM F 1526 (Atomic Spectroscopy Testing and Materials), SEMI Standard M33 (Semiconductor), and UCS Standard (Ultra-Clean Society). A cleanroom of better than class 4 is recommended, a glancing angle between 0.25% and 0.75% of the critical angle is suggested, and a fine vacuum with about 10 Pa or a flush with helium or nitrogen gas shall be used. A reference sample with known amounts of iron or nickel (about 1×10^{12} atoms/cm^2) and a blank reference material (BRM) shall be stored in the same container and used for analysis. The reference material shall be analyzed by a reliable method of quantitative analysis. Dip- and spin-coating as well as microdrop specimens are available. A linear calibration with relative sensitivity factors is obligatory.

An international interlaboratory test program was carried out by 15 different laboratories in Japan, Europe, and USA in 1997. The test program was based on this ISO standard. Four test specimens and one reference material were distributed. The results were already discussed in Section 6.1.5.

The mentioned standard ISO 14706:2000 is helpful for quality assurance (QA) in the semiconductor industry. A second standard complements the first one by collection of elements deposited on silicon wafers and used as reference materials. It is called ISO 17331:2004—Surface Chemical Analysis—Chemical methods for the collection of elements from the surface of silicon-wafer working reference materials and their determination by total-reflection X-ray fluorescence (TXRF) spectrometry. All standards of ISO/TC 201 are commercially available and can be purchased from the ISO Central Secretariat in Geneva, Switzerland. New standards for calibration of ultrathin layers are produced by AXO (Applied X-ray Optics), Dresden, Germany. By means of physical vapor deposition (PVD), submonolayers with a density of 10^{13} atoms/cm^2 could be produced, which is equivalent to an areal deposition of 1%. The thickness of a single atom is about 0.25 nm for most elements while the atomic density of a complete monolayer is about 10^{15} atoms/cm^2.

Since 2012, a group of experts has been developing a prenormative standard to promote the use of TXRF in biological and environmental analysis. The project was accepted by the ISO/TC201 committee and a first draft on "Technical specification for the use of TXRF spectroscopy in biological and environmental analysis" was written, providing guidelines for qualitative and quantitative analyses [113]. A number of different samples were included: with environmental origin (different kinds of water, soils, bioindicators, and aerosols), with biological importance (wine, honey, milk, vegetables, and cereal), and with medical/clinical aspects (blood, serum, plasma, and tissue samples).

In this context, an interlaboratory test with more than 70 laboratories has been approved by the Versailles Project on Advanced Materials and Standards (VAMAS). This international organization is stimulating cross-border consensus and collaboration in order to support international trade. The "1st TXRF round robin test: water samples" was organized with 19 labs from 10 different countries. Two certified labs with experience in ICP-MS were involved. Four different kinds of water samples had to be analyzed: drinking water, waste water, leaching water, and purified water. Much work remains to demonstrate advantages and limits of TXRF as a standard tool for elemental analysis of environmental and biological materials.

A German equivalent of ISO is DIN (Deutsche Industrie Norm), which acts as a national mirror committee. The "Normenausschuss" NA 062-08-16 AA is responsible for standards of "Chemische Oberflächenanalyse und Rastersondenmikroskopie" and corresponds to ISO/TC 201. Already in 2004, the working group NMP 815 (Normenausschuss Materialprüfung) around A. Prange developed the standard DIN 51003:2004-05; Total reflection X-ray fluorescence—Principles and definitions. It is available from Beuth-Verlag, under http://www.beuth.de/de/norm/din-51003/69485626. It may also be mentioned that the VDI/VDE in Germany (Verein Deutscher Ingenieure/Verband der Elektronik Informationstechnik) has developed and published a guideline "Röntgenoptische Systeme" or "X-Ray optical systems." It has the designation VDI/VDE 5575 and is also available from Beuth-Verlag. Page 1 is related to terms and definitions, page 2 is concerned with assembly and measuring methods for the characterization of X-ray optical systems including X-ray sources and X-ray spectral analysis, and page 3 describes X-ray capillaries.

Finally, it may be underlined that good laboratory practice (GLP) is a must for professional analytical determinations in a laboratory. Quality management and assurance are a matter of course. Analytical determinations should be performed in accord with a standard operating procedure (SOP). Such an instruction describes the respective sequence of operations exactly. Any operation has to be documented in a protocol that has to be underlined by two persons. The staff members carrying out the operation have to be informed and trained regularly, and alterations have to be approved and documented. SOPs are normally used in clinical studies and pharmaceutical processes. They are the base for an ISO or DIN standard. Industrial laboratories, manufacturers, and supplying companies can be certified according to ISO/TS 16949.

6.3.5 International Cooperation and Activity

Especially in the semiconductor industry, new materials and components are produced with ever-decreasing structures for the production of nanoscaled integrated circuits. According to Moore's law from 1965, the number of transistors on a chip is doubled every 18 to 24 months. Consequently, the dimensions of microelectronic structures are decreased exponentially, which is put down in the International Technology Roadmap of Semiconductors (ITRS). By contrast, the dimensions of substrates, especially of wafers, are increasing in order to reduce the manufacturing costs. Wafers with a diameter of 310 mm and even of 450 mm are produced today.

A number of analytical methods (optical, electrical, X-ray, electron beam, and ion beam techniques) are suitable for investigations of such materials and components. Since 2009, efforts were made to bundle several complementary competencies within Europe in order to control material properties down to atomic dimensions. Several partners joined the European Integrated Activity of Excellence and Networking for Nano- and Microelectronics Analysis (ANNA). Particular common tasks were: (1) the production of a large set of wafers intentionally contaminated with various transition metals and light elements; and (2) the determination of these elements by different analytical techniques including TXRF [114]. A uniform distribution of the impurities, low detection limits, and a reliable quantification were sought. Aluminum and sodium as light elements and Ti, Fe, Cu, Ni, and Mo as transition metals were determined as impurities. "Impurity free" surfaces could be guaranteed for these elements with a level below 10^9 atoms/cm^2 (except for aluminum with 10^{10} atoms/cm^2).

Complementary competencies of different methods and laboratories can be revealed by intercomparisons and round-robin tests. Organic contaminants, such as plasticizers, disinfectants, and fire retardants, were deposited on silicon strips and analyzed, for example, by XANES [115]. Nano layers such as ultrashallow junctions of boron and arsenic implanted in Si wafers were characterized by GI-XRF, ellipsometry, and SIMS (see Section 6.2.3). Additionally, thin layers of high-*k* materials, hafnium-based, and only 1–10 nm thin, were characterized. ANNA as joint laboratory with 18 infrastructures offered transnational access and reference samples. The establishment of "Golden Laboratories" working as reference laboratories is a goal [115].

Complementary methods were also applied to the characterization of thin high-*k* layers made of Al_2O_3. Measurements by GI-XRF were evaluated with a fundamental-parameter approach, which is completely reference-free [116]. Besides, a model was designed for an analytical platform that integrates seven analytical methods into one sample chamber suitable for UHV and 450 mm wafers. A novel positioning system was patented with seven or even nine free axes. Five X-ray methods are planned for integration: XRF, TXRF, GI-XRF, XRR, and XRD. Ellipsometry and VUV reflectometry complete this palette. A detailed table summarizes applicability, detection limits, concentration range, spatial and depth resolution, time consumption, costs, and strengths and weaknesses of all these methods [116].

A novel instrument for nanoanalytical investigations has been designed and constructed by the TUB (Technische Universität Berlin) and by the PTB (Physikalische-Technische Bundesanstalt) in Berlin [117]. The PTB is Germany's metrology institute. A schematic view of the device is shown in Figure 6.13. The instrument consists of an UHV chamber mounted on a base frame. This chamber has to be aligned to one of two beamlines of the synchrotron BESSY II. Samples to be investigated can be transferred to the sample manipulator inside the chamber via a load-lock.

The patent-pending sample manipulator with nine axes is represented as a 3D view in Figure 6.14. It allows three rotational and three translational movements of the sample, and a further three movements of the detectors, all of them driven by computer-controlled stepper motors. A translation is possible between ±12 and ±55 mm, a tilt between −5° and 110°. The resolution is about 0.0005° or 2 arcsec and ±2 μm, the scan velocity is about 0.25 mm/s and 0.3°/s. Two energy-dispersive detectors oriented perpendicularly to the incident beam and to each other can be flanged to the chamber (DN 160 CF). They enable the performance of six different X-ray analytical methods: XRF, TXRF, GI-XRF, XRD, XRR, and GI-SAXS. Various types of samples can be investigated: small pieces of a few mm^3, thin foils, flat and even curved samples, samples with micro- and nanostructures, and 4 inch wafers (diameter of about 100 mm).

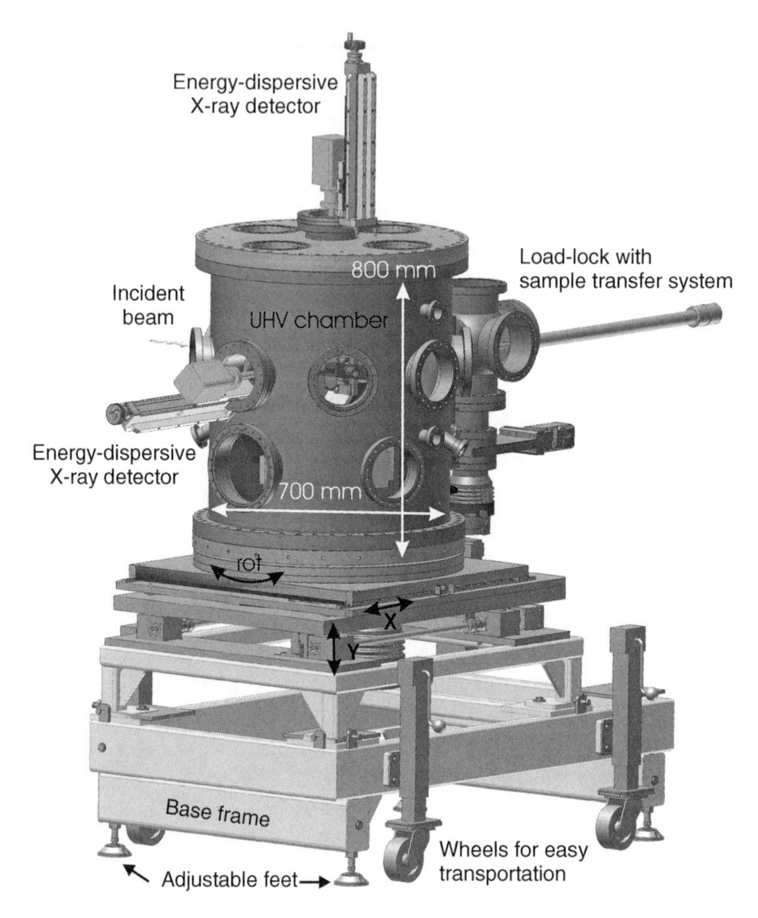

Figure 6.13. Schematic view of a novel UHV sample chamber placed on a movable base frame. It can be attached to a synchrotron beamline, samples are transferred through a load-lock to the sample manipulator inside the chamber, and two energy-dispersive detectors can be used for analysis. Figure from Ref. [117], reproduced with permission. Copyright © 2013, American Institute of Physics.

The instrument was commissioned and applied to the characterization of deposited and implanted or buried layers, and of interfaces in the nanometer regime [117]. Light elements, like B, C, N, and Al, could be detected. Measurements were carried out using the linear polarization of the synchrotron beam. In particular, it was proved again that XRR and GI-XRF are complementary techniques. A simultaneous analysis can lead to a corrected model for a layered sample and can allow an improved calculation avoiding systematic errors of both methods.

Until 2014, three work places could be provided with chamber and patent-pending manipulator for $100 \times 100\,\text{mm}^2$ samples. The first work place was opened at the PTB beamline of BESSY II in Berlin, Germany, in 2012 and

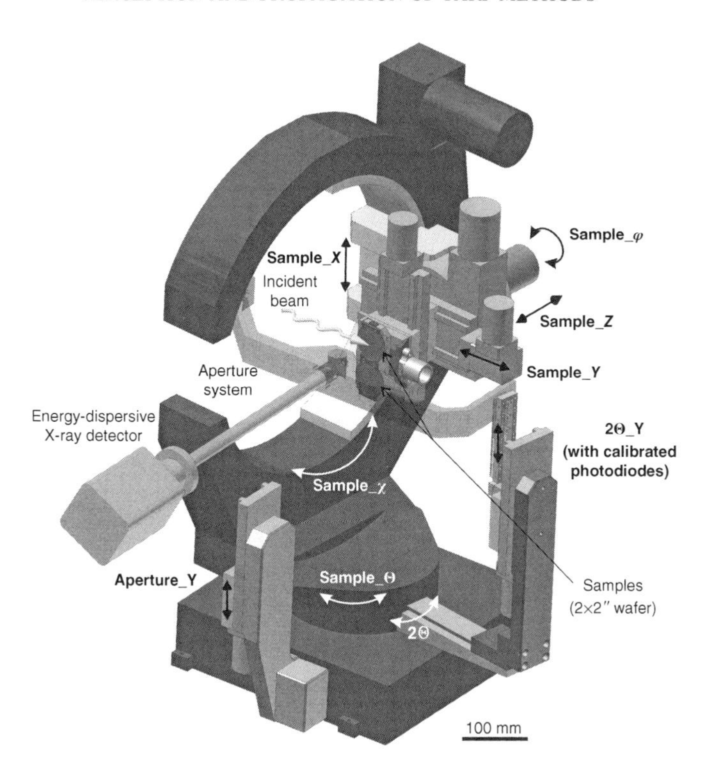

Figure 6.14. 3D view of the sample manipulator included in the sample chamber. A nine-axis sample manipulator can be used for adjustment and movement of sample and detector. Sample chamber and manipulator are already installed at three different work places. Figure from Ref. [117], reproduced with permission. Copyright © 2013, American Institute of Physics.

installed in BLiX (Berlin Laboratory for Innovative X-ray Technology) of the TU in Berlin. A further transportable station initiated by IAEA in 2011 was constructed and tested in 2013. The installation of the chamber is completed now for operation at the XRF beamline of Elettra Sincrotrone in Trieste, Italy. It can also be used as a stand-alone instrument if a high- or low-power X-ray tube is connected to it. The equipment will be fully commissioned and accessible to IAEA members in 2014 [118]. A third station is designated for a beamline of the CEA at SOLEIL in Saint-Aubin, France.

Beyond this European cooperation, a working group was created in 2012 as an open discussion forum for international communication in the field of TXRF and related methods. Facebook was chosen as social network due to its widespread availability and its ease of use. The main Facebook page describes the group as an "Open forum for specialists in X-ray spectroscopic techniques, TXRF and GI-XRF mainly, where you can share knowledge, experiences, questions, and lines of the future of this technology. Scientific knowledge should be shared by all." Everybody can become a member of the group after a simple verification of the applicant's interests in TXRF by the group administrator. The number of group members was 50 at the beginning of 2014; the administrator is Ramon Fernandez-Ruiz [119].

Two pages show the platform of the group. The first is named "TXRF Workgroup Forum" (http://www.facebook.com/groups/TXRFSpectrometry). It is mainly restricted to accepted members. Other people can only read the description of the group and can take note of its members. The members themselves have access to all information shared within the group. The second page "TXRF Workgroup Public Page" is open to everyone, to members, and nonmembers (http://facebook.com/TXRFSpectrometry). Everybody can find news and reports here, on events and activities of the TXRF group with a public character.

REFERENCES

1. Török, S.B. and Van Grieken, R.E. (1994). X-ray spectrometry. *Anal. Chem.*, **66**, 186R–206R.
2. Török, S.B., Labar, J., Injuk, J., and Van Grieken, R.E. (1996). X-ray spectrometry. *Anal. Chem.*, **68**, 467R–485R.
3. Potts, P.J., Ellis, A.T., Kregsamer, P., Marshall, J., Streli, C., West, M., and Wobrauschek, P. (2002). Atomic spectrometry update. X-ray fluorescence spectrometry. *J. Anal. At. Spectrom.*, **17**, 1439–1455.
4. West, M., Kregsamer, P., Potts, P.J., Streli, C., Vanhoof, C., and Wobrauschek, P. (2007). Atomic spectrometry update. X-ray fluorescence spectrometry. *J. Anal. At. Spectrom.*, **22**, 1304–1332.
5. International Union of Pure and Applied Chemistry (1976). Nomenclature, symbols, units and their usage in spectrochemical analysis - II. Data interpretation. *Pure Appl. Chem.*, **45**, 99.
6. Reus, U., Freitag, K., Haase, A., and Alexandre, J.F. (1989). Total-reflection X-ray fluorescence spectrometry. Application methods. *Spectra 2000*, **143**, 42–46.
7. Prange, A. and Schwenke, H. (1992). Trace element analysis using total-reflection X-ray fluorescence spectrometry. *Adv. X-Ray Anal.*, **35B**, 899–923.
8. Klockenkämper, R. (1997). *Total-Reflection X-ray Fluorescence Analysis*, 1st ed., John Wiley & Sons, Inc.: New York.
9. Michaelis, W., Fanger, H.U., Niedergesäß, R., and Schwenke, H. (1985). Intercomparison of the multielement analytical methods TXRF, NAA and ICP with regard to trace element determinations in environmental samples. In: Sansoni, B. (editor) *Instrumentelle Multielementanalyse*, VCH Verlagsgemeinschaft: Weinheim, pp. 693–710.
10. Michaelis, W. (1986). Multielement analysis of environmental samples by total-reflection X-ray fluorescence spectrometry, neutron activation analysis and inductively coupled plasma optical emission spectroscopy. *Fresenius Z. Anal. Chem.*, **324**, 662–671.
11. Burba, P., Willmer, P.G., and Klockenkämper, R. (1988). Elementspuren-Bestimmungen (AAS, ICP-OES, TRFA) in natürlichen Wässern nach Voranreicherung: ein Vergleich. *Vom Wasser*, **71**, 179–194.

12. Stößel, R.-P. and Prange, A. (1985). Determination of trace elements in rainwater by total reflection X-ray fluorescence. *Anal. Chem.*, **57**, 2880–2885.
13. Prange, A., Knöchel, A., and Michaelis, W. (1985). Multielement determination of dissolved heavy metal traces in sea water by total-reflection X-ray fluorescence spectrometry. *Anal. Chim. Acta*, **172**, 79–100.
14. Prange, A., Knoth, J., Stößel, R.-P., Böddeker, H., and Kramer, K. (1987). Determination of trace elements in the water cycle by total-reflection X-ray fluorescence spectrometry. *Anal. Chim. Acta*, **195**, 275–287.
15. Freimann, P. and Schmidt, D. (1988). Application of total reflection X-ray fluorescence analysis for the determination of trace metals in the North Sea. *Spectrochim. Acta*, **44B**, 505–511.
16. Egorov, A.I., Kabina, L.P., Kondurov, I.A., Korotkikh, E.M., Martynov, V.V., Shchebetov, A.F., and Sushkov, P.A. (1992). *Adv. X-Ray Anal.*, **35B**, 959.
17. Prange, A., Böddeker, H., and Kramer, K. (1993). Determination of trace elements in riverwater using total-reflection X-ray fluorescence. *Spectrochim. Acta*, **48B**, 207–215.
18. Reus, U., Markert, B., Hoffmeister, C., Spott, D., and Guhr, H. (1993). Determination of trace metals in river water and suspended solids by TXRF spectrometry. A methodical study on analytical performance and sample homogeneity. *Fresenius J. Anal. Chem.*, **347**, 430–435.
19. Prange, A. (1983). Entwicklung eines spurenanalytischen Verfahrens zur Bestimmung von gelösten Schwermetallen in Meerwasser mit Hilfe der Totalreflexions-Röntgenfluoreszenzanalyse, PhD Thesis. University of Hamburg.
20. Prange, A., von Tümpling Jr., W., Niedergesäß, R., and Jantzen, E. (1995). Die gesamte Elbe auf einen Blick – Elementverteilungsmuster der Elbe von der Quelle bis zur Mündung. *Wasserwirtschaft - Wassertechnik*, **7**, 22–33.
21. Prange, A. und Mitarbeiter (2001). Erfassung und Beurteilung der Belastung der Elbe mit Schadstoffen, Band 2/3, Anhang zum Abschlussbericht Band 1/3, GKSS-Forschungszentrum; Helmholtz-Zentrum Geesthacht; Institute for Coastal Research.
22. Schmidt, D., Gerwinski, W., and Radke, I. (1993). Trace metal determinations by total-reflection X-ray fluorescence analysis in the open Atlantic Ocean. *Spectrochim. Acta*, **48B**, 171–181.
23. Haarich, M., Schmidt, D., Freimann, P., and Jakobsen, A. (1993). North Sea research projects ZISCH and PRISMA: application of total reflection X-ray fluorescence in sea-water analysis. *Spectrochim. Acta*, **48B**, 183–192.
24. Freimann, P., Schmidt, D., and Neubauer-Ziebarth, A. (1993). Reference materials for quality assurance in sea-water analysis: performance of total-reflection X-ray fluorescence in the intercomparison and certification stages. *Spectrochim. Acta*, **48B**, 193–198.
25. Farías, S.S., Casa, V.A., Vázquez, C., Ferpozzi, L., Pucci, G.N., and Cohen, I.M. (2003). Natural contamination with arsenic and other trace elements in ground waters of Argentine Pampean Plain. *Sci. Total Environ.*, **309**, 187–199.
26. Custo, G., Litter, M.I., Rodríguez, D., and Vázquez, C. (2006). Total reflection X-ray fluorescence trace mercury determination by trapping complexation: Application in advanced oxidation technologies. *Spectrochim. Acta*, **B61**, 1119–1123.

27. Hegedüs, F., and Winkler, P. (1992). Uranium concentration measurement in water samples with XRF. *Adv. X-Ray Anal.*, **35B**, 965–967.
28. Barros, H., Marcó Parra, L.M., Bennun, L., and Greaves, E.D. (2010). Determination of arsenic in water samples by Total Reflection X-Ray Fluorescence using preconcentration with alumina. *Spectrochim. Acta*, **B65**, 489–492.
29. Marguí, E., Kregsamer, P., Hidalgo, M., Tapias, J., Queralt, I., and Streli, C. (2010). Analytical approaches for Hg determination in wastewater samples by means of total reflection X-ray fluorescence spectrometry. *Talanta*, **82**, 821–827.
30. Reus, U. and Prange, A. (1993). Atomika Instruments GmbH. *Application Notes.*
31. Barreiros, M.A., Carvalho, M.L., Costa, M.M., Marques, M.I., and Ramos, M.T. (1997). Application of total-reflection XRF to elemental studies of drinking water. *X-Ray Spectrom.*, **26**, 165–168.
32. Pashkova, G.V., Revenkov, A.G., and Finkelshtein, A.L. (2013). Study of factors affecting the results of natural water analyses by total reflection X-ray fluorescence. *X-Ray Spectrom.*, **42**, 524–530.
33. Muia, L.M., Razafindramisa, F.L., and Van Grieken, R.E. (1991). Total reflection X-ray fluorescence analysis using an extended focus tube for the determination of dissolved elements in rain water. *Spectrochim. Acta*, **B46**, 1421–1427.
34. Miesbauer, H. (1997). Multielement determination in sediments, pore water and river water of Upper Austrian rivers by total-reflection X-ray fluorescence, *Spectrochim. Acta*, **B52**, 1003–1007.
35. Gatari; M.J., Maina, D.M., Wagner, A., Boman, J., Gaita, S.M., and Bartilol, S. (2013). *Trace metals in waters of Lake Victoria basin*, 15th International Conference on TXRF and related methods, Book of Abstracts, P15, p. 48.
36. Gerwinski, W. and Schmidt, D. (1998). Automated solid-phase extraction for tracemetal analysis of seawater: sample preparation for total-reflection X-ray fluorescence measurements. *Spectrochim. Acta*, **B53**, 1355–1364.
37. Reus, U., Freitag. K., and Fleischhauer, J. (1989). Trace analysis in ultrapure acids at levels below 1 ng/ml. *Fresenius Z. Anal. Chem.*, **334**, 674.
38. Prange, A., Kramer, K., and Reus, U. (1991). Determination of trace element impurities in ultrapure reagents by total reflection X-ray spectrometry. *Spectrochim. Acta*, **46B**, 1385–1393.
39. Bilbrey, D.B., Leland, D.J., Leyden, D.E., Wobrauschek, P., and Aiginger, H. (1987). Determination of metals in oil using total reflection X-ray fluorescence spectrometry. *X-Ray Spectrom.*, **16**, 161–165.
40. Freitag, K., Reus, U., and Fleischhauer, J. (1989). The application of TXRF spectrometry for the determinations of trace metals in lubricating oils. *Fresenius Z. Anal. Chem.*, **334**, 675-??
41. Reus, U. (1991). Determination of trace elements in oils and greases with total reflection X-ray fluorescence: sample preparation methods. *Spectrochim. Acta*, **46B**, 1403–1411.
42. Schirrmacher, M., Freimann, P., Schmidt, D., and Dahlmann, G. (1993). Trace elemental determination by total-reflection X-ray fluorescence (TXRF) for the differentiation between pure fuel oil (bunker oil) and waste oil (sludge) in maritime shipping legal cases. *Spectrochim. Acta*, **48B**, 199–205.

43. Ojeda, N., Greaves, E.D., Alvarado, J., and Sajo-Bohus, L. (1993). Determination of vanadium, iron, nickel and sulphur in petroleum crude oil by total-reflection X-ray fluorescence. *Spectrochim. Acta*, **48B**, 247–253.
44. Knoth, J., Schwenke, H., Marten, R., and Glauer, J. (1977). Total-reflection X-ray fluorescence spectrometric determination of elements in nanogram amounts. *J. Clin. Chem. Clin. Biochem.*, **15**, 557–560.
45. Yap, C.T. (1988). X-ray total reflection fluorescence analysis of iron, copper, zinc, and bromine in human serum. *Appl. Spectrosc.*, **42**, 1250–1253.
46. Prange, A., Böddeker, H., and Michaelis, W. (1989). Multielement determination of trace elements in whole blood and blood serum by TXRF. *Fresenius J. Anal. Chem.*, **335**, 914–918.
47. Knöchel, A., Bethel, U., and Hamm, V. (1989). Multielement analysis of trace elements in blood serum by TXFA. *Fresenius Z. Anal. Chem.*, **334**, 673.
48. Ayala, R.E., Alvarez, E.M., and Wobrauschek, P. (1991). Direct determination of lead in whole human blood by total reflection X-ray fluorescence spectrometry. *Spectrochim. Acta*, **46B**, 1429–1432.
49. Dogan, P., Dogan, M., and Klockenkämper, R. (1993). Determination of trace elements in blood serum of patients with Behcet disease by total reflection X-ray fluorescence analysis. *Clin. Chem. (Washington, D. C.)*, **39**, 1039–1041.
50. Greaves, E.D., Meitín, J., Sajo-Bohus, L., Castelli, C., Liendo, J., and Borgerg, C. (1995). Trace element determination in amniotic fluid by total reflection X-ray fluorescence. *Adv. X-Ray Chem. Anal. Jpn.*, **26s**, 47–52.
51. Greaves, E.D., Marco-Parra, L.M., Rojas, A., and Sajo-Bohus, L. (2000). Determination of platinum levels in serum and urine samples from pediatric cancer patients by TXRF. *X-Ray Spectrom.*, **29**, 349–353.
52. Telgmann, L., Holtkamp, M., Künnemeyer, J., Gelhard, C., Hartmann, M., Klose, A., Sperling, M., and Karst, U. (2011). Simple and rapid quantification of gadolinium in urine and blood plasma samples by means of total reflection X-ray fluorescence (TXRF). *Metallomics*, **3**, 1035–1040.
53. Ostachowitz, B., Lankosz, M., Tomik, B., Adamek, D., Wobrauschek, P., Streli, C., and Kregsamer, P. (2006). Analysis of some chosen elements of cerebrospinal fluid and serum in amyotrophic lateral sclerosis patients by total reflection X-ray fluorescence. *Spectrochim. Acta*, **B61**, 1210–1213.
54. Liendo, J.A., Gonzalez, A.C., Castelli, C., Gomez, J., Jimenez, J., Marco, L., Sajo-Bohus, L., Greaves, E.D., Fletcher, N.R., and Bauman, S. (1999). Comparison between proton-induced X-ray emission (PIXE) and total reflection X-ray fluorescence (TXRF) spectrometry for the elemental analysis of human amniotic fluid. *X-Ray Spectrom.*, **28**, 3–8.
55. Carvalho, M.L., Custodio, P.J., Reus, U., and Prange, A. (2001). Elemental analysis of human amniotic fluid and placenta by total-reflection X-ray fluorescence and energy-dispersive X-ray fluorescence: child weight and maternal age dependence. *Spectrochim. Acta*, **B56**, 2175–2180.
56. Savage, I., and Haswell, S.J. (1998) The development of analytical methodology for simultaneous trace elemental analysis of blood plasma samples using total reflection X-ray fluorescence spectrometry. *J. Anal. At. Spectrom.*, **13**, 1119–1122.

57. Khuder, A., Bakir, M.A., Karjou, J., and Sawan, M.K. (2007). XRF and TXRF techniques for multi-element determination of trace elements in whole blood and human hair samples. *J. Radioanal. Nucl. Chem.*, **273**, 435–442.

58. Marco, L.M., Greaves, E.D., and Alvarado, J. (1999). Analysis of human blood serum and human brain samples by total reflection X-ray fluorescence spectrometry applying Compton peak standardization. *Spectrochim. Acta*, **B54**, 1469–1480.

59. Ketelsen, P. und Knöchel, A. (1984). Multielementanalyse von Aerosolen mit Hilfe der Röntgenfluoreszenzanalyse mit totalreflektierendem Probenträger (TRFA). *Fresenius Z. Anal. Chem.*, **317**, 333–342.

60. Ketelsen, P., und Knöchel, A. (1985). Multielementanalyse von größenklassierten Luftstaubproben. *Staub Reinhalt. Luft*, **45**, 175–178.

61. Gerwinski, W., Goetz, D., Koelling, S., und Kunze, J. (1987). Multielement-Analyse von Müllverbrennungs-Schlacke mit der Totalreflexions-Röntgenfluoreszenz (TRFA). *Fresenius Z. Anal. Chem.*, **327**, 293–296.

62. Gerwinski, W., and Goetz, D. (1987). Multielement analysis of standard reference materials with total reflection X-ray Fluorescence (TXRF). *Fresenius Z. Anal. Chem.*, **327**, 690–693.

63. Burba, P., Willmer, P.G., Becker, M., and Klockenkämper, R. (1989). Determination of trace elements in high-purity aluminium by total reflection X-ray fluorescence after their separation on cellulose loaded with hexamethylenedithiocarbamates. *Spectrochim. Acta*, **44B**, 525–532.

64. Reus, U. (1989). Total reflection X-ray fluorescence spectrometry: matrix removal procedures for trace analysis of high-purity silicon, quartz and sulphuric acid. *Spectrochim. Acta*, **44B**, 533–542.

65. Chen, J.S., Berndt, H., Klockenkämper, R., and Tölg, G. (1990). Trace analysis of high-purity iron by total reflection X-ray fluorescence spectrometry. *Fresenius J. Anal. Chem.*, **338**, 891–894.

66. Koopmann, C., and Prange, A. (1991). Multielement determination in sediments from the German Wadden Sea - investigations on sample preparation techniques. *Spectrochim. Acta*, **46B**, 1395–1402.

67. Gatari, M.J., Mugoh, F., Njenga, L.W., Shepherd, K.D., Sila, A., Kamau, M.N., and Maina, D.M. (2013). *Prediction of soil properties: An experiment using Mt Kenya forest soils*, 15th International Conference on TXRF and related methods, Book of Abstracts, O7, p. 23.

68. Klockenkämper, R., Becker, M., and Bubert, H. (1991). Determination of the heavy-metal ion-dose after implantation in silicon-wafers by total-reflection X-ray fluorescence analysis. *Spectrochim. Acta*, **46B**, 1379–1383.

69. Battiston, G.A., Gerbasi, R., Degetto, S., and Sbrignadello, G. (1993). Heavy metal speciation in coastal sediments using total-reflection X-ray fluorescence spectrometry. *Spectrochim. Acta*, **48B**, 217–221.

70. Freiburg, C., Krumpen, W., and Troppenz, U. (1993). Determinations of cerium, europium and terbium in the electroluminescent materials gadolinium oxysulfide (Gd_2O_2S) and lanthanum oxysulfide (La_2O_2S) by total-reflection x-ray spectrometry. *Spectrochim. Acta*, **48B**, 263–267.

71. von Bohlen, A., Klockenkämper, R., Otto, H., Tölg, G., and Wiecken, B. (1987). Qualitative survey analysis of thin layers of tissue samples - Heavy metal traces in human lung tissue. *Int. Arch. Occup. Environ. Health*, **59**, 403–411.
72. Eller, R. und Weber, G. (1987). Bestimmung der durch Flavonoide komplexierten Metalle in Zwiebeln mit Totalreflexions-Röntgenfluoreszenzanalyse (TRFA). *Fresenius Z. Anal. Chem.*, **328**, 492–494.
73. von Bohlen, A., Klockenkämper, R., Tölg, G., and Wiecken, B. (1988). Microtome sections of biomaterials for trace analyses by TXRF. *Fresenius Z. Anal. Chem.*, **331**, 454–458.
74. Klockenkämper, R., von Bohlen, A., and Wiecken, B. (1989). Quantification in total reflection X-ray fluorescence analysis of microtome sections. *Spectrochim. Acta*, **44B**, 511–518.
75. Günther, K. and von Bohlen, A. (1990). Simultaneous multielement determination in vegetable foodstuffs and their respective cell fractions by total-reflection X-ray fluorescence (TXRF). *Z. Lebensm. Unters. Forsch.*, **190**, 331–335.
76. Günther, K. and von Bohlen, A. (1991). Multielement speciation in vegetable foodstuffs by gel permeation chromatography (GPC) and total-reflection X-ray fluorescence (TXRF). *Spectrochim. Acta*, **46B**, 1413–1419.
77. Kulesh, N.A., Muela, A., Fdez-Gubieda, L., Novoselova, Y.P., Denisova, T.P., Safranov, A.P., Zarubina, K., and Kurlyandskaya, G.V. (2013). *Study of trace iron concentration in biological samples by total reflection X-ray fluorescence spectrometry*, 15th International Conference on TXRF and related methods, Book of Abstracts, O12, p. 99.
78. Prange, A. (1989). Total reflection X-ray spectrometry: method and applications. *Spectrochim. Acta*, **44B**, 437–452.
79. International Atomic Energy Agency, Austria (1993). Intercomparison run on trace and minor elements in candidate lichen research material (IAEA-336) and AQCS cabbage material (IAEA-359), Progress report No. 1.
80. Schmeling, M., Alt, F., Klockenkämper, R., and Klockow, D. (1997). Multielement analysis by total reflection X-ray fluorescence spectrometry for the certification of lichen research material. *Fresenius J. Anal. Chem.*, **357**, 1042–1044.
81. Rink, I., Rostam-Khani, P., Knoth, J., Schwenke, H., de Gendt, S., and Wortelboer, R. (2001). Calibration of straight total reflection X-ray fluorescence spectrometry - results of a European Round Robin test. *Spectrochim. Acta*, **56B**, 2283–2292.
82. Papadakis, I., Vendelbo, E., Van Nevel, L., and Taylor, P. (2000). IMEP-14 Trace Elements in Sediment, Report to Participants, EUR 19595 EN, European Commission JRC IRMM, Geel, Belgium.
83. Klockenkämper, R., Alt, F., Brandt, R., Jakubowski, N., Messerschmidt, J., and von Bohlen, A. (2001). Results of proficiency testing with regard to sediment analysis by FAAS, ICP-MS and TXRF. *J. Anal. At. Spectrom.*, **16**, 658–663.
84. Tölg, G. and Klockenkämper, R. (1993). The role of total-reflection X-ray fluorescence in atomic spectroscopy. *Spectrochim. Acta*, **48B**, 111–127.
85. Michaelis, W. (1986). Multielement analysis of environmental samples by total-reflection X-ray fluorescence spectrometry, neutron activation analysis and inductively coupled plasma optical emission spectroscopy. *Fresenius Z. Anal. Chem.*, **324**, 662–671.

86. Krivan, V., Schneider, G., Baumann, H., and Reus, U. (1994). Multi-element characterization of tobacco smoke condensate. *Fresenius J. Anal. Chem.*, **348**, 218–225.

87. Pepelnik, R., Prange, A., and Niedergesäß, R. (1994). Comparative study of multi-element determination using inductively coupled plasma mass spectrometry, total reflection X-ray fluorescence spectrometry and neutron activation analysis. *J. Anal. At. Spectrom.*, **9**, 1071–1074.

88. Slavin, W. (1992). *Spectrosc. Intern.*, **4**, 22.

89. Ehmann, W.D. and Vance, D.E. (1991). *Radiochemistry and Nuclear Methods of Analysis*, John Wiley & Sons: New York.

90. Lieser, K.H., Flakowski, M., and Hoffmann, P. (1994). Determination of trace elements in small water samples by total reflection X-ray fluorescence (TXRF) and by neutron activation analysis (NAA). *Fresenius J. Anal. Chem.*, **350**, 135–138.

91. Tabuchi, Y., Kaku, S., and Tsuji, K. (2013). *TXRF analysis of halogen in environmental and biological samples*, 15th International Conference on TXRF and related methods, Book of Abstracts, P45, p. 116.

92. Global Assessment Report, (2004). *The Laboratory Life Science and Analytical Instrumentation Industry*, 8th ed., Strategic Directions International (SDi), Inc.: Los Angeles, CA, USA.

93. Klockenkämper, R. (2006). Challenges of total reflection X-ray fluorescence for surface- and thin-layer analysis. *Spectrochim. Acta*, **61B**, 1082–1090.

94. CAS (Chemical Abstracts Service), online. © 2005 by the American Chemical Society (ACS). Search for keywords: *xrs* or *xrf* or *x ray spectro; txrf* or *total reflection x ray fluorescence; faas* or *flame atomic absorption spectro; et aas* or *electrothermal aas; icp aes* or *inductively coupled plasma oes; icp ms* or *icp mass spectro.*

95. von Bohlen, A. (2009). Total reflection X-ray fluorescence and grazing incidence X-ray spectrometry — Tools for micro and surface analysis. A review. *Spectrochim. Acta*, **64B**, 821–832.

96. Krämer, M., von Bohlen, A., Sternemann, C., Paulus, M., and Hergenröder, R. (2006). X-ray standing waves: a method for thin layered systems. *J. Anal. At. Spectrom.*, **21**, 1136–1142.

97. Krämer, M., von Bohlen, A., Sternemann, C., Paulus, M., and Hergenröder, R. (2007). Synchrotron radiation induced X-ray standing waves analysis of layered structures. *Appl. Surf. Sci.*, **253**, 7, 3533–3542.

98. Penka, V. and Hub, W. (1989). Application of total reflection X-ray fluorescence in semiconductor surface analysis. *Spectrochim. Acta*, **44B**, 483–490.

99. Klockenkämper, R., Becker, H.W., Bubert, H., Jenett, H., and von Bohlen, A. (2002). Depth profiles of a shallow implanted layer in a Si wafer determined by different methods of thin layer analysis. *Spectrochim. Acta*, **57B**, 1593–1599.

100. Kunimura, S., Kudo, S., Suzuki, K., and Hojo, Y. (2013). *Methods for improving detection limits of total reflection X-ray fluorescence spectrometry using weak white X-rays*, 15th International Conference on TXRF and related methods, Book of Abstracts, O10, p. 93.

101. Kunimura, S., Watanabe, D., and Kawai, J. (2009). Optimization of a glancing angle for simultaneous trace element analysis by using a portable total reflection X-ray fluorescence spectrometer. *Spectrochim. Acta*, **64B**, 288–290.

102. Kunimura, S. and Kawai, J. (2013). Trace elemental determination by portable total reflection X-ray fluorescence spectrometer with low wattage X-ray tube. *X-Ray Spectrom.*, **42**, 171–173.

103. Imashuku, S., Ping Tee, D., and Kawai, J. (2012). Improvement of total reflection X-ray fluorescence spectrometer sensitivity by flowing nitrogen gas. *Spectrochim. Acta*, **73B**, 75–78.

104. Pahlke, A., Bachmann, M., Eggert, T., Fojt, R., Graczek, M., Höllt, L., Knobloch, J., Miyakawa, N., Pahlke, S., Pahlke, St., Rumpff, J., Scheid, O., Simsek, A., Stötter, R., Wennemuth, I., and Wiest, F. (2013). *Large area silicon drift detectors for TXRF applications*, 15th International Conference on TXRF and related methods, Book of Abstracts, I25, p. 229.

105. Krämer, M., Beckhoff, B., Dietsch, R., Holz, T., Hönicke, P., and Weißbach, D. (2013). *New multilayer tools for TXRF measurements*, 15th International Conference on TXRF and related methods, Book of Abstracts, O20, p. 221.

106. Wobrauschek, P. (2013). *The role of the IAEA in the spreading of TXRF*, The 15th International conference on total reflection X-ray fluorescence analysis and related methods (TXRF2013) and The 49th Annual conference on X-ray chemical analysis, Book of Abstracts, I17, 111–112.

107. Streli, C., Wobrauschek, P., Pepponi, G., and Zoeger, N. (2004). A new total reflection X-ray fluorescence vacuum chamber with sample changer analysis using a silicon drift detector for chemical analysis. *Spectrochim. Acta*, **59B**, 1199–1203.

108. Wobrauschek, P., Streli, C., Kregsamer, P., Meirer, F., Jokubonis, C., Markowicz, A., Wegrzynek, D., and Chineao Cano, E. (2008). Total reflection X-ray fluorescence attachment module modified for analysis in vacuum. *Spectrochim. Acta*, **63B**, 1404–1407.

109. Huang, Y. and Yu, H. (2013). *Grazing incidence X-ray technique and its application at SSRF*, 15th International Conference on TXRF and related methods, Book of Abstracts, I10, p. 89.

110. Zaitz, M.A. (2013). Personal information at a round table discussion of the 15th international conference on TXRF 2013, Osaka, Japan.

111. Gross, A. (2014). Personal information after a round table discussion of the 15th international conference on TXRF 2013, Osaka, Japan.

112. Gohshi, Y. (1997). Summary of ISO/TC 201 Standard: VI ISO 14706:2000–Surface chemical analysis – Determination of surface elemental contamination on silicon wafers by total reflection X-ray fluorescence (TXRF) spectroscopy. *Surf. Interface Anal.*, **33**, 369–370.

113. Depero, L.E. (2013). *TXRF as a tool for environmental and biological analysis*, The 15th International conference on total reflection X-ray fluorescence analysis and related methods (TXRF2013) and The 49th Annual conference on X-ray chemical analysis, Book of Abstracts, I13, 103–104.

114. Beckhoff, B., Nutsch, A., Altmann, R., Borionetti, G., Pello, C., Polignano, M.L., Codegoni, D., Grasso, S., Cazzini, E., Bersani, M., Lazzeri, P., Gennaro, S., Kolbe, M., Müller, M., Kregsamer, P., and Posch, F. (2009). Highly sensitive detection of inorganic contamination. *Solid State Phenom.*, **145**, 101–104.

115. Nutsch, A., Beckhoff, B., Altmann, R., Van Den Berg, J.A., Giubertoni, D., Hoenicke, P., Bersani, M., Leibold, A., Meirer, F., Müller, M., Pepponi, G.,

Otto, M., Petrik, P., Reading, M., Pfitzner, L., and Ryssel, H. (2009). Complementary metrology within a European joint laboratory. *Solid State Phenom.*, **145**, 97–100.

116. Holfelder, I., Beckhoff, B., Fliegauf, R., Hönicke, P., Nutsch, A., Petrik, P., Roeder, G., and Weser, J. (2013). Complementary methodologies for thin film characterization in one tool – a novel instrument for 450 mm wafers. *J. Anal. At. Spectrom.*, **28**, 549–557.
117. Lubeck, J., Beckhoff, B., Fliegauf, R., Holfelder, I., Hönicke, P., Müller, M., Pollakowski, B., Reinhardt, F., and Weser, J. (2013). A novel instrument for quantitative nanoanalytics involving complementary X-ray methodologies. *Rev. Sci. Instrum.*, **84**, 045106-1, 7 pages.
118. Karydas, A.G., Beckhoff, B., Bogovac, M., Eichert, D., Fliegauf, R., Gambitta, A., Grötzsch, D., Herzog, C., Jark, W., Kaiser, R.B., Kanngießer, B., Leani, J.J., Lühl, L., Kiskinova, M., Lubeck, J., Malzer, W., Diawara, Y., Migliori, A., Vakula, N., and Weser, J. (2013). *A multipurpose experimental facility for advanced X-ray spectrometry applications*, 15th International Conference on TXRF and related methods, Book of Abstracts, I16, p. 109.
119. Fernandez-Ruiz, R. (2013). *TXRF workgroup and social networks: An alternative tool for scientific collaboration*, The 15th International conference on total reflection X-ray fluorescence analysis and related methods (TXRF2013) and The 49th Annual conference on X-ray chemical analysis, Book of Abstracts, I26, 230–231.

CHAPTER

7

TRENDS AND FUTURE PROSPECTS

In the preceding chapters, total reflection X-ray fluorescence (TXRF) analysis was represented as a novel variant of energy-dispersive X-ray fluorescence. It was described with respect to its principles, instrumentation, performance, applications, efficiency, and competitiveness. The perception and propagation of TXRF and related methods was illuminated. Several reviews in journals and contributions in books have been given with brief or partial reports on TXRF and related methods [1–16]. Annual updates on X-ray fluorescence spectrometry appearing in the Journal of Analytical Atomic Spectrometry report on the progress of TXRF and related methods, for example, Refs [17,18]. Reports on activities in XRF laboratories of the International Atomic Energy Agency (IAEA) are published periodically in IAEA newsletters, for example, Ref. [19].

In this final chapter, present trends and future prospects will be considered. The efficiency of TXRF and grazing incidence X-ray fluorescence (GI-XRF) may well be increased by various recent instrumental and methodical

Total-Reflection X-ray Fluorescence Analysis and Related Methods, Second Edition.
Reinhold Klockenkämper and Alex von Bohlen.

developments. Particular trends will be described and combinations of X-ray methods performed in total reflection geometry and related to TXRF will be highlighted especially.

7.1 INSTRUMENTAL DEVELOPMENTS

Modifications or extensions of the customary instrumentation commonly aim at improving the sensitivity. Instead of conventional X-ray tubes, new variants of X-ray sources have been developed. Particularly, synchrotron radiation has been used for a highly effective excitation. A worldwide dissemination of synchrotron beamlines dedicated to X-ray excitation and an easier access to such facilities have been the trend of the last decade. Capillaries are employed in order to focus the exciting beam for line-scans and area maps. Waveguides are applied as adaptors and intensifiers of the exciting beam. In order to improve the spectral resolution, new types of cryogenic detectors have been developed.

7.1.1 Excitation by Synchrotron Radiation

The lowest detection limits can be achieved by using a synchrotron beam for excitation. Respective large-scale storage rings for electrons or positrons necessary for this purpose are accessible; 81 beamlines are listed in Table 3.2 of Section 3.3. Three advantages of using this ideal excitation source should be emphasized again:

1. The high photon flux and brilliance of a synchrotron beam gives a drastically increased primary X-ray intensity. The gain is 5–15 orders of magnitude in comparison with conventional X-ray tubes.
2. Synchrotron radiation covers a broad energy range of the X-ray spectrum and is highly suitable for energy tuning. A double crystal monochromator (DCM) or a plane grating monochromator (PGM) is preferable for the selection of a particular small energy band (<30 eV) (see Section 3.4.3). The photon energy can easily be adjusted to an optimum excitation energy of the elements sought.
3. The spectral background already reduced by total reflection is lowered further because of the polarization. The synchrotron beam is linearly polarized at 100% in the horizontal plane of the storage ring. Consequently, the arrangement shown in Figure 7.1 should be set up [20]. With this configuration, the background can reach its ultimate lowest limit determined by the “bremsstrahlung” of emitted photoelectrons [21].

High primary intensity, adjusted excitation energy, and the lowest possible background have resulted in detection limits decrease by as much as three

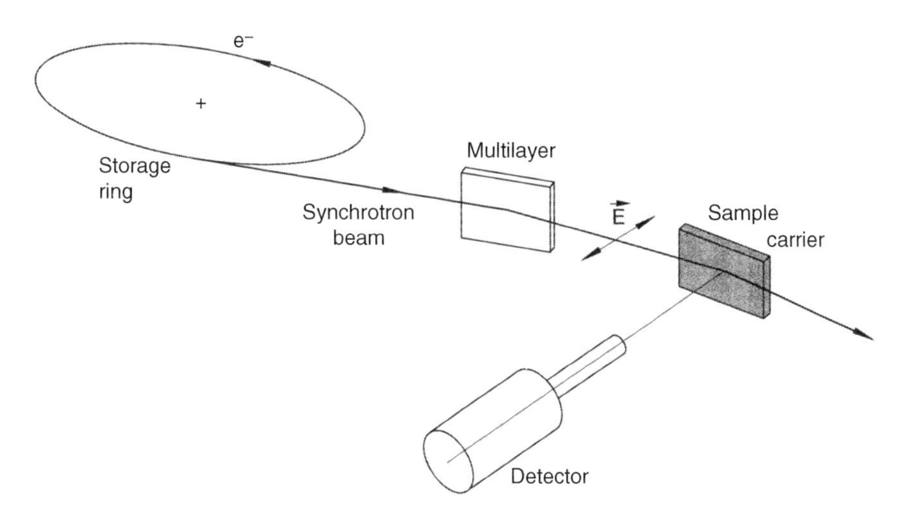

Figure 7.1. Arrangement for TXRF with excitation by a synchrotron beam. This beam is linearly polarized, that is, the electrical field strength ***E*** lies in the plane of the electron (or positron) orbit. For a low spectral background, the detector axis must be placed in this plane and directed perpendicular to the beam itself (side-looking). In order to use the advantage of polarization, the sample carrier has to be positioned vertically. Figure from Ref. [22], reproduced with permission. Copyright © 1996, John Wiley and Sons.

orders of magnitude compared to conventional excitation. The fg level has been reached, with 20 fg strontium, 30 fg nickel, 150 fg cadmium [23], and 200 fg even for magnesium [24]. In wafer analysis, detection limits between 3×10^8 and 1×10^9 atoms/cm^2 have been obtained for metallic contaminations [25,26]. Furthermore, basic principles have been described and practical examples given for the depth analysis of impurities and for the characterization of multilayered structures [27–31]. The easy access to such storage rings and the installation of suitable experimental working stations will lead to a widespread use of synchrotron radiation in the future.

Beckhoff *et al.* have given a short overview of different methods of X-ray spectrometry applied to the characterization of wafer contaminations and nanolayers at BESSY II in Berlin, Germany [32]. The Physikalische-Technische Bundesanstalt (PTB) operates a laboratory with the possibility for conventional XRF, TXRF, GI-XRF, and near-edge X-ray absorption fine structure (NEXAFS) (see Section 7.3.3). Outstanding features are an UHV sample chamber for 15 mm up to 75 mm wafers, a second larger chamber with a load-lock for 200 and 300 mm wafers, a mobile cleanroom of class 100, a high-vacuum load-lock, an electrostatic chuck, an eight-axis manipulator, calibrated photodiodes, and Si(Li) detectors with ultrathin windows but without a supporting grid. A rough vacuum is reached by scroll pumps and high-vacuum down to 2×10^{-5} Pa is achieved by turbomolecular pumps. The different axes of the manipulator are driven by stepper motors with a maximum resolution of 0.0001° and 0.25 μm and with a minimum speed of 0.3°/s and 0.25 mm/s.

Excitation is possible by a plane grating monochromator in the range of about 100 eV up to 1860 eV so that light elements, such as C, N, O, Na, Mg, and Al, can be excited by their K peaks, and transition elements, such as Fe, Cu, and Zn, by their L peaks. The instrumental assembly allows the direct TXRF of light element contaminations with the lowest detection limits that reach down to 100 fg or 10^9 atoms/cm^2. Nanolayers can be characterized by GI-XRF and composition, thickness, and density can be determined by a reference-free quantification. Finally, NEXAFS can reveal information about the chemical speciation of light elements by tuning the excitation energy.

7.1.2 New Variants of X-Ray Sources

New types of X-ray sources were constructed and used for excitation in X-ray spectral analysis aiming at a high X-ray flux.

The first variant was constructed that represents a special X-ray tube with a spiral orbit of electrons [33]. It is a tabletop instrument of lab-size based on a microtron injector for high-energetic electrons, which are guided on an orbit of 8 cm radius in a betatron-like motion. With energies up to 1 MeV, the electrons are focused on a target and decelerated. Because of multipulse resonance injection, the X-radiation is pulsed with a repetition rate of 1 kHz and a pulse width of 10 μs. White radiation is produced and emitted from a focus of 10 μm diameter at a wide divergence of some 100 mrad (5.7°). Kα peaks of heavy elements, such as tungsten, can be excited (59.3 keV). A high X-ray flux of 10^{14} photons/s/mm^2/mrad^2/0.1% BW is achieved so that NEXAFS measurements can be carried out within 30 min.

This device is a sophisticated X-ray source where highly accelerated electrons of a microtron are decelerated by collisions with a target (bremsstrahlung). It is not a synchrotron where highly accelerated electrons are forced to run in a circular orbit and are decelerated by a magnetic deflection (synchrotron radiation). The new MIRRORCLE X-ray source was applied to two examples, but both examples did not make use of total reflection geometry. First, X-ray absorption near-edge structure (XANES) measurements were carried out in absorption mode [34]. A flat Si(111) crystal was used for an energy scan of 18–22 keV, an ellipsoidally curved Si crystal focused the X-ray beam onto a sample consisting of a Mo foil or a K_2MoO_4 powder. The transmitted X-rays were recorded by an X-ray charge-coupled device (CCD) camera and the XANES spectra were Fourier-transformed. They showed radial distributions with four oscillations and main maxima at 0.20 and 0.33 nm.

The second application of the MIRRORCLE X-ray source demonstrated residual stress by XRD measurements [35]. A gear made of chromium–molybdenum steel after carbonizing was analyzed. At an energy of 57 and 97 keV, XRD spectra of Fe (110) with $d = 0.205$ nm were recorded at 2θ angles of 5.9° and 3.7°, respectively. Compressive stress of some 100 MPa (negative values) could be determined in a depth up to 0.5 mm.

Another type of a new X-ray tube has been available since 2008. The usual solid anode is exchanged by a jet of liquid metal, which is continuously renewed. Gallium or Ga compounds are used for the liquid jet and this anode is continuously regenerated in a high-pressure loop. At a jet velocity of 100 m/s, new target material is always present and the power loading is not restricted by the melting point of the anode. Gallium with its low melting point of only 30 °C is ideally suited for that purpose. A thin gallium layer is condensed on the beryllium exit window but disappears continuously in the form of a few droplets.

The companies Bruker, Excillum, and Incoatec have jointly developed this lab-sized X-ray tube (MetalJet D1 and D2). A 10-fold increase of brightness was achieved and a microfocus could be realized leading to a very low spectral background [36]. The spot size is between 5 and 50 μm at a power between 50 and 500 W, respectively. The operating voltage can be increased to 70 keV or even 160 keV and detection limits in the low pg range are obtained. The D1 type won the R&D 100 award in 2011 and is integrated with the Bruker Nanostar. It is mainly used for structural biological applications.

Another promising type is a plasma X-ray source (PXS) jointly developed by the Max-Born Institut (MBI) für Nichtlineare Optik und Kurzzeitspektroskopie and the Institute for Scientific Instruments (IfG) in Berlin, Germany [37,38]. A femtosecond-laser-beam is focused onto a solid target with a high repetition rate. A thin copper tape of 20 μm thickness is used as the target and continuously moved in order to get new fresh material for all laser pulses. A line of craters each with a spot size of 5 μm is generated and a hot plasma appears above. The ionization of the target material and the acceleration of electrons in the electrical field of the laser produce characteristic X-rays as well as X-ray bremsstrahlung between 0.1 and 10 keV. The target is enclosed in an aluminum chamber with entrance and exit windows. The windows are shielded by continuously moved plastic tapes in order to protect them against the deposition of copper debris by laser shots. The laser energy is >5 mJ, the pulse duration is 35 fs, the repetition rate is 1 kHz, and the power load is 10^{18} W/cm^2. The emitted X-radiation is focused by X-ray optics onto the sample to be investigated and enables an ultrafast time-resolved X-ray analysis.

A glow-discharge tube of the Grimm-type usually applied for optical emission spectrometry was rebuilt by addition of an iron anode and by application of a high-voltage of 10 keV [39]. An X-ray beam transmitted by a beryllium window was collimated and directed as a flat parallel beam to a sample under total reflection. Simultaneously, a second X-ray beam from a rotating anode tube with a molybdenum target and a voltage of 40 keV was turned to the sample—perpendicular to the first X-ray beam. Both beams excite the sample; one beam is more suitable for lighter elements while the other one is more effective for heavier elements.

7.1.3 Capillaries and Waveguides for Beam Adapting

For microanalytical tasks, a focusing device may be used in order to strengthen the incoming-beam intensity. A single capillary can be applied, but more commonly polycapillary optics are used, such as Kumakhov lenses [40], which are commercially available as small pens or pins. Up to 100 000 small hallow glass tubes are bundled in a glass sleeve and this array is used for guiding X-rays along the stretched tubes. A divergent beam of a mm wide X-ray source is totally reflected at the inner walls of the gently bent capillaries several times. It can be collimated or focused, respectively, and either leads to a parallel outgoing beam or to a small focal spot with high intensity, respectively. A focus of 10–100 μm diameter can be attained with a gain of intensity that is 100–1000-fold in comparison to pinhole collimation [40]. The sample is excited, for example, in the (30°/45°) geometry as is common practice for energy-dispersive spectrometry (EDS). Scanning of the sample allows an elemental mapping with a spatial resolution down to 10 μm as shown, for example, by Haschke *et al.* [41] or Mages [42]. Capillaries are available from IFG (Institute for Scientific Instruments), Berlin or from XOS (X-Ray Optical Systems, Inc.), East Greenbush, NY, or from Center for X-Ray Optics, Albany, NY.

Such a focused beam, however, is not suitable for TXRF or GI-XRF where the (0°/90°) geometry is necessary. In this total reflection geometry, flat samples have to be excited by a parallel beam with a height of about 50 μm. A single or double reflector (cutoff reflector) or a monochromator is frequently used as a beam adapting unit, which simultaneously works as a filtering unit. A similar arrangement was applied by Cheburkin and Shotyk in 1996 [43]. A collimator was built by two rectangular microscope slides that were pressed together between steel plates. An incoming beam of a line-focus X-ray tube passes the collimator by several total reflections at the upper ($50 \times 25 \times 1\ mm^3$) and lower microscope slide ($70 \times 25 \times 1\ mm^3$). This lower glass plate also serves as a sample carrier demonstrated in Figure 7.2a. The divergence of the beam is <0.03°. For 15 elements, the detection limits were found to be about 150 pg.

Further studies of X-ray waveguides were carried out by Egorov *et al.* [44–47]. Their waveguide consisted of two flat and even rectangular quartz-glass plates, which were mated together. The waveguide was called a "slitless" collimator because the slit between both plates could not be observed visually. However, an X-ray beam can propagate through the slit with several 100–1000 zigzag lines after total reflections at both plates. A standing wave arises within the slit and the incoming beam is confined in one direction with $0 \leq \theta \leq \theta_{crit}$ and guided through the collimator. Only high-energetic X-ray photons with $E \geq E_{crit}$ leave the slit, but with almost no loss of intensity.

Commonly used TXRF spectrometers with low-pass filters or monochromators have a vertical divergence of ± 0.001°. This quantity is increased to ± 0.1° for a spectrometer equipped with a waveguide collimator and the intensity is increased by a factor of 100. An early design of Egorov *et al.* [44] is shown in Figure 7.2b. The waveguide plates as well as the sample carrier are made of

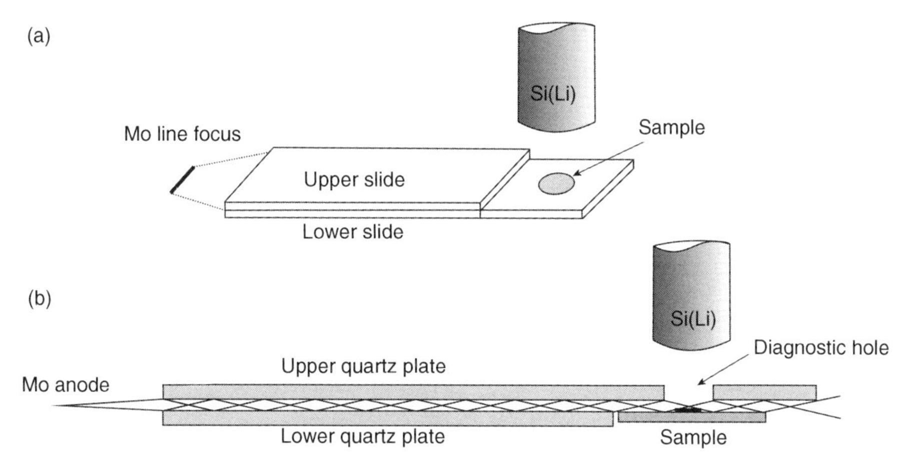

Figure 7.2. (a) Waveguide collimator after Ref. [43]. Two rectangular microscope slides, ~1 mm thick and made of simple glass, are pressed together. The lower slide is 70 mm × 25 mm, the upper slide is only 50 mm × 25 mm. A molybdenum tube with a line focus of 10 mm × 0.1 mm is used for excitation. The sample is placed on the right side of the lower slide. Figure from Ref. [43], reproduced with permission. Copyright © 1996, John Wiley and Sons. (b) Waveguide collimator after Ref. [44]. Two flat and even quartz-glass plates are mated together with a clearance of 30 nm. The upper plate is 100 mm long with a "diagnostic" hole on the right with 1 cm diameter. The sample carrier is placed right of the lower plate just beyond the "diagnostic" hole. Figure from Ref. [44], reproduced with permission. Copyright © 2000, ICDD.

quartz glass. The sample carrier is placed next to the lower and shorter plate and pressed against the upper plate beyond a "diagnostic" hole. A simple Mo tube with 25 keV was chosen for excitation and a Si(Li) detector was used for spectra registration within 1000 s. Detection limits for contaminations on a silicon wafer showed a geometric mean of 2×10^{10} atoms/cm^2 for the Fe-group elements.

In a review article, the peculiarities of waveguides for X-ray propagation were clarified [46]. Two optically polished quartz-glass plates of $100 \times 33 \times 2.5$ mm^3 with a roughness of 5 nm were kept apart by small strips of titanium and uniformly pressed together. The thickness of the strips was controlled by Rutherford backscattering. A set of 14 different waveguides with a slit width between 43 nm and 120 μm was studied. Furthermore, a "slitless" collimator was built without Ti strips and its width was determined by attenuated internal total reflection (AITR) with a laser beam ($\lambda = 680$ nm). Measurements at slit-widths between 43 and 300 nm led to a value of ~ 30 nm for the "slitless" collimator by extrapolation.

The intensity of the Kα peak of a Cu X-ray tube was measured after passing the different waveguides with a divergence of $\pm 0.2°$. It was shown that an X-ray beam can propagate through a narrow slit between two plates of 100 mm length and 33 mm height in two extreme cases [45]: (1) If the slit width is greater than 3 μm, a direct propagation of the incoming beam with multiple total reflections is possible. The beam is totally reflected at the inner walls of the two plates and follows a zigzag line. The intensity of the exit beam increases linearly with the slit width. (2) If the slit width is smaller than about 200 nm, the intensity of the exit beam remains constant. An X-ray standing wave is built up in the space of the extended slit with a uniform field of interference. This case based on extremely narrow slits is called "waveguide resonance."

The minimum width for the clearance necessary for a real waveguide is the smallest period of the standing wave at total reflection [47]. In accord with Equation 2.12, this lower limit is

$$s_{\min} = \frac{\lambda}{2\sin \alpha_{\text{crit}}} \tag{7.1}$$

The maximum width is determined by the spatial coherence necessary for a standing wave field. It is approximately half the longitudinal coherence length of the incoming X-ray beam [45,46], which is determined by Equation 4.21:

$$s_{\max} = \frac{\lambda}{4} \cdot \frac{\lambda}{\Delta\lambda} \tag{7.2}$$

For quartz-glass reflectors and Mo-Kα radiation with $\lambda/\Delta\lambda \approx 2000$, the respective limits are 20 and 36 nm; for quartz glass and Cu-Kα radiation, the range is between 21 and 77 nm. The roughness of the reflectors has to be smaller than a sixth of the lower limit, that is, less than 4 nm [45].

The waveguide resonator demonstrated in Figure 7.2 was used with a clearance of 200 nm. The exit beam of Mo-Kα with a divergence of about $\pm$ 0.1° was used for excitation. Analysis of a monocrystal film of $Ge_{0.24}Si_{0.76}$ showed metallic impurities, such as Fe, Ni, Cu, and Zn with detection limits of about 3.6×10^{10} at/cm^2 [47]. Besides, the authors investigated a 100 nm thin film of $YBa_2Cu_3O_{6.9}/ZrO_2(Y)$. They claimed detection limits to be lowered by two orders of magnitude and maintained that an upgrade of the device could even compete with synchrotron excitation [46].

The lateral intensity distribution behind such a planar X-ray waveguide was investigated by theory and experiment [48]. For theory, the electromagnetic field within the planar waveguide was calculated. The results indicate that a waveguide resonator cannot be replaced by a slit model. The intensity pattern shows several modes different to that of a single or double slit. The Fresnel number was introduced in order to simplify the problem and to differ between Fraunhofer and Fresnel diffraction.

In contrast to the mentioned "waveguide," Sanchez used two parallel rectangular wafer pieces, distinctly separated by two aluminum spacers of 20 μm. This device was referred to as a "beam guide" instead of a "waveguide" [49]. As illustrated in Figure 7.3a, the sample is deposited on the free end of the lower plate with the detector above. Because of the large clearance of both plates, only a few reflections are possible. For an incident angle of

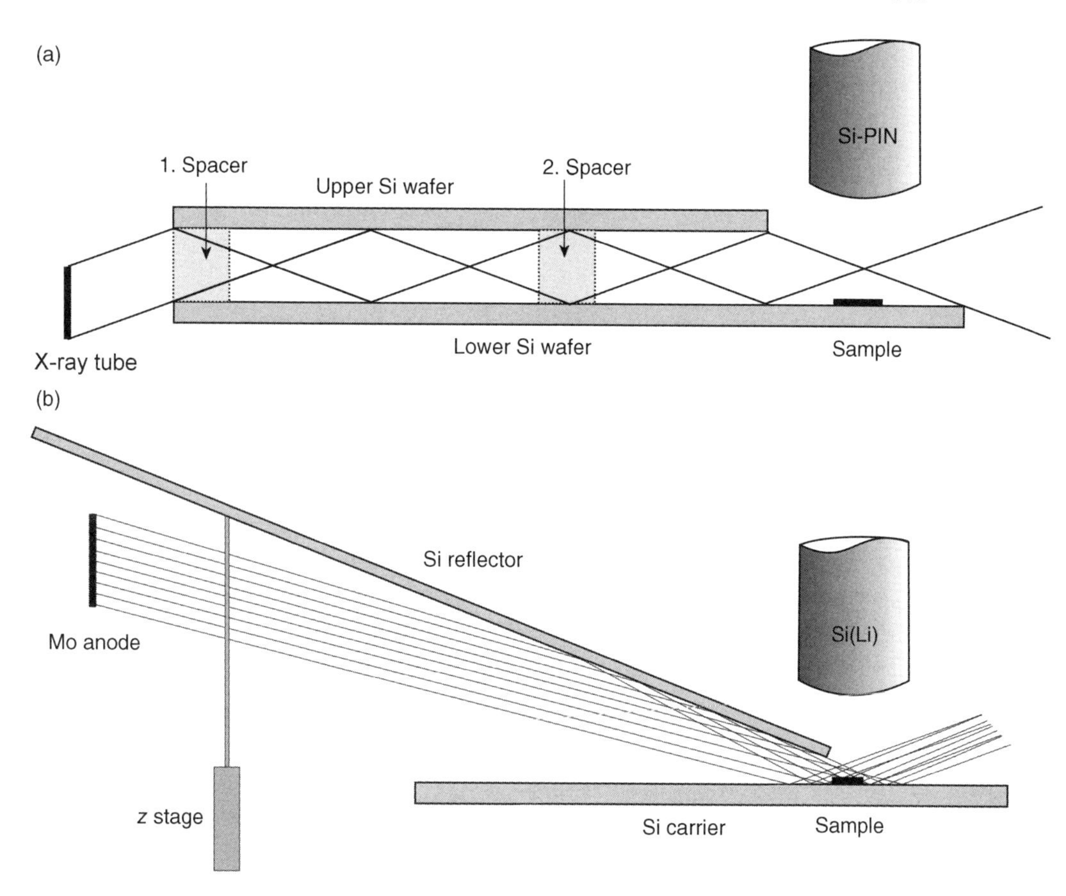

Figure 7.3. (a) Beamguide collimator after Ref. [49]. The upper rectangular part of a silicon wafer is 90 mm long and 10 mm wide, the lower part is 100 mm long. An aluminum spacer of 20 μm keeps the plates separate. The sample is deposited on the right side of the lower plate. Figure from Ref. [49], reproduced with permission. Copyright © 2002, Elsevier. (b) Beamguide collimator after Ref. [50]. The sample carrier made of a Si wafer with 70 mm × 20 mm is placed below a Si reflector with 80 mm × 30 mm. This reflector is resting on two pieces of a 30 μm aluminum foil left and right of the sample. Figure from Ref. [50], reproduced with permission. Copyright © 2002, John Wiley and Sons.

0.069° and a cutoff energy of 26 keV, only five reflections fill up the beam guide. A soil sample was analyzed with 1000 s live-time. In comparison with conventional XRF (at 45°/45° geometry), the detection limits for eight elements were improved by a factor between 4 and 1000 with a geometric mean of 15. With respect to Egorov and Egorov [45], it may be remarked that the beam guide does not act as a waveguide because the clearance is greater than 3 μm. Standing waves cannot be built because the coherence condition is not fulfilled.

An arrangement by Tsuji and Wagatsuma [50] differs distinctly because the upper plate is tilted by a small angle of about 0.1° by means of a z stage as shown in Figure 7.3b. The upper plate rests on two thin aluminum foils (30 μm) to the right and to the left of the sample. The intensity of the primary beam is increased by its total reflection at the sample carrier and at the tilted reflector as well. A gold layer with a diameter of 10 mm and a thickness of 5 or 10 nm, respectively, was used as a sample, and an intensity/angle profile was recorded by tilting the reflector. The spectral peaks of Au-Lα, Lβ, and Mα were shown to be increased by a factor of about 12, while Si-Kα was once more duplicated in comparison to the arrangement without the reflector. The annoying peak of Ar-Kα emitted from ambient air was reduced strongly.

7.1.4 New Types of X-ray Detectors

Beyond silicon-drift detectors (SDD), new types of X-ray detectors have been developed in the last two decades. They are based on superconducting tunnel junctions (STJ) or on adiabatic microcalorimeters (AMC). Both types of detectors belong to the class of cryogenic detectors because they work at temperatures near the absolute zero point. Their principle is very easy. STJs measure a tunnel current induced by a broken Cooper pair of a superconducting material. AMCs measure the rise of temperature caused by an incident photon, which is absorbed. The radiant energy of a photon is converted into electrical or thermal energy, and vice versa the current or temperature rise is a measure for the photon energy.

Superconducting tunnel junctions consist of two superconductive layers separated by a thin layer of an insulating material. Aluminum oxide is frequently used as a barrier because it can be deposited as a very thin layer of only 2 nm thickness with no defects so that a short circuit of the neighboring superconducting layers is prevented. A bias voltage is applied with a voltage of about 2 mV and the resulting tunneling current is on the order of a few nA.

Adiabatic microcalorimeters consist of an X-ray absorber, a very sensitive thermometer, a thermal link, and a heat sink or thermal bath kept at a very low temperature. A 6 keV photon may produce a rise of 5 mK in the thermometer if operated at 500 mK. Such detectors have a high efficiency of nearly 100% in a broad energy band and show an excellent spectral resolution of about 2 eV for photons of 6 keV, that is, they have a resolving power $E/\Delta E \approx 3000$. Consequently, they can clearly resolve the Kα doublet of manganese.

There are three different types of cryogenic X-ray detectors:

1. A typical niobium-based STJ detector consists of a stack of five thin layers: Nb–Al–Al_2O_3–Al–Nb with an area of about 200 µm × 200 µm [51,52]. The superconductive niobium layers (critical temperature 9.2 K) are nearly 200 nm thick and serve as electrodes, the aluminum layers are 50 nm thick, separated by a 2 nm thin Al_2O_3 barrier (altogether about 500 nm or 0.5 µm). Cooper pairs of electrons in niobium can tunnel the thin barrier according to the Josephson effect. Moreover, an incident X-ray photon can break up Cooper pairs and produce "quasiparticles" as free charge carriers, which are similar to electron-hole pairs. It needs only

about 0.002 eV to overcome the superconducting energy gap of niobium and to create a quasiparticle. It can tunnel the junction barrier in the direction of the biased voltage and generate a current pulse. The sensor is coupled with a cooled JFET front-end amplifier for a read-out with a low electronic noise.

In accord with Equation 3.25, with a value e_{pair} of 0.002 eV and a Fano factor of about 1, the peak width determined by quantum statistics is about 8 eV for a photon energy of 6 keV. The electronic noise at such low temperature is only about 4 eV, so that the total peak width is 9 eV resulting in a resolving power of 670. For a low photon energy of 1 keV, the peak width is 5 eV and the resolving power is 160. The count rate capability is about 10^4 cps. On the other side, STJs need liquid helium at a temperature of 4.2 K and this temperature has even to be lowered down to 100 mK by adiabatic demagnetization. Unfortunately, this prerequisite of cryogenic operation is rather troublesome. Furthermore, ghost peaks and peak splittings in the spectra are severe drawbacks.

2. A representative AMC is composed of a transition edge sensor (TES) and uses superconducting layers, for example, Al/Ti/Au, biased for a normal-superconducting transition. It is coupled with a current sensing SQUID (superconducting quantum interference device) for amplification. The temperature of the heat sink has to be about 100 mK. Unfortunately, this detector has only a low count rate capability and the drawback of an unstable energy scaling.

A new TES microcalorimeter was developed by NIST (National Institute for Standards and Technology). It was cooled down to a temperature below 4 K by a so-called pulse tube cooling device that works mechanically; liquid nitrogen or helium are not necessary [53]. In order to reach a final temperature below 100 mK, an ADR (adiabatic demagnetization refrigerator) was applied. This system is suitable for industrial applications. It has a working time of only about 10 h, which can be extended to 24 h.

This cryogenic detector shows a high resolution for low-energy photons. It was shown to be especially advantageous for the analysis of silicon wafers when nitrogen has to be determined simultaneously with titanium, tungsten, or tantalum. Conventional Si(Li)s or SDDs are incapable of this task. Figure 7.4 shows two spectral sectors recorded by a TES microcalorimeter. The Kα peaks of nitrogen and silicon have a peak width of only 7 or 19 eV, respectively. Neighboring Lα peaks of titanium, and Mα or Mβ peaks of tungsten and tantalum are clearly separated. Unfortunately, the count rate capability of this detector is strongly restricted to about 10 cps due to a small sensor area of only 20 µm × 20 µm. It may be improved for the next generation of microcalorimeters.

3. Metallic magnetic calorimeters (MMC) are made of a planar metallic sensor that is paramagnetic [54]. X-ray photons are first absorbed by a thin pad, made of gold with an area of 180 µm × 180 µm and a height of

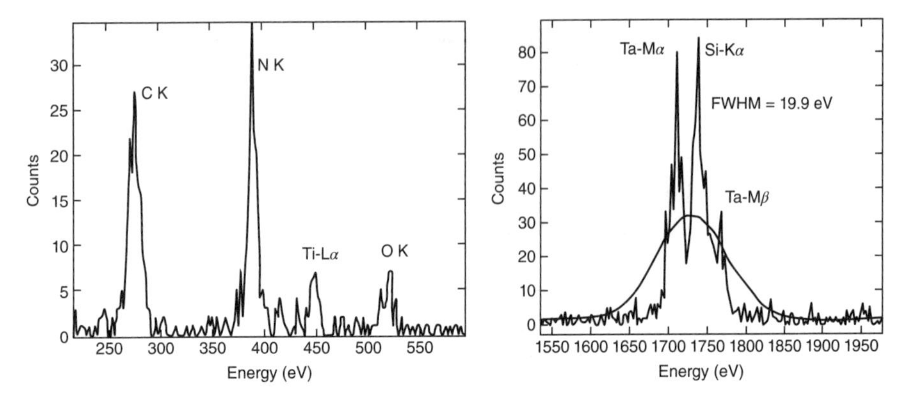

Figure 7.4. Two parts of an X-ray spectrum taken with a TES microcalorimeter. A Ti–N spectrum is shown left, and a Si–Ta spectrum is displayed right. All peaks are clearly separated by the TES detector. A conventional Si(Li) or SDD showing the thin profile (right) is unable to resolve these peaks. Figure from Ref. [53], reprinted with permission from the National Institute of Standards and Technology.

8 μm. As a result of absorption, the cryogenic temperature is increased a little and the magnetization of the paramagnetic sensor is decreased. The planar sensor has a diameter of 50 μm and a height of 8 μm. Usually, a solid solution of gold doped with about 500 ppm erbium (Au : Er) is chosen. The magnetization of this alloy is caused by the magnetic moment of erbium ions and is due to 4f electrons of the N-shell, which is only partly filled and is well below the outer O- and P-shell of the ion. The 4f band with only 11 instead of 14 electrons can be treated as a doublet with a gap of 0.0015 eV.

The magnetization increases with the magnetic induction and decreases with the temperature. At a low temperature of 50 mK and a small magnetic induction of only 6 mT the alloy is magnetized with 150 A/m (equivalent with an increase of the magnetic field strength). An incident photon with energy of 6 keV raises the temperature by nearly 1 mK. Because of its low impedance, the sensor is coupled with a SQUID. This detector also needs a cryostat supplied with pumped liquid helium. The final temperature is reached by means of an adiabatic demagnetization refrigerator with paramagnetic salt pills. The complete device is protected against external magnetic fields by a superconducting shield of lead. The energy resolution amounts to about 3 eV for photons of 6 keV leading to a resolving power of 2000. Unfortunately, only a low count rate of 2 cps can be treated so that a counting time of more than 1 h is necessary in order to reach a measurement precision better than 1%.

Microcalorimeters can be used for building large arrays with small pixels, which can serve as 2D detectors. Such devices have manifold applications for X-ray astronomy in space when the temperature is near the absolute zero point *a priori*. However, none of these detectors is ideally suited for energy-dispersive detection of X-rays in common laboratories to date.

An X-ray color camera may also be regarded as a new type of X-ray detector. It was developed as a new product of the Institute for Scientific Instruments in Berlin and PNSensor in Munich, Germany. The SL-cam® is a compact instrument for X-ray mapping and was already mentioned in Section 4.1.3.3. It is based on a polycapillary lens combined with an energy-dispersive array detector or CCD and visualizes the elemental composition of a sample in real time. An area of $12 \times 12\,\text{mm}^2$ is depicted with 264×264 pixels; the spatial resolution is 8 μm at best. Because all spectra of the individual pixels are taken simultaneously, the elemental distribution of a sample can be displayed online representing individual elements by different colors. Images at a 1 : 1 ratio are produced with a parallel polycapillary lens; magnifications are possible with a conical lens and both can easily be exchanged. Real-time observations enable or simplify *in situ* investigations by X-ray analyses as mentioned in Section 7.2.5.

7.2 METHODICAL DEVELOPMENTS

Some special subjects of methodical development belong to continuous and recent trends. The detection of light elements is a problem of X-ray fluorescence analysis in general but a challenge for TXRF in particular. Further developments are related to ablation and deposition techniques. They aim at thin-layer analysis while a destruction of the samples is accepted as is also the case for several methods of thin-layer analysis. A continuing trend of investigations is tied up with the variant of grazing exit X-ray fluorescence (GE-XRF) because of particular strong points.

Reliable quantification is a continuing trend that aims at avoiding matrix effects or taking them into account. A recently revived trend is a reference-free quantification only based on physical models and parameters. For most methods of spectrochemical analyses, such an aim is considered to be unattainable because of extremely complex chemical processes. For TXRF and related methods, however, this aim is within reach. Another difficult task for TXRF is a time-resolved *in situ* analysis, but efforts have been made in this direction.

7.2.1 Detection of Light Elements

Several obstacles hamper the detection of light elements especially. The eight elements C, N, O, F, Ne, Na, Mg, and Al have their Kα peaks in the low-energy region between 0.25 and 1.5 keV. They can be overlapped with, and therefore interfered by, 16 Lα and 16 Lβ peaks of medium-Z elements [55]. A spectral resolution of 100 eV of the detector is not sufficient to resolve all these peaks (see Figure 4.12). Further problems are related to fluorescence excitation, energy-dispersive detection, and quantitative evaluation.

The fluorescence signals for light elements are three to four orders of magnitude smaller compared with medium-Z elements. The low sensitivity determined by Equation 2.37 is caused by a reduced fluorescence yield in accord with Equation 1.7, and by a decreased photoelectric mass-absorption coefficient in accord with Equation 1.43. This drawback is an inherent restriction. Besides, it has to be kept in mind that the information depth of low-energy photons is quite low. It is less than 10 μm for aluminum, about 1 μm for sodium, and about 0.1 μm for carbon [55].

The spectral background in the low-energy region is high due to the scattered radiation of the X-ray tube. Its continuum is especially responsible, with a natural steep rise at low energies in accord with Equation 1.10. For synchrotron excitation, the spectral background mainly consists of the "brems" continuum of photoelectrons emitted by the sample material, by elastic Rayleigh and inelastic Compton scattering, and by resonant inelastic Raman scattering [56,57].

The detector efficiency is bad in the low-energy region, as demonstrated in Figure 3.24. The spectral resolution is extensively limited by the electric noise, in accord with Equation 3.27. The new generation of silicon-drift detectors with ultrathin windows, however, shows a higher sensitivity and an improved spectral resolution as can be seen from Figure 3.29. To improve the spectral resolution within the low-energy range, it has been proposed that the energy-dispersive detector be replaced by a wavelength-dispersive device [57,58]. On the other side, the improvement of resolution leads to a loss in sensitivity. This loss, however, can be compensated by the high brilliance of synchrotron radiation used for excitation.

The absorption of the primary X-rays and the fluorescence X-rays with low photon energies is comparatively high. The absorption by ambient air and by the windows of X-ray tubes and detectors leads to a reduced intensity. This problem can be settled by means of a closed chamber, which can be evacuated, and furthermore by new polymeric materials, which can replace beryllium windows [59]. The ultrathin foils can be reinforced with a grid of silicon, silicon nitride, aluminum, or a nylon mesh.

The high absorption of low-energy photons within the sample matrix can lead to matrix effects that give rise to further difficulties. Quantification suffers generally from absorption and enhancement effects [58] and these effects are more severe in the low-energy range, that is, for low-Z elements. Usually, standard reference samples are not available or not really suitable. For micro- and trace analyses by TXRF, the standard-addition technique might not be applicable. Consequently, calculations that consider matrix effects become absolutely necessary. For surface and thin-layer analyses by GI-XRF, angle-dependent intensity profiles must be recorded and the fundamental parameter approach considered in Section 4.6.4 has to be applied [55,60]. A reference-free quantification is aimed at by Beckhoff and coworkers [61–63].

Most of the mentioned problems are inherent to energy-dispersive X-ray fluorescence and cannot be alleviated. The others have been tackled in several attempts, especially by the group of Streli [55,56,60,64–70]. Specifically designed instrumentation is absolutely essential. Excitation by the *K* radiation of a Si, Cr, or a Sc anode in an X-ray tube is recommended, and the tube window should be as thin as possible (8 μm Kapton® foil). Ultimately, a windowless X-ray tube is recommended. Instead of a low-pass filter, a double multilayer or Bragg crystal should be used as monochromator. An HPGe detector is preferable to a Si(Li) detector, possibly equipped with an ultrathin Norwar window of 0.3 μm thickness [55]. A modern SDD with integrated JFET and a thin polyimide foil was already recommended in Section 3.6.2. A vacuum chamber is of course mandatory, but rough vacuum with a pressure of 1 Pa is sufficient for the detection of light elements.

All these efforts have produced absolute detection limits in the low-ng region. Streli *et al.* [56] reported 13 ng carbon, 10 ng oxygen, 3.4 ng fluorine, 0.5 ng sodium, and 0.4 ng magnesium. Relative detection limits for concentrations were estimated from a spectrum of NIST SRM 1643 c "Water." They are at the level of 0.1 μg/ml for a counting time of 1000 s. Again, Streli *et al.* [69] determined 0.5 μg/ml for sodium, 0.4 μg/ml for magnesium, and 0.03 μg/ml for aluminum.

In order to reduce the detection limits further, synchrotron beams were used for excitation. Because of the high spectral brilliance, detection limits of only a few pg were reached: 9 pg fluorine, 2 pg magnesium, and 0.5 pg aluminum according to Ref. [69]. In comparison to the excitation by a simple Cr tube, the detection limits for sodium, manganese, and aluminum could be reduced from the 100 pg range down to the 100 fg range (about 5×10^9 atoms/cm^2) concerning contamination on silicon wafers [55].

A new version of the low-cost Wobrauschek module was developed at the ATI (Atominstitut) in Vienna, Austria, illustrated in Figure 7.5. The spectrometer is commercially available as already mentioned in Section 6.3.1. The compact module consists of a vacuum chamber, a Mo/Si multilayer, a sample carousel, and a down-looking SDD from KETEK [70,71]. The chamber, which is closed by an entrance and an exit window with a thin Kapton foil, can be evacuated so that a medium vacuum of 100 hPa results. Consequently, the spectral background due to X-ray scattering in air is reduced, the absorption of low-energy X-rays by ambient air is diminished, and the annoying Ar peak does not appear. An X-ray tube with a Cr anode for light elements and a Mo anode for heavy elements can be attached for excitation. The reflected X-ray beam can be observed by a CCD camera and the sample carrier can be adjusted by translation and rotation.

It was tested for the detection of elements in residues on silicon substrates. Aluminum as a light element could be determined with a detection limit of 52 pg; heavy elements, such as rubidium, showed detection limits of about 1 pg [71]. Furthermore, carbon was determined in natural freshwater biofilms while titanium served as internal standard [72]. The accuracy was checked by an independent combustive carbon analyzer. A recovery rate of 91% ± 8% demonstrated a reliable quantification.

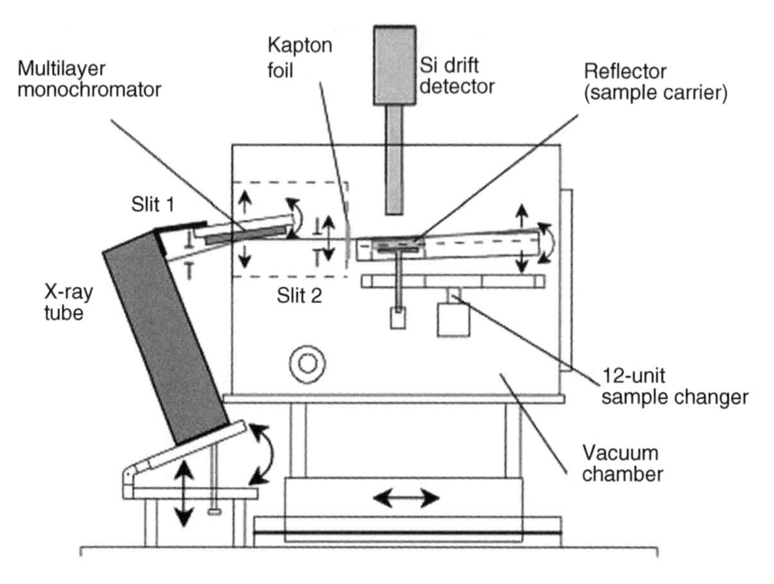

Figure 7.5. Novel spectrometer of the Atominstitut, Vienna, Austria [70]. The X-ray tube with a Mo or Cr anode is attached to a sample chamber, which can be evacuated by a rotating pump. An adjustable W/C or Mo/Si multilayer is used for monochromatization. The sample carousel with 12 positions is computer-controlled and can keep up to 12 different sample carriers. A Si(Li) or a silicon drift detector equipped with an electron trap and a thin window is applied for the registration of X-ray spectra. Figure from Ref. [70], reproduced with permission. Copyright © 2004, Elsevier.

In a further study, the new chamber was employed at a BESSY II beamline with a side-looking detector perpendicular to the upright wafer [56,73]. The light elements C, N, O, and F as well as Na, Mg, and Al were deposited on silicon wafers via standard droplets containing the respective elements in the ng range. In order to avoid excitation of silicon ($K_{\text{edge}} = 1.838$ keV), the wiggler and undulator mode was chosen and excitation energies were tuned by means of a PGM. The light elements could be determined in the dried residue with detection limits between 0.5 pg (carbon) and 6 pg (aluminum) after tuning the excitation energy. These values correspond with 5.0×10^{10} carbon atoms/cm^2 and 2.2×10^{10} aluminum atoms/cm^2 fulfilling the requirements of the SEMATECH roadmap [56].

In addition to traces of light elements detected by TXRF, thin shallow layers of light elements can be characterized by GI-XRF. Two implanted layers of aluminum in silicon could be investigated with a depth and thickness of 10 and 50 nm, respectively [55]. In addition, a single carbon layer (5 nm) and a C–Ni–C multilayer (1.6 nm carbon, 2.1 nm nickel, 1.6 nm carbon) were each deposited on a silicon wafer and studied by GI-XRF [73]. Intensity/angle scans for carbon and nickel were fitted with theoretical curves, which were based on a model with nominal thickness and density of layers or sublayers.

If a synchrotron is not available, excitation of light elements can be done by the Mα peak of a W tube (1.775 keV). Unfortunately, the background below the aluminum peak is dominated anyway by resonant inelastic Raman scattering [56,57,74]. Intensity and width of the Raman band depend on the glancing angle of the incident beam and disappear for glancing angles below 0.06° [73]. For the detection of aluminum contaminants, either a small glancing angle has to be chosen or the intensity of the Raman band has to be subtracted carefully. It can also be eliminated by deconvolution of the spectrum, which can lead to a detection limit of about 3×10^9 Al atoms/cm^2 [57,74]. The residue of droplets with 1 μg of boron could be determined leading to a detection limit of some ng [73].

7.2.2 Ablation and Deposition Techniques

Apart from instrumental modifications, special techniques of sample treatment were developed with the aim of trace enrichment or improved *spatial* resolution. Trace elements from an aqueous solution can be deposited on a TXRF carrier by an electrochemical reaction. The carrier should serve as a cathode in a DC cell with a continuous flow of the electrolyte. A conductive material, such as glassy carbon, is suitable as both the cathode and the TXRF carrier. Such electrochemical enrichment has been in its infancy for a long time [75–77].

The opposite process is the anodic decomposition of a metallic surface, also called electropolishing. It may be combined with a subsequent TXRF analysis of the electrolyte. A high current density and a low temperature are beneficial to the fine etching of thin metallic layers. However, the depth resolution seems to be restricted to a layer thickness of about 0.1 μm [78].

Native oxide layers on silicon surfaces can be decomposed by hydrofluoric acid, as noted in Section 5.4.7.2. The aqueous converted solution can be analyzed by TXRF. A continually repeated oxidation, ablation, and analysis of the uppermost layer can give a depth resolution of 1–3 nm (see Section 4.5.2.1).

In addition to such chemical or electrochemical methods used for depth profiling, physical methods have been proposed. An initial suggestion has come from Schwenke *et al.* [58], and a patent application has been filed [79]. Figure 7.6 depicts the procedure. The layered sample is first etched or eroded by an ion beam of a sputter device, usually by an Ar^+ beam. This process is commonly used in combination with methods of surface and thin-layer analyses, for example, Auger electron spectrometry (AES), X-ray photoelectron spectrometry (XPS), or secondary ion mass spectrometry (SIMS). The sputtered material emitted as an atomic vapor partly passes a shielding slit positioned above the sample. The vapor is deposited as a thin layer on a substrate suitable as a TXRF carrier.

This substrate is horizontally shifted during the sputter process, and thereby the vertical concentration profile is transformed to a horizontal one. Afterward, the thin layer with its lateral distribution of atoms is subjected to TXRF

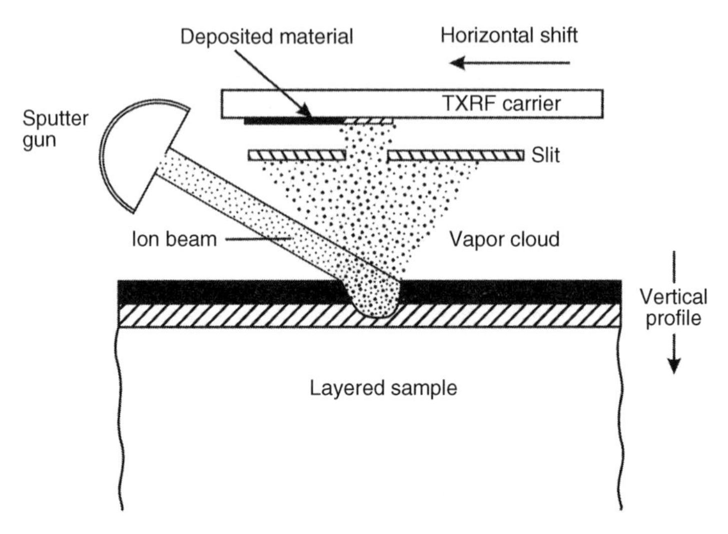

Figure 7.6. Sputter-etching device for vertical depth profiling. The layered sample is etched by an ion beam, and the sampled material is deposited on a carrier shifted horizontally behind a slit. Then, the carrier with the deposited thin layer is placed in a TXRF instrument and analyzed step by step. Figure from Ref. [58], reproduced with permission. Copyright © 1993, Elsevier.

analysis. The substrate is placed under the aperture (<5 mm) of the energy-dispersive detector and again horizontally shifted, step by step. A lateral scan is recorded, which after calibration gives the depth profile as desired.

This method is recommended because it is applicable to multilayer structures with a thickness of some 10 nm up to even several hundred nm. The samples need not to be flat and even, but may have rough or wavy surfaces. The depth resolution can be on the order of 5 nm. An additional variant is offered by use of a collimated beam instead of a broad ion beam or even by a focused laser beam [80]. Material can be sputtered from a small spot, deposited on a carrier, and analyzed by TXRF. In this way, a *micro-distribution* analysis becomes possible. By successive ablations from neighboring spots, a line scan can be recorded. The lateral resolution might be about 5 µm. Above all, the instrumental device is rather simple and high vacuum is not necessary.

A second approach to depth-profile analysis is also based on sputtering. The subsequent analysis, however, is not applied to the sample material removed but to that which remains. The method was first suggested by Knoth *et al.* [81] and promptly applied by Wiener *et al.* [82] and by Frank *et al.* [83]. First, the samples are sputter-etched by a wide ion source of 4–5 cm diameter. Clean argon gas is used, adjusted to a pressure of 5×10^{-2} Pa, which allows for high ion flux up to 1 mA/cm^2 with a relatively low energy of 500 eV. As illustrated in Figure 7.7, the wide ion beam is directed perpendicular to the sample.

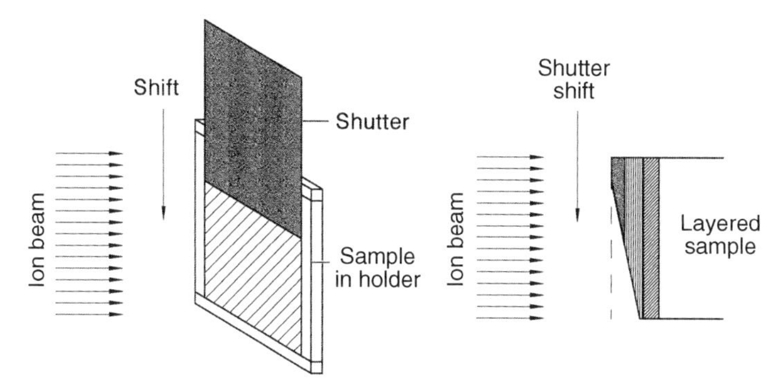

Figure 7.7. Ion beam sputtering for a bevel cut of a layered sample: (a) during the sputter process, a tantalum shutter is steadily moved down; (b) the sample is etched to a bevel shape and then subjected to a lateral scan with TXRF. Figure from Ref. [82], reproduced with permission. Copyright © 1995, American Institute of Physics.

During sputtering, the sample is more and more shielded by a tantalum shutter, which is steadily moved in front of the sample [82]. In a modified technique, the sample is moved along a straight line behind a shield with an aperture of about 1 cm^2 [83]. Both techniques can etch the sample to a bevel with an inclination angle of smaller than 0.0001°. The vertical depth scale is thereby transformed to a horizontal length scale at a magnification of about 10^6, that is, layers of 10 nm thickness might give stripes of 1 cm width. Layers of such a thickness are sputtered within 1–10 min.

After the sputter process, the beveled samples are moved in steps below a TXRF detector with a diaphragm of 0.2–1 mm width. The angle of incidence is chosen to be well below the critical angle of the relevant elements (about 0.1–0.3°). The intensity for the individual elements is recorded against the lateral position, producing a laterally resolved line scan. To derive a concentration versus depth profile, both axes have to be calibrated. The length axis or abscissa is calibrated by determining the different thicknesses of the individual layers. This can easily be done by an additional measurement with a profilometer [83], or it may be done by an additional experiment with TXRF. This time, however, the incident angle should be set far above the critical angle [82]. The intensity axis or ordinate can be calibrated by the algorithm described in Section 4.5.2.2. The fluorescence intensity is then calculated for the remaining layers, dependent on the sputter depth.

The foregoing method is not only applicable to flat layered samples with sharp interfaces as is a prerequisite for nondestructive depth-profiling by TXRF, but it is also applicable to layered samples with diffuse interfaces. A depth resolution of 2–3 nm and a sampling depth up to several hundred nanometers may be obtained. Of course, the method is destructive and time-consuming so that applications are few.

7.2.3 Grazing Exit X-ray Fluorescence

Developments in the 1990s featured TXRF analysis at the grazing exit either *instead of* the grazing incidence [84–86] or *in addition to* the grazing incidence [86–89]. The new option has been called "grazing exit" or "grazing emission" X-ray fluorescence (GE-XRF). The three different variants are highlighted in Figure 7.8. The conventional method is illustrated in Figure 7.8a. The primary beam is incoming at grazing incidence and totally reflected at the sample carrier. The fluorescence radiation is detected perpendicular to the carrier. This (0°, 90°) configuration is changed into a (90°, 0°) configuration in Figure 7.8b. The angle of incidence and the takeoff angle are interchanged. The sample is excited at normal incidence while the fluorescence radiation is

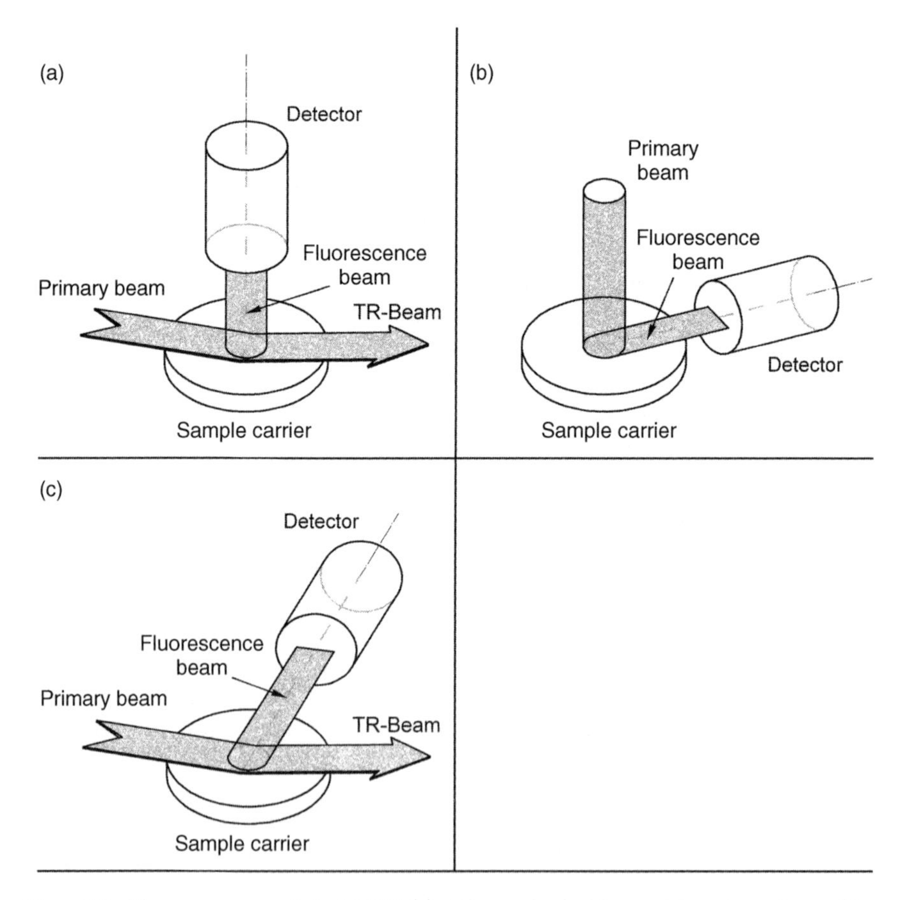

Figure 7.8. Three arrangements for TXRF: (a) at the grazing incidence of the primary beam; (b) at the grazing exit of the fluorescence beam; and (c) at the grazing incidence and exit of both beams. TR beam = totally reflected beam. Figure from Ref. [22], reproduced with permission. Copyright © 1996, John Wiley and Sons.

detected under grazing exit angles, that is, at (90°, 0°) geometry. In Figure 7.8c, the primary beam is directed toward the carrier at grazing incidence and the fluorescence beam is detected at the grazing exit. Both angles are about zero at this (0°, 0°) arrangement, while the angle between both beams is 90°. The three modifications may be called GI-XRF (grazing incidence X-ray fluorescence), GE-XRF (grazing exit or grazing emission XRF), and GIE-XRF (grazing incidence and grazing exit XRF). Any further distinction between a fixed and a variable grazing angle seems to be unnecessary.

All three techniques can involve the effect of total reflection if the relevant angle is below the corresponding critical angle. This is shown for GE experiments by Figure 7.9. The atoms excited to fluorescence by the primary beam may be placed above a substrate or within a thin layer on top of a substrate. In any case, different coherent fluorescence beams may be parallel after reflection at the substrate and so interfere with each other. The interference pattern of a standing wave will be highly distinctive if total reflection occurs at or below the critical angle.

The analytical procedure for all three techniques is the same. Micro- or trace analyses are performed after a small sample amount is put onto a suitable glass carrier at a fixed angle. Surface and thin-layer analyses are carried out by tilting the flat sample and by recording angle-dependent intensity profiles. The following differences should be considered:

- Under the GE condition the critical angle is determined by the fluorescence radiation, whereas under the GI condition it is determined by the

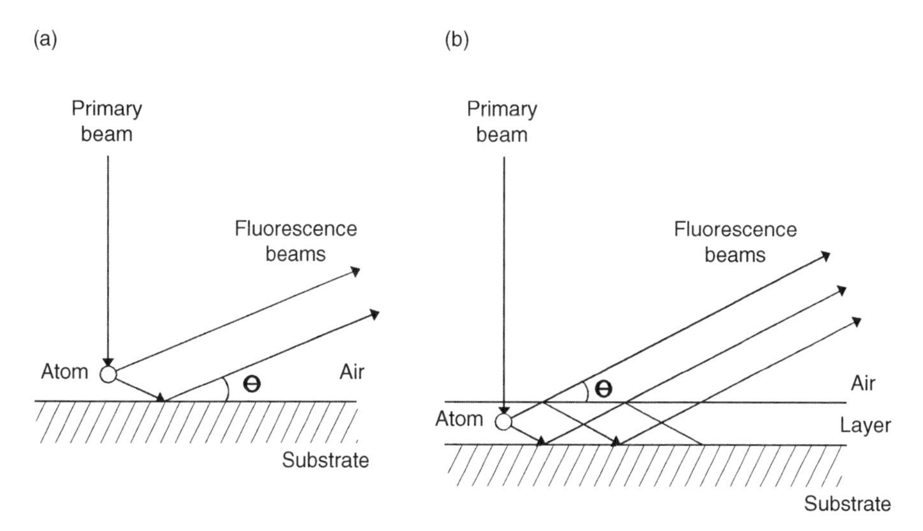

Figure 7.9. Beam paths for X-ray fluorescence at grazing exit. The atom excited by the primary beam is placed (a) above the substrate or (b) within a layer on top of a substrate. Different beams emitted from the atom may be reflected at the substrate and may interfere with each other. Figure from Ref. [22], reproduced with permission. Copyright © 1996, John Wiley and Sons.

primary radiation. According to Equation 1.68, the critical angle for GE is larger than that for GI inasmuch as the respective photon energy is smaller.

- The information depth under the GE condition corresponds to the *emergence* depth of the fluorescence beam and is somewhat smaller than that under GI condition, which corresponds to the *penetration* depth of the primary beam.

Back in 1983, Becker *et al.* performed GE experiments for surface analysis [90]. Thereafter, the GE technique was used to characterize submonolayers of adsorbates [91] and implantation profiles in wafers [92]. Angle-dependent profiles with an oscillation structure were observed for thin layers on a substrate [85–87]. The results obtained for GE-XRF were shown to agree quite well with those obtained by GI-XRF. For a comparison, the reciprocity theorem of optics named for H.L.F. von Helmholtz can be applied (see Ref. [93]). This theorem simplifies the calculation of fluorescence intensities, which can be based on the recursion formalism (Section 2.4). A straightforward calculation is also possible but needs an asymptotic analysis of the Maxwell equations in a stratified layer [94].

The new variants of GE- or GIE-XRF were mostly employed in nondestructive surface or thin-layer analysis using a synchrotron beam for excitation. They showed important advantages but also drawbacks. A monochromatic excitation, such as is needed for GI-XRF, is not necessary. This is the first distinct advantage of the GE configuration. Furthermore, the normal incidence of GE-XRF allows a spatially resolved analysis by the focused microbeam of synchrotron radiation, offering a lateral resolution of a few μm. Concentration profiles along straight lines or area maps can be recorded by scanning the sample. This possibility is excluded for GI-XRF because grazing incidence is an inherent obstacle to high spatial resolution. On the other side, the sensitivity of GE-XRF is reduced due to a much smaller solid angle of the emitted radiation. Detection limits are worse by two orders of magnitude.

A further advantage of GE-XRF is the possibility of simply replacing the energy-dispersive detector by a wavelength-dispersive detector. The choice of a crystal spectrometer enables a more reliable detection of light elements because of far better resolution in the low-energy spectrum. Such a combination is recommendable for GE-XRF but less suitable for GI-XRF due to intensity limitations. On the other hand, absorption effects become more severe under GE conditions and may diminish the advantage. While self-absorption of the exciting *incident* beam is strongly reduced because of a much shorter path length at normal incidence, the path length of the *emergent* beam with photons of lower energy is strongly increased and causes matrix effects that are detrimental to the quantitative determination of elements.

In comparison to GI and GE instrumentation, GIE-XRF makes still higher experimental demands. The incidence and takeoff angles have to be controlled with an accuracy of 0.001°. Furthermore, the fluorescence intensity is minimal because the incident and the exit beams are strongly restricted in their divergence. On the other hand, GIE-XRF allows the variation of a second decisive angle. The incident angle can be varied at different fixed takeoff angles and vice versa. The various combinations allow a cross-checking and may yield additional information [86].

A compact spectrometer for GE-XRF can easily be constructed as shown by Ashida *et al.* [95]. An X-ray tube (Magnum of MOXTEK) was attached to a Z-stage for positioning and adjustment of the sample carrier, and an SDD (AMPTEK) was installed for X-ray detection. GE-XRF geometry (90°/0°) was realized with 8 and 10 mm distances. A close aluminum collimator was set between sample carrier and detector only a few tens of μm apart from the carrier. Detection limits of about 0.2 ng or 20 ng/ml (ppb), respectively, were achieved for cobalt and gallium in aqueous solutions but conventional TXRF is 100fold more sensitive.

The determination of trace elements in drinking water and environmental samples carried out by GE-XRF was compared with other analytical techniques [96]. GE-XRF with wavelength-dispersive detection was applied to determine silicon in biological matrices such as cellulose, serum, urine, spinach, and pork liver [97]. The samples spiked with traces of silicon were digested in a microwave oven and these traces were determined by GE-XRF with acceptable precision. A special procedure was developed for sample preparation [98]. Polycarbonate discs were first siliconized within a central ring of about 30 mm diameter. Thereafter, the silicone layer was oxidized by a plasma asher. Finally, drops of about 1000 μl could be pipetted onto the ring area and analyzed by GE-XRF. In a further study, GE-XRF was applied to the analysis of artist's pigments [99]. A quantification of GE-XRF analyses was performed with emphasis on light elements and their wavelength-dispersive detection [100].

Grazing exit XRF has a special position in the field of X-ray fluorescence analysis as demonstrated by a review [101]. The variant benefits from the effect of the total reflection, but uses normal incidence and wavelength-dispersive detection. The review claims detection limits of 10^{10} atoms/cm^2 for wafer analysis and of ng fractions for environmental samples. However, it can be stated that usual TXRF is superior to GE-XRF for the detection of surface contamination on silicon wafers. Detection limits of TXRF for light elements like aluminum are three orders of magnitude lower (2×10^9/cm^2 instead of 3.7×10^{12}/cm^2 [101]).

A first study on aluminum ions implanted in silicon wafers and characterized by GE-XRF was carried out at a beamline of ESRF in Grenoble, France [102]. Different samples were implanted with different energies between 1 and 100 keV but with the same nominal dose or fluence of 10^{16}

aluminum atoms/cm^2. In order to prevent an overlapping of Al-Kα and Si-Kα, the samples were excited at different selected photon energies of 1582 and 2000 eV, respectively. A von Hamos spectrometer equipped with an ADP crystal of high spectral resolution and a position-sensitive detector was applied for analyses. Figure 7.10 demonstrates three different angular intensity profiles for three different implantations after fitting. From here, concentration/depth profiles were derived as described in Sections 4.6.4 and 5.4.8. They are shown in Figure 7.11 for four different implantations representing Gaussian distributions with a maximum in a depth between 4 and 170 nm. The deviations between experimentally measured and theoretically calculated profiles according to TRIM (Transport and Range of Ions in Matter) were about ±5% for the maxima, and about ±1% for the widths [102].

Within the scope of the European Integrated Activity of Excellence and Networking for Nano- and Microelectronics Analysis (ANNA), the two methods of GI-XRF and GE-XRF were compared regarding the characterization of the aforementioned aluminum implants in silicon wafers [103]. GI-XRF was performed at the PGM beamline (plane grating mirror) at BESSY II with a silicon drift detector, while GE-XRF was carried out at the beamline of ESRF mentioned earlier. The results are shown in Figure 7.12 for three samples. They were cross-checked with

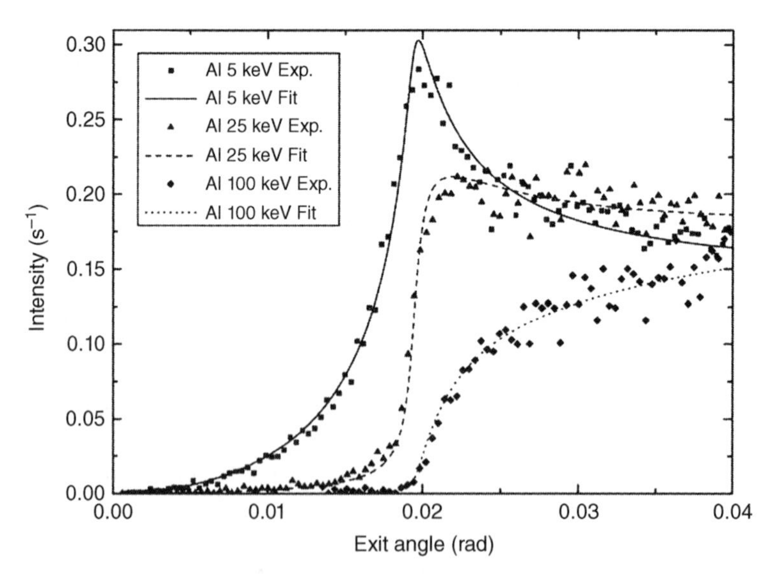

Figure 7.10. Intensity/angle scan for aluminum impurities implanted in silicon and determined by GE-XRF. The profiles belong to energies of 5, 25, and 100 keV (from left to right) whereas the dose or fluence was always 10^{16}/cm^2. The critical angle for total reflection amounts to 0.02 rad or 1.13°, respectively. Theoretical calculations (solid lines) were fitted to the experimental values (dots) and agree quite well. Figure from Ref. [102], reproduced with permission. Copyright © 2010, Elsevier.

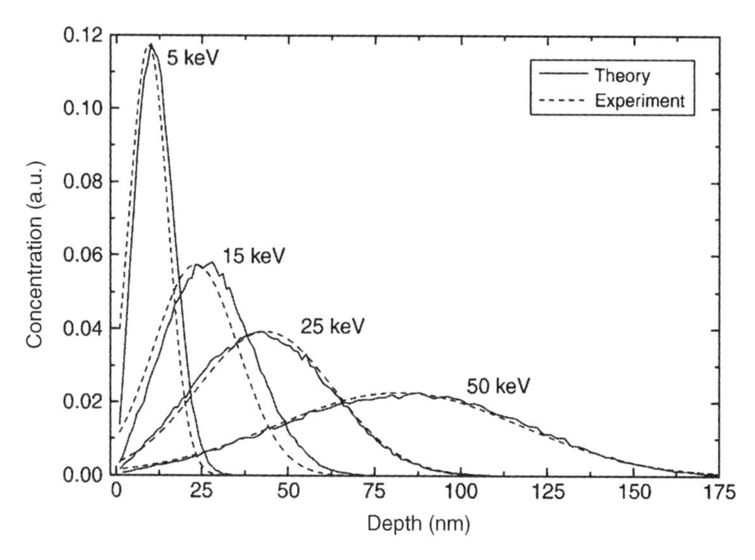

Figure 7.11. Concentration/depth profiles for aluminum implanted in silicon wafers with different energies: 5, 15, 25, and 50 keV. With increasing energy, deeper and broader profiles were generated. Experimental measurements by GE-XRF (dotted line) and theoretical calculations after TRIM (solid line) correspond extensively. Figure from Ref. [103], reproduced with permission. Copyright © 2012, Royal Society of Chemistry.

profiles determined experimentally by SIMS measurements and theoretically by TRIM calculations and showed good agreement. SIMS, assessed as a well-established method of depth profiling, suffers from systematic errors for the first 5 nm just below the surface (Fig. 7.12b and c). But it can be concluded that GE-XRF as well as GI-XRF are capable for depth-profiling of ultrashallow layers in the regime from 100 nm down to a few nm.

In general, GE-XRF as well as GI-XRF can compete very well with established methods of depth-profiling, like SIMS, Rutherford backscattering spectrometry (RBS), AES, and XPS. It should be emphasized that both methods of X-ray fluorescence are nondestructive and need no ultrahigh vacuum. A medium vacuum is sufficient for the detection of light elements. They furthermore allow reference-free quantification. SIMS is destructive and suffers from systematic errors in the near surface region. RBS is also nondestructive but shows a low depth resolution in general and a low efficiency for light elements. Both methods of electron spectrometry, AES and XPS, are confined to layers of some nm, while the X-ray methods reach also deeper layers from a few nm up to some μm. Finally, all methods of surface and thin-layer analyses apart from X-ray methods need ultrahigh vacuum produced by special pumps, for example, by cryopumps. X-ray methods can work without vacuum or only need a medium vacuum achieved by simple mechanical pumps.

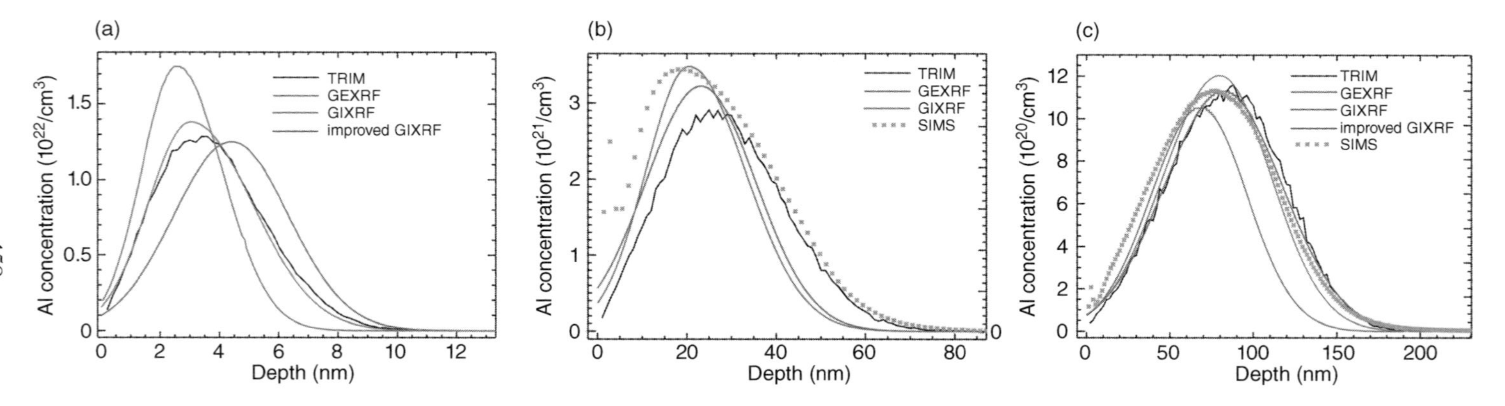

Figure 7.12. GI-XRF and GE-XRF results of shallow implants in silicon after reference-free quantification. The depth profiles were recorded for aluminum layers implanted in silicon wafers with different energies: (a) 5 keV, (b) 15 keV; and (c) 50 keV. The profiles determined by GI-XRF (dark gray) and GE-XRF (gray) were compared with those determined by SIMS (dots) and TRIM (black). GI-XRF profiles (Figure 7.12a and c) were corrected assuming a thin layer of SiO_2. The improved profiles match very well. Figure from Ref. [103], reproduced with permission. Copyright © 2012, Royal Society of Chemistry.

7.2.4 Reference-Free Quantification

In general, appropriate standard reference materials are used for a quantification of experimental results. Their matrix should be very similar to the sample that has to be analyzed. However, for a number of samples, especially for new materials, such reference materials are lacking. The theoretical dependence of X-ray fluorescence intensity and concentration or area-related mass of an element is well-known (e.g., Ref. [104]). Secondary and tertiary excitation leading to notorious matrix-effects of absorption and enhancement can be included. For excitation in total-reflection geometry, a standing wave of incident and reflected beam has to be taken into account. This way, X-ray spectra of TXRF, GI-XRF, and also of NEXAFS can be calculated. A reference-free quantification is a demanding aim of a research group with Beckhoff at the PTB in Berlin [62].

A theoretical calculation requires a response function of excitation and detection. Two sets of different parameters have to be known: experimental instrument parameters as well as fundamental atomic parameters. The first set of experimental parameters involves (a) X-ray excitation of the sample with radiant power and spectral distribution; (b) detection of X-ray fluorescence with spectral resolution and efficiency; and (c) beam geometry with angles, distances, and divergences. The second set of fundamental parameters includes (a) photoelectric mass-absorption coefficients of pure elements for the exciting photons; (b) total mass attenuation coefficient of the elements for the photons to be detected, (c) transition probabilities of different spectral peaks or lines; and (d) fluorescence yields and jump ratios at absorption edges.

All these values are listed in different tables, for example, in the X-ray data booklet [105], or can be read from a database, www.cxro.lbl.gov [106]. However, many of them are based on experiments made in the 1970s. Their uncertainty or reliability is not sufficient in many cases. For a better reliability, they have to be determined experimentally as was done in a special case by the PTB [62]. A synchrotron beam was used for excitation with intense and tunable radiation. For detection with high efficiency and spectral resolution, a wavelength dispersive spectrometer based on an SGM (spherical grating monochromator) was built. Mass absorption coefficient, fluorescence yield, and transition probability could be determined experimentally for the light elements B, C, and Al. The cross-sections for resonant Raman scattering below the K edges of Si, V, and Ni were also measured.

A spectrum of a 200 mm silicon wafer was recorded at an excitation energy of 1622 eV and a glancing angle of 0.9° (critical angle 1.10°). A considerable footprint effect was first corrected, and afterward the spectral background was taken into account. Figure 7.13 shows fluorescence peaks of a few elements (0.1 to 2 ng/cm^2) above a continuous background, which results from different components, such as Rayleigh and Compton scatter, resonant Raman scatter, and continuous "bremsstrahlung" of emitted photoelectrons at synchrotron excitation [32].

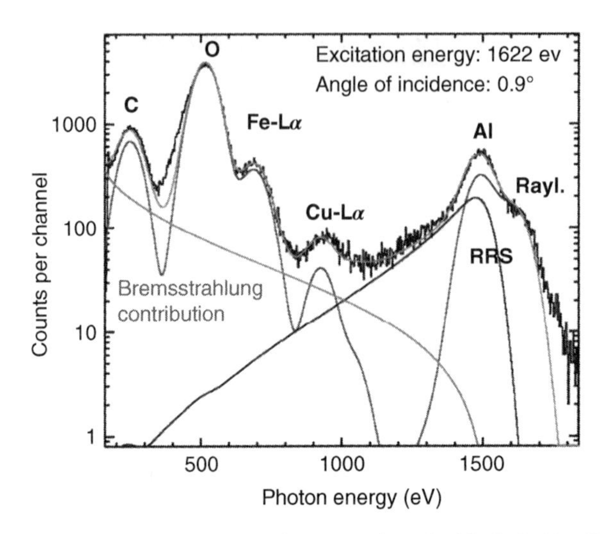

Figure 7.13. TXRF spectrum of a silicon wafer contaminated with C, O, Fe, Cu, and Al up to 0.2 ng/cm^2. The spectral peaks of these elements appear above a continuous background, which was deconvoluted with respective fundamental parameters of the sample and experimental parameters of the detector (its "response function"). Different components were taken into account, such as Rayleigh and Compton scattering, resonant Raman scattering, and "bremsstrahlung" of photoelectrons emitted by aluminum atoms with 0.2 ng/cm^2. Figure from Ref. [32], reprinted with permission from the American Chemical Society. Copyright © 2008, Trans Tech Publications.

The method of reference-free quantification was performed for the analysis of particulate contaminations on a substrate by GI-XRF [63]. In order to simulate aerosol particles, artificial nanoscaled structures were deposited on silicon wafers. The surface was first covered with a photoresist and a defined pattern of "dots" was imprinted by lithography. The dots had a diameter of 2.7 μm and were arranged in seven rows of about 50 μm horizontal distance. Altogether, 2250 dots with a minimal distance of 50 μm were deposited on an area of about 350 μm × 20 mm. The total area was covered with a thin metallic layer. Afterward, the photoresist was removed and the dots appeared as isolated flat cylinders or "pads" with a diameter of 2.7 μm and a height of only 20–100 nm. Only 0.2% of the total area was deposited with metallic pads. The diameter was checked by scanning electron microscopy (SEM) images, the height was determined by a profilometer, and the density was derived from weighing the samples.

Two sets of samples were studied, covered with chromium and with permalloy (66% iron and 34% nickel), respectively, and both sets had four samples with a small nominal height of 10, 20, 50, or 100 nm. One additional sample was produced with a 2 nm gold layer on top of chromium pads. These artificial structures were investigated by GI-XRF. Measurements were carried out at the FCM beamline in the PTB laboratory at the electron storage ring of BESSY II [63].

For the chromium pads with gold layers on top of them, the photon energy of the incident beam was selected to be 6.8 keV. The critical angle of total reflection for chromium was 0.45°, for gold it was 0.68°, and for the silicon substrate 0.26°. The K peaks of chromium and gold were observed with a silicon drift detector and a reference-free quantification was performed. As expected, the angular-dependent fluorescence intensity showed periodic oscillations for Cr-Kα below 0.26°, and for Au-Mα below 0.68°, while the intensity of silicon increased steeply at 0.26°. As demonstrated in Figure 7.14a, theory and experiment generally agree; they only differ a little because of shadowing effects. That means that the X-ray standing wave (XSW) field caused by total reflection at the silicon substrate is hardly disturbed by the nanoscaled structures because only 0.2% of the total area is covered. For the chromium and permalloy structures without gold layer, an incident beam of 9.3 keV was selected; the critical angle for chromium was 0.32°, for permalloy it was 0.35°, and for silicon 0.19°. This time, the K peaks of Cr, Fe, and Au were observed with a silicon drift detector and a reference-free quantification was performed again. As demonstrated in Figure 7.14b, the measured intensity/angle profiles showed only smeared and weak oscillations; however, theory and experiment correspond at a rough estimate [63].

The mass-deposition of chromium pads, which was in the range of 30–125 pg/cm^2, could also be determined by reference-free quantification. Assuming a flat top of the pads, the average values fit quite well with the nominal values of the manufacturer with a deviation of only 6%. For the permalloy-pads, however, the results showed deviations of about 55%. The heights of these pads were determined afterward by SEM/energy dispersive X-ray spectrometry (EDX) and corrected by a factor 0.42. The correction led to a nominal mass

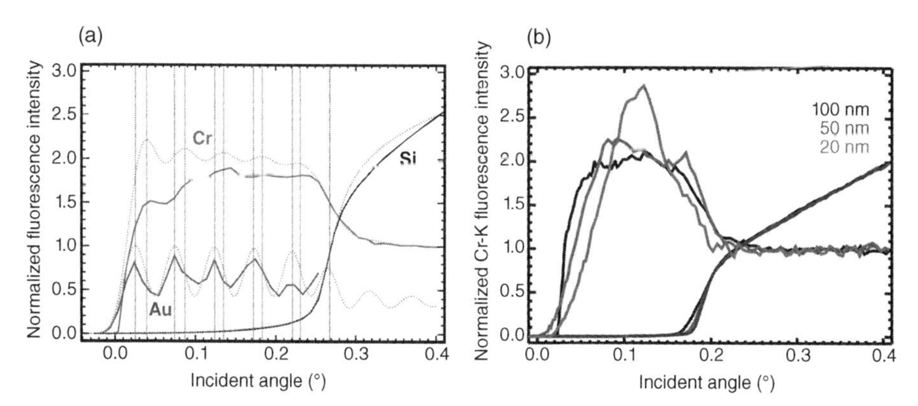

Figure 7.14. (a) Intensity/angle scan recorded by GI-XRF for an artificial park of cylindrical metallic pads with a height of 100 nm. The pads consisting of chromium were deposited on a Si wafer and capped with a gold layer. The solid curves represent measurements for Cr-Kα, Au-Mα, and Si-Kα; the dotted curves were calculated reference-free. The oscillations indicate nodes and antinodes of the XSW field above the Si substrate. (b) Intensity/angle scan for Cr pads without gold layer but with different heights of 20, 50, and 100 nm. Figure from Ref. [63], reproduced with permission. Copyright © 2012, Royal Society of Chemistry.

deposition that corresponds quite well with the measured values at a deviation of 14%. In summary, it can be stated that GI-XRF with reference-free quantification is a reliable method for nanoparticles provided that the XSW field is undisturbed.

Experiments described earlier were repeated for two nanoscaled objects deposited on a silicon wafer. Monodisperse NaCl nanocrystals were investigated at the PGM beamline of BESSY II and artificial chromium pads were studied again at the FCM beamline [107]. In contrast to calculations with the formalism of XSW, however, a novel approach was tested with the concept of geometrical optics (GO). Results for the NaCl nanocrystals showed good agreement (better than ±5%) for both concepts at glancing angles above 66% of the critical angle. Results for the chromium pads are similar; however, the GO approach considers an absorption and shadowing effect of neighboring pads and consequently gives a better adjustment at the critical angle and at zero angle. Furthermore, interference oscillations were smeared out if a roughness of 5 nm at the roof of the pads was taken into account.

7.2.5 Time-Resolved *In Situ* Analysis

Different analytical investigations can be distinguished with respect to external sample conditions. Studies can be carried out *in vivo* for living organisms, or *in vitro* for samples in an artificial environment, for example, in a test tube. Studies conducted *in situ* observe the samples on their spot or place of their origin. Most spectrochemical analyses take place in the artificial environment of a laboratory, but only sometimes *in situ*. A process-related *in situ* analysis is necessary for the exploration of ores, coal, and oil, or for the production of industrial components. Spectrochemical analyses at different points of the process have to be carried out within minutes. Conventional X-ray fluorescence is one of the most applied methods for such assignments. The observation and control of physico-chemical reactions, for example, by heat treatment, is also called *in situ*. Additionally, time-resolved investigations can be necessary with a time resolution of milliseconds.

For that purpose, a real-time measurement has to be carried out. Recently, a special example was given by Sakurai [108]. It was made possible for XRR and GI-XRF by improvements of the time resolution. For XRR, a multilayer monochromator was placed between a rotating X-ray tube and sample stage in the narrowest geometry. A fairly wide angular divergence of the beam allowed simultaneous data collection by a fast Si-strip detector, which is position-sensitive. Spectral peaks could be recorded within some 0.1 s over a total angular range of 2° with a sufficient precision of about 1%. A further attempt was made by synchrotron excitation with a bandwidth of 5% and a simple wavelength-dispersive spectrometer with a spectral resolution of about 0.1%.

Another simple but typical example will be described in Section 7.3.4 for XRD studies of layers during evaporation and chemical vapor coating. Process analysis is a promising supplement and may be trend-setting in the future of

TXRF and related methods. The pulsed X-ray source of a laser plasma with a repetition time of 1 ms and the new X-ray color camera, both mentioned in Section 7.1.4, may be very helpful for this task [37,38].

7.3 FUTURE PROSPECTS BY COMBINATIONS

Irradiation of a sample by X-rays in total reflection geometry can also be combined with several other methods of spectral analysis. All of these combinations have basic advantages in common, such as a large variety of samples that can be analyzed, a microscopic sample amount necessary for analyses, nondestructive, and a simple and fast performance. Besides, all combinations can profit from excitation by synchrotron radiation with variable photon energy.

Figure 7.15 gives an overview of the different combinations of TR of X-rays with different methods of spectroscopy [109]. The original and most widely applied combination is that of total reflection X-ray fluorescence or TXRF. Related combinations are GI-XRF and GE-XRF (see Sections 6.2.3 and 7.2.3). The combination of total reflection with X-ray reflectivity (TR-XRR) was already mentioned as an accompanying method of GI-XRF in Section 4.6.4.

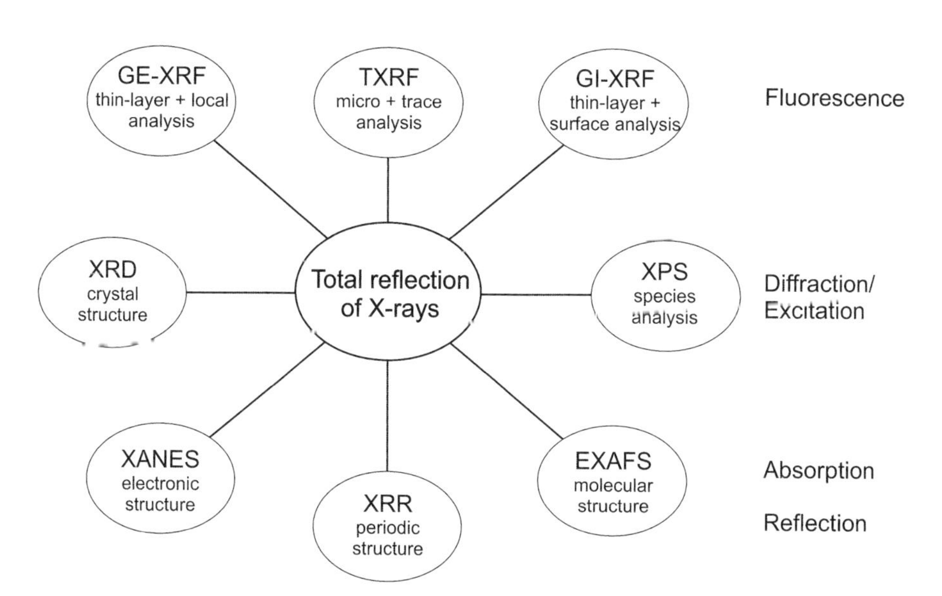

Figure 7.15. Different methods of spectral analysis based on fluorescence, diffraction, absorption, and reflection of X-rays, and emission of photoelectrons can be combined with excitation by X-rays in total reflection geometry. The combinations are preferably applicable to specific tasks, such as the analysis of microsamples, ultratraces, surfaces and thin near-surface films or layers, element species, and molecular and even atomic (electronic) structures. Figure from Ref. [109], reproduced with permission. Copyright © 2006, Elsevier.

Total reflection combined with X-ray diffraction (TR-XRD) and with X-ray photoelectron spectrometry (TR-XPS) seems to be less frequently applied. However, the combination of TR with X-ray absorption fine structure (TR-XAFS) increased in use over the last decade. Special attention was turned to a particular case of TR-XAFS, which is related to the near-edge of X-ray absorption and which is called TR-XANES or TR-NEXAFS. All of these combined methods have special capabilities; the improved sensitivity by excitation in TR geometry can be used for micro- and trace analysis, for thin-layer and for local analysis. Moreover, the combinations can be capable for species analysis, for the determination of crystal structures or periodic structures of stratified layers, and for detection of molecular structures of chemical compounds or even of electronic structures of atoms.

Local analysis has to be distinguished from microanalysis as it aims at the determination of elements at defined positions of microstructured samples. On the other side, species analysis is a completion of element analysis. It is the differentiation of similar chemical compounds of the same elements but with different valences. Species analysis is important if it is necessary to determine the different element species in addition to the elemental composition of a sample. This happens especially for biological and environmental samples when problems of physical tolerability, bioavailability, and environmental mobility have to be solved. Some methods of spectrochemical analyses, such as XRD and ICP-MS, are suitable for such a species analysis because different species can be distinguished simply by different lines. Other methods, for example, AAS and ICP-OES, can be combined with a chromatographic separation of species. TXRF itself is not capable of such a differentiation because the spectral peaks of different species do not differ enough apart from some examples in the low-energy region of X-rays. However, some of the mentioned combinations in total reflection or grazing incidence geometry are capable of species analysis even at trace levels.

Several studies on the combination with X-ray diffraction, X-ray absorption, and X-ray photoelectron emission have already been published in the 1990s [110–119]. All these methods benefit from a reduced background at grazing incidence and lead to an improved species or structure analysis. A condensed consideration of all the combinations will be given in the following sections.

7.3.1 Combination with X-ray Reflectometry

An ideal supplement to GI-XRF is the method of X-ray reflectometry [9,31,120–122]. Both methods allow the absolute determination of density, thickness, and even roughness of layered materials. However, the element composition of the layers can only be determined by GI-XRF, whereas the layer thickness can be measured more precisely by XRR. If the results of the individual fits are not in accord but ambiguous at first, both methods can work

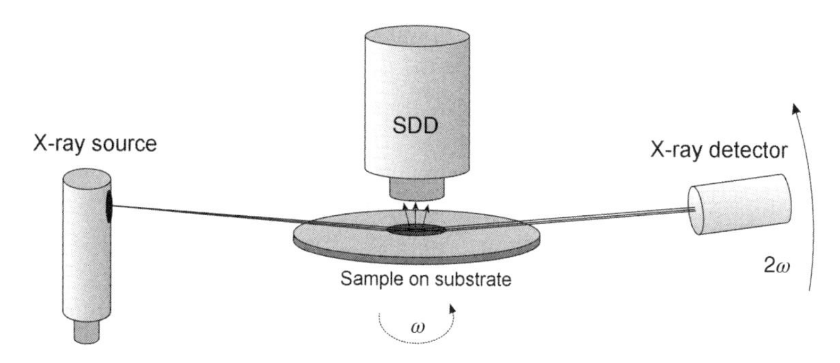

Figure 7.16. Experimental arrangement for X-ray reflectivity measurements in parallel to GI-XRF. The sample is tilted by a small angle α, the detector tilted by 2α for a geometry of $(-\alpha/180°-2\alpha)$ in the total reflection regime. A simple gas-flow detector or a Si-Pin diode is used for the detection of the reflected monochromatic beam. The fluorescence radiation of the sample is detected in $(-\alpha/90°)$ geometry by a Si(Li) detector or by an SDD.

in tandem to find the true model of a layered material. A combined piece of equipment that allows for both TXRF and XRR measurements can easily be assembled as demonstrated in Figure 7.16. It is commercially available from Rigaku and ATI.

The additional experimental effort is small; only a second X-ray detector is needed, positioned in the direction of the reflected beam. The detector can be a simple ionization chamber, a gas-flow detector, and even a Si-PIN diode has already been used. Only the intensity of the strong Kα or Lα peak of the exciting X-ray beam has to be recorded depending on the glancing angle of the incident beam. Weaker peaks, such as Kβ or Lβ peaks, can be filtered out by a thin metal foil.

Multilayered systems of inorganic and organic materials show certain oscillations of the intensity called Kiessig fringes. Their period is related to the number N and thickness d of layers: $\Delta\alpha = \lambda/(2Nd)$. The topic was actually dealt with in Section 2.1.2. Three examples have been demonstrated and the results were found simply and fast. Figure 2.2 shows the XRR profile for a 70 nm silicon layer on a gold substrate. The period $\Delta\alpha$ of the Kiessig maxima or minima of 0.028° leads to a thickness of 72.5 nm in accord with Equation 2.7. Figure 2.3 represents the profile for a 30 nm cobalt layer on a silicon substrate. The period of 0.065° corresponds to a thickness of 31.3 nm. Figure 2.6 was recorded for a stack of 15 Pt/Co bilayers each 2.1 nm thick. The profile has 15 Kiessig fringes between two successive Bragg reflections with a period of 0.133°, which corresponds to a thickness of 2.2 nm. The results confirm that number and thickness of several layers can simply be determined by the number and period of Kiessig fringes. The results have an accuracy of about 4% and can help for different kinds of thin-layer analysis by GI-XRF as mentioned in Section 5.4.8.2.

7.3.2 EXAFS and Total Reflection Geometry

X-ray absorption fine structure was applied by Glocker and Frohnmeyer already in 1923 [123]. It was developed by several scientists after 1970, and even more so in the 1980s when synchrotron beam lines became available and revolutionized the experimental situation by a drastic increase of intensity and reduction of measuring times. A comprehensive account of the development of XAFS was given by Stern [124].

X-ray absorption fine structure is especially suitable for determining different species of an element, that is, the chemical state or valence of an element in different compounds, and to distinguish them. Some famous examples may be enumerated: (i) different oxides and other compounds of titanium, such as TiO, Ti_2O_3, TiO_2, TiN, TiC, and TiS_2; (ii) different oxides of arsenic, such as As trioxide and pentoxide, As_2O_3 and As_2O_5, called arsenite and arsenate; (iii) different sulfides of arsenic, such as As_4S_4 and As_2S_3, that is, the orange–red pigment realgar and the yellow orpiment; (iv) different organic compounds of nitrogen, such as cyanate and thio-cyanate (-CNO and -SCN); and (v) several compounds of boron carbonitrides with the formula $B_xC_yN_z$. XAFS was initially performed in classical excitation geometry but it can take advantage of total reflection geometry.

Two special variants of XAFS are distinguished, namely, extended X-ray absorption fine structure (EXAFS) and XANES. In 1980, the latter acronym was already used by Bianconi [125]. XAFS means the determination of X-ray absorption already treated in Section 1.4.1. If an X-ray beam strikes a thin layer of a certain material with intensity N_0, a part of it is absorbed so that only a reduced intensity N is transmitted. The reduction is caused by the photoelectric effect, when core electrons from deeper shells are first raised to unoccupied higher shells or are even set free above the vacuum level. In accord with the law of Lambert–Beer in Equation 1.40, the absorption of a material is determined by the mass-absorption coefficient (τ/ρ) in cm^2/g. This specific coefficient follows the Bragg–Pierce law of Equation 1.43. The double logarithmic plot of Figure 1.24 shows a straight line with slope –8/3 dependent on the primary photon energy. It furthermore shows discontinuities, called jumps, positioned at the absorption edges of the respective elements. These discontinuities are the subject of XAFS, a method that is applicable to liquids, to solids in the form of thin plates, briquettes, powders, thin films or layers, and even to gases.

Absorption spectra can be recorded by an incident primary beam passing a thin sheet or layer. The *transmitted* intensity is measured in relation to and dependence on the photon energy of the incident beam. This mode of operation may be called *absorption mode*. As a result of X-ray absorption by the photoelectric effect, the ionized atom can rebuild its original state leading to the emission either of Auger electrons (radiationless) or of X-ray fluorescence photons. The *fluorescence* intensity of a thin layer and also of a thick sample induced by the incident primary beam can be recorded in relation to and dependence on the primary photon energy. In comparison with the *absorption*

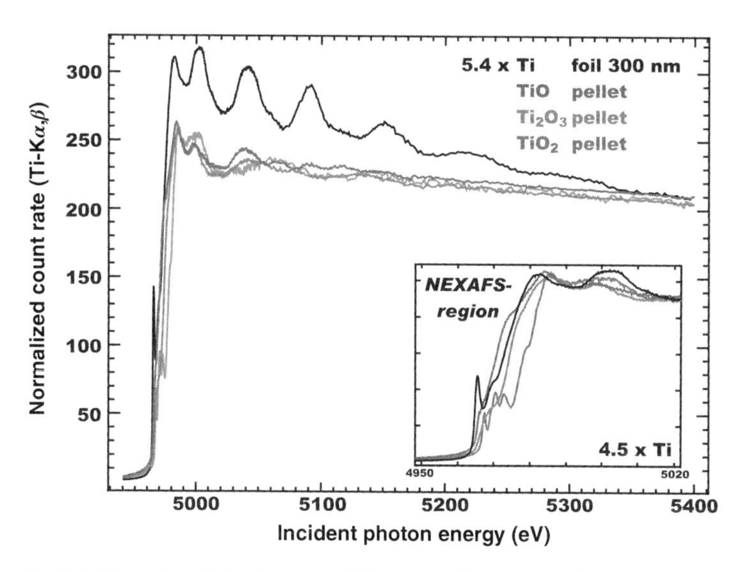

Figure 7.17. XAFS spectra of titanium and different titanium oxides. The K absorption edge of titanium appears at 4.964 keV, the pure foil of titanium and three different oxides show different peaks at the steep flank and oscillations at the sloping flank. The NEXAFS region is nearly 30 eV above an absorption edge (see inset), while the EXAFS region reaches up to 300 eV above this edge. Figure from Ref. [126], reproduced with permission. Copyright © 2009, American Chemical Society.

mode, XAFS spectra are mainly recorded in this *fluorescence mode*. The fluorescence intensity is plotted versus the primary photon energy, typically in a range up to ~500 eV above an absorption edge (EXAFS). If the range is restricted to less than 50 eV, the method is called XANES or NEXAFS (NE means near edge) (see Section 7.3.3). A typical example for EXAFS is given in Figure 7.17 for a titanium foil (300 nm) and three different pellets of titanium oxides [126]. TiO, Ti_2O_3, and TiO_2 with different valences (bi-, tri-, and tetra-valent) can clearly be distinguished from metallic titanium (zero or nonvalent).

In principle, XAFS spectra of a metal show a first steep rise at their respective K, L, and M edges. It is attributed to elevation of a K, L, or M electron above the Fermi level. Monoatomic gases can show a single peak or two peaks within 5 eV above the high-energy side of the respective edges. They are due to electron transitions into unoccupied higher shells. Molecular gases and chemical compounds of liquids and solids show several fluctuations at the sloping flank due to transitions into a valence band of neighboring atoms. These structures occur in the range of 5–50 eV. They are attributed to multiple reflections of the emitted photons even at more distant atoms (short-range order effects) and are called XANES [127]. Further and lower oscillations can be found in the range of 50–500 eV above the absorption edge. They are related to reflections by nearest neighbors of the central atom (long-range order) and also depend on the crystal lattice, temperature, and pressure. These structures are called EXAFS [125]. A theory based on the essential physics showed that

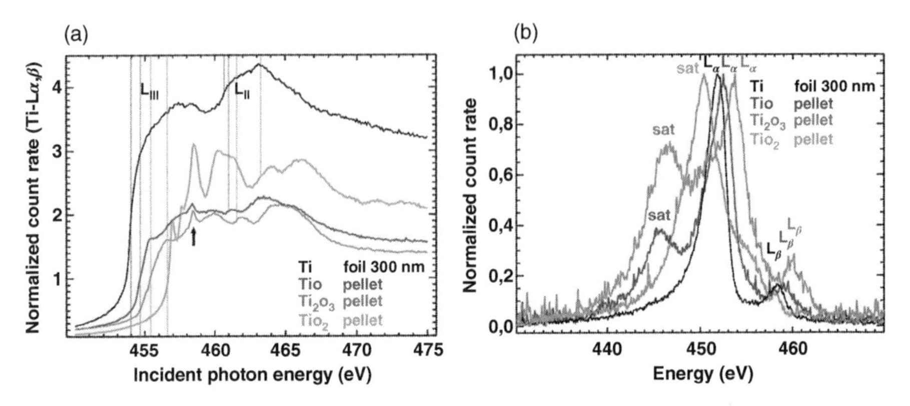

Figure 7.18. (a) L_{II} and L_{III} edge of a pure titanium foil and pellets of three different titanium oxides. The peak at 458.5 eV is characteristic for TiO_2 and is less distinct for TiO and Ti_2O_3. (b) $L\alpha$ and $L\beta$ peaks of the above mentioned samples show a shift of only 1 or 2 eV whereas the satellite peaks differ mainly in their height. Figure from Ref. [126], reproduced with permission. Copyright © 2009, American Chemical Society. (See colour plate section)

the different position of atoms near the center atom can even be read from Fourier-transformed XAFS data [124].

Binary compounds of titanium have an increasing technological use as ceramics, semiconductors, and superconductors. Tri- and tetravalent titanium, such as TiO, Ti_2O_3, TiO_2, TiN, TiC, and TiS_2, were compressed to pellets and characterized by NEXAFS [126]. Experiments were carried out for soft X-rays at BESSY II in Berlin and for hard X-rays at SPring-8 in Japan in (45°/45°) geometry. Absorption spectra were recorded for the L_{II} and L_{III} edges by a Si(Li) detector (above 454 eV) and for the K edge by an SDD (above 4966 eV). Additionally, emission spectra were observed for the $L\alpha$ and $L\beta$ peak by a wavelength-dispersive spectrometer with $E/\Delta E = 430$ (at about 455 eV) and for several $K\beta$ peaks with $E/\Delta E > 1000$ (at about 4960 eV).

Figure 7.18 demonstrates absorption spectra of the L_{II} and L_{III} edges (Figure 7.18a) and emission spectra of $L\alpha$ and $L\beta$ peaks (Figure 7.18b) of metallic titanium and different titanium oxides [126]. The L edges of titanium showed the multiplet structure typical for 3D transition metals. Besides, several complementary results were obtained. Clues regarding which titanium species are present include: (i) a chemical shift of the L_{II} and L_{III} edges and of the corresponding $L\alpha$ and $L\beta$ peaks of about 1 eV per oxidation state; (ii) shifts of a rather intense $K\beta_5$ peak of about 5 eV for the oxides; (iii) an additional low-energy satellite of the $L\alpha$ peak for the oxides; and (iv) quite different K absorption spectra for TiO_2, TiS_2, and TiC. Differences depend on the oxidation state of titanium as well as on its ligand if it differs from oxygen.

7.3.3 Combination with XANES or NEXAFS

The measurement of XAFS spectra follows after excitation of a sample by primary X-rays, generally by the intense X-ray source of a tunable synchrotron beam. A small band of excitation energy is scanned beyond the absorption edge in several small steps and the fluorescence intensity is measured. NEXAFS is a synonym for XANES [127]; however, while XANES is coupled with higher absorption edges (above 1 keV), NEXAFS is linked with lower absorption edges (below 1 keV), that is, of the low-*Z* elements, like carbon, nitrogen, and oxygen. NEXAFS is especially suitable for surface studies and for molecular science. A monograph on NEXAFS was written by Stöhr [128].

The energetic position of the main absorption edges of 92 elements are listed in Table 7.1. They are equivalent with the excitation energy of the respective X-ray peaks and can also be used for X-ray photoelectron spectroscopy (see Section 7.3.5). This simple table does not contain satellite peaks of different compound elements or energy shifts by nearest neighbors.

Novel materials with nanoscaled and complex structures require a chemical characterization of layers and interfaces. X-ray excitation at 45° incidence and under total reflection was coupled with XANES in several studies. Usually, AES, and SIMS are employed for speciation analysis; however, XANES is better suitable for that task. Nondestructive methods like XPS can suffer from sensitivity, and destructive techniques like SIMS can modify the chemical properties during analysis. XANES is not only nondestructive, but also highly sensitive and suitable for a characterization of layers and interfaces on the nanoscale. With respect to XPS and AES with an information depth of only 5 to 8 nm, the X-ray analytical methods can reach far deeper layers. The penetration depth of X-rays is much larger than that of photo- and Auger electrons and also of secondary ions. Consequently, XANES is preferred for μm-deep buried layers, μm-thick foils, and μm-thick pellets of powders. Meanwhile XANES is well established and routinely used for chemical speciation in many laboratories.

Fourteen different examples will be given here in order to explain the capabilities of TR-NEXAFS and TR-XANES.

Copper contamination was detected on Si wafers and shown to be from oxides but not silicides [74]. Copper ions were implanted into a 20 nm thick layer of SiO_2 grown on a silicon wafer. The oxide layer was removed by a HF solution and a XANES spectrum of Cu 1s was recorded as a fingerprint at the Stanford Synchrotron Radiation Lightsource (SSRL) of Stanford University. Copper with a high area deposition showed a spectrum similar to the reference spectra of Cu oxides but different to the spectra of metallic copper and Cu silicide.

A speciation of arsenic was shown to be possible by SR-XANES. Cucumber plants were fed by nutrient solutions with small amounts of two inorganic arsenic compounds: arsenite and arsenate. The As (III) species (arsenites) is the most toxic, followed by the As (V) species (arsenates) and much less toxic organic compounds. Experiments were carried out at HASYLAB, beamline L, in Hamburg, Germany [129]. Xylem saps of the plants were analyzed by TXRF at trace levels down to 0.2 ng/ml (0.2 ppb). For XANES, the K edge of arsenic at 11.86 keV was studied by tuning the energy of the synchrotron radiation with a crystal monochromator. The spectra showed clear differences between the inorganic compounds and an organic standard of dimethyl arsenic acid.

TABLE 7.1. Excitation Potentials or Absorption Edges (K, three L, five M edges) from Elements with Atomic Number Z

At. No. Z	Symbol	K edge 1s	L_1 edge 2s	L_2 edge $2p_{1/2}$	L_3 edge $2p_{3/2}$	M_1 edge 3s	M_2 edge $3p_{1/2}$	M_3 edge $3p_{3/2}$	M_4 edge $3d_{3/2}$	M_5 edge $3d_{5/2}$
1	H	0.0136								
2	He	0.0246								
3	Li	0.0547								
4	Be	0.1115								
5	B	0.1880								
6	C	0.2842								
7	N	0.4099								
8	O	0.5431								
9	F	0.6967								
10	Ne	0.8702	0.0485	0.02170	0.02160					
11	Na	1.0708	0.0635	0.03065	0.03081					
12	Mg	1.3030	0.0887	0.04978	0.04950					
13	Al	1.5596	0.1178	0.07295	0.07255					
14	Si	1.8390	0.1497	0.09982	0.09942					
15	P	2.1455	0.1890	0.1360	0.1350					
16	S	2.4720	0.2309	0.1636	0.1625					
17	Cl	2.8224	0.2700	0.2020	0.2000					
18	Ar	3.2059	0.3263	0.2506	0.2484	0.0293	0.0159	0.0157		
19	K	3.6084	0.3786	0.2973	0.2946	0.0348	0.0183	0.0183		
20	Ca	4.0385	0.4384	0.3497	0.3462	0.0443	0.0254	0.0254		
21	Sc	4.492	0.4980	0.4036	0.3987	0.0511	0.0283	0.0283		
22	Ti	4.966	0.5609	0.4602	0.4538	0.0587	0.0326	0.0326		
23	V	5.465	0.6267	0.5198	0.5121	0.0663	0.0372	0.0372		

24	Cr	5.989	0.6960	0.5838	0.5741	0.0741	0.0422	0.0422		
25	Mn	6.539	0.7691	0.6499	0.6387	0.0823	0.0472	0.0472		
26	Fe	7.112	0.8446	0.7199	0.7068	0.0913	0.0527	0.0527		
27	Co	7.709	0.9251	0.7932	0.7781	0.1010	0.0589	0.0599		
28	Ni	8.333	1.0086	0.8700	0.8527	0.1108	0.0680	0.0662		
29	Cu	8.979	1.0967	0.9523	0.9327	0.1225	0.0773	0.0751		
30	Zn	9.659	1.1962	1.0449	1.0218	0.1398	0.0914	0.0886	0.0102	0.0101
31	Ga	10.367	1.2990	1.1432	1.1164	0.1595	0.1035	0.1000	0.0187	0.0187
32	Ge	11.103	1 4146	1.2481	1.2170	0.1801	0.1249	0.1208	0.0298	0.0292
33	As	11.867	1.527	1.3591	1.3236	0.2047	0.1462	0.1412	0.0417	0.0417
34	Se	12.658	1.652	1.4743	1.4339	0.2296	0.1665	0.1607	0.0555	0.0546
35	Br	13.474	1.782	1.5960	1.5500	0.257	0.189	0.182	0.0700	0.0690
36	Kr	14.326	1.921	1.7309	1.6784	0.2928	0.2222	0.2144	0.0950	0.0938
37	Rb	15.200	2.065	1.864	1.804	0.3267	0.2487	0.2391	0.1130	0.1120
38	Sr	16.105	2.216	2.007	1.940	0.3587	0.2803	0.2700	0.1360	0.1342
39	Y	17.038	2.373	2.156	2.080	0.3920	0.3106	0.2988	0.1577	0.1558
40	Zr	17.998	2.532	2.307	2.223	0.4303	0.3435	0.3298	0.1811	0.1788
41	Nb	18.986	2.698	2.465	2.371	0.4666	0.3761	0.3606	0.2050	0.2023
42	Mo	20.000	2.866	2.625	2.520	0.5063	0.4116	0.3940	0.2311	0.2279
43	Tc	21.044	3.043	2.793	2.677	0.544	0.4476	0.4177	0.2576	0.2539
44	Ru	22.117	3.224	2.967	2.838	0.5861	0.4835	0.4614	0.2842	0.2800
45	Rh	23.220	3.412	3.146	3.004	0.6281	0.5213	0.4965	0.3119	0.3072
46	Pd	24.350	3.604	3.330	3.173	0.6716	0.5599	0.5323	0.3405	0.3352
47	Ag	25.514	3.806	3.524	3.351	0.7190	0.6038	0.5730	0.3740	0.3683
48	Cd	26.711	4.018	3.727	3.538	0.7720	0.6526	0.6184	0.4119	0.4052
49	In	27.940	4.238	3.938	3.730	0.8272	0.7032	0.6653	0.4514	0.4439
50	Sn	29.200	4.465	4.156	3.929	0.8847	0.7565	0.7146	0.4932	0.4849

(continued)

TABLE 7.1. *(Continued)*

At. No. Z	Symbol	K edge 1s	L_1 edge 2s	L_2 edge $2p_{1/2}$	L_3 edge $2p_{3/2}$	M_1 edge 3s	M_2 edge $3p_{1/2}$	M_3 edge $3p_{3/2}$	M_4 edge $3d_{3/2}$	M_5 edge $3d_{5/2}$
51	Sb	30.491	4.698	4.380	4.132	0.946	0.8127	0.7664	0.5375	0.5282
52	Te	31.814	4.939	4.612	4.341	1.006	0.8708	0.8200	0.5834	0.5730
53	I	33.169	5.188	4.852	4.557	1.072	0.931	0.875	0.6308	0.6193
54	Xe	34.561	5.453	5.107	4.786	1.1487	1.0021	0.9406	0.6890	0.6764
55	Cs	35.985	5.714	5.359	5.012	1.211	1.071	1.003	0.7405	0.7266
56	Ba	37.441	5.989	5.624	5.247	1.293	1.137	1.063	0.7957	0.7805
57	La	38.925	6.266	5.891	5.483	1.362	1.209	1.128	0.8530	0.8360
58	Ce	40.443	6.549	6.164	5.723	1.436	1.274	1.187	0.9024	0.8838
59	Pr	41.991	6.835	6.440	5.964	1.511	1.337	1.242	0.9483	0.9288
60	Nd	43.569	7.126	6.722	6.208	1.575	1.403	1.297	1.0033	0.9804
61	Pm	45.184	7.428	7.013	6.459	-	1.471	1.357	1.0520	1.0270
62	Sm	46.834	7.737	7.312	6.716	1.723	1.541	1.420	1.1109	1.0834
63	Eu	48.519	8.052	7.617	6.977	1.800	1.614	1.481	1.1586	1.1273
64	Gd	50.239	8.376	7.930	7.243	1.881	1.688	1.544	1.2219	1.1896
65	Tb	51.996	8.708	8.252	7.514	1.968	1.768	1.611	1.2769	1.2411
66	Dy	53.789	9.046	8.581	7.779	2.047	1.842	1.676	1.3330	1.2926
67	Ho	55.618	9.394	8.918	8.071	2.128	1.923	1.741	1.392	1.351
68	Er	57.486	9.751	9.264	8.358	2.207	2.006	1.812	1.453	1.409
69	Tm	59.390	10.116	9.617	8.648	2.307	2.090	1.885	1.515	1.468
70	Yb	61.332	10.486	9.978	8.944	2.398	2.173	1.950	1.576	1.528
71	Lu	63.314	10.870	10.349	9.244	2.491	2.264	2.024	1.639	1.589
72	Hf	65.351	11.271	10.739	9.561	2.601	2.365	2.108	1.716	1.662
73	Ta	67.416	11.682	11.136	9.881	2.708	2.469	2.194	1.793	1.735

74	W	69.525	12.100	11.544	10.207	2.820	2.575	2.281	1.872	1.809
75	Re	71.676	12.527	11.959	10.535	2.932	2.682	2.367	1.949	1.883
76	Os	73.871	12.968	12.385	10.871	3.049	2.792	2.457	2.031	1.960
77	Ir	76.111	13.419	12.824	11.215	3.174	2.909	2.551	2.116	2.040
78	Pt	78.395	13.880	13.273	11.564	3.296	3.027	2.645	2.202	2.122
79	Au	80.725	14.353	13.734	11.919	3.425	3.148	2.743	2.291	2.206
80	Hg	83.102	14.839	14.209	12.284	3.562	3.279	2.847	2.385	2.295
81	Tl	85.530	15.347	14.698	12.658	3.704	3.416	2.957	2.485	2.389
82	Pb	88.005	15.861	15.200	13.035	3.851	3.554	3.066	2.586	2.484
83	Bi	90.524	16.388	15.711	13.419	3.999	3.696	3.177	2.688	2.580
84	Po	93.105	16.939	16.244	13.814	4.149	3.854	3.302	2.798	2.683
85	At	95.730	17.493	16.785	14.214	4.317	4.008	3.426	2.909	2.787
86	Rn	98.404	18.049	17.337	14.619	4.482	4.159	3.538	3.022	2.892
87	Fr	101.137	18.639	17.907	15.031	4.652	4.327	3.663	3.136	3.000
88	Ra	103.922	19.237	18.484	15.444	4.822	4.490	3.792	3.248	3.105
89	Ac	106.755	19.840	19.083	15.871	5.002	4.656	3.909	3.370	3.219
90	Th	109.651	20.472	19.693	16.300	5.182	4.830	4.046	3.491	3.332
91	Pa	112.601	21.105	20.314	16.733	5.357	5.001	4.174	3.611	3.442
92	U	115.606	21 757	20.948	17.166	5.548	5.182	4.303	3.728	3.552

Source: From Refs [105,106], reproduced with permission from the Center for X-Ray Optics and the Advanced Light Source, Lawrence Berkeley National Laboratory.

Nitrate and sulfate compounds could be distinguished in spectra of aerosols collected at different sites. The aerosols were characterized by NEXAFS after collection on silicon wafers in a cascade impactor [130,131]. Sampling times of only some 10 min and small sampling volumes of only 2 m^3 air were sufficient for analyses. Experiments were carried out in the PTB laboratory of BESSY II in Berlin with a plane grating monochromator. The exciting energy was varied in a range of 15 eV above the K edge at 410 eV of nitrogen; the incident angle was set to 2.5° (1.5° below the critical angle of total reflection). One hundred twenty steps of 0.125 eV at a measuring time of only 20 s lead to a period of 40 min for the total spectrum. The $K\alpha\beta$ peak of nitrogen was recorded and normalized to the radiant power. The spectra of ammonium sulfate and sodium nitrate were taken as reference.

The detection of nitrogen was possible down to 10^{11} atoms/cm^2 or even lower. Molar ratios of ammonium and nitrate were determined for aerosols with different aerodynamic diameters (0.25, 0.5, and 1 μm) and from different sampling sites (the sea shore of Antarctica and Sardinia and a suburban and rural region in Hungary).

Organic contaminations on silicon wafers were detected and the results gave a hint to their source and avoidance [132]. Small pieces of a silicon wafer were intentionally but only slightly contaminated with droplets of a stock solution of four chemical compounds of high industrial importance: sodium cyanate, sodium thiocyanate, stearic acid, and naphthalene trisulfonic acid. Droplets of 25 μl each with 1 μg of carbon were dried under a laminar flow. The wafer pieces were fixed in a vertical position and a side-looking detector was attached to a vacuum chamber.

Spectra were taken at the PTB beamline with a PGM at BESSY II in Berlin in the fluorescence mode of NEXAFS [132]. As demonstrated by Figure 7.19, they showed up to five peaks or resonances, which were found to be different for the four organic compounds mentioned earlier. The spectra were used as fingerprints and showed that the real contamination of a blank wafer at positions off a droplet was similar to the intentional contamination of stearic acid detected at positions on a droplet. This contaminant could have come from a lubricant used for cutting the wafer pieces but not completely removed by cleaning.

Chemical speciation of different compounds of light elements used as lubricants or detergents for wafer production is made possible by NEXAFS. Two different compounds of light elements, such as sodium cyanate and sodium thiocyanate as already mentioned in the previous example, were analyzed in the fluorescence mode [61]. For samples of 500 pg, the K absorption edges of C, N, and O were recorded as illustrated in Figure 7.20. Up to four different peaks or resonances characterize the different chemical species. However, a "background" of unintended omnipresent contamination of carbon and oxygen on the silicon wafer interferes with the relevant spectra.

Layers above a substrate containing a particular element in different chemical neighborhoods can be distinguished. A self-assembled monolayer was built by

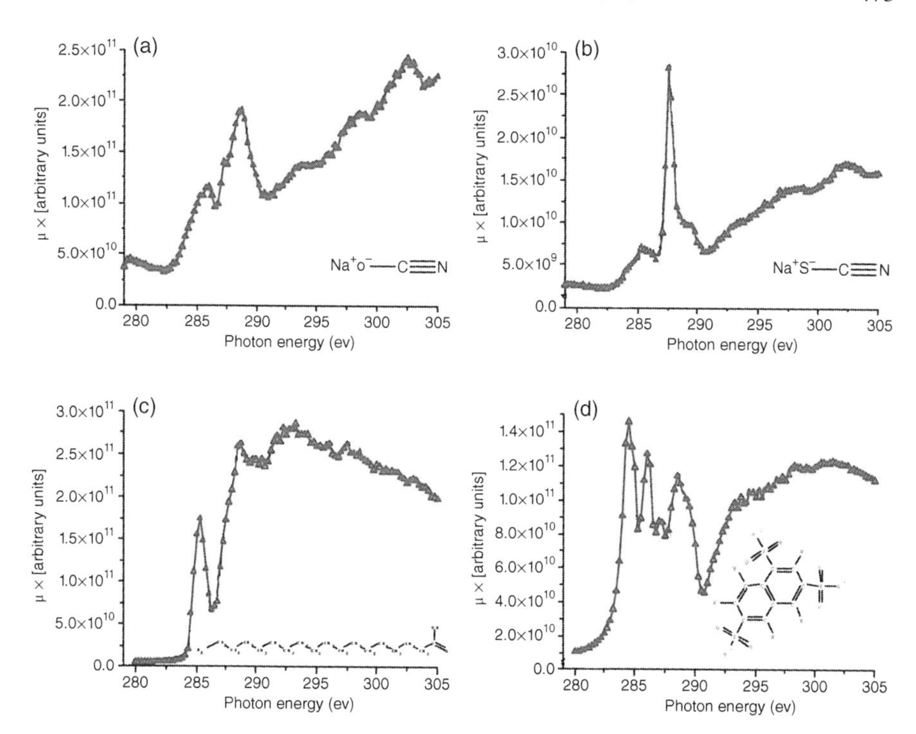

Figure 7.19. NEXAFS spectra of four different organic sodium compounds near the K edge of carbon. With a step width of 0.25 eV and a counting time of 40 s, the total range was scanned with 80 steps in about 1 h. The K edge of pure carbon is at 284 eV; but different peaks appear between 286 and 302 eV, dependent on the carbon bonding. The spectra can be used as a fingerprint of the different compounds. The spectra are similar to the carbon spectra of Figure 7.20a, but have a higher spectral resolution. Figure from Ref. [132], reproduced with permission. Copyright © 2003, Elsevier.

organic molecules on germanium surfaces [133]. A flat germanium wafer with a roughness below 0.3 nm was first dipped in concentrated HF (50%) for 40 s in order to remove the oxide layer. TXRF analysis showed a residual fluorine contamination of 2×10^{14} atoms/cm^2. After drying, the wafer was immersed in a solution of an alcylthiol [HS$-$(CH_2)$_{11}$$-$$OCOF_3$] of 1 mmol/l. The hydrogen-terminated surface of the germanium wafer was bonded to molecules with the HS-group as an anchor while the long alkyl chain with a head of $OCOF_3$ was erected with a certain angle. A monolayer was built within several hours, starting as islands that grew together gradually. GI-XRF was applied to such a layer and fluorine was determined as a "marker ion" (1) in a 0.2 nm lower layer covering the surface, (2) in a thin upper layer, 1.4–1.6 nm above the surface, and (3) in clusters that remained throughout the whole layer. Additionally, NEXAFS of fluorine showed a difference of the lower lying and the upper standing layer of thiol molecules.

Different iron compounds in biogenic samples can be distinguished by XANES. Human cancer cells were grown, harvested, washed, centrifuged,

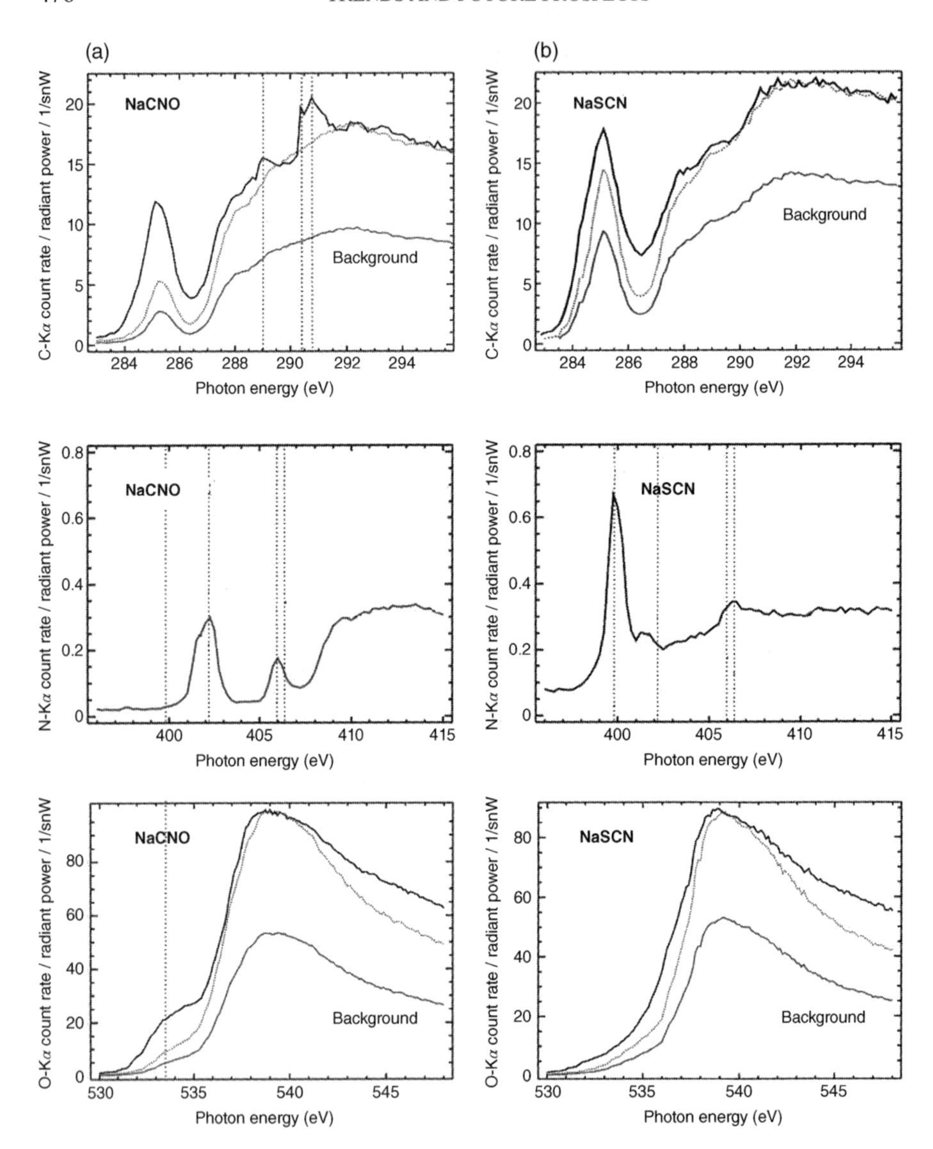

Figure 7.20. NEXAFS spectra of two different organic compounds of sodium. (a) Spectra of sodium cyanate NaCNO; (b) spectra of sodium thiocyanate NaSCN, with a mass of 500 pg each. The spectra represent the K edges of the light elements C, N, and O from top to bottom. Relevant resonances (oscillations) are indicated by dotted lines. Figure from Ref. [61], reprinted with permission. Copyright © 2007, American Chemical Society.

and resuspended in an isotonic NaCl solution [134]. Five microliters of the cell suspension were pipetted onto quartz carriers, dried, and analyzed by TXRF and by XANES at the beamline L of HASYLAB at DESY, Hamburg, Germany. Nine different organic and two inorganic iron compounds with oxidation state II and III and different ligands were used as standards. For XANES, the excitation energy was varied by a double Si (111) monochromator in 200 steps around the K edge of iron at 7112 eV. At a fixed glancing angle of 0.15° (well below the critical angle), the Fe-Kα intensity was measured by an SDD with an acquisition time of a few seconds, and normalized to 0.0 in a preedge position and to 1.0 in a postedge position. The XANES spectra of cancer cells were very similar to the spectrum of ferritin, which is an intracellular protein that binds and stores Fe (III) as FeO(OH). An oxidative stress of the cells by administration of 5-fluorouracil shifted the absorption spectra to a higher energy.

Binary and ternary compounds of the light elements boron, carbon, and nitrogen, which are basic materials for new solids, nanolayers, and nanotubes, were characterized using NEXAFS. Thin layers of boron carbonitride compounds were grown by CVD (chemical vapor deposition) with trimethylamine borane only and with N_2 or NH_3 at different partial pressures [135]. The layers were deposited on silicon wafers with a thickness of about 200 nm. The elemental composition of these layers was examined by TXRF and particular species were determined by NEXAFS. Reference samples of BN and of B_4C were used for fingerprint analysis.

Samples were investigated in the PTB laboratory at BESSY II in Berlin, Germany. The glancing angle of the silicon substrate was fixed at 2.5°, which ensured total reflection for all experiments. The excitation energy was tuned in a range of about 25 eV above the absorption edges of boron at 188 eV, of carbon at 284 eV, and of nitrogen at about 400 eV. The intensity of the K$\alpha\beta$ peaks of B, C, and N was measured in a counting time of 10 s. NEXAFS spectra were recorded with 125 steps of a width of 0.2 eV in a period of 20 min. The spectra provided evidence that pure boron carbide B_4C is the main component of the layer produced in the absence of NH_3. When NH_3 was added to the process, a layer of B_2C_2N could be detected and when the partial pressure of NH_3 was increased, a compound of BCN was produced with a hexagonal structure.

Compounds of boron carbonitrides can be analyzed by NEXAFS. Ternary compounds of boron, carbon, and nitrogen are novel materials of high technical interest used as extremely hard layers in electronic, optoelectronic, and luminescent devices. Boron carbonitride layers were produced by CVD with a single-source precursor, such as trimethylamine borane $(CH_3)_3NBH_3$, triethylamine borane $(C_2H_5)_3NBH_3$, and trimethylborazine $(CH_3)_3N_3B_3H_3$ with acronyms TMAB, TEAB, and TMB, respectively [136]. The layers were synthesized in a reaction chamber at temperatures up to 700 °C and a low pressure of 1–4 Pa. The chamber was flooded with hydrogen, helium, nitrogen, and partly with ammonia at a partial pressure of 0.5–2 Pa. At a growth rate of 5 nm/min, several different layers of boron carbonitride were produced with a thickness of about 200 nm. All of the layers were deposited on polished Si (100) plates.

Several methods of X-ray spectral analysis were applied for a characterization of these thin layers. SEM/EDX was used to determine the main components B, C, and N between 7% and 45% leading to a stoichiometric formula $B_xC_yN_z$. For a sum $z + x + y = 10$, the uncertainty of the indices was estimated to be ±0.5. In a triangle diagram, two main clusters of samples could be identified as demonstrated in Figure 7.21. The first cluster was found for six samples of TMAB and TEAB without the addition of ammonia. The respective center lies in the carbonic region and leads to the formula $B_{3.7}C_{4.4}N_2$. The second cluster was built by the remaining 10 samples of TMAB and TEAB with the addition of ammonia and for four samples of TMB without or with ammonia. The center lies in the nitridic region with the stoichiometric formula $B_{3.6}C_{1.5}N_{4.9}$. While the fraction of boron is nearly the same for all samples, carbon and nitrogen substitute each other in samples from the two different clusters.

Near-edge X-ray absorption fine structure spectra of B, C, and N were recorded above 185, 280, and 395 eV, respectively, as shown in Figure 7.22.

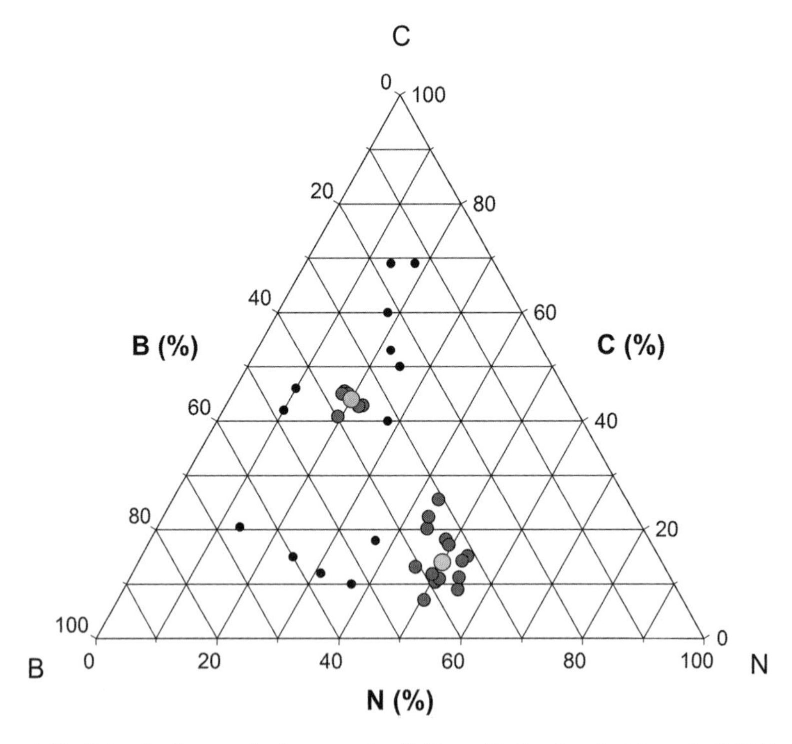

Figure 7.21. Triangle diagram for boron carbonitrides with the chemical formula $B_xC_yN_z$. The elemental composition of such layers is represented in at.%. Pure boron, pure carbon, or pure nitrogen has a position at the three edges of the triangle; binary compounds can be found in the center of the three sides. The different compounds analyzed by EPMA build two clusters around $B_{3.7}C_{4.4}N_2$ and $B_{3.6}C_{1.5}N_{4.9}$. Several data from literature are marked as small black dots. Figure from Ref. [136], reproduced with permission. Copyright © 2012, John Wiley and Sons.

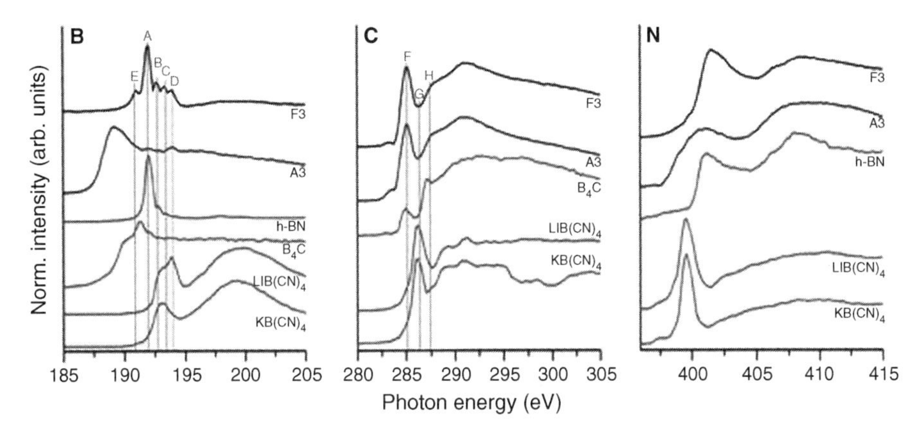

Figure 7.22. NEXAFS spectra near the K edge of boron, carbon, and nitrogen. Spectra were recorded for different layers of borane compounds produced with ammonia (F3) and without (A3). Reference spectra of boron carbide and hexagonal boron nitride and of Li and K cyanoborate can be used as fingerprints for the different bondings. Figure from Ref. [136], reproduced with permission. Copyright © 2012, John Wiley and Sons.

B-Kα spectra of borane compounds reveal B–N bonds as the main structure of hexagonal BN_3 configurations with additional BN and BN_2 components. C-Kα spectra represent a hexagonal bonded carbon while N-Kα spectra are similar to the hexagonal crystalline form of boron nitride, h-BN. The broad peaks point to more amorphous and less crystalline structures.

Layers of boron carbonitride covering nickel layers can be analyzed by NEXAFS or XANES. These methods combined with GI-XRF were applied to such double layers on top of a silicon substrate [137]. A nickel layer of 5 nm resulted from the deposition of nickel on a silicon wafer by PVD (physical vapor deposition). The layer $B_xC_yN_z$ was deposited by CVD of borane (TMAB) with ammonia at 200, 400, and 500 °C yielding a layer with a thickness between 4 and 6 nm. The Kα spectra of B, C, and N as well as the L$\alpha\beta$ spectra of nickel were recorded by varying the excitation energy above the respective K or L edges. Simultaneously, the glancing angle was changed between 0.5° and 15°. It could be shown that Ni–C bonds occur at a temperature of 200 °C in an interface between $B_xC_yN_z$ and nickel. This bond vanishes if the temperature is increased so that nickel is only existent as a metal. Simultaneously, Ni–Si bonds appear in an interface between nickel layer and silicon substrate.

Different compounds built as interfaces between different layers can be identified by XANES. A very special kind of layers, namely, aluminum doped ZnO layers (ZnO:Al) protected by a polycrystalline silicon layer (poly-Si), are highly relevant for solar cells in photovoltaic devices [138]. Panes of glass were coated with a transparent and conductive layer of ZnO:Al covered by poly-Si. High-temperature treatments up to 1050 °C followed by annealing procedures were used. A chemical speciation of the buried interface was performed by NEXAFS [138]. Zn silicates and Al oxides were found within an interface of 10 nm.

Organic complexes used as cytostatic agents can be distinguished by XANES. Platinum complexes, such as cis- and carbo-platinum inhibit cell growth and are used as strong chemotherapeutic substances; however, they have unpleasant or even toxic side effects. In order to develop new antitumor drugs, platinum was replaced by platinoids, such as ruthenium, rhodium, and rhenium [139]. Human cancer cell lines treated with a highly active rhodium agent were investigated by XANES in fluorescence mode. At grazing incidence, the excitation energy was shifted across the K edge of Rh (23.220 keV) and different types of rhodium species were detected in the spectrum. They could be distinguished by comparing spectra of different rhodium complexes as fingerprints. Complexes similar to $RhCl_3$ were shown to be the most effective cytostatic agents.

Cryosections of brain tissue can be characterized by TXRF and XANES with μm local resolution. Samples were cut from tumor tissue or Parkinson's-affected tissue of a human brain during autopsy [140]. The tissue samples were cryosectioned to a thickness of 5–20 μm and the sections were placed and stretched on polymer discs. TXRF as well as XANES were carried out with synchrotron excitation, while 2D mapping was made possible by focusing the monochromatized beam with polycapillary lenses or Kirkpatrick–Baez mirrors to a diameter of 15 μm. Particular elements, such as Cl, Fe, Ca, and Zn, with high levels of concentration can be used for a general discrimination of malignancy. Different oxidation states between 1+ and 3+ of iron, copper, and zinc in tissues were determined. It could be shown that the distance of atoms within the first coordination zone from a central atom is correlated with the grade of malignancy.

Chemical speciation by XANES or NEXAFS is also possible in grazing exit geometry as shown by Meirer et al. [141]. It was tested for the analysis of arsenic traces in water droplets and compared with grazing incidence geometry. The combination of XANES with grazing exit geometry showed distinct oscillations above the K edge of arsenic at 11.867 keV. They were not damped or smeared out as was the case for the combination with grazing incident geometry. Due to a short path length of the primary beam, GE-XRF hardly suffered from self-absorption effects in concentrated samples (>20 ng arsenic/μl). Furthermore, the sample could be excited by a focused microbeam of 40 μm spot size, and area maps of 1200×1000 μm^2 became possible. However, GE-XRF is much less sensitive for traces (factor 1/30) so that small amounts of arsenic (<4 ng arsenic/μl) could only be characterized by XANES with grazing incidence geometry.

7.3.4 X-ray Diffractometry at Total Reflection

X-ray diffraction analysis is a method parallel to the wavelength dispersive mode of X-ray fluorescence and both methods are based on Bragg's reflection (see Section 1.5.2). For wavelength-dispersive spectrometry, the sample excited by X-rays emits X-rays that are reflected at an analyzer crystal under a certain reflection angle. The relevant interplanar spacing of the analyzer crystal is known, the glancing angle is measured by means of a goniometer, and the wavelength of the fluorescence radiation is determined by Bragg's Equation 1.62. Accordingly, the energy of the emitted photons is calculated and the element emitting this radiation is identified. For X-ray diffraction, the opposite path is treated.

Two experimental adjustments are possible. The Debye–Scherrer disposition is suitable for powdered samples filled into small and thin capillaries, while the Bragg–Brentano disposition is used for flat samples, such as sheets, plates, films, layers, and pellets. In both cases, the analyzer crystal is replaced by the sample, which is irradiated for analysis. Figure 7.23 shows the Bragg–Brentano adjustment where the sample is tangential to the measuring circle and distances of sample–X-ray tube and sample–detector are equal and remain constant. For this para-focusing geometry, the peripheral angle is constant (180°–2α) for all positions of the detector; however, the focusing condition of Rowland is fulfilled only for one certain angle α. Two kinds of scanning are possible: (1) for (α/α) rotation, the sample is fixed while the X-ray tube and detector rotate toward one another with the same angular velocity ω; and (2) for ($\alpha/2\alpha$) rotation, the X-ray tube is fixed and the sample is rotated with angular velocity ω while the detector rotates with 2ω.

Usually, the main Kα peak of an X-ray tube is selected by a simple metal foil acting as an attenuation filter. The monochromatic X-rays are reflected at certain glancing angles, which correspond to particular interplanar spacings

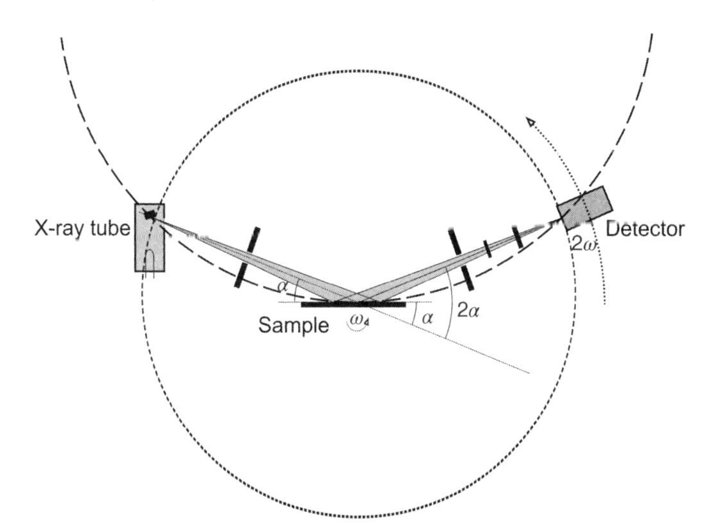

Figure 7.23. XRD spectrometer in Bragg–Brentano adjustment. A flat crystalline sample is irradiated by a monochromatic X-ray beam. It is reflected at the lattice planes under certain glancing angles. An intensity/angle scan is carried out by means of a goniometer with a simple gas-filled or scintillation detector representing a diffraction spectrum of the sample. For a polychromatic X-ray beam, a fixed glancing angle can be chosen; but in this case, a Si(Li) or SDD detector with a good spectral resolution has to record the total energy spectrum.

of the crystalline sample. The photon energy or wavelength of the incident beam is known, and the glancing angle is varied by means of a goniometer. Consequently, the interplanar spacing of the corresponding crystal lattice plane can be calculated by the reverse Bragg equation for the *m*th order of diffraction:

$$d = m\frac{\lambda}{2} \cdot \frac{1}{\sin\alpha} \quad (7.3)$$

Usually, a crystalline sample provides several reflections with different sets of Miller indices, *hkl*, even for one single crystal and moreover for different crystals of a sample. For a rhombic lattice with lattice constants *a*, *b*, and *c*, several interplanar spacings are possible for different combinations of the Miller indices:

$$\frac{1}{d_{hkl}^2} = \frac{h^2}{a^2} + \frac{k^2}{b^2} + \frac{l^2}{c^2} \quad (7.4)$$

For a cubic lattice, the three constants *a*, *b*, and *c* are equal. Interplanar spacings are listed for about 50 000 substances in an extensive data file of JCPDS (Joint Committee on Powder Diffraction Standards). As an example, Figure 7.24 shows a spectrum of a car finish with white pigments of TiO_2, ZnO, and Ba_2SO_4. XRD can be applied for the determination of the crystal structures of known samples but also for the identification of chemical compounds possibly in addition to element-specific X-ray fluorescence analysis. Consequently, XRD can be used for phase analyses and the study of phase transitions, for their dependence of temperature and pressure, for the investigation of the texture of samples, and of mechanical stress and strain.

X-ray diffraction and total reflection were already combined by Yoneda and Horiuchi in 1971 as described by Horiuchi 22 years later [142]. First, they determined different elements in seawater, in blood, and in spring water. Moreover, an ultrathin organic film of *n*-$C_{33}H_{68}$ (an alkane) was investigated. Such a Langmuir–Blodgett film has a molecular structure with vertical axes perpendicular to the substrate. Diffraction occurred at the crystal lattice planes (110) and (200) of the film. Background noise from scattering within the substrate was removed or significantly reduced by total reflection. In accord with Bragg's law, weak diffraction peaks were found at two particular angles corresponding to the *d* values. The evaporated film with a thickness up to 100 nm was studied during a heating process between 30 and 60 °C *in vacuo*.

Between 1993 and 1995, the combination of total reflection and diffraction of X-rays was developed mainly by Horiuchi and coworkers [143–148]. It was called TR-XRD (total reflection X-ray diffraction) or GI-XRD (grazing incidence X-ray diffraction). Because of the low penetration depth at total reflection, it is a powerful tool for evaluating thin near-surface layers, such

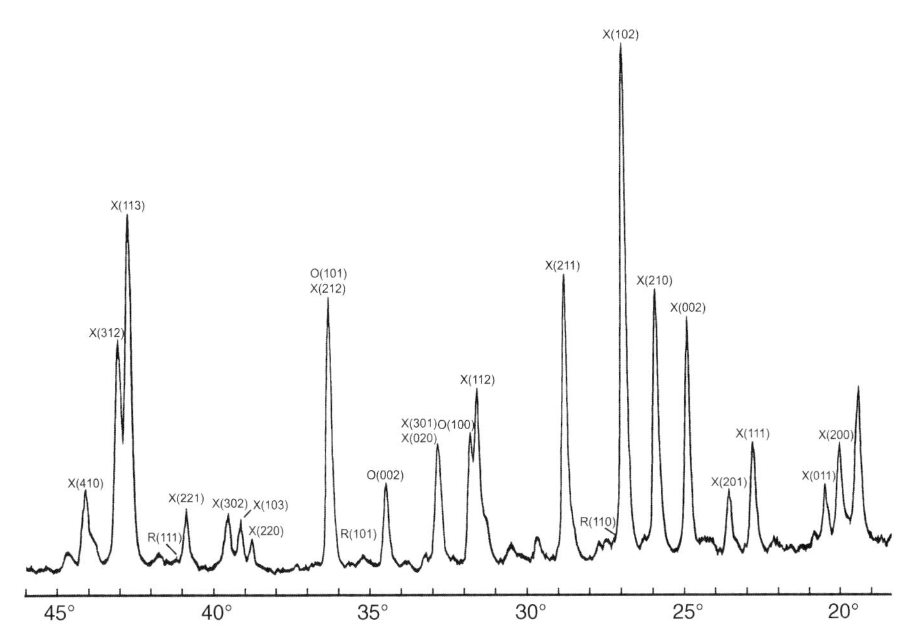

Figure 7.24. XRD spectrum of a white car paint containing different pigments. The peaks correspond to the pigments R = TiO_2, O = ZnO, and X = Ba_2SO_4. Each pigment shows several reflections according to the different lattice planes with Miller indices *hkl*.

as molecular thin films. Figure 7.25a shows a diffractometer with total reflection geometry of a circular substrate, which is rotated around a vertical axis for in-plane diffractometry [143]. The X-ray beam of a Mo or a W anode was used for diffraction and the irradiated area was about $8 \times 15\,mm^2$. The substrate with a thin film was rotated around its vertical axis and a Si(Li) detector was used for the detection of the diffracted beam. Figure 7.25b shows a diagram of this arrangement where the sample substrate is coated with a thin alkane layer. This layer was characterized by XRD *in vacuo* and *in situ* [144]. A crystalline quartz glass and a single crystal of KCl were used as substrates. The thickness of the layer increased from 15 and 100 nm during evaporation and decreased during heating from 30 to 60 °C.

A further study reported on an organometallic thin film of ferrocene, $(C_5H_5)_2Fe$, deposited on crystalline quartz glass and on a KCl crystal [145]. A slurry of ferrocene powder in acetone was pipetted on the substrate and evaporated at room temperature so that a thin film was cast on the substrate. It was irradiated under total reflection of the incident beam. A diffractometer was applied with either in-plane or vertical geometry, respectively. The substrate with the thin film and the detector, can be rotated azimuthally around a vertical axis or altitudinally around a horizontal axis, respectively. However, to avoid time-consuming rotation during the evaporation process, a single particular angle of reflection was chosen and fixed but "white" polychromatic

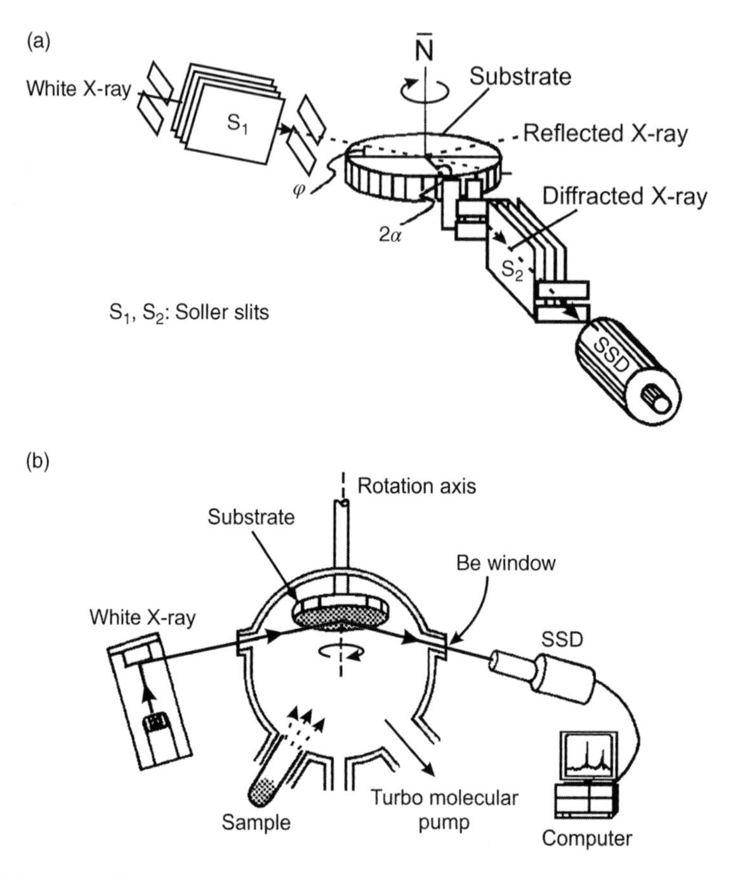

Figure 7.25. X-ray diffraction at total reflection. (a) The experimental arrangement with total reflection geometry shows a circular substrate that can be rotated around a vertical axis for in-plane diffractometry. The Soller slits have a divergence of about 0.15°. (b) The schematic diagram of this instrumentation shows the sample substrate placed in an evaporation chamber for *in situ* investigations. The Soller slits have been omitted for clarity. Figure from Ref. [143], reproduced with permission. Copyright © 1993, Elsevier.

instead of monochromatic radiation was applied. A total energy-dispersive spectrum was recorded by a Si(Li) detector displaying fluorescence and scattering peaks as well as diffraction peaks of the sample. As shown in Figure 7.26, the Kα and Kβ peaks of iron and furthermore the diffraction peaks (001) of ferrocene and (200) of KCl arise without any disturbing noise from scattering at the substrate. The diffraction peaks suggest that ferrocene crystals grow epitaxially and orient their (110) plane parallel to the KCl surface.

Furthermore, thin films of copper phthalocyanine (CuPc) were studied by *in situ* observations during the vapor coating process [146]. The thickness of the films growing by evaporation could be determined by TXRF. The structure of the thin films was studied by TR-XRD simultaneously. It could be revealed that the crystalline film changed its structure at about 9 nm thickness.

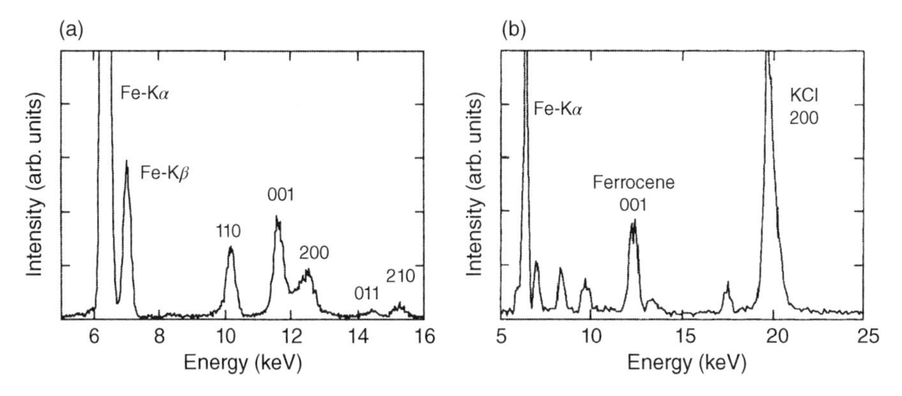

Figure 7.26. X-ray diffraction spectrum of a thin layer of ferrocene recorded by (a) an in-plane diffractometer and (b) a vertical type diffractometer with total reflection geometry at a fixed angle of diffraction. The spectra represent the Fe-Kα and Kβ peak and diffraction peaks of ferrocene and KCl. Figure from Ref. [145], reproduced with permission from the author.

In addition to the preceding experiments with grazing incidence, grazing exit geometry with a synchrotron beamline was applied and called GE-XRD [147]. The synchrotron beam was monochromatized by a W/Si multilayer and focused by Kirkpatrick–Baez optics with a pair of elliptical mirrors. The beam size was reduced to about $7 \times 6\,\mu m^2$ and local microanalysis was made possible. The spectra were recorded via a Si(Li) detector for the investigation of thin multilayers. Thickness, density, and surface roughness could be determined by variation of the exit angle.

For GE-XRD measurements, the sample was rotated around a vertical axis and the spectra were recorded with a position-sensitive proportional counter (PSPC). A thin layer of palladium with 45 nm thickness was deposited on a quartz substrate, afterward oxidized in air, and finally reduced by exposing the palladium layer to diluted hydrogen gas. Hydrogen reduction leads to many dark circular spots of some 10 μm diameter in a brown-colored film. The Pd (111) and the PdO (101) peak appeared in the brown areas while only the Pd (111) peak was observed in the dark spots. These spots were shown to consist of pure palladium completely reduced by hydrogen from the oxide, PdO.

In a further study, quantum well structures used for high-power light emitting diodes (LEDs) were characterized by high-resolution XRD [148]. Quaternary layers of $Al_xGa_yIn_{1-x-y}N$ with a thickness of 10 to 25 nm were characterized with a special six-axis diffractometer for in-plane diffraction but without total reflection geometry. TXRF was additionally performed to get some particular missing data.

The characterization of laterally structured surface gratings of GaAs was carried out by X-ray diffraction under total external reflection geometry [149]. Measurements and model calculations developed by a dynamical theory of scattering were compared for five samples and shown to be quantitatively correspondent. The mesoscopic parameters of the grating and also microscopic imperfections of the surface were considered in the extensive theoretical calculations.

7.3.5 Total Reflection and X-ray Photoelectron Spectrometry

In 1957, Kai Siegbahn developed the photoelectron spectrometry (PES), which he referred to as ESCA (electron spectroscopy for chemical analysis). In 1981, he received the Nobel Prize for his successful efforts. The method is based on the photoelectric effect when core electrons are ejected from atoms or molecules by photons of UV or X-rays. In accord with the excitation, the two variants are called ultraviolet and X-ray photoelectron spectrometry, UPS and XPS, respectively. The emitted electrons may stem from gases, liquids, and preferably from solids. They have to be detected and their kinetic energy has to be measured [150,151]. Afterward, the binding energy of the ejected electron can be determined and the original electronic state of atoms and their vibrational and rotational levels in molecules can be derived. Consequently, the electronic structure of atoms and the chemical structure of molecules in a material can be revealed. For solids, the method of XPS is strongly surface-sensitive because photoelectrons can only escape from near-surface layers of a few nm.

The samples are irradiated by a monochromatic X-ray beam with a known energy, $h\nu$, focused to a beam of about 100 μm, mostly of Al-Kα at 1486.7 eV. The spectrum of the electrons is measured either by a cylindrical mirror analyzer (CMA) or by a concentric hemispherical analyzer (CHA). The first consists of two coaxial cylinders and the second consists of two concentric hemispheres. Both analyzers are operated at UHV, while the electrons are detected by an electron multiplier, specifically by a channeltron or a channelplate. The CHA has a better energy resolution (down to 0.05 eV instead of 1.5 eV). Further details can be found in the literature [151,152].

Photoelectron emission requires that a bound electron of an atom or molecule is hit by a photon and ejected. It leaves a hole in the electron shell (which can be filled in a following process with X-ray emission) and is ejected with a certain kinetic energy. The process is controlled by the conservation of energy according to Einstein's relation:

$$h\nu = E_{\text{binding}} + E_{\text{kin}} + \Phi \tag{7.5}$$

where $h\nu$ is the photon energy, E_{binding} is the binding energy of the electron, E_{kin} is its kinetic energy after ejection, and Φ is the work function of the analyzer. The photon energy has to be known and the kinetic energy of the electron has to be measured by the analyzer. Consequently, the binding energy can be calculated from this formula provided that the work function is known, which usually has a small, constant value (about 2–6 eV). Figure 7.27 illustrates

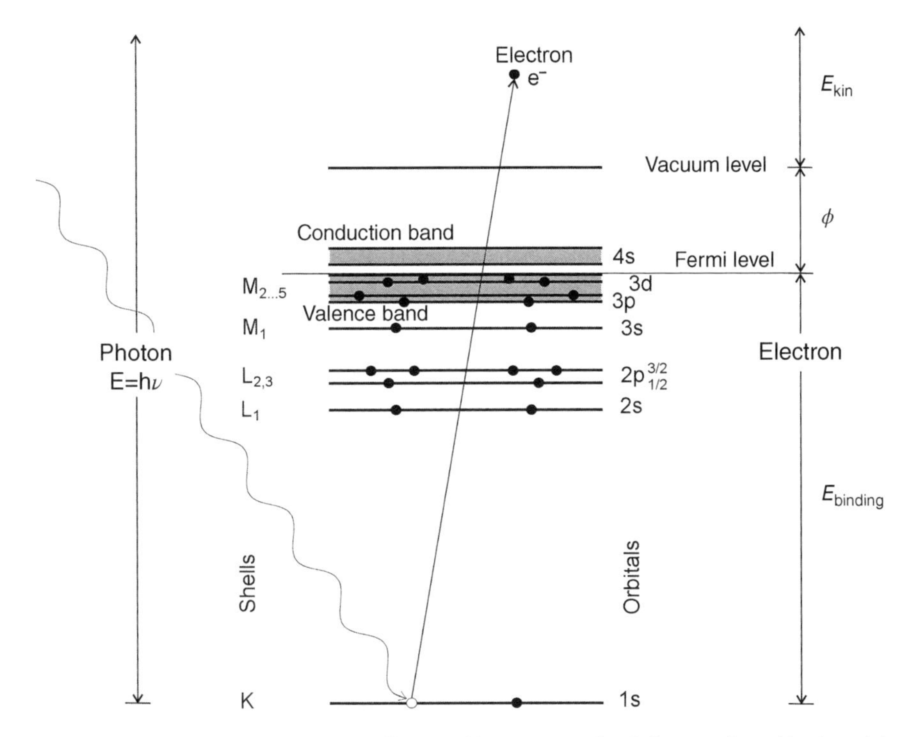

Figure 7.27. The photoelectric process illustrated in an energy-level diagram (logarithmic scale). The incident photon with energy $E = h\nu$ penetrates the intrinsic shells of an atom and hits a core electron. If its binding energy E_{binding} is below the energy E of the incident photon, the electron is ejected from the atom and leaves it with the remaining energy as kinetic energy. The different orbitals in Siegbahn notation are indicated. Fermi and vacuum level have a small difference Φ, which is called work function of the spectrometer.

different energetic levels of an atom, the valence band, and the conduction band with Fermi and vacuum level.

X-ray photoelectron spectra are recorded as intensity/energy diagrams with peaks but also with bands. Examples for the element silver and for the organic compound cystine are illustrated in Figure 7.28a and b, respectively. The abscissa can be represented by the kinetic energy or the binding energy, respectively. Different transitions lead to different peaks and are named in accord with the core electrons. The notation follows the specific quantum numbers of the ejected electron: 1s; 2s, $2p_{1/2}$ and $2p_{3/2}$; 3s, $3p_{1/2}$ and $3p_{3/2}$, $3d_{3/2}$ and $3d_{5/2}$; 4s, $4p_{1/2}$ and $4p_{3/2}$; $4d_{3/2}$ and $4d_{5/2}$, $4f_{5/2}$ and $4f_{7/2}$, and so on. The first number is the principal quantum number n; the following letter represents the azimuthal quantum number l; and the index gives the total angular momentum j equal to the fraction $l \pm 1/2$. All peaks with $n \geq 2$ arise as doublets, only the s-peaks are singular. These peaks are listed in tables or plotted in diagrams. Their energetic position roughly follows Moseley's law treated in Section 1.3.2.1. Their positions are equal to the photon energies of X-ray edges used for

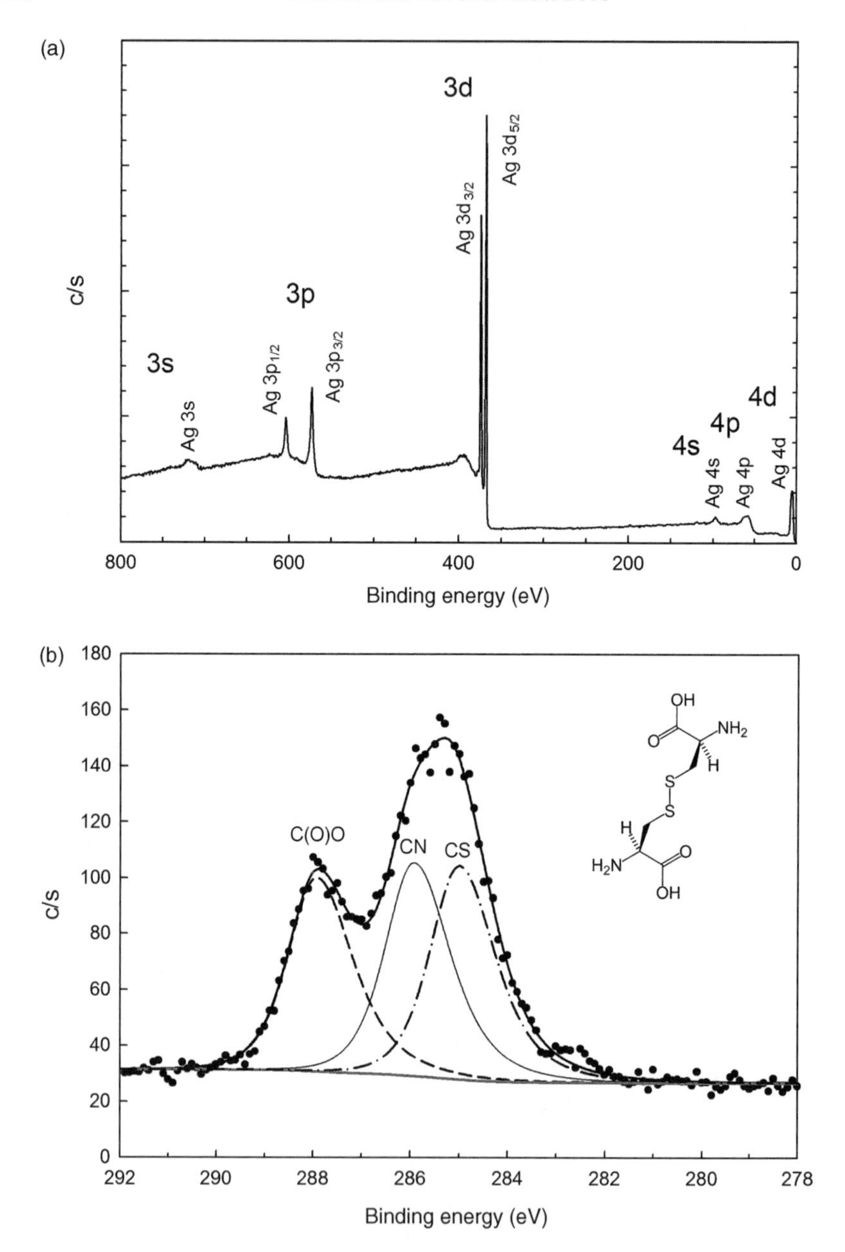

Figure 7.28. Example of an XPS spectrum of the element silver (a) and of the organic compound cystine (b), here in dependance on the binding energy. Cystine is an amino acid with a disulfide bond and the molecular formula $[SCH_2\,CH\,(NH_2)\,COOH]_2$. Several peaks of silver are marked in the spectrum in Figure 7.28a: 3s, 3p and 3d, 4s, 4p, and 4d. Three valence bands of the organic compound appear in the XPS spectrum in Figure 7.28b representing three different bonds of carbon. Both spectra were recorded at Leibniz-Institut für Analytische Wissenschaften – ISAS – e.V., Dortmund, Germany.

EXAFS and XANES and consequently can also be read from Table 7.1. The spectral width or FWHM of these peaks is about 0.3 eV up to 1.5 eV. Oxides usually have broader peaks. A chemical shift depends on nearest neighbors of an atom and usually depends on the kind of bonding.

All conducting solid samples can simply be analyzed by XPS. Insulators may need a charge referencing or even a neutralization by a low-voltage electron flood gun during analysis. Basically, all elements of the periodic table can be detected and a semiquantitative analysis can be carried out with sensitivity factors obtained from reference samples. The method is suited for major and minor components, but not for traces. Detection limits are in the range of mg/g. Today, XPS is a well-established and widely used analytical tool in a number of different fields of industry [150]. It is routinely used for the analysis of a variety of materials, such as metals, alloys, semiconductors, superconductors, inorganic and organic compounds, catalysts, glasses, glues, oils, implants, wafers, paints, woods, and polymers. Problems of adhesion, coating, corrosion, lubrication, oxidation, passivation, plating, sliding, and welding can be treated. Frequently, the method is applied to examine the modification of a material after a chemical treatment of its surface, such as activation and passivation.

In contrast to XRF and TXRF, photoelectrons are detected instead of X-ray photons and their energy is determined. However, the method of XPS, like XRF, needs an X-ray tube or a synchrotron for the excitation of a sample. Consequently, an improved excitation with a total reflection arrangement can also be exploited for XPS as already mentioned earlier [117–120]. It should be called TR-XPS, and some examples of this combination are treated here.

In 1972, Henke was the first regarding the total reflection effect for photoelectrons [117]. Twenty years later, Kawai *et al.* presented a numerical simulation of TR-XPS [116]. X-rays at total reflection can penetrate into a sample as an evanescent wave. For an Al-Kα beam and a copper layer, the critical angle of total reflection is $\alpha_{crit} = 2.17^\circ$. The depth of the evanescent wave was calculated according to Equation 1.79. It gives values between 1.5 and 4 nm for α values between 0° and 2.17°, respectively. The penetration depth of X-rays for larger glancing angles goes up to 250 nm (factor of $1/\sqrt{\beta} \approx 61$). Photoelectrons coming from these depths and going to the surface lose energy by scattering and increase the background. The escape depth or the mean free path of the original Cu 2p photoelectrons, however, is only about 3 nm. Consequently, the background in the XPS spectrum caused by inelastically scattered photoelectrons may be halved by TR-XPS. Simultaneously, the peak intensity can be increased by a factor of 2–4 for thin layers at the surface.

As predicted in his paper [116], Kawai *et al.* presented a TR-XPS spectrum for the first time in 1995 [153]. It is represented in Figure 7.29 showing a spectrum of silver that was evaporated on a silicon wafer and excited under TR geometry. The spectral background is reduced to about half due to the reduction of inelastic scattering of photoelectrons. The spectral peaks are increased by a factor of about 1.5.

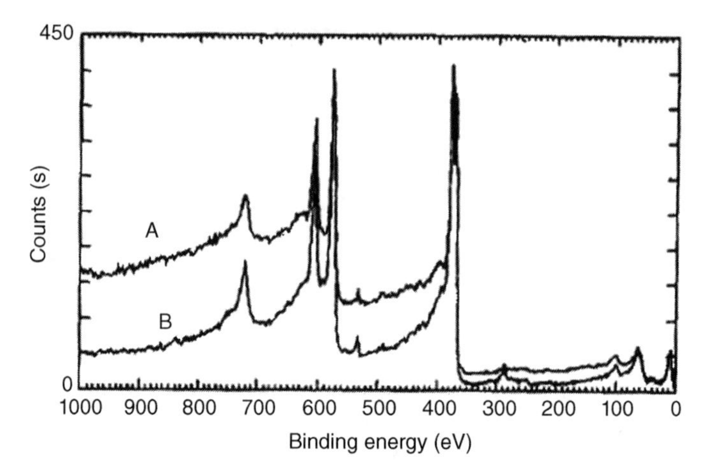

Figure 7.29. First photoelectron spectra of silver after vapor deposition on a silicon wafer. Spectrum A was recorded at a glancing angle above the critical angle while spectrum B was taken at a glancing angle below the critical angle of total reflection. The high background of spectrum A is reduced in spectrum B due to total reflection of the incident beam. The spectral peaks from left to right are 3s, $3p_{1/2}$ and $3p_{3/2}$, $3d_{1/2}$ plus $3d_{3/2}$. The latter doublet with a distance of 6 eV is not resolved. Figure from Ref. [119], reproduced with permission from the author.

XPS spectra can elucidate chemical bonds of molecular species. For example, thin boron carbonitride layers mentioned in Section 7.3.3 were investigated by TR-XPS [136]. Figure 7.30 confirms B–N and B–C bonds, and additional C–N bonds of such a layer in sp^2 and sp^3 configurations. Total reflection XPS was furthermore applied to a bilayer of copper phthalocyanine ($C_{32}H_{18}CuN_8$)

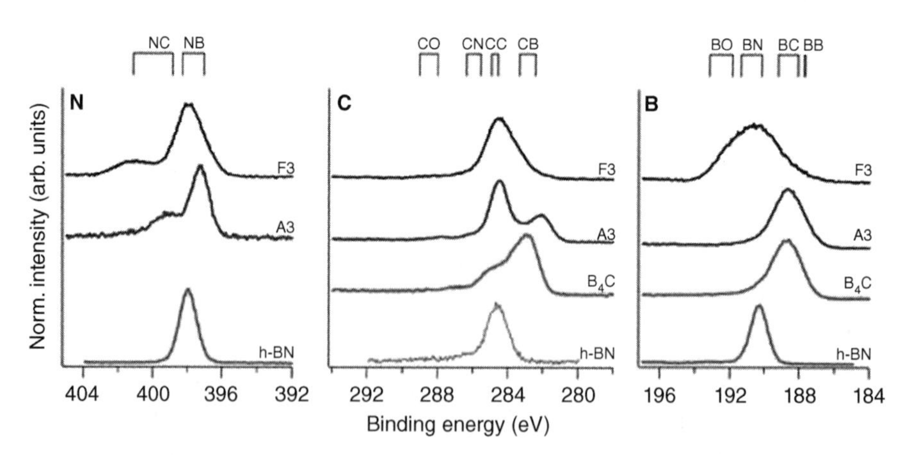

Figure 7.30. XPS spectra of B1s, C1s, and N1s, recorded from two different layers of borane, which were produced without and with adding of NH_3 (F3 and A3). These spectra are compared with XPS spectra of B_4C and hexagonal boron-nitride (below). The different bondings of NC, NB, CC, CB, BN, and BC are responsible for the peaks and marked on top. Figure from Ref. [136], reproduced with permission. Copyright © 2012, John Wiley and Sons.

with 5 nm thickness on a 200 nm thick layer of iron, both deposited on a silicon wafer [154]. Because of the total reflection geometry, the KMN Auger peaks of iron were reduced, the O 1s and the Fe 2p peaks remained constant, while the KLL peak of carbon, the Cu 2s and Cu 2p peaks, the N 1s and the C 1s peaks were increased. The O 1s peak was shown to originate from surface oxides on the iron layer.

REFERENCES

1. Prange, A. (1989). Total reflection X-ray spectrometry: method and applications. *Spectrochim. Acta*, **44B**, 437–452.
2. Klockenkämper, R. (1991). Totalreflexions-Röntgenfluoreszenzanalyse. In: H. Günzler, R. Borsdorf, W. Fresenius, W. Huber, H. Kelker, I. Lüderwald, G. Tölg, and H. Wisser (editors) *Analytiker Taschenbuch*, **Bd. 10**, Springer: Berlin, p. 111–152.
3. Klockenkämper, R., and von Bohlen, A. (1992). Total reflection X-ray fluorescence - An efficient method for micro-, trace- and surface-layer-analysis. *J. Anal. At. Spectrom.*, **7**, 273–279.
4. Prange, A. and Schwenke, H. (1992). Trace element analysis using total-reflection X-ray fluorescence spectrometry. *Adv. X-Ray Anal.*, **35B**, 899–923.
5. Klockenkämper, R., Knoth, J., Prange, A., and Schwenke, H. (1992). Total reflection X-ray fluorescence. *Anal. Chem.*, **64**, 1115A–1123A.
6. Schwenke, H. and Knoth, J. (1993). Total reflection XRF. In: R. Van Grieken and A. Markowicz (editors) *Handbook on X-Ray Spectrometry, Practical Spectroscopy Series*, Vol. 14, Dekker: New York, pp. 453.
7. van den Hoogenhof, W.W. and de Boer, D.K.G. (1993). Glancing-incidence X-ray analysis. *Spectrochim. Acta*, **48B**, 277–284.
8. Taniguchi, K. (1991). *Bunseki*, **3**, 168.
9. de Boer, D.K.G., Leenaers, A.J.G., and van den Hoogenhof, W.W. (1995). Glancing-incidence X-ray analysis of thin-layered materials: a review. *X-Ray Spectrom.*, **24**, 91–102.
10. Kregsamer, P., Streli, C., and Wobrauschek, P. (2002). Total-reflection X-ray fluorescence. *in* Van Grieken, R.E. and Markowicz, A.A. (Eds.) *Handbook on X-ray Spectrometry, Practical Spectroscopy Series*, Vol. 29, 2nd edition, Marcel Dekker: New York, 559–602.
11. Beckhoff, B., Kanngiesser, B., Langhoff, N., Wedell, R., and Wolff, H. (2006). (editors), *Handbook of Practical X-ray Fluorescence Analysis*, Springer, New York.
12. Streli, C. (2006). Recent advances in TXRF. *Appl. Spectrosc. Rev.*, **41**, 473–489.
13. Klockenkämper, R. (2006). Challenges of total reflection X-ray fluorescence for surface- and thin-layer analysis. *Spectrochim. Acta*, **61B**, 1082–1090.
14. Wobrauschek, P. (2007). Total reflection X-ray fluorescence analysis – a review. *X-Ray Spectrom.*, **36**, 289–300.
15. von Bohlen, A. (2009). Total reflection X-ray fluorescence and grazing incidence X-ray spectrometry — Tools for micro and surface analysis. A review. *Spectrochim. Acta*, **64B**, 821–832.

16. Alov, N.V. (2011). Total reflection X-ray fluorescence analysis: physical foundations and analytical application (a review). *Inorganic Materials*, **47**, 1487–1499.
17. West, M., Ellis, A.T., Potts, P.J., Streli, C., Vanhoof, C., Wegrzynek, D., and Wobrauschek, P. (2010). Atomic spectrometry update–X-ray fluorescence spectrometry. *J. Anal. At. Spectrom.*, **25**, 1503–1545.
18. West, M., Ellis, A.T., Potts, P.J., Streli, C., Vanhoof, C., Wegrzynek, D., and Wobrauschek, P. (2012). Atomic spectrometry update–X-ray fluorescence spectrometry. *J. Anal. At. Spectrom.*, **27**, 1603–1644.
19. Karydas, A.G., Beckhoff, B., Bogovac, M., Darby, I., Eichert, D., Fliegauf, R., Gambitta, A., Grötzsch, D., Herzog, C., Jark, W., Kaiser, R.B., Kanngießer, B., Kiskinova, M., Lubeck, J., Malzer, W., Markowicz, A., Migliori, A., Sghaier, H., and Weser, J. (2013). Activities in the IAEA XRF laboratory. *XRF Newsletter of IAEA*, editor A.G. Karydas, No. **24**, 1–3.
20. Wobrauschek, P., Kregsamer, P., Ladisich, W., Rieder, R., and Streli. C. (1993). Total-reflection X-ray fluorescence analysis using special X-ray sources. *Spectrochim. Acta*, **48B**, 143–151.
21. Takaura, N., Brennan, S., Pianetta, P., Laderman, S.S., Fischer-Colbrie, A., Kortright, J.B., Wherry, D.C., Miyazaki, K., and Shimazaki, A. (1995). Quantitative consideration of background contributions to TXRF spectra for the case of a Synchrotron radiation X-ray source. *Adv. X-ray Chem. Anal. Jpn.*, **26s**, 113–118.
22. Klockenkämper, R. (1997). *Total-Reflection X-Ray Fluorescence Analysis*, 1st ed., John Wiley & Sons, Inc., New York.
23. Rieder, R., Wobrauschek, P., Ladisich, W., Streli, C., Aiginger, H., Garbe, S., Gaul, G., Knöchel, A., and Lechtenberg, F. (1995). Total reflection X-ray fluorescence analysis with synchrotron radiation monochromatized by multilayer structures. *Nucl. Instr. Meth.*, **A355**, 648–653.
24. Streli, C., Wobrauschek, P., Ladisich, W., Rieder, R., Aiginger, H., Ryon, R.W., and Pianetta, P. (1994). Total reflection X-ray fluorescence analysis of light elements using synchrotron radiation. *Nucl. Instr. Meth.*, **A345**, 399–403.
25. Laderman, S.S., Fischer-Colbrie, A., Shimazaki, A., Miyazaki, K., Brennan, S., Takaura, N., Pianetta, P., and Kortright, J.B. (1995). High sensitivity total reflection X-ray fluorescence spectroscopy of silicon wafers using synchrotron radiation. *Adv. X-ray Chem. Anal. Jpn.*, **26s**, 91–96.
26. Liu, K.Y., Kojima, S., Kawado, S., and Iida, A. (1995). SR-TXRF analysis of metallic impurities on silicon surface. *Adv. X-ray Chem. Anal. Jpn.*, **26s**, 107–112.
27. Iida, A., Yoshinaga, A., Sakurai, K., and Gohshi, Y. (1986). Synchrotron radiation excited X-ray fluorescence analysis using total reflection of X-rays. *Anal. Chem.*, **58**, 394–397.
28. Matsushita, T., Iida, A., Ishikawa, T., Nakagiri, T., and Sakai, K. (1986). X-ray standing waves excited in multilayered structures. *Nucl. Instr. Meth.*, **A246**, 751–754.
29. Iida, A. (1991). *Adv. X-Ray Anal.*, **34**, 23.
30. Iida, A. (1992). *Adv. X-Ray Anal.*, **35B**, 795.
31. Krämer, M. (2007). Potentials of synchrotron radiation induced X-ray standing waves and X-ray reflectivity measurements in material analysis, PhD Thesis, University of Dortmund, 128 pages.

32. Beckhoff, B., Fliegauf, R., Kolbe, M., Müller, M., Pollakowski, B., Weser, J., and Ulm, G. (2008). X-ray spectrometry for wafer contamination analysis and speciation as well as for reference-free nanolayer characterization. *Solid State Phenom.*, **134**, 277–280.
33. Yamada, H., Hasegawa, D., Yamada, T., and Maeo, S. (2013). *High power X-ray beam realized by the MIRRORCLE-type tabletop synchrotron*, 15th International Conference on TXRF and related methods, Book of Abstracts, O18, p. 217.
34. Maeo, S., Hasegawa, D., Yamada, T., and Yamada, H. (2013). *XANES installation with MIRRORCLE X-ray source*, 15th International Conference on TXRF and related methods, Book of Abstracts, P27, p. 65.
35. Maeo, S., Katori, N., Kawahara, M., Tsujita, M., and Yamada, H. (2013). *High energy X-ray residual stress analysis by using MIRRORCLE X-ray source*, 15th International Conference on TXRF and related methods, Book of Abstracts, P64, p. 147.
36. Maderitsch, A., Smolek, S., Wobrauschek, P., Takman, P., and Streli, C. (2013). *TXRF using liquid metal jet tube*, 15th International Conference on TXRF and related methods, Book of Abstracts, P13, p. 44.
37. Zamponi, F., Ansari, Z., Schmising, C.K., Rothhardt, P., and Zhavoronkov, N. (2009). Femtosecond hard X-ray plasma sources with a kilohertz repetition rate. *Appl. Phys.*, **A96**, 1, 51–58.
38. Mantouvalou, I., Jung, R., Tuemmler, J., Legall, H., Bidu, T., Stiel, H., Malzer, W., Kanngiesser, B., and Sandner, W. (2011). Note: study of extreme ultraviolet and soft X-ray emission of metal targets produced by laser-plasma-interaction. *Rev. Sci. Instrum.*, **82**, 066103.
39. Tsuji, K., Sato, T., Wagatsuma, K., Claes, M., and Van Grieken, R. (1999). Preliminary experiment of total reflection X-ray fluorescence using two glancing X-ray beams excitation. *Rev. Sci. Instrum.*, **70**, 1621–1623.
40. MacDonald, C.A. (2010). Focusing polycapillary optics and their applications, Review article. *X-ray Optics and Instrum.*, **2010**, article ID 867049, 17 pages.
41. Haschke, M., Pfannekuch, P., and Scruggs, B. (2000). Ultra-trace analysis by micro X-ray fluorescence spectroscopy. *Adv. X-Ray Anal.*, **43**, 435–441.
42. Mages M. (2013). *Elementbestimmungen in aquatischen Biofilmen und Zooplankton mittels Totalreflexions-Röntgenfluoreszenzanalytik*, Dissertation, Universität Lüneburg, Institut für Ökologie und Umweltchemie, 127 Seiten.
43. Cheburkin, A. and Shotyk, W. (1996). Double-plate sample carrier for a simple total reflection X-ray fluorescence analyzer. *X-Ray Spectrom.*, **25**, 175–178.
44. Egorov, V.K., Kondratiev, O.S., Zuev, A.P., and Egorov, E.V. (1999). The modification of TXRF-method by use of X-ray slitless collimator. *Adv. X-Ray Anal.*, **43**, 406–417.
45. Egorov, V.K. and Egorov, E.V. (2003). Waveguide resonance mechanism for X-ray beam propagation: physics and experimental background. *Adv. X-Ray Anal.*, **46**, 307–313.
46. Egorov, V.K. and Egorov, E.V. (2004). The experimental background and the model description for the waveguide-resonance propagation of X-ray radiation through a planar narrow extended slit, Review. *Spectrochim. Acta*, **59B**, 1049–1069.
47. Egorov, V.K. and Egorov, E.V. (2007). Background of X-ray nanophotonics based on the planar air waveguide resonator. *X-Ray Spectrom.*, **36**, 381–397.

48. Iwasaki, H. and Kawai, J. (2013). *Electromagnetic analysis of a planar X-ray waveguide*, 15th International Conference on TXRF and related methods, Book of abstracts, P16, p. 50.

49. Sanchez, H.J. (2002). Total reflection X-ray fluorescence analysis using plate beam-guides. *Nucl. Instr. and Meth. in Phys. Res.*, **B194**, 90–95.

50. Tsuji, K. and Wagatsuma, K. (2002). Enhancement of TXRF intensity by using a reflector. *X-Ray Spectrom.*, **31**, 358–362.

51. Cesareo, R. (2000). X-ray physics: Interaction with matter, production, detection. *La Rivista del Nuovo Cimento*, vol. 23, serie 4, numero 7, 232 pages.

52. Höhne, J., Altmann, M., Angloer, G., Buehler, M., von Feilitzsch, F., Frank, T., Hettl, P., Hertrich, T., Jochum, J., Nuessle, T., Pfnuer, S., Schnagl, J., and Waenninger, S. (1999). *Cryogenic microcalorimeters and tunnel junctions for high-resolution energy dispersive X-ray spectrometry*, Proc. SPIE 3743, In-line Characterization, Yield Reliability, and Failure Analyses in Microelectronic Manufacturing, 162; doi: 10.1117/12.346911.

53. Redfern, D., Nicolosi, J., Höhne, J., Weiland, R., Simmacher, B., and Hollerich, C. (2002). The microcalorimeter for industrial applications. *J. Res. Natl. Inst. Stand. Technol.*, **107**, 621–626.

54. Fleischmann, L., Linck, M., Burck, A., Domesl, C., Höhn, C., Kempf, S., Lausberg, S., Pabinger, A., Pies, C., Porst, J.P., Rotzinger, H., Schäfer, S., Weldle, R., Fleischmann, A., Enss, C., and Deidel, G.M. (2009). Metallic magnetic calorimeters for X-ray spectroscopy. *IEEE Trans. Appl. Supercond.*, **19**, 63–68.

55. Streli, C., Kregsamer, P., Wobrauschek, P., Gatterbauer, H., Pianetta, P., Pahlke, S., Fabry, L., Palmetshofer, L., and Schmeling, M. (1999). Low Z total reflection X-ray fluorescence analysis – challenges and answers. *Spectrochim. Acta*, **54B**, 1433–1441.

56. Streli, C., Wobrauschek, P., Beckhoff, B., Ulm, G., Fabry, L., and Pahlke, S. (2001). First results of TXRF measurements of low-Z elements on Si-wafer surfaces at the PTB plane grating monochromator beamline for undulator radiation at BESSY II. *X-Ray Spectrom.*, **30**, 24–31.

57. Baur, K., Brennan, S., Burrow, B., Werho, D., and Pianetta, P. (2001). Laboratory and synchrotron radiation total-reflection X-ray fluorescence: new perspectives in detection limits and data analysis. *Spectrochim. Acta*, **56B**, 2049–2056.

58. Schwenke, H., Bormann, R., Knoth, J., and Prange, A. (1993). Some potential developments for trace element and surface analysis using a grazing incident X-ray beam. *Spectrochim. Acta*, **48B**, 293–299.

59. Holfelder, I., Beckhoff, B., Fliegauf, R., Hönicke, P., Nutsch, A., Petrik, P., Roeder, G., and Weser, J. (2013). Complementary methodologies for thin film characterization in one tool – a novel instrument for 450 mm wafers. *J. Anal. At. Spectrom.*, **28**, 549–557.

60. Streli, C., Wobrauschek, P., Randolf, G., Rieder, R., Ladisich, W., and Aiginger, H. (1995). Light element analysis with TXRF at different excitation energies: theory and experiment. *Adv. X-ray Chem. Anal. Jpn.*, **26s**, 63–68.

61. Beckhoff, B., Fliegauf, R., Kolbe, M., Müller, M., Weser, J., and Ulm, G. (2007). Reference-free total reflection X-ray fluorescence analysis of semiconductor surfaces with synchrotron radiation. *Anal. Chem.*, **79**, 7873–7882.

62. Beckhoff, B. (2008). Reference-free X-ray spectrometry based on metrology using synchrotron radiation. *J. Anal. At. Spectrom.*, **23**, 845–853.
63. Reinhardt, F., Osan, J., Török, S., Pap, A.E., Kolbe, M., and Beckhoff, B. (2012). Reference-free quantification of particle-like surface contaminations by grazing incidence X-ray fluorescence analysis. *J. Anal. At. Spectrom.*, **27**, 248–255.
64. Streli, C., Aiginger, H., and Wobrauschek, P. (1989). Total reflection X-ray fluorescence analysis of low-Z elements. *Spectrochim. Acta*, **44B**, 491–498.
65. Streli, C., Wobrauschek. P., and Aiginger, H. (1992). Light element analysis with TXRF. *Adv. X-Ray Anal.*, **35B**, 947–952.
66. Hein, M., Hoffmann, P., Lieser, K.H., and Ortner, H.M. (1992). Measurement of low-Z elements by TXRF. *Fresenius J. Anal. Chem.*, **343**, 760.
67. Streli, C., Aiginger, H., and Wobrauschek, P. (1993). Light element determination with a new spectrometer for total-reflection X-ray fluorescence. *Spectrochim. Acta*, **48B**, 163–170.
68. Streli, C., Aiginger, H., and Wobrauschek, P. (1993). A new spectrometer for total reflection X-ray fluorescence analysis of light elements. *Nucl. Instr. Meth.*, **A334**, 425–429.
69. Streli, C., Wobrauschek, P., Ladisich, W., Rieder, R., and Aiginger, H. (1995). Total reflection X-ray fluorescence analysis for light elements under various excitation conditions. *X-Ray Spectrom.*, **24**, 137–142.
70. Streli, C., Wobrauschek, P., Pepponi, G., and Zoeger, N. (2004). A new total reflection X-ray fluorescence vacuum chamber with sample changer analysis using a silicon drift detector for chemical analysis. *Spectrochim. Acta*, **59B**, 1199–1203.
71. Wobrauschek, P., Streli, C., Kregsamer, P., Meirer, F., Jokubonis, C., Markowicz, A., Wegrzynek, D., and Chineao Cano, E. (2008). Total reflection X-ray fluorescence attachment module modified for analysis in vacuum. *Spectrochim. Acta*, **63B**, 1404–1407.
72. Ovari, M., Streli, C., Wobrauschek, P., and Zaray, G. (2009). Determination of carbon in natural freshwater biofilms with total reflection X-ray fluorescence spectrometry. *Spectrochim. Acta*, **64B**, 802–804.
73. Streli, C., Pepponi, G., Wobrauschek, P., Beckhoff, B., Ulm, G., Pahlke, S., Fabry, L., Ehmann, Th., Kanngießer, B., Malzer, W., and Jark, W. (2003). Analysis of low Z elements on Si wafer surfaces with undulator radiation induced total-reflection X-ray fluorescence at the PTB beamline at BESSY. *Spectrochim. Acta*, **58B**, 2113–2121.
74. Baur, C., Brennan, S., Pianetta, P., and Opila, R. (2002). Looking at trace impurities on silicon wafers with synchrotron radiation. *Anal. Chem.*, **74**, 609A–616A.
75. Kollotzek, D. (1980). Beitrag zur Elementspurenbestimmung durch Röntgenfluoreszenzspektrometrie mit Anregung unter Totalreflexion der Primärstrahlung. *Diploma Thesis*, University of Stuttgart.
76. Eller, R., Alt, F., Tölg, G., and Tobschall, H.J. (1989). An efficient combined procedure for the extreme trace determination of gold, platinum, palladium and rhodium with the aid of graphite furnace atomic absorption spectrometry and total-reflection X-ray fluorescence analysis. *Fresenius Z. Anal. Chem.*, **334**, 723–739.

77. Fan, Q. and Gohshi, Y. (1993). Enhancement of total reflection X-ray fluorescence spectroscopy with electrochemical deposition. *Appl. Spectrosc.*, **47**, 1742–1756.
78. Krämer, K. (1982). *Tiefenprofilanalyse an Metallen im Nanometerbereich mit elektrochemischer Schichtabtragung durch Elektropolieren*, Ph. D. Thesis, University of Stuttgart.
79. Bormann, R. and Schwenke, H. (1992). Offenlegungsschrift DE 4028044 A1, Int. Cl. G 01 N23/203, Deutsches Patentamt.
80. Bredendiek-Kämper, S., von Bohlen, A., Klockenkämper, R., Quentmeier, A., and Klockow., D. (1996). Microanalysis of solid samples by laser-ablation and total-reflection X-ray fluorescence. *J. Anal. At. Spectrom.*, **11**, 537–541.
81. Knoth, J., Bormann, R., Gutschke, R., Michaelsen, C., and Schwenke, H. (1993). Examination of layered structures by total-reflection X-ray fluorescence analysis. *Spectrochim. Acta*, **48B**, 285–292.
82. Wiener, G., Michaelsen, C., Knoth, J., Schwenke, H., and Bormann, R. (1995). Concentration-depth profiling using total-reflection X-ray fluorescence spectrometry in combination with ion-beam microsectioning techniques. *Rev. Sci. Instrum.*, **66**, 20–23.
83. Frank, W., Thomas, H.-J., and Schindler, A. (1995). Depth profiling by means of the combination of glancing incidence x-ray fluorescence spectrometry with low energy ion beam etching technique. *Spectrochim. Acta*, **50B**, 265–270.
84. Sasaki, Y. and Hirokawa, K. (1990). Refraction effect of scattered X-ray fluorescence at surface. *Appl. Phys. A*, **50**, 397–404.
85. Noma, T. and Iida, A. (1994). Surface analysis of layered thin films using a synchrotron x-ray microbeam combined with a grazing-exit condition. *Rev. Sci. Instrum.*, **65**, 837–844.
86. Noma, T., Iida, A., and Sakurai, K. (1993). Fluorescent-X-ray-interference effect in layered materials. *Phys. Rev.*, **B48**, 17524–17526.
87. Tsuji, K. and Hirokawa, K. (1994). Takeoff angle-dependent X-ray fluorescence of layered materials using a glancing incident X-ray beam. *J. Appl. Phys.*, **A75**, 7189–7194.
88. Tsuji, K., Sato, S., and Hirokawa, K. (1995). X-ray fluorescence analysis of thin films at glancing-incident and –take-off angles. *Adv. X-Ray Chem. Anal. Jpn.*, **26s**, 151–156.
89. Sasaki, Y. (1995). Fluorescent X-ray interference from a metal monolayer and metal-labeled proteins. *Adv. X-Ray Chem. Anal. Jpn.*, **26s**, 193–198.
90. Becker, S., Golovchenko, J.A., and Patel, J.R. (1983). X-ray evanescent-wave absorption and emission. *Phys. Rev. Lett.*, **50**, 153–156.
91. Hasegawa, S., Ino, S., Yamamoto, Y., and Daimon, H. (1985). Chemical analysis of surfaces by total-reflection-angle X-ray spectroscopy in RHEED experiments (RHEED-TRAXS). *Jpn. J. Appl. Phys.*, **24**, L387–L390.
92. Sasaki, Y.C. and Hirokawa, K. (1991). New non-destructive depth profile measurement by using a refracted X-ray fluorescence method. *Appl. Phys. Lett.*, **58**, 1384–1386.
93. Born, M. and Wolf, E. (1975). *Principles of Optics*, Pergamon Press: New York.

94. de Bokx, P.K. and Urbach, H.P. (1995). Calculation of fluorescence intensities in grazing-emission X-ray fluorescence spectrometry. *Adv. X-ray Chem. Anal. Jpn.*, **26s**, 199–204.
95. Ashida, T., Sawamura, T., and Tsuji, K. (2013). *Development of a compact GE-XRF spectrometer for fast trace element analysis*, 15th International Conference on TXRF and related methods, Book of Abstracts, P4, p. 29.
96. Holynska, B., Olko, M., Ostachowicz, B., Ostachowicz, J., Wegrzynek, D., Claes, M., Van Grieken, R., and De Bokx, P. (1998). Performance of total reflection and grazing emission X-ray fluorescence spectrometry for the determination of trace metals in drinking water in relation to other analytical techniques. *Fresenius J. Anal. Chem.*, **362**, 294–298.
97. Claes, M., Van Dyck, K., Deelstra, H., and Van Grieken, R. (1999). Determination of silicon in organic matrices with grazing-emission X-ray fluorescence spectrometry. *Spectrochim. Acta*, **54B**, 1517–1524.
98. Claes, M., de Bokx, P., Willard, N., Veny, P., and Van Grieken, R. (1997). Optimization of sample preparation for grazing emission X-ray fluorescence in micro- and trace analysis applications. *Spectrochim. Acta*, **52B**, 1063–1070.
99. Claes, M., Van Ham, R., Janssens, K., Van Grieken, R., Klockenkämper, R., and von Bohlen, A. (1999). Micro-analysis of artists' pigments by grazing-emission X-ray fluorescence spectrometry. *JCPDS-International Centre for Diffraction Data*, **41**, 262–277.
100. Spolnik, Z.M., Claes, M., Van Grieken, R.E., de Bokx, P.K., and Urbach, H.P. (1999). Quantification in grazing-emission X-ray fluorescence spectrometry. *Spectrochim. Acta*, **54B**, 1525–1537.
101. Claes, M., de Bokx, P., and Van Grieken, R. (1999). Progress in laboratory grazing emission X-ray fluorescence spectrometry. *X-Ray Spectrom.*, **28**, 224–229.
102. Kayser, Y., Banas, D., Cao, W., Dousse, J.C., Hoszowska, J., Jagodzinski, P., Kavcic, M., Kubala-Kukus, A., Nowak, S., Pajek, M., and Szlachetko, J. (2010). Depth profiles of Al impurities in Si wafers determined by means of the high resolution grazing emission x-ray fluorescence technique. *Spectrochim. Acta*, **65B**, 445–449.
103. Hönicke, P., Kayser, Y., Beckhoff, B., Müller, M., Dousse, J.C., Hoszowska, J., and Nowak, S. H. (2012). Characterization of ultra-shallow aluminum implants in silicon by grazing incidence and grazing emission X-ray fluorescence spectroscopy. *J. Anal. At. Spectrom.*, **27**, 1432–1438.
104. Bertin, E.P. (1975). *Principles and Practice of Quantitative X-ray Fluorescence Analysis*, 2nd ed., Plenum Press: New York.
105. Author collective (1986). *X-ray Data Booklet*, 1st ed., editor D. Vaughan, Lawrence Berkeley Laboratory: Berkeley; 3rd ed. (2009), editor A.C. Thompson; here chapter 1.1 Electron binding energies, Table 1-1.
106. www.cxro.lbl.gov. U.S. Department of Energy, National Laboratory Operated by the University of California 2014, X-ray Data Base, 2014.
107. Nowak, S.H., Reinhardt, F., Beckhoff, B., Dousse, J.C., and Szlatchetko, J. (2013). Geometrical optics modelling of grazing incidence x-ray fluorescence of nano-scaled objects. *J. Anal. At. Spectrom.*, **28**, 689–696.

108. Sakurai, K. (2013). *Recent instrumentation in grazing-incidence X-ray analysis*, 15th International Conference on TXRF and related methods, Book of abstracts, I03, p. 6.
109. Klockenkämper, R. (2006). Challenges of total reflection X-ray fluorescence for surface- and thin-layer analysis, Review. *Spectrochim. Acta*, **61B**, 1082–1090.
110. Marra, W.C., Eisenberger, P., and Cho, A.Y. (1979). X-ray total-external-reflection – Bragg diffraction: A structural study of the GaAs-Al interface. *J. Appl. Phys.*, **50**, 6927–6933.
111. Segmüller, A. (1987). Characterization of epitaxial films by grazing-incidence X-ray diffraction. *Thin Solid Films*, **154**, 33–42.
112. Feidenhansl, R. (1989). Surface structure determination by X-ray diffraction. *Surface Science Rep.*, **10**, 105–188.
113. Huang, T.C. (1992). *Adv. X-Ray Anal.*, **35A**, 143.
114. Wulff, H., Klimke, J., and Quade, A. (1995). *GIT Fachz. Lab.*, **39**, 1063.
115. Horiuchi, T. and Matsushige, K. (1993). Total-reflection X-ray diffractometry and its applications to evaporated organic thin films. *Spectrochim. Acta*, **48B**, 137–142.
116. Kawai, J., Takami, M., Fujinami, M., Hashigushi, Y., Hayakawa, S., and Gohshi, Y. (1992). A numerical simulation of total reflection X-ray photoelectron spectroscopy (TRXPS). *Spectrochim. Acta*, **47B**, 983–991.
117. Henke, B.L. (1972). Ultrasoft X-ray reflection, refraction and production of photoelectrons (100-1000 eV region). *Phys. Rev.*, **A6**, 94.
118. Kawai, J., Adachi, H., Hayakawa, S., Zheng, Z., Kobayashi, K., Gohshi, Y., Maeda, K., and Kitayima, Y. (1994). Depth selective X-ray absorption fine structure spectrometry. *Spectrochim. Acta*, **49B**, 739–743.
119. Kawai, J., Hayakawa, S., Kitajima, Y., and Gohshi, Y. (1995). Total reflection X-ray absorption and photoelectron spectroscopy. *Adv. X-ray Chem. Anal. Jpn.*, **26s**, 97–102.
120. Huang, T.C. and Parrish, W. (1992). *Adv. X-Ray Anal.*, **35A**, 137.
121. Lengeler, B. (1992). X-ray reflection, a new tool for investigating layered structures and interfaces. *Adv. X-Ray Anal.*, **35**, 127.
122. Hüppauf, M. (1993). Charakterisierung von dünnen Schichten und von Gläsern mit Röntgenreflexion und Röntgenfluoreszenzanalyse bei streifendem Einfall, PhDThesis, RWTH Aachen, and JÜL-report JÜL-2730, ISSN 0366-0885.
123. Glocker, R. and Frohnmeyer, W. (1923). Use of X-rays in quantitative chemical analysis. *Fortschr. Geb. Röntgenstr.*, **31**, 90–92.
124. Stern, E.A. (2001). Musings about the development of XAFS. *J. Synchrotron. Rad.*, **8**, 49–54.
125. Bianconi, A. (1980). Surface X-ray absorption spectroscopy: Surface EXAFS and Surface XANES. *Appl. Surf. Sci.*, **6**, 392–418.
126. Reinhardt, F., Beckhoff, B., Eba, H., Kanngiesser, B., Kolbe, M., Mizusawa, M., Müller, M., Pollakowski, B., Sakurai, K., and Ulm, G. (2009). Evaluation of high-resolution X-ray absorption and emission spectroscopy for the chemical speciation of binary titanium compounds. *Anal. Chem.*, **81**, 1770–1776.
127. http://en.wikipedia.org/wiki/XANES
128. Stöhr, J. (1992). NEXAFS Spectroscopy. *Springer Series in Surface Science*, **25**, Springer, Berlin-Heidelberg.

129. Streli, C., Pepponi, G., Wobrauschek, P., Jocubonis, C., Falkenberg, G., Zaray, G., Broekaert, J., Fittschen, U., and Peschel, B. (2006). Recent results of synchrotron radiation induced total reflection X-ray fluorescence analysis at HASYLAB, beamline L. *Spectrochim. Acta*, **61B**, 1129–1134.

130. Török, S., Osan, J., Beckhoff, B., and Ulm, G. (2003). Ultra-trace speciation of nitrogen compounds in aerosols collected on silicon wafer surfaces by means of TXRF-NEXAFS. *Powder Diffraction*, **19**, 81–86.

131. Osan, J., Török, S., Beckhoff, B., Ulm, G., Hwang, H., Ro, C.U., Abete, C., and Fuoco, R. (2006). Nitrogen and sulphur compounds in coastal Antarctic fine aerosol particles – an insight using non-destructive X-ray microanalytical methods. *Atmospheric Environment*, **40**, 4691–4702.

132. Pepponi, G., Beckhoff, B., Ehmann, T., Ulm, G., Streli, C., Fabry, L., Pahlke, S., and Wobrauschek, P. (2003). Analysis of organic contaminants on Si wafers with TXRF-NEXAFS. *Spectrochim. Acta*, **58B**, 2245–2253.

133. Lommel, M., Hönicke, P., Kolbe, M., Müller, M., Reinhardt, F., Mobus, P., Mankel, E., Beckhoff, B., and Kolbesen, B.O. (2009). Preparation and characterization of self-assembled monolayers on germanium surfaces. *Solid State Phenom.*, **145**, 169–172.

134. Polgari, Z., Meirer, F., Sasamori, S., Ingerle, D., Pepponi, G., Streli, C., Rickers, K., Reti, A., Budai, B., Szoboszlai, N., and Zaray, G. (2011). Iron speciation in human cancer cells by K-edge total reflection X-ray fluorescence –X-ray absorption near edge structure analysis. *Spectrochim. Acta*, **66B**, 274–279.

135. Baake, O., Hoffmann, P.S., Klein, A., Pollakowski, B., Beckhoff, B., Ensinger, W., Kosinova, M., Fainer, N., Sulyaeva, V.S., and Trunova, V. (2008). Chemical character of BC_xN_y layers grown by CVD with trimethylamine borane. *X-Ray Spectrom.*, **38**, 68–73.

136. Hoffmann, P.S., Baake, O., Kosinova, M.L., Beckhoff, B., Klein, A., Pollakowski, B., Trunova, V.A., Sulyaeva, V.S., Kuznetsov, F.A., and Ensinger W. (2011). Chemical bonds and elemental compositions of BC_xN_y layers produced by chemical vapour deposition with trimethylamine borane, triethylamine borane, or trimethylborazine. *X Ray Spectrom.*, **41**, 240–246.

137. Pollakowski, B., Hoffmann, P., Kosinova, M., Baake, O., Trunova, V., Unterumsberger, R., Ensinger, W., and Beckhoff, B. (2012). Nondestructive and non-preparative chemical nanometrology of internal material interfaces at tunable high information depths. *Anal. Chem.*, **85**, 193–200.

138. Becker, C., Pagels, M., Zachäus, C., Pollakowski, B., Beckhoff, B., et al. (2013). Chemical speciation at buried interfaces in high-temperature processed polycrystalline silicon thin-film solar cells on ZnO:Al. *J. Appl. Phys.*, **113**, 044519, 8 pages

139. Szoboszlai, N., Gaal, A., Bosze, S., Majer, Z., Pepponi, G., Meirer, F., Ingerle, D., and Streli, C. (2013). *Determination and speciation of Rh in cancer cells by TXRF and K-edge SR TXRF*, 15th International Conference on TXRF and related methods, Book of Abstracts, P50, p. 125.

140. Lancosz, M., Szcerbowska-Boruchowska, M., Wandzilak, A., Czyzycki, M., Adamek, D., and Radwanska, E. (2013). *Application of X-ray fluorescence and X-ray absorption micro-spectroscopies in biomedical research*, 15th International Conference on TXRF and related methods, Book of Abstracts, I20, p. 206.

141. Meirer, F., Pepponi, G., Streli, C., Wobrauschek, P., and Zoeger, N. (2009). Grazing exit versus grazing incidence geometry for x-ray absorption near edge structure analysis of arsenic traces. *J. Appl. Phys.*, **105**, 074906-1–074906-7. http://dx.doi.org/10.1063/1.3106086.

142. Yoneda, Y. and Horiuchi, T. (1971). Optical flats for use in X-ray spectro-chemical microanalysis. *Rev. Sci. Instrum.*, **42**, 1069–1070.

143. Horiuchi, T. (1993). Initial idea to use optical flats for X-ray fluorescence analysis and recent applications to diffraction studies. *Spectrochim. Acta*, **48B**, 129–136.

144. Horiuchi, T. and Matsushige, K. (1993). Total-reflection X-ray diffractometry and its applications to evaporated organic thin films. *Spectrochim. Acta*, **48B**, 137–142.

145. Ishida, K., Horiuchi, T., and Matsushige, K. (1995). Epitaxial growth of organo-metallic thin films studied by total reflection X-ray diffraction. *Adv. X-Ray Chem. Anal. Jpn.*, **26s**, 157–162.

146. Hayashi, K., Horiuchi, T., and Matsushige, K. (1995). Simultaneous analysis of total reflection X-ray diffraction and fluorescence from copperphthalocyanine thin films during evaporation process. *Jpn. J. Appl. Phys.*, **Part 1 34** (12A), 6478–6482.

147. Noma, T., Takada, K., and Iida, A. (1999). Surface-sensitive X-ray fluorescence and diffraction analysis with grazing-exit geometry. *X-Ray Spectrom.*, **28**, 433–439.

148. Groh, L., Hums, C., Bläsing, J., Krost, A., and Dadgar, A. (2011). Characterization of AlGaInN layers using X-ray diffraction and fluorescence. *Phys. Status Solidi*, **B 248**, 622–626.

149. Tolan, M., Press, W., Brinkop, F., and Kotthaus, J.P. (1995). X-ray diffraction from laterally structured surfaces: Total external reflection. *Phys. Rev.*, **B51**, 2239–2251.

150. http://en.wikipedia.org/wiki/X-ray_photoelectron_spectroscopy

151. Bubert, H., Riviere, J.C., and Werner, W.S.M. (2011). X-ray photoelectron spectroscopy (XPS). in *Surface and Thin Film Analysis*, Friedbacher, G., and Bubert, H., (editors) 2nd ed., Wiley-VCH, Weinheim, 9–41.

152. Hüfner, S. (2003). *Photoelectron Spectroscopy*, 3rd ed., Springer, Berlin, 662 pages.

153. Kawai, J., Hayakawa, S., Kitajima, Y., and Gohshi, Y. (1995). X-ray absorption and photoelectron spectroscopies using total reflection X-rays. *Anal. Sciences*, **11**, 519–524.

154. Imashuku, S. and Kawai, J. (2013). *Total reflection X-ray photoelectron spectroscopy of thin film*, 15th International Conference on TXRF and related methods, Book of Abstracts, O5, p. 21.

INDEX

Total-Reflection X-ray Fluorescence Analysis and Related Methods, Second Edition.
Reinhold Klockenkämper and Alex von Bohlen.

T

Y

Z